W9-AQZ-765

READINGS IN THE PHILOSOPHY OF SCIENCE

edited by
Baruch A. Brody
Massachusetts Institute of Technology

READINGS IN THE PHILOSOPHY OF SCIENCE

Prentice-Hall, Inc.,
Englewood Cliffs, New Jersey

Library of Congress Catalog Card Number: 71–98091

Printed in the United States of America

13–760702–4

Current Printing (last digit):

10 9 8 7 6 5

PRENTICE-HALL INTERNATIONAL, INC., *London*
PRENTICE-HALL OF AUSTRALIA, PTY. LTD., *Sydney*
PRENTICE-HALL OF CANADA, LTD., *Toronto*
PRENTICE-HALL OF INDIA PRIVATE LIMITED, *New Delhi*
PRENTICE-HALL OF JAPAN, INC., *Tokyo*

To Dena

about whom Proverbs 31
was written

PREFACE

No area in philosophy has grown as rapidly in recent years as has the philosophy of science. There are an increasing number of specialists in this area, and a great many schools have added courses in the philosophy of science to their undergraduate and graduate curriculum. Unfortunately, there has not been a corresponding increase in the number of texts and anthologies suitable for these courses, and teachers have been forced to require their students to purchase numerous books and to spend a good deal of time in their library reserve rooms reading additional material. The purpose of this anthology is to alleviate this unsatisfactory situation by providing sufficient current material of high quality in one book so that a good undergraduate (or first year graduate) philosophy of science course can be built entirely around it.

This anthology is divided into three sections: Section I concerns scientific explanation and prediction; Section II, the structure and function of scientific theories; and Section III, the confirmation of scientific hypotheses. These problems have been chosen because they are clearly central to the philosophy of science; no worthwhile course in this field can possibly omit them. While further sections could have been added, it seemed more desirable to treat the central problems at some depth rather than to treat every problem superficially.

Three principles have guided me in choosing the selections: (a) as far as possible, every major position on a particular issue should be represented; (b) the various selections, besides presenting their own point of view, should also analyze other points of view; (c) selections should presuppose as little philosophy, logic, and science as is possible. It is hoped, therefore, that this book will give even a student relatively new to philosophy a genuine sense of the difficulties and potentialities inherent in a variety of positions, rather than a mere superficial knowledge of what several philosophers have said about a particular issue.

Each section begins with a substantial introduction that structures the section by explaining its problems, by stating the possible solutions to these problems that have been offered, and by briefly indicating the roles of the particular selections in the discussion of these problems. It is hoped that these introductions will enable the student to understand by himself what is said in the selections, thereby freeing the teacher from the task of exposition and enabling him to devote his time in class to the critical analysis and evaluation of the positions presented in the selections.

No book could reasonably hope to include all the worthwhile material that has been published in the philosophy of science in recent years. Since the serious student will certainly want access to this material, this book therefore concludes with a detailed bibliographical essay that introduces the student to material not included in this volume.

This book would probably never have appeared were it not for the constant aid and advice of Alan Lesure of Prentice-Hall. I would like to take this opportunity, therefore, to express my appreciation to him. I am also indebted to several of my colleagues, particularly to Sylvain Bromberger and Jerry Katz for their advice on many points, and to my secretary, Martha Sullivan, for her invaluable aid in producing the manuscript for this book. Finally, in dedicating this book to my wife, I am merely expressing my awareness of the many debts that I owe to her, the least of which is for her encouragement that enabled me to complete this book.

GENERAL INTRODUCTION

Philosophers of science are primarily concerned with three kinds of questions. One kind deals with the implications of new scientific findings for traditional philosophical issues; thus, philosophers of science question whether the principle of indeterminacy in quantum mechanics shows that human actions are not entirely determined or whether recent research on computers and artificial intelligence supports the thesis that human beings are merely very complex machines. A second kind of question deals with the analysis of the fundamental concepts of the diverse scientific disciplines; thus, philosophers of science have analyzed such concepts as number, space, force, goal, and living organism. Finally, a third kind of question deals with the nature of the goals of the scientific enterprise and the methods the scientist employs to attain these goals; thus, philosophers of science have argued about whether science should attempt to explain, or merely describe, the data observed in the laboratory or whether the scientist must postulate the existence of unobserved entities in order to deal effectively with his data. The selections in this anthology are directed primarily to this last type of problem, although in the process they frequently deal with the other types as well.

Some philosophers believe that this last type of problem, the nature of the goals and methods of science, cannot be dealt with since there are no permanent goals or methods of science. Much as scientists give up one scientific theory in favor of another, so they formulate new conceptions of their goals and of the methods to be employed in reaching them. Accordingly, the best we can hope for is a detailed account of the many methods for reaching these goals favored by scientists throughout history.

This view is clearly based upon the following assumptions: (a) different scientists at different points in the development of science have had different conceptions of the goals and methods of science; (b) the task of the philosopher of science is to describe the conception of the goals and methods of science actually held by scientists. Those who disagree with this view, while generally in agreement about (a), challenge (b) on the grounds that it confuses a normative discipline like the philosophy of science with descriptive disciplines like the history and sociology of science. After all, the existence of disciplines like the history and sociology of science that describe these many conceptions of the goals and methods of science in no way rules out the legitimacy of a discipline that is aimed at determining which of these conceptions are correct.

This normative conception of the philosophy of science goes back at least as far as the Rennaissance philosophers of science, such as Bacon, Galileo, and Descartes, who claimed that earlier incorrect scientific methods caused the sterility of science before their time. Moreover, many of the authors represented here hold this conception. It is not our purpose at this point to decide whether this normative account, or the earlier descriptive account, of the philosophy of science is correct; we merely want to warn the reader that the following works were written by men with differing viewpoints about this fundamental issue.

The philosophy of science, as conceived of in this book, is not a new discipline. Plato and Aristotle both wrote accounts of the goals and methods of the scientist. Extremely important and valuable work was produced in this area during the seventeenth century by Bacon, Galileo, Descartes, Newton, and Leibniz and during the nineteenth century by Herschel, Whewell, Mill, Jevons, Peirce, Mach, Herz, Poincare, Duhem, and Pearson. However, all the selections included in this book were written in the present century. These earlier writings, although still informative and provocative, contain many ideas which are today considered outmoded; it therefore seemed advisable to let the student begin his study of the philosophy of science with more recent writings. Those interested in acquiring a more thorough understanding of the subject must eventually become familiar with these classical texts. The bibliography contains several references that will enable the student to begin work on this task.

The most important school in twentieth-century philosophy of science is logical positivism (or logical empiricism). This movement began in Austria and Germany in the 1920's, spread to England and America in the 1930's, and continues to have a great influence on the philosophic community in these countries. This school held that most of the traditional philosophical issues are meaningless pseudo-problems and that philosophers ought, therefore, to turn their attention to more legitimate problems such as the nature of the goals and methods of the empirical sciences. Given this great interest in the philosophy of science and a variety of highly sophisticated analytical tools developed by modern logicians, these logical positivists produced a powerful and persuasive conception of the scientific enterprise that has been adopted by many philosophers and scientists who do not agree with the basic presuppositions of logical positivism. Nearly all the sections in this book begin with selections by important logical empiricists, like Professors Rudolph

Carnap, Carl G. Hempel, and Ernest Nagel, who helped develop this conception of science.

More recently, several younger philosophers of science have argued that the logical empiricist's view of science should be supplemented or supplanted. Most of the selections that follow deal with their suggestions. Although no text could possibly include all the interesting material that has appeared in recent years, I have tried to include in this text enough of this material so that the reader can obtain a balanced view of the present state of the subject.

CONTENTS

part 1

EXPLANATION AND PREDICTION: GOALS OF THE SCIENTIFIC ENTERPRISE

INTRODUCTION

Many people believe that the scientist's major concern is with collecting information about particular objects and events. Anyone who believes this must be extremely surprised when, upon reading a science textbook, he finds instead many laws and theories and far fewer items of information about particular objects and events than he had supposed. It then should be come clear to him that the scientist's main task is to formulate laws and theories and to adduce evidence for them.

Why is this the scientist's prime concern? A popular answer is that these laws and theories are needed to fulfill two of the important goals of the scientific enterprise: the explanation of what has been observed and the prediction of what will be observed. This answer clearly presupposes that scientific explanation and prediction necessarily involve laws and theories. Professor Carl G. Hempel defends this presupposition in the first two selections.

Hempel has formulated two models for explanation and prediction. Both are called "covering-law" models because both involve scientific laws and theories. One model, the deductive-nomological model, portrays a type of explanation or prediction where the *explanandum*, the sentence describing the data to be explained, or the *praedicendum*, the sentence describing the predicted occurrence, follows deductively from the *explanans*, the sentence describing that which is being offered as an explanation, or the *praedicens*, the sentence that describes the grounds for the prediction. Moreover, the explanans or praedicens necessarily contains at least one non-statistical law or theory. The other model, the probabilistic model, portrays a type of explanation or prediction where the explanans or praedicens necessarily contains at least one statistical law or theory and where the explanans or praedicens merely offers high inductive support for, but does not entail the truth of, the explanandum or praedicendum.

Hempel claims that all scientific explanations and predictions are either deductive-nomological or probabilistic. What type of evidence does he offer for this claim? To begin with, he gives several examples of scientific explanations and predictions and shows that they fit his model. Secondly, he tries to show that many explanations and predictions that do not seem to fit these models actually do fit them. Finally, and most importantly, he argues that an explanation or prediction must offer us adequate grounds for believing in the truth of the explanandum or of the praedicendum, and only explanations and predictions that fit these models can do this.

The distinctive feature of both models, as stated earlier, is their presupposition that laws and theories are necessary for scientific explanations and predictions. But how may we distinguish between laws and theories on the one hand and ordinary statements of fact on the other? One might think that the difference between a law like "all objects attract each other with a force inversely proportional to the square of the distance between them" and a mere statement of fact like "John went home and had dinner" is that the former is a universal statement whereas the latter is a particular statement. This way of drawing the distinction will not do, however, because there are many true universal statements that are not laws of nature. One example is "All the coins in my pocket at this moment are nickels." Hempel draws the distinction in terms of the potentially infinite scope of a law of nature and in terms of its purely qualitative predicates.

Many philosophers believe that the difference between laws of nature and mere statements of fact is that the former express "necessary" truths while the latter merely express "contingent" truths. Although these philosophers are not entirely clear about what they mean by "necessary," they seem to be postulating a new type of necessity, physical or natural necessity, which is quite different from the formal necessity discussed by the logician. Hempel, in disagreement with these philosophers, is trying to distinguish between laws of nature and mere statements of fact without using this problematic notion of physical necessity.

Several philosophers sympathetic to Hempel's goal have nevertheless objected to his method. Professor Alfred J. Ayer offers several examples of laws of nature that do not meet Hempel's requirements and of mere statements of fact that do meet them. He therefore concludes that Hempel's way of drawing the distinction is inadequate and attempts to draw the distinction in terms of the differences in our attitudes to laws of nature and to mere statements of fact. Unfortunately, he provides us with no justification for having these different attitudes. Professor Richard Braithwaite attempts to draw the distinction in terms of the place the statement occupies in an established scientific system and the consequent difference in the nature of the evidence for the statement. The reader will have to decide for himself whether these alternative approaches are superior to Hempel's and whether these attempts to draw the distinction without invoking the notion of physical necessity are successful.

The covering-law models of explanation and prediction have provoked much discussion. Initially, the discussion centered around whether these models are applicable to explanations outside of the physical sciences; these discussions will be considered below. More recently, several philosophers have argued that these models to not provide us with a correct account of explanations in the physical sciences.

Professor Michael Scriven has criticized the models because of the role they assign to laws and theories in scientific explanation. He argues that: (a) The models mistakenly include in the explanation itself some of the grounds (the laws) for the explanation while they correctly do not include in the explanation itself the rest of the grounds for the explanation; (b) since there are cases where laws are not needed even as grounds for the explanation, the models are mistaken in their presupposition that each explanation presupposes the existence of a scientific law; (c) even when laws are used as grounds for an explanation, they are not, as the model requires, necessarily true. As a matter of fact, most laws are known to be inaccurate; (d) there are explanations that meet all of the requirements of the models but which are nevertheless, for reasons not even considered in the models, invalid.

As Scriven points out, these arguments are based on his own conception of a scientific explanation, viz., the conveyance of an understanding of a phenomenon. Any information that can provide such an understanding, even if it does not involve a law or a theory, is adequate as the explanans of some explanation. An explanation can be challenged in at least three ways: It may be argued that the information conveyed is false, that it has no bearing on the phenomenon being explained, or that it is not the sort of thing that is being looked for in this particular context. Laws and theories, even when inaccurate, can be used as a defense against this second type of challenge, so they are possible (but not necessary) partial grounds for defending an explanation that has been challenged—but this is all that they are.

Unlike Scriven, Professor Sylvain Bromberger does not challenge the need for laws and theories in that type of scientific explanation (answers to certain types of "why-questions") that he discusses. He does, however, challenge Hempel's account of the type of laws that are adequate for this kind of explanation and he describes the type that must be used. He also shows how his account avoids certain ingenious counter-examples to Hempel's models.

Bromberger, however, is not merely concerned with altering the requirements of the model. He wants to raise the more fundamental objection that Hempel and others have discussed under the general rubric of the theory of explanation answers to why-questions, how-questions, what-is-the-cause-of-questions, etc., without realizing that the structure of the explanations that are answers to these very different types of questions is not the same in every case. Consequently, even Bromberger's model is only meant to cover the answer, to one type of why-question. Elsewhere, Bromberger argues for a distinction between the general theory of explanation and a theory of the structure of the answers to these very different types of questions.

The reader should carefully decide whether Scriven and Bromberger have demonstrated that the covering-law models are inadequate as accounts of explanations in the physical sciences. They have, however, clearly raised some doubts about the adequacy of the models.

From the very beginning, proponents of the covering-law models have had to face the challenge that the biological, social, and historical sciences have their own patterns of explanation, ones that are not analyzable as covering-law explanations. In this book, we will not be able to consider all the other modes of explanation that have been proposed; we will only consider functional explanations and certain forms of historical explanations.

Consider the following examples of functional explanations that have been offered in the social and biological sciences:

(1) Why do the Hopi Indians continue to engage in rain-making ceremonies even though these ceremonies do not normally produce the desired result? They do so because these ceremonies serve to reinforce group identity by providing occasions for the group to engage in a common activity, and this reinforcement is vital to the survival and well-being of the group.

(2) Why do all vertebrates have hearts that beat? The heartbeat is needed to circulate blood throughout the organism and this blood is necessary for the survival of the organism.

There are several points that we should note about these examples. First, although they involve goals, the survival and well-being of the group or organism, they do not presuppose that these goals are consciously or unconsciously held by anyone. In this respect, these explanations differ from purposive explanations of human behavior. Second, these explanations still have some explanatory import. And finally, these explanations do not involve any laws, so they are counter-examples to Hempel's claim that all explanations must involve laws or theories.

Professor Ernest Nagel, on the one hand, attempts to avoid this objection by showing how functional or teleological explanations can be translated, without any loss of meaning, into non-functional explanations that fit the covering-law models. Hempel, on the other hand, claims that the above explanations are not satisfactory, and that when they are modified so as to be satisfactory explanations, they will involve laws or theories and will fit the models. These two replies are not incompatible, for Hempel's satisfactory functional explanations are just those that seem to be the most likely candidates for translatability, following the pattern suggested by Nagel, into nonfunctional terms. These replies raise fundamental issues, both about the meaning of functional language and about the role of such explanations in the sciences; thus the reader should carefully consider their adequacy.

We come now to some of the modes of explanation used in the historical sciences. One must keep in mind that history, in the strict sense of that term, is not the only historical science. Biologists, when studying the evolutionary history of a species; geologists, when studying the development of a certain rock formation; and astronomers, when tracing the history of the solar system, are all engaged in historical investigations. And there are some philosophers, like Professor Gallie, who claim that there is a special method of explanation, used in all of these historical sciences, which does not fit the covering-law models.

Gallie has in mind genetic explanations in which the explanans describe a necessary, but not a sufficient, condition for the occurence of the event described in the explanandum. Consequently, the explanans do not entail, nor do they necessarily offer high inductive support for, the explanandum. So we have here a legitimate mode of explanation that does not fit either of the covering-law models.

An obvious objection to Gallie's account is that the existence of a condition necessary for an event does not by itself explain the occurrence of that event. After all, the presence of oxygen in the air is necessary for the occurrence of any human event, like your reading this book, but it hardly explains why that event occurred. Gallie is aware of this objection, but he feels that he can meet it by stipulating that one can use in a genetic explanation only those necessary conditions that are part of a continuing development in a certain direction or that are part of a persisting set of elements within a succession of events. Gallie also stipulates that in the case of human actions, the condition must render the action intelligible or justifiable. Once again, the reader should decide for himself whether this special mode of explanation is legitimate, and, if so, whether Gallie has given an adequate account of the conditions for its legitimate use.

Professor William Dray, in his discussion of Collingwood's theory of historical understanding, is concerned with a method used by some historians to explain the actions of thinking individuals. This type of explanation is therefore used only in history, in the strict sense of that term, and not in the historical sciences in general. These historians explain the actions in question by showing that they were the rational thing for the agent to do in his circumstances. There is, to be sure, a general principle employed in these explanations, viz., that under these conditions, the rational thing to do was what the agent did do, but this is a normative principle rather than a descriptive empirical law. Consequently, the covering-law models do not cover this mode of historical explanation.

Is this a legitimate mode of explanation? If the agent in question is usually an irrational individual, are his actions in this case explainable on the grounds that they were the rational thing to do? Don't we need, instead of Dray's normative principle, some descriptive laws that tell us how agents like the one in question behave under these circumstances? These are questions that Dray must answer before his special mode of explanation can be accepted as satisfactory.

In conclusion, then, the following issues central to the problems of scientific explanation and prediction remain in dispute: the need for laws and theories to explain phenomena, the worth of the covering-law models as a general theory of scientific explanation, and the role that laws and theories play in the development of a science.

The Covering-Law Models and the Nature of Scientific Laws

Carl G. Hempel and
*Paul Oppenheim**

STUDIES IN THE LOGIC OF EXPLANATION

§1. *Introduction*

To explain the phenomena in the world of our experience, to answer the question "Why?" rather than only the question "What?", is one of the foremost objectives of all rational inquiry; and especially, scientific research in its various branches strives to go beyond a mere description of its subject matter by providing an explanation of the phenomena it investigates. While there is rather general agreement about this chief objective of science, there exists considerable difference of opinion as to the function and the essential characteristics of scientific explanation. In the present essay, an attempt will be made to shed some light on these issues by means of an elementary survey of the basic pattern of scientific explanation and a subsequent more rigorous analysis of the concept of law and of the logical structure of explanatory arguments.

The elementary survey is presented in part I of this article; part II contains an analysis of the concept of emergence; in part III, an attempt is made to exhibit and to clarify in a more rigorous manner some of the peculiar and perplexing logical problems to which the familiar elementary analysis of explanation gives rise.

---

*This paper represents the outcome of a series of discussions among the authors; their individual contributions cannot be separated in detail. The technical developments contained in Part IV, however, are due to the first author, who also put the article into its final form. [Parts II and IV omitted in this reprinting.]

We wish to express our thanks to Dr. Rudolf Carnap, Dr. Herbert Feigl, Dr. Nelson Goodman, and Dr. W. V. Quine for stimulating discussions and constructive criticism.

PART I. ELEMENTARY SURVEY OF SCIENTIFIC EXPLANATION

§2. *Some Illustrations*

A mercury thermometer is rapidly immersed in hot water; there occurs a temporary drop of the mercury column, which is then followed by a swift rise. How is this phenomenon to be explained? The increase in temperature affects at first only the glass tube of the thermometer; it expands and thus provides a larger space for the mercury inside, whose surface therefore drops. As soon as by heat conduction the rise in temperature reaches the mercury, however, the latter expands, and as its coefficient of expansion is considerably larger than that of glass, a rise of the mercury level results. This account consists of statements of two kinds. Those of the first kind indicate certain conditions which are realized prior to, or at the same time as, the phenomenon to be explained; we shall refer to them briefly as antecedent conditions. In our illustration, the antecedent conditions include, among others, the fact that the thermometer consists of a glass tube which is partly filled with mercury, and that it is immersed into hot water. The statements of the second kind express certain general laws; in our case, these include the laws of the thermic expansion of mercury and of glass, and a statement about the small thermic conductivity of glass. The two sets of statements, if adequately and completely formulated, explain the phenomenon under consideration: They entail the consequence that the mercury will first drop, then rise. Thus, the event under discussion is explained by subsuming it under general laws, i.e., by showing that it occurred in accordance with those laws, by virtue of the realization of certain specified antecedent conditions.

Consider another illustration. To an observer in a row boat, that part of an oar which is under water appears to be bent upwards. The phenomenon is explained by means of general laws—mainly the law of refraction and the law that water is an optically denser medium than air—and by reference to certain antecedent conditions—especially the facts that part of the oar is in the water, part in the air, and that the oar is practically a straight piece of wood. Thus, here again, the question "*Why* does the phenomenon happen?" is construed as meaning "According to what general laws, and by virtue of what antecedent conditions, does the phenomenon occur?"

So far, we have considered exclusively the explanation of particular events occurring at a certain time and place. But the question "Why?" may be raised also in regard to general laws. Thus, in our last illustration, the question might be asked: Why does the propagation of light conform to the law of refraction? Classical physics answers in terms of the undulatory theory of light, i.e., by stating that the propagation of light is a wave phenomenon of a certain general type, and that all wave phenomena of that type satisfy the law of refraction. Thus, the explanation of a general regularity consists in subsuming it under another, more comprehensive regularity, under a more general law. Similarly, the validity of Galileo's law for the free fall of bodies near the earth's surface can be explained by deducing it from a more comprehensive set of laws, namely Newton's laws of motion and his law of gravitation, together with some statements about particular facts, namely the mass and the radius of the earth.

§3. *The Basic Pattern of Scientific Explanation*

From the preceding sample cases let us now abstract some general characteristics of scientific explanation. We divide an explanation into two major constituents, the *explanandum* and the *explanans*.[1] By the *explanandum*, we understand the sentence describing the phenomenon to be explained (not that phenomenon itself); by the *explanans*, the class of those sentences which are adduced to account for the phenomenon. As was noted before, the explanans falls into two subclasses; one of these contains certain sentences $C_1, C_2, \ldots, C_k$ which state specific antecedent conditions; the other is a set of sentences $L_1, L_2, \ldots, L_r$ which represent general laws.

If a proposed explanation is to be sound, its constituents have to satisfy certain conditions of adequacy, which may be divided into logical and empirical conditions. For the following discussion, it will be sufficient to formulate these requirements in a slightly vague manner; in part III, a more rigorous analysis and a more precise restatement of these criteria will be presented.

I. *Logical conditions of adequacy*

(R_1) The explanandum must be a logical consequence of the explanans; in other words, the explanandum must be logically deducible from the information contained in the explanans, for otherwise, the explanans would not constitute adequate grounds for the explanandum.

(R_2) The explanans must contain general laws, and these must actually be required for the derivation of the explanandum. We shall not make it a necessary condition for a sound explanation, however, that the explanans must contain at least one statement which is not a law; for, to mention just one reason, we would surely want to consider as an explanation the derivation of the general regularities governing the motion of double stars from the laws of celestial mechanics, even though all the statements in the explanans are general laws.

(R_3) The explanans must have empirical content; i.e., it must be capable, at least in principle, of test by experiment or observation. This condition is implicit in (R_1); for since the explanandum is assumed to describe some empirical phenomenon, it follows from (R_1) that the explanans entails at least one consequence of empirical character, and this fact confers upon it testability and empirical content. But the point deserves special mention because, as will be seen in section 4, certain arguments which have been offered as explanations in the natural and in the social sciences violate this requirement.

II. *Empirical condition of adequacy*

(R_4) The sentences constituting the explanans must be true.

That in a sound explanation, the statements constituting the explanans have to satisfy some condition of factual correctness is obvious. But it might seem more appropriate to stipulate that the explanans has to be highly confirmed by all the relevant evidence available rather than that it should be true. This stipulation, however, leads to awkward consequences. Suppose that a certain phenomenon was explained at an earlier stage of science, by means of an explanans which was well supported by the evidence then at hand, but which had been highly disconfirmed by more recent empirical findings. In such a case, we would have to say that originally the explanatory account was a correct explanation, but that it ceased to be one later, when unfavorable evidence was discovered. This does not appear to accord with sound common usage, which directs us to say that on the basis of the limited initial evidence, the truth of the explanans, and thus the soundness of the explanation, had been quite probable, but that the ampler evidence now available made it highly probable that the explanans was not true, and hence that the account in question was not—and had never been—a correct explanation. (A similar point will be made illustrated, with respect to the requirement of truth for laws, in the beginning of section 6.)

Some of the characteristics of an explanation which have been indicated so far may be summarized in the following schema:

$$
\text{Logical deduction}
\begin{array}{lll}
\left\{\begin{array}{l} C_1, C_2, \cdots, C_k \\ \\ L_1, L_2, \cdots, L_r \end{array}\right. &
\begin{array}{l} \text{Statements of antecedent} \\ \text{conditions} \\ \text{General Laws} \end{array} &
\left.\begin{array}{l} \\ \\ \\ \end{array}\right\} \text{Explanans} \\
\hline
\qquad\quad E &
\begin{array}{l} \text{Description of the} \\ \text{empirical phenomenon} \\ \text{to be explained} \end{array} &
\left.\begin{array}{l} \\ \\ \\ \end{array}\right\} \text{Explanandum}
\end{array}
$$

Let us note here that the same formal analysis, including the four necessary conditions, applies to scientific prediction as well as to explanation. The difference between the two is of a pragmatic character. If E is given, i.e., if we know that the phenomenon described by E has occurred, and a suitable set of statements $C_1, C_2, \cdots, C_k, L_1, L_2, \cdots, L_r$ is provided afterwards, we speak of an explanation of the phenomenon in question. If the latter statements are given and E is derived prior to the occurrence of the phenomenon it describes, we speak of a prediction. It may be said, therefore, that an explanation is not fully adequate unless its explanans, if taken account of in time, could have served as a basis for predicting the phenomenon under consideration.[2] Consequently, whatever will be said in this article concerning the logical characteristics of explanation or prediction will be applicable to either, even if only one of them should be mentioned.

It is this potential predictive force which gives scientific explanation its importance: Only to the extent that we are able to explain empirical facts can we attain the major objective of scientific research, namely not merely to record the phenomena of our experience, but to learn from them, by basing upon them theoretical generalizations which enable us to anticipate new occurrences and to control, at least to some extent, the changes in our environment.

Many explanations which are customarily offered, especially in prescientific discourse, lack this predictive character, however. Thus, it may be explained that a car turned over on the road "because" one of its tires blew out while the car was travelling at high speed. Clearly, on the basis of just this information, the accident could not have been predicted, for the explanans provides no explicit general laws by means of which the prediction might be effected, nor does it state adequately the antecedent conditions which would be needed for the prediction. The same point may be illustrated by reference to W. S. Jevons's view that every explanation consists in pointing out a resemblance between facts, and that in some cases this process may require no reference to laws at all and "may involve nothing more than a single identity, as when we explain the appearance of shooting stars by showing that they are identical with portions of a comet."[3] But clearly, this identity does not provide an explanation of the phenomenon of shooting stars unless we presuppose the laws governing the development of heat and light as the effect of friction. The observation of similarities has explanatory value only if it involves at least tacit reference to general laws.

In some cases, incomplete explanatory arguments of the kind here illustrated suppress parts of the explanans simply as "obvious"; in other cases, they seem to involve the assumption that while the missing parts are not obvious, the incomplete explanans could at least, with appropriate effort, be so supplemented as to make a strict derivation of the explanandum possible. This assumption may be justifiable in some cases, as when we say that a lump of sugar disappeared "because" it was put into hot tea, but it is surely not satisfied in many other cases. Thus, when certain peculiarities in the work of an artist are explained as outgrowths of a specific type of neurosis, this observation may contain significant clues, but in general it does not afford a sufficient basis for a potential prediction of those peculiarities. In cases of this kind, an incomplete explanation may at best be considered as indicating some positive correlation between the antecedent conditions adduced and the type of phenomenon to be explained, and as pointing out a direction in which further research might be carried on in order to complete the explanatory account.

The type of explanation which has been considered here so far is often referred to as causal explanation. If E describes a particular event, then the antecedent circumstances described in the sentences $C_1, C_2, \cdots, C_k$ may be said jointly to "cause" that event, in the sense that there are certain empirical regularities, expressed by the laws $L_1, L_2, \cdots, L_r$, which imply that whenever conditions of the kind indicated by $C_1, C_2, \cdots, C_k$ occur, an event of the kind described in E will take place. Statements such as $L_1, L_2, \ldots, L_r$, which assert general and unexceptional connections between specified characteristics of events, are customarily called causal, or deterministic laws. They are to be distinguished from the so-called statistical laws which assert that in the long run, an explicitly stated percentage of all cases satisfying a given set of conditions are accompanied by an event of a certain specified kind. Certain cases of scientific explanation involve "subsumption" of the explanandum under a set of laws of which at least some are statistical in character. Analysis of the peculiar logical structure of that type of subsumption involves difficult special problems. The present essay will be restricted to an examination of the causal type of explanation, which has retained its significance in large segments of contemporary science, and even in some areas where a more adequate account calls for reference to statistical laws.[4]

§4. *Explanation in the Non-Physical Sciences. Motivational and Teleological Approaches*

Our characterization of scientific explanation is so far based on a study of cases taken from the physical sciences. But the general principles thus obtained apply also outside this area.[5] Thus, various types of behavior in laboratory animals and in human subjects are explained in psychology by subsumption under laws or even general theories of learning or conditioning; and while frequently, the regularities invoked cannot be stated with the same generality and precision as in physics or chemistry, it is clear, at least, that the general character of those explanations conforms to our earlier characterization.

Let us now consider an illustration involving sociological and economic factors. In the fall of 1946, there occurred at the cotton exchanges of the United States a price drop which was so severe that the exchanges in New York, New Orleans, and Chicago had to suspend their activities temporarily. In an attempt to explain this occurrence, newspapers traced it back to a large-scale speculator in New Orleans who had feared his holdings were too large and had therefore begun to liquidate his stocks; smaller speculators had then followed his example in a panic and had thus touched off the critical decline. Without attempting to assess the merits of the argument, let us note that the explanation here suggested again involves statements about antecedent conditions and the assumption of general regularities. The former include the facts that the first speculator had large stocks of cotton, that there were smaller speculators with considerable holdings, that there existed the institution of the cotton exchanges with their specific mode of operation, etc. The general regularities referred to are—as often in semi-popular explanations—not explicitly mentioned; but there is obviously implied some form of the law of supply and demand to account for the drop in cotton prices in terms of the greatly increased supply under conditions of practically unchanged demand; besides, reliance is necessary on certain regularities in the behavior of individuals who are trying to preserve or improve their economic position. Such laws cannot be formulated at present with satisfactory precision and generality, and therefore, the suggested explanation is surely incomplete, but its intention is unmistakably to account for the phenomenon by integrating it into a general pattern of economic and socio-psychological regularities.

We turn to an explanatory argument taken from the field of linguistics.[6] In Northern France, there exists a large variety of words synonymous with the English "bee," whereas in southern France, essentially only one such word is in existence. For this discrepancy, the explanation has been suggested that in the Latin epoch, the South of France used the word "apicula," the North the word "apis." The latter, because of a process of phonologic decay in northern France, became the monosyllabic word "é"; and monosyllables tend to be eliminated, especially if they contain few consonantic elements, for they are apt to give rise to misunderstandings. Thus, to avoid confusion, other words were selected. But "apicula," which was reduced to "abelho," remained clear enough and was retained, and finally it even entered into the standard language, in the form "abbeille." While the explanation here described is incomplete in the sense characterized in the previous section, it clearly exhibits reference to specific antecedent conditions as well as to general laws.[7]

While illustrations of this kind tend to support the view that explanation in biology, psychology, and the social sciences has the same structure as in the physical sciences, the opinion is rather widely held that in many instances, the causal type of explanation is essentially inadequate in fields other than physics and chemistry, and especially in the study of purposive behavior. Let us examine briefly some of the reasons which have been adduced in support of this view.

One of the most familiar among them is the idea that events involving the activities of humans singly or in groups have a peculiar uniqueness and irrepeatability which makes them inaccessible to causal explanation because the latter, with its reliance upon uniformities, presupposes repeatablity of the phenomena under consideration. This argument which, incidentally, has also been used in support of the contention that the experimental method is inapplicable in psychology and the social sciences, involves a misunderstanding of the logical character of causal explanation. Every individual event, in the physical sciences no less than in psychology or the social sciences, is unique in the sense that it, with all its peculiar characteristics, does not repeat itself. Nevertheless, individual events may conform to, and thus be explainable by means of, general laws of the causal type. For all that a causal law asserts is that any event of a specified kind, i.e., any event having certain specified characteristics, is accompanied by another event which in turn has certain specified characteristics; for example, that in any event involving friction, heat is developed. And all that is needed for the testability and applicability of such laws is the recurrence of events with the antecedent characteristics, i.e., the repetition of those characteristics, but not of their individual instances. Thus, the argument is inconclusive. It gives occasion, however, to emphasize an important point concerning our earlier analysis: When we spoke of the explanation of a single event, the term "event" referred to the occurrence of some more or less complex characteristic in a specific spatio-temporal location or in a certain individual object, and not to *all* the characteristics of that object, or to all that goes on in that space-time region.

A second argument that should be mentioned here[8] contends that the establishment of scientific generalizations—and thus of explanatory principles—for human behavior is impossible because the reactions of an individual in a given situation depend not only upon that situation, but also upon the previous history of the individual. But surely, there is no a priori reason why generalizations should not be attainable which take into account this dependence of behavior on the past history of the agent. That indeed the given argument "proves" too much, and is therefore a *non sequitur*, is made evident by the existence of certain physical phenomena, such as magnetic hysteresis and elastic fatigue, in which the magnitude of a specific physical effect depends upon the past history of the system involved, and for which nevertheless certain general regularities have been established.

A third argument insists that the explanation of any phenomenon involving purposive behavior calls for reference to motivations and thus for teleological rather than causal analysis. Thus, for example, a fuller statement of the suggested explanation for the break in the cotton prices would have to indicate the large-scale speculator's motivations as one of the factors determining the event in question. Thus, we have to refer to goals sought, and this, so the argument runs, introduces a type of explanation alien to the physical sciences. Unquestionably, many of the—frequently incomplete—explanations which are offered for human actions involve reference to goals and motives; but does this make them essentially different from the causal explanations of physics and chemistry? One difference which suggests itself lies in the circumstance that in motivated behavior, the future appears to affect the present in a manner which is not found in the causal explanations of the physical sciences. But clearly, when the action of a person is motivated, say, by the desire to reach a certain objective, then it is not the as yet unrealized future event of attaining that goal which can be said to determine his present behavior, for indeed the goal may never be actually reached; rather—to put it in crude terms—it is (a) his desire, present before the action, to attain that particular objective, and (b) his belief, likewise present before the action, that such and such a course of action is most likely to have the desired effect. The determining motives and beliefs, therefore, have to be classified among the antecedent conditions of a motivational explanation, and there is no formal difference on this account between motivational and causal explanation.

Neither does the fact that motives are not accessible to direct observation by an outside observer constitute an essential difference between the two kinds of explanation; for also the determining factors adduced in physical explanations are very frequently inaccessible to direct observation. This is the case, for instance, when opposite electric charges are adduced in explanation of the mutual attraction of two metal spheres. The presence of those charges, while eluding all direct observation, can be ascertained by various kinds of indirect test, and that is sufficient to guarantee the empirical character of the explanatory statement. Similarly, the presence of certain motivations may be ascertainable only by indirect methods, which may include reference to linguistic utterances of the subject in question, slips of the pen or of the tongue, etc.; but as long as these methods are "operationally determined" with reasonable clarity and precision, there is no essential difference in this respect between motivational explanation and causal explanation in physics.

A potential danger of explanation by motives lies in the fact that the method lends itself to the facile construction of *ex post facto* accounts without predictive force. It is a widespread tendency to "explain" an action by ascribing it to motives conjectured only after the action has taken place. While this procedure is not in itself objectionable, its soundness requires that (1) the motivational assumptions in question be capable of test, and (2) that suitable general laws be available to lend explanatory power to the assumed motives. Disregard of these requirements frequently deprives alleged motivational explanations of their cognitive significance.

The explanation of an action in terms of the motives of the agent is sometimes considered as a special kind of teleological explanation. As was pointed out above, motivational explanation, if adequately formulated, conforms to the conditions for causal explanation, so that the term "teleological" is a misnomer if it is meant to imply either a non-causal character of the explanation or peculiar determination of the present by the future. If this is borne in mind, however, the term "teleological" may be viewed, in this context, as referring to causal explanations in which some of the antecedent conditions are motives of the agent whose actions are to be explained.[9]

Teleological explanations of this kind have to be distinguished from a much more sweeping type, which has been claimed by certain schools of thought to be indispensable especially in biology. It consists in explaining characteristics of an organism by reference to certain ends or purposes which the characteristics are said to serve. In contradistinction to the cases examined before, the ends are not assumed here to be consciously or subconsciously pursued by the organism in question. Thus, for the phenomenon of mimicry, the explanation is sometimes offered that it serves the purpose of protecting the animals endowed with it from detection by its pursuers and thus tends to preserve the species. Before teleological hypotheses of this kind can be appraised as to their potential explanatory power, their meaning has to be clarified. If they are intended somehow to express the idea that the purposes they refer to are inherent in the design of the universe, then clearly they are not capable of empirical test and thus violate the requirement (R_3) stated in section 3. In certain cases, however, assertions about the purposes of biological characteristics may be translatable into statements in non-teleological terminology which assert that those characteristics function in a specific manner which is essential to keeping the organism alive or to preserving the species.[10] An attempt to state precisely what is meant by this latter assertion—or by the similar one that without those characteristics, and other things being equal, the organism or the species would not survive—encounters considerable difficulties. But these need not be discussed here. For even if we assume that biological statements in teleological form can be adequately translated into descriptive statements about the life-preserving function of certain biological characteristics, it is clear that (1) the use of the concept of purpose is not essential in these contexts, since the term "purpose" can be completely eliminated from the statements in question, and (2) teleological assumptions, while now endowed with empirical content, cannot serve as explanatory principles in the customary contexts. Thus, e.g., the fact that a given species of butterflies displays a particular kind of coloring cannot be inferred from—and therefore cannot be explained by means of—the statement that this type of coloring has the effect of protecting the butterflies from detection by pursuing birds, nor can the presence of red corpuscles in the human blood be inferred from the statement that those corpuscles have a specific function in assimilating oxygen and that this function is essential for the maintenance of life.

One of the reasons for the perseverance of teleological considerations in biology probably lies in the fruitfulness of the teleological approach as a heuristic device: Biological research which was psychologically motivated by a teleological orientation, by an interest in purposes in nature, has frequently led to important results which can be stated in non-teleological terminology and which increase our scientific knowledge of the causal connections between biological phenomena.

Another aspect that lends appeal to teleological considerations is their anthropomorphic character. A teleological explanation tends to make us feel that we really "understand" the phenomenon in question, because it is accounted for in terms of purposes, with which we are familiar from our own experience of purposive behavior. But it is important to distinguish here understanding in the psychological sense of a feeling of empathic familiarity from understanding in the theoretical, or cognitive, sense of exhibiting the phenomenon to be explained as a special case of some general regularity. The frequent insistence that explanation means the reduction of something unfamiliar to ideas or experiences already familiar to us is indeed misleading. For while some scientific explanations do have this psychological effect, it is by no means universal: The free fall of a physical body may well be said to be a more familiar phenomenon than the law of gravitation, by means of which it can be explained; and surely the basic ideas of the theory of relativity will appear to many to be far less familiar than the phenomena for which the theory accounts.

"Familiarity" of the explicans is not only not necessary for a sound explanation (as we have just tried to show), but it is not sufficient either. This is shown by the many cases in which a proposed explicans sounds suggestively familiar, but upon closer inspection proves to be a mere metaphor, or an account lacking testability, or a set of statements which includes no general laws and therefore lacks explanatory power. A case in point is the neovitalistic attempt to explain biological phenomena by reference to an entelechy or vital force. The crucial point here is not—as it is sometimes made out to be—that entelechies cannot be seen or otherwise directly observed; for that is true also of gravitational fields, and yet, reference to such fields is essential in the explanation of various physical phenomena. The decisive difference between the two cases is that the physical explanation provides (1) methods of testing, albeit indirectly, assertions about gravitational fields, and (2) general laws concerning the strength of gravitational fields, and the behavior of objects moving in them. Explanations by entelechies satisfy the analogue of neither of these two conditions. Failure to satisfy the first condition represents a violation of (R_3); it renders all statements about entelechies inaccessible to empirical test and thus devoid of empirical meaning. Failure to comply with the second condition involves a violation of (R_2). It deprives the concept of entelechy of all explanatory import; for explanatory power never resides in a concept, but always in the general laws in which it functions. Therefore, notwithstanding the flavor of familiarity of the metaphor it invokes, the neovitalistic approach cannot provide theoretical understanding.

The preceding observations about familiarity and understanding can be applied, in a similar manner, to the view held by some scholars that the explanation, or the understanding, of human actions requires an empathic understanding of the personalities of the agents.[11] This understanding of another person in terms of one's own psychological functioning may prove a useful heuristic device in the search for general psychological principles which might provide a theoretical explanation; but the existence of empathy on the part of the scientist is neither a necessary nor a sufficient condition for the explanation, or the scientific understanding, of any human action. It is not necessary, for the behavior of psychotics or of people belonging to a culture very different from that of the scientist may sometimes be explainable and predictable in terms of general principles even though the scientist who establishes or applies those principles may not be able to understand his subjects empathically. And empathy is not sufficient to guarantee a sound explanation, for a strong feeling of empathy may exist even in cases where we completely misjudge a given personality. Moreover, as the late Dr. Zilsel has pointed out, empathy leads with ease to incompatible results; thus, when the population of a town has long been subjected to heavy bombing attacks, we can understand, in the empathic sense, that its morale should have broken down completely, but we can understand with the same ease also that it should have developed a defiant spirit of resistance. Arguments of this kind often appear quite convincing; but they are of an *ex post facto* character and lack cognitive significance unless they are supplemented by testable explanatory principles in the form of laws or theories.

Familiarity of the explanans, therefore, no matter whether it is achieved through the use of teleological terminology, through neovitalistic metaphors, or through other means, is no indication of the cognitive import and the predictive force of a proposed explanation. Besides, the extent to which an idea will be considered as familiar varies from person to person and from time to time, and a psychological factor of this kind certainly cannot serve as a standard in assessing the worth of a proposed explanation. The decisive requirement for every sound explanation remains that it subsume the explanandum under general laws.

. . .

PART III. LOGICAL ANALYSIS OF LAW AND EXPLANATION

§6. *Problems of the Concept of General Law*

From our general survey of the characteristics of scientific explanation, we now turn to a closer examination of its logical structure. The explanation of a phenomenon, we noted, consists in its subsumption under laws or under a theory. But what is a law? What is a theory? While the meaning of these concepts seems intuitively clear, an attempt to construct adequate explicit definitions for them encounters considerable difficulties. In the present section, some basic problems of the concept of law will be described and analyzed; in the next section, we intend to propose, on the basis of the suggestions thus obtained, definitions of law and of explanation for a formalized model language of a simple logical structure.

The concept of law will be construed here so as to apply to true statements only. The apparently plausible alternative procedure of requiring high confirmation rather than truth of a law seems to be inadequate: It would lead to a relativized concept of law, which would be expressed by the phrase, "Sentence *S* is a law relatively to the evidence *E*." This does not seem to accord with the meaning customarily assigned to the concept of law in science and in methodological inquiry. Thus, for example, we would not say that Bode's general formula for the distance of the planets from the sun was a law relatively to the astronomical evidence available in the 1770's, when Bode propounded it, and that it ceased to be a law after the discovery of Neptune and the determination of its distance from the sun; rather, we would say that the limited original evidence had given a high probability to the assumption that the formula was a law, whereas more recent additional information reduced that probability so much as to make it practically certain that Bode's formula is not generally true, and hence not a law.[12]

Apart from being true, a law will have to satisfy a number of additional conditions. These can be studied independently of the factual requirement of truth, for they refer, as it were, to all logically possible laws, no matter whether factually true or false. Adopting a convenient term proposed by Goodman,[13] we will say that a sentence is lawlike if it has all the characteristics of a general law, with the possible exception of truth. Hence, every law is a lawlike sentence, but not conversely.

Our problem of analyzing the concept of law thus reduces to that of explicating the meaning of *lawlike sentence*. We shall construe the class of lawlike sentences as including analytic general statements, such as "A rose is a rose," as well as the lawlike sentences of empirical science, which have empirical content.[14] It will not be necessary to require that each lawlike sentence permissible in explanatory contexts be of the second kind; rather, our definition of explanation will be so constructed as to guarantee the factual character of the totality of the laws—though not of every single one of them—which function in an explanation of an empirical fact.

What are the characteristics of lawlike sentences? First of all, lawlike sentences are statements of universal form, such as "All robins' eggs are greenish-blue," "All metals are conductors of electricity," "At constant pressure, any gas expands with increasing temperature." As these examples illustrate, a lawlike sentence usually is not only of universal, but also of conditional form; it makes an assertion to the effect that universally, if a certain set of conditions, C, is realized, then another specified set of conditions, E, is realized as well. The standard form for the symbolic expression of a lawlike sentence is therefore the universal conditional. However, since any conditional statement can be transformed into a non-conditional one, conditional form will not be considered as essential for a lawlike sentence, while universal character will be held indispensable.

But the requirement of universal form is not sufficient to characterize lawlike sentences. Suppose, for example, that a certain basket, b, contains at a certain time t a number of red apples and nothing else.[15] Then the statement

(S_1) Every apple in basket b at time t is red.

is both true and of universal form. Yet the sentence does not qualify as a law; we would refuse, for example, to explain by subsumption under it the fact that a particular apple chosen at random from the basket is red. What distinguishes (S_1) from a lawlike sentence? Two points suggest themselves, which will be considered in turn, namely, finite scope, and reference to a specified object.

First, the sentence (S_1) makes, in effect, an assertion about a finite number of objects only, and this seems irreconcilable with the claim to universality which is commonly associated with the notion of law.[16] But are not Kepler's laws considered as lawlike although they refer to a finite set of planets only? And might we not even be willing to consider as lawlike a sentence such as the following?

(S_2) All the sixteen ice cubes in the freezing tray of this refrigerator have a temperature of less than 10 degrees centigrade.

This point might well be granted; but there is an essential difference between (S_1) on the one hand and Kepler's laws as well as (S_2) on the other: The latter, while finite in scope, are known to be consequences of more comprehensive laws whose scope is not limited, while for (S_1) this is not the case.

Adopting a procedure recently suggested by Reichenbach,[17] we will therefore distinguish between fundamental and derivative laws. A statement will be called a derivative law if it is of universal character and follows from some fundamental laws. The concept of fundamental law requires further clarification; so far, we may say that fundamental laws, and similarly fundamental lawlike sentences, should satisfy a certain condition of non-limitation of scope.

It would be excessive, however, to deny the status of fundamental lawlike sentence to all statements which, in effect, make an assertion about a finite class of objects only, for that would rule out also a sentence such as "All robins' eggs are greenish-blue," since presumably the class of all robins' eggs—past, present, and future—is finite. But again, there is an essential difference between this sentence and, say, (S_1). It requires empirical knowledge to establish the finiteness of the class of robins' eggs, whereas, when the sentence (S_1) is construed in a manner which renders it intuitively unlawlike, the terms "basket *b*" and "apple" are understood so as to imply finiteness of the class of apples in the basket at time *t*. Thus, so to speak, the meaning of its constitutive terms alone—without additional factual information—entails that (S_1) has a finite scope. Fundamental laws, then, will have to be construed so as to satisfy what we have called a condition of non-limited scope; our formulation of that condition however, which refers to what is entailed by "the meaning" of certain expressions, is too vague and will have to be revised later. Let us note in passing that the stipulation here envisaged would bar from the class of fundamental lawlike sentences also such indesirable candidates as "All uranic objects are spherical," where "uranic" means the property of being the planet Uranus; indeed, while this sentence has universal form, it fails to satisfy the condition of non-limited scope.

In our search for a general characterization of lawlike sentences, we now turn to a second clue which is provided by the sentence (S_1). In addition to violating the condition of non-limited scope, this sentence has the peculiarity of making reference to a particular object, the basket *b*; and this, too, seems to violate the universal character of a law.[19] The restriction which seems indicated here should however again be applied to fundamental lawlike sentences only; for a true general statement about the free fall of physical bodies on the moon, while referring to a particular object, would still constitute a law, albeit a derivative one.

It seems reasonable to stipulate, therefore, that a fundamental lawlike sentence must be of universal form and must contain no essential—i.e., uneliminable—occurrences of designations for particular objects. But this is not sufficient; indeed, just at this point, a particularly serious difficulty presents itself. Consider the sentence:

(S_3) Everything that is either an apple in basket *b* at time *t* or a sample of ferric oxide is red.

If we use a special expression, say "*x* is ferple," as synonymous with "*x* is either an apple in *b* at *t* or a sample of ferric oxide," then the content of (S_3) can be expressed in the form:

(S_4) Everything that is ferple is red.

The statement thus obtained is of universal form and contains no designations of particular objects, and it also satisfies the condition of non-limited scope; yet clearly, (S_4) can qualify as a fundamental lawlike sentence no more than can (S_3).

As long as "ferple" is a defined term of our language, the difficulty can readily be met by stipulating that after elimination of defined terms, a fundamental lawlike sentence must not contain essential occurrences of designations for particular objects. But this way out is of no avail when "ferple," or another term of the kind illustrated by it, is a primitive predicate of the language under consideration. This reflection indicates that certain restrictions have to be imposed upon those predicates, i.e., terms for properties or relations, which may occur in fundamental lawlike sentences.[19]

More specifically, the idea suggests itself of permitting a predicate in a fundamental lawlike sentence only if it is purely universal, or, as we shall say, purely qualitative, in character; in other words, if a statement of its meaning does not require reference to any one particular object or spatio-temporal location. Thus, the terms "soft," "green," "warmer than," "as long as," "liquid," "electrically charged," "female," "father of" are purely qualitative predicates, while "taller than the Eiffel Tower," "medieval," "lunar," "arctic," "Ming" are not.[20]

Exclusion from fundamental lawlike sentences of predicates which are not purely qualitative would at the same time ensure satisfaction of the condition of non-limited scope; for the meaning of a purely qualitative predicate does not require a finite extension; and indeed, all the sentences considered above which violate the condition of non-limited scope make explicit or implicit reference to specific objects.

The stipulation just proposed suffers, however, from the vagueness of the concept of purely qualitative predicate. The question whether indication of the meaning of a given predicate in English does or does not require reference to some one specific object does not always permit an unequivocal answer since English as a natural language does not provide explicit definitions or other clear explications of meaning for its terms. It seems therefore reasonable to attempt definition of the concept of law not with respect to English or any other natural language, but rather with respect to a formalized language—let us call it a model language, *L*,—which is governed by a well-determined system of logical rules, and in which every term either is characterized as primitive or is introduced by an explicit definition in terms of the primitives.

This reference to a well-determined system is customary in logical research and is indeed quite natural in the context of any attempt to develop precise criteria for certain logical distinctions. But it does not by itself suffice to overcome the specific difficulty under discussion. For while it is now readily possible to characterize as not purely qualitative all those among the defined predicates in *L* whose definiens contain an essential occurrence of some individual name, our problem remains open for the primitives of the language, whose meanings are not determined by definitions within the language, but rather by semantical rules of interpretation. For we want to permit the interpretation of the primitives of *L* by means of such attributes as blue, hard, solid, warmer, but not by the properties of being a descendant of Napoleon, or an arctic animal or a Greek statue; and the difficulty is precisely that of stating rigorous criteria for the distinction between the permissible and the non-permissible interpretations. Thus the problem of setting up an adequate definition for purely qualitative attributes now arises again; namely for the concepts of the meta-language in which the semantical interpretation of the primitives is formulated. We may postpone an encounter with the difficulty by presupposing formalization of the semantical meta-language, the meta-meta-language, and so forth; but somewhere, we will have to stop at a non-formalized meta-language, and for it a characterization of purely qualitative predicates will be needed and will present much the same problems as non-formalized English, with which we began. The characterization of a purely qualitative predicate as one whose meaning can be made explicit without reference to any one particular object points to the intended meaning but does not explicate it precisely, and the problem of an adequate definition of purely qualitative predicates remains open.

There can be little doubt, however, that there exists a large number of property and relation terms which would be rather generally recognized as purely qualitative in the sense here pointed out, and as permissible in the formulation of fundamental lawlike sentences; some examples have been given above, and the list could be readily enlarged. When we speak of purely qualitative predicates, we shall henceforth have in mind predicates of this kind.

. . .

NOTES

1. These two expressions, derived from the Latin *explanare*, were adopted in preference to the perhaps more customary terms "explicandum" and "explicans" in order to reserve the latter for use in the context of explication of meaning, or analysis. On explication in this sense, cf. Carnap [Concepts], p. 513. Abbreviated titles in brackets refer to the bibliography at the end of this article.

2. The logical similarity of explanation and prediction, and the fact that one is directed towards past occurrences, the other towards future ones, is well expressed in the terms "postdictability" and "predictability" used by Reichenbach in [Quantum Mechanics], p. 13.

3. [Principles], p. 533.

4. The account given above of the general characteristics of explanation and prediction in science is by no means novel; it merely summarizes and states explicitly some fundamental points which have been recognized by many scientists and methodologists.

Thus, e.g., Mill says: "An individual fact is said to be explained by pointing out its cause, that is, by stating the law or laws of causation of which its production is an instance," and "a law of uniformity in nature is said to be explained when another law or laws are pointed out, of which that law itself is but a case, and from which it could be deduced." ([Logic], Book III, Chap. xii, Sec. 1). Similarly, Jevons, whose general characterization of explanation was critically discussed above, stresses that "the most important process of explanation consists in showing that an observed fact is one case of a general law or tendency." ([Principles], p. 533.) Ducasse states the same point as follows: "Explanation essentially consists in the offering of a hypothesis of fact, standing to the fact to be explained as case of antecedent to case of consequent of some already known law of connection." ([Explanation], pp. 150–51.) A lucid analysis of the fundamental structure of explanation and prediction was given by Popper in [Forschung], Sec. 12, and, in an improved version, in his work [Society], especially in Chap. xxv and in n. 7 referring to that chapter. For a recent characterization of explanation as subsumption under general theories, cf., for example, Hull's concise discussion in [Principles], Chap. 1. A clear elementary examination of certain aspects of explanation is given in Hospers [Explanation], and a concise survey of many of the essentials of scientific explanation which are considered in the first two parts of the present study may be found in Feigl [Operationism], pp. 284 ff.

5. On the subject of explanation in the social sciences, especially in history, cf. also the following publications, which may serve to supplement and amplify the brief discussion to be presented here: Hempel ["Laws"]; Popper [*Society*]; White ["Explanation"]; and the articles "Cause" and "Understanding" in Beard and Hook [Terminology].

6. The illustration is taken from Bonfante [Semantics], Sec. 3.

7. While in each of the last two illustrations, certain regularities are unquestionably relied upon in the explanatory argument, it is not possible to argue convincingly that the intended laws, which at present cannot all be stated explicitly, are of a causal rather than a statistical character. It is quite possible that most or all of the regularities which will be discovered as sociology develops will be of a statistical type. Cf., on this point, the suggestive observations by Zilsel in [Empiricism] Sec. 8, and [Laws]. This issue does not affect, however, the main point we wish to make here, namely that in the social no less than in the physical sciences, subsumption under general regularities is indispensable for the explanation and the theoretical understanding of any phenomenon.

8. Cf., for example, F. H. Knight's presentation of this argument in [Limitations], pp. 251–52.

9. For a detailed logical analysis of the character and the function of the motivation concept in psychological theory, see Koch [Motivation]. A stimulating discussion of teleological behavior from the standpoint of contemporary physics and biology is contained in the article [Teleology] by Rosenblueth, Wiener and Bigelow. The authors propose an interpretation of the concept of purpose which is free from metaphysical connotations, and they stress the importance of the concept thus obtained for a behavioristic analysis of machines and living organisms. While our formulations above intentionally use the crude terminology frequently applied in philosophical arguments concerning the applicability of causal explanation to purposive behavior, the analysis presented in the article referred to is couched in behavioristic terms and avoids reference to "motives" and the like.

10. An analysis of teleological statements in biology along these lines may be found in Woodger [Principles], especially pp. 432 ff.; essentially the same interpretation is advocated by Kaufmann in [Methodology], Chap. 8.

11. For a more detailed discussion of this view on the basis of the general principles outlined above, cf. Zilsel [Empiricism], Secs. 7 and 8, and Hempel [Laws], Sec. 6.

12. The requirement of truth for laws has the consequence that a given empirical statement S can never be definitely known to be a law; for the sentence affirming the truth of S is logically equivalent with S and is therefore capable only of acquiring a more or less high probability, or degree of confirmation, relatively to the experimental evidence available at any given time. On this point, cf. Carnap [Remarks]. For an excellent non-technical exposition of the semantical concept of truth, which is here applied, the reader is referred to Tarski [Truth].

13. [Counterfactuals]. p. 125.

14. This procedure was suggested by Goodman's approach in [Counterfactuals]. Reichenbach, in a detailed examination of the concept of law, similarly construes his concept of nomological statement as including both analytic and synthetic sentences; cf. [Logic], Chap. viii.

15. The difficulty illustrated by this example was stated concisely by Langford [Review], who referred to it as the problem of distinguishing between universals of fact and causal universals. For further discussion and illustration of this point, see also Chisholm [Conditional], especially pp. 301 f. A systematic analysis of the problem was given by Goodman in [Counterfactuals], especially Part III. While not concerned with the specific point under discussion, the detailed examination of counterfactual conditionals and their relation to laws of nature, in Chap. viii of Lewis's work [Analysis], contains important observations on several of the issues raised in the present section.

16. The view that laws should be construed as not being limited to a finite domain has been expressed, among others, by Popper [Forschung], Sec. 13 and by Reichenbach [Logic], p. 369.

17. [Logic], p. 361. Our terminology as well as the definitions to be proposed later for the two types of law do not coincide with Reichenbach's, however.

18. In physics, the idea that a law should not refer to any particular object has found its expression in the maxim that the general laws of physics should contain no reference to specific space-time points, and that spatio-temporal coordinates should occur in them only in the form of differences or differentials.

19. The point illustrated by the sentences (S_3) and (S_4) above was made by Goodman, who has also emphasized the need to impose certain restrictions upon the predicates whose occurrence is to be permissible in lawlike sentences. These predicates are essentially the same as those which Goodman calls projectible. Goodman has suggested that the problems of establishing precise criteria for projectibility, of interpreting counterfactual conditionals, and of defining the concept of law are so intimately related as to be virtually aspects of a single problem. (Cf. his articles [Query] and [Counterfactuals].) One suggestion for an analysis of projectibility has recently been made by Carnap in [Application]. Goodman's note [Infirmities] contains critical observations on Carnap's proposals.

20. That laws, in addition to being of universal form, must contain only purely universal predicates was clearly argued by Popper ([Forschung], Sec. 14, 15). Our alternative expression "purely qualitative predicate" was chosen in analogy to Carnap's term "purely qualitative property" (cf. [Application]). The above characterization of purely universal predicates seems preferable to a simpler and perhaps more customary one, to the effect that a statement of the meaning of the predicate must require no reference to particular objects. For this formulation might be too exclusive since it could be argued that stating the meaning of such purely qualitative terms as "blue" or "hot" requires illustrative reference to some particular object which has the quality in question. The essential point is that no one specific object has to be chosen; any one in the logically unlimited set of blue or of hot objects will do. In explicating the meaning of "taller than the Eiffel Tower," "being an apple in basket *b* at the time *t*," "medieval," etc., however, reference has to be made to one specific object or to some one in a limited set of objects.

BIBLIOGRAPHY

Throughout the article, the abbreviated titles in brackets are used for reference.

Beard, Charles A., and Hook, Sidney, [Terminology], "Problems of terminology in historical writing." Chap. iv of Theory and practice in historical study: A report of the Committee on Historiography. New York: Social Science Research Council, 1946.

Bergmann, Gustav, [Emergence], "Holism, historicism, and emergence." *Philosophy of Science* II, (1944), 209–21.

Bonfante, G., [Semantics], Semantics, language. An article in P. L. Harriman, ed., *The Encyclopedia of Psychology*. Philosophical Library, New York, 1946.

Broad, C. D., [Mind], *The mind and its place in nature*. New York, 1925.

Carnap, Rudolf, [Semantics], *Introduction to Semantics*. Harvard University Press, 1942.

———. [Inductive Logic], "On Inductive Logic." *Philosophy of science*, XII (1945), 72–97.

———. [Concepts], "The Two Concepts of Probability." *Philosophy and Phenomenological Research*, V (1945), 513–32.

———. [Remarks], "Remarks on Induction and Truth." *Philosophy and Phenomenological Research*, VI (1946), 590–602.

———. [Application], "On the Application of Inductive Logic." *Philosophy and Phenomenological Research*, VIII (1947), 133–47.

Chisholm, Roderick M., [Conditional], "The Contrary-to-Fact Conditional." *Mind*, IV (1946), 289–307.

Church, Alonzo, [Logic], "Logic, Formal." An article in Dagobert D. Runes, ed. *The Dictionary of Philosophy*. Philosophical Library, New York, 1942.

Ducasse, C. J., [Explanation], "Explanation, Mechanism, and Teleology." *The Journal of Philosophy*, XXII (1925), 150–55.

Feigl, Herbert, [Operationism], "Operationism and Scientific Method." *Psychological Review*, LII (1945), 250–59, 284–88.

Goodman, Nelson, [Query], "A Query on Confirmation." *The Journal of Philosophy*, XLIII (1946), 383–85.

———. [Counterfactuals], "The Problem of Counterfactual Conditionals." *The Journal of Philosophy*, XLIV (1947), 113–28.

———. [Infirmities], "On Infirmities of Confirmation Theory." *Philosophy and Phenomenological Research*, VIII (1947), 149–51.

Grelling, Kurt, and Oppenheim, Paul, [Gestaltbegriff], "Der Gestaltbegriff im Lichte der neuen Logik." *Erkenntnis*, VII (1937–38), 211–25, 357–59.

Grelling, Kurt, and Oppenheim, Paul, [Functional Whole], "Logical Analysis of Gestalt as Functional Whole." Preprinted for distribution at Fifth Internat. Congress for the Unity of Science, Cambridge, Mass., 1939.

Helmer, Olaf, and Oppenheim, Paul, [Probability], "A Syntactical Definition of Probability and of Degree of Confirmation." *The Journal of Symbolic Logic*, X (1945), 25–60.

Hempel, Carl G., [Laws], "The Function of General Laws in History." *The Journal of Philosophy*, XXXIX (1942), 35–48.

———. [Studies], "Studies in the Logic of Confirmation." *Mind*, LIV (1945); Part I: 1–26, Part II: 97–121.

Hempel, Carl G., and Oppenheim, Paul , [Degree], "A Definition of Degree of Confirmation." *Philosophy of Science*, XII (1945), 98–115.

Henle, Paul, [Emergence], "The Status of Emergence." *The Journal of Philosophy*, XXXIX (1942), 486–93.

Hospers, John, [Explanation], "On Explanation." *The Journal of Philosophy*, XLIII (1946), 337–56.

Hull, Clark L., [Variables], "The Problem of Intervening Variables in Molar Behavior Theory." *Psychological Review*, L (1943), 273–91.

———. [Principles] *Principles of Behavior*. New York, 1943.

Jevons, W. Stanley, [Principles], *The Principles of Science*. London, 1924. (1st ed. 1874.)

Kaufmann, Felix, [Methodology], *Methodology of the Social Sciences*. New York, 1944.

Knight, Frank H., [Limitations], "The Limitations of Scientific Method in Economics." In Tugwell, R., ed., *The Trend of Economics*. New York, 1924.

Koch, Sigmund, [Motivation], "The Logical Character of the Motivation Concept." *Psychological Review*, XLVIII (1941). Part I: 15–38, Part II: 127–54.

Langford, C. H., [Review], "Review" *in The Journal of Symbolic Logic*, VI (1941), 67–68.

Lewis, C. I., [Analysis], *An Analysis of Knowledge and Valuation*. La Salle, Ill., 1946.

McKinsey, J. C. C., [Review], "Review of Helmer and Oppenheim" [Probability]. *Mathematical Reviews*, VII (1946), 45.

Mill, John Stuart, [Logic], *A System of Logic*.

Morgan, C. Lloyd, *Emergent Evolution*. New York, 1923.

———. *The Emergence of Novelty*. New York, 1933.

Popper, Karl, [Forschung], *Logik der Forschung*. Wien, 1935.

———. [Society], *The Open Society and its Enemies*. London, 1945.

Reichenbach, Hans, [Logic], *Elements of Symbolic Logic*. New York, 1947.

———. [Quantum mechanics], *Philosophic Foundations of Quantum Mechanics*. University of California Press, 1944.

Rosenblueth, A., Wiener, N., and Bigelow, J., [Teleology], "Behavior, Purpose, and Teleology." *Philosophy of Science*, X (1943), 18–24.

Stace, W. T., [Novelty], "Novelty, Indeterminism and Emergence." *Philosophical Review*, XLVIII (1939), 296–310.

Tarski, Alfred, [Truth], "The Semantical Conception of Truth, and the Foundations of Semantics." *Philosophy and Phenomenological Research*, IV (1944), 341–76.

Tolman, Edward Chase, [Behavior], "Purposive Behavior in Animals and Men." New York, 1932.

White, Morton G., [Explanation], Historical Explanation. *Mind*, LII (1943), 212–29.

Woodger, J. H., [Principles], *Biological Principles*. New York, 1929.

Zilsel, Edgar, [Empiricism], "Problems of Empiricism." In *International Encyclopedia of Unified Science*, II, No. 8. The University of Chicago Press, 1941.

———. [Laws], "Physics and the Problem of Historico-Sociological Laws." *Philosophy of Science*, VIII (1941), 567–79.

Carl G. Hempel

PROBABILISTIC EXPLANATION

. . .

5.4 PROBABILISTIC EXPLANATION: FUNDAMENTALS

Not all scientific explanations are based on laws of strictly universal form. Thus, little Jim's getting the measles might be explained by saying that he caught the disease from his brother, who had a bad case of the measles some days earlier. This account again links the explanandum event to an earlier occurrence, Jim's exposure to the measles; the latter is said to provide an explanation because there is a connection between exposure to the measles and contracting the disease. That connection cannot be expressed by a law of universal form, however; for not every case of exposure to the measles produces contagion. What can be claimed is only that persons exposed to the measles will contract the disease with high probability, i.e., in a high percentage of all cases. General statements of this type, which we shall soon examine more closely, will be called *laws of probabilistic form or probabilistic laws*, for short.

In our illustration, then, the explanans consists of the probabilistic law just mentioned and the statement that Jim was exposed to the measles. In contrast to the case of deductive-nomological explanation, these explanans statements do not deductively imply the explanandum statement that Jim got the measles; for in deductive inferences from true premisses, the conclusion is invariably true, whereas in our example, it is clearly possible that the explanans statements might be true and yet the explanandum statement false. We will say, for short, that the explanans implies the explanandum, not with "deductive certainty," but only with near-certainty or with high probability.

The resulting explanatory argument may be schematized as follows:

> The probability for persons exposed to the measles
> to catch the disease is high.
> Jim was exposed to the measles.
> ════════════════════ [makes highly probable]
> Jim caught the measles.

In the customary presentation of a deductive argument, which was used, for example, in the schema (*D-N*) above, the conclusion is separated from the premisses by a single line, which serves to indicate that the premisses logically imply the conclusion. The double line used in our latest schema is meant to indicate analogously that the "premisses" (the explanans) make the "conclusion" (the explanandum sentence) more or less probable; the degree of probability is suggested by the notation in brackets.

Arguments of this kind will be called *probabilistic explanations*. As our discussion shows, a probabilistic explanation of a particular event shares certain basic characteristics with the corresponding deductive-nomological type of explanation. In both cases, the given event is explained by reference to others, with which the explanandum event is connected by laws. But in one case, the laws are of universal form; in the other, of probabilistic form. And while a deductive explanation shows that, on the information contained in the explanans, the explanandum was to be expected with "deductive certainty," an inductive explanation shows only that, on the information contained in the explanans, the explanandum was to be expected with high probability, and perhaps with "practical certainty"; it is in this manner that the latter argument meets the requirement of explanatory relevance.

5.5. STATISTICAL PROBABILITIES AND PROBABILISTIC LAWS

We must now consider more closely the two differentiating features of probabilistic explanation that have just been noted: the probabilistic laws they invoke and the peculiar kind of probabilistic implication that connects the explanans with the explanandum.

Suppose that from an urn containing many balls of the same size and mass, but not necessarily of the same color, successive drawings are made. At each drawing, one ball is removed, and its color is noted. Then the ball is returned to the urn, whose contents are thoroughly mixed before the next drawing takes place. This is an example of a so-called random process or random experiment, a concept that will soon be characterized in more detail. Let us refer to the procedure just described as experiment *U*, to each drawing as one performance of *U*, and to the color of the ball produced by a given drawing as the result, or the outcome, of that performance.

If all the balls in an urn are white, then a statement of strictly universal form holds true of the results produced by the performance of U: Every drawing from the urn yields a white ball, or yields the result W, for short. If only some of the balls—say, 600 of them—are white, whereas the others—say 400—are red, then a general statement of probabilistic form holds true of the experiment: The probability for a performance of U to produce a white ball, or outcome W, is .6; in symbols:

(5a) $$P(W, U) = .6$$

Similarly, the probability of obtaining heads as a result of the random experiment C of flipping a fair coin is given by

(5b) $$P(H, C) = .5$$

and the probability of obtaining an ace as a result of the random experiment D of rolling a regular die is

(5c) $$P(A, D) = \frac{1}{6}$$

What do such probability statements mean? According to one familiar view, sometimes called the "classical" conception of probability, the statement (5a) would have to be interpreted as follows: Each performance of the experiment U effects a choice of one from among 1,000 basic possibilities, or basic alternatives, each represented by one of the balls in the urn; of these possible choices, 600 are "favorable" to the outcome W; and the probability of drawing a white ball is simply the ratio of the number of favorable choices available to the number of all possible choices, i.e., 600/1,000. The classical interpretation of the probability statements (5b) and (5c) follows similar lines.

Yet this characterization is inadequate; for if before each drawing, the 400 red balls in the urn were placed on top of the white ones, then in this new kind of urn experiment—let us call it U'—the ratio of favorable to possible basic alternatives would remain the same, but the probability of drawing a white ball would be smaller than in the experiment U, in which the balls are thoroughly mixed before each drawing. The classical conception takes account of this difficulty by requiring that the basic alternatives referred to in its definition of probability must be "equipossible" or "equiprobable"—a requirement presumably violated in the case of experiment U'.

This added proviso raises the question of how to define equipossibility or equiprobability. We will pass over this notoriously troublesome and controversial issue, because—even assuming that equiprobability can be satisfactorily characterized—the classical conception would still be inadequate, since probabilities are assigned also to the outcomes of random experiments for which no plausible way is known of marking off equiprobable basic alternatives. Thus, for the random experiment D of rolling a regular die, the six faces might be regarded as representing such equiprobable alternatives; but we attribute probabilities to such results as rolling an ace, or an odd number of points, etc., also in the case of a loaded die, even though no equiprobable basic outcomes can be marked off here.

Similarly—and this is particularly important—science assigns probabilities to the outcomes of certain random experiments or random processes encountered in nature, such as the step-by-step decay of the atoms of radioactive substances, or the transition of atoms from one energy state to another. Here again, we find no equiprobable basic alternatives in terms of which such probabilities might be classically defined and computed.

To arrive at a more satisfactory construal of our probability statements, let us consider how one would ascertain the probability of the rolling of an ace with a given die that is not known to be regular. This would obviously be done by making a large number of throws with the die and ascertaining the *relative frequency*, i.e., the proportion, of those cases in which an ace turns up. If, for example, the experiment D' of rolling the given die is performed 300 times and an ace turns up in 62 cases, then the relative frequency, 62/300, would be regarded as an approximate value of the probability $p(A, D')$ of rolling an ace with the given die. Analogous procedures would be used to estimate the probabilities associated with the flipping of a given coin, the spinning of a roulette wheel, and so on. Similarly, the probabilities associated with radioactive decay, with the transitions between different atomic energy states, with genetic processes, etc., are determined by ascertaining the corresponding relative frequencies; however, this is often done in highly indirect ways rather than by simply counting individual atomic or other events of the relevant kinds.

The interpretation in terms of relative frequencies applies also to probability statements such as (5b) and (5c), which concern the results of flipping a fair (i.e., homogeneous and strictly cylindrical) coin or tossing a regular (homogeneous and strictly cubical) die: What the scientist (or the gambler, for that matter) is concerned with in making a probability statement is the relative frequency with which a certain outcome O can be expected in long series of repetitions of some random experiment R. The counting of "equiprobable" basic alternatives and of those among them which are "favorable" to O may be regarded as a heuristic device for guessing at the relative frequency of O. And indeed when a regular die or a fair coin is tossed a large number of times, the different faces tend to come up with equal frequency. One might expect this on the basis of symmetry considerations of the kind frequently used in forming physical hypotheses, for our empirical knowledge affords no grounds on which to expect any of the faces to be favored over any other. But while such considerations often are heuristically useful, they must not be regarded as certain or as self-evident truths: Some very plausible symmetry assumptions, such as the principle of parity, have been found not to be generally satisfied at the subatomic level. Assumptions about equiprobabilities are therefore always subject to correction in the light of empirical data concerning the actual relative frequencies of the phenomena in question. This point is illustrated also by the statistical theories of gases developed by Bose and Einstein and by Fermi and

Dirac, respectively, which rest on different assumptions concerning what distributions of particles over a phase space are equiprobable.

The probabilities specified in the probabilistic laws, then, represent relative frequencies. They cannot, however, be strictly defined as relative frequencies in long series of repetitions of the relevant random experiment. For the proportion, say, of aces obtained in throwing a given die will change, if perhaps only slightly, as the series of throws is extended; and even in two series of exactly the same length, the number of aces will usually differ. We do find, however, that as the number of throws increases, the relative frequency of each of the different outcomes tends to change less and less, even though the results of successive throws continue to vary in an irregular and practically unpredictable fashion. This is what generally characterizes a random experiment R with outcomes O_1, O_2, . . . O_n: Successive performances of R yield one or another of those outcomes in an irregular manner; but the relative frequencies of the outcomes tend to become stable as the number of performances increases. And the probabilities of the outcomes, $p(O_1, R), p(O_2, R), \ldots, p(O_n, R)$, may be regarded as ideal values that the actual frequencies tend to assume as they become increasingly stable. For mathematical convenience, the probabilities are sometimes defined as the mathematical *limits* toward which the relative frequencies converge as the number of performances increases indefinitely. But this definition has certain conceptual shortcomings, and in some more recent mathematical studies of the subject, the intended empirical meaning of the concept of probability is deliberately, and for good reasons, characterized more vaguely by means of the following so-called *statistical interpretation of probability:*[1]

The statement

$$p(O, R) = r$$

means that in a long series of performances of random experiment R, the proportion of cases with outcome O is almost certain to be close to r.

The concept of *statistical probability* thus characterized must be carefully distinguished from the concept of *inductive or logical probability*, which we considered in section 4.5. Logical probability is a quantitative logical relation between definite *statements;* the sentence

$$c(H, K) = r$$

asserts that the hypothesis H is supported, or made probable, to degree r by the evidence formulated in statement K. Statistical probability is a quantitative relation between repeatable *kinds of events*: a certain kind of outcome, O, and a certain kind of random process, R; it represents, roughly speaking, the relative frequency with which the result O tends to occur in a long series of performances of R.

What the two concepts have in common are their mathematical characteristics: Both satisfy the basic principles of mathematical probability theory:

(a) The possible numerical values of both probabilities range from 0 to 1:

$$0 \leqslant p(O, R) \leqslant 1$$
$$0 \leqslant c(H, K) \leqslant 1$$

(b) The probability for one of two mutually exclusive outcomes of R to occur is the sum of the probabilities of the outcomes taken separately; the probability, on any evidence K, for one or the other of two mutually exclusive hypotheses to hold is the sum of their respective probabilities:

If O_1, O_2 are mutually exclusive, then

$$p(O_1 \text{ or } O_2, R) = p(O_1, R) + p(O_2, R)$$

If H_1, H_2 are logically exclusive hypotheses, then

$$c(H_1 \text{ or } H_2, K) = c(H_1, K) + c(H_2, K)$$

(c) The probability of an outcome that necessarily occurs in all cases—such as O or not O—is 1; the probability, on any evidence, of a hypothesis that is logically (and in this sense necessarily) true, such as H or not H, is 1:

$$p(O \text{ or not } O, R) = 1$$
$$c(H \text{ or not } H, K) = 1$$

Scientific hypotheses in the form of statistical probability statements can be, and are, tested by examining the long-run relative frequencies of the outcomes concerned; and the confirmation of such hypotheses is then judged, broadly speaking, in terms of the closeness of the agreement between hypothetical probabilities and observed frequencies. The logic of such tests, however, presents some intriguing special problems, which call for at least brief examination.

Consider the hypothesis, H, that the probability of rolling an ace with a certain die is .15; or briefly, that $p(A, D) = .15$, where D is the random experiment of rolling the given die. The hypothesis H does not deductively imply any test implications specifying how many aces will occur in a finite series of throws of the die. It does not imply, for example, that exactly 75 among the first 500 throws will yield an ace, nor even that the number of aces will lie between 50 and 100, say. Hence, if the proportion of aces actually obtained in a large number of throws differs considerably from .15, this does not refute H in the sense in which a hypothesis of strictly universal form, such as "All swans are white, "can be refuted, in virtue of the *modus tollens* argument, by reference to one counter-instance, such as a black swan. Similarly, if a long run of throws of the given die yields a proportion of aces very close to .15, this does not confirm H in the sense in which a hypothesis is confirmed by the finding that a test sentence I that it logically implies is in fact true. For in this latter case, the hypothesis asserts I by logical implication, and the test result is thus confirmatory in the sense of showing that a certain part of what the hypothesis asserts is indeed true; but nothing strictly analogous is shown for *H* by confirmatory frequency data; for *H* does *not* assert by implication that the frequency of aces in some long run will definitely be very close to .15.

But while *H* does not logically preclude the possibility that the proportion of aces obtained in a long series of throws of the given die may depart widely from .15, it does logically imply that such departures are highly improbable in the statistical sense, i.e., that if the experiment of performing a long series of throws (say, 1,000 of them per series) is repeated a large number of times, then only a tiny proportion of those long series will yield a proportion of aces that differs considerably from .15. For the case of rolling a die, it is usually assumed that the results of successive throws are "statistically independent"; this means roughly that the probability of obtaining an ace in a throw of the die does not depend on the result of the preceding throw. Mathematical analysis shows that in conjunction with this independence assumption, our hypothesis *H* deductively determines the statistical probability for the proportion of aces obtained in *n* throws to differ from .15 by no more than a specified amount. For example, *H* implies that for a series of 1,000 throws of the die here considered, the probability is about .976 that the proportion of aces will lie between .125 and .175; and similarly, that for a run of 10,000 throws the probability is about .995 that the proportion of aces will be between .14 and .16. Thus, we may say that if *H* is true, then it is practically certain that in a long trial run the observed proportion of aces will differ by very little from the hypothetical probability value .15. Hence, if the observed long-run frequency of an outcome is not close to the probability assigned to it by a given probabilistic hypothesis, then that hypothesis is very likely to be false. In this case, the frequency data count as disconfirming the hypothesis, or as reducing its credibility; and if sufficiently strong disconfirming evidence is found, the hypothesis will be considered as practically, though not logically, refuted and will accordingly be rejected. Similarly, close agreement between hypothetical probabilities and observed frequencies will tend to confirm a probabilistic hypothesis and may lead to its acceptance.

If probabilistic hypotheses are to be accepted or rejected on the basis of statistical evidence concerning observed frequencies, then appropriate standards are called for. These will have to determine (1) what deviations of observed frequencies from the probability stated by a hypothesis are to count as grounds for rejecting the hypothesis, and (2) how close an agreement between observed frequencies and hypothetical probability is to be required as a condition for accepting the hypothesis. The requirements in question can be made more or less strict, and their specification is a matter of choice. The stringency of the chosen standards will normally vary with the context and the objectives of the research in question. Broadly speaking, it will depend on the importance that is attached, in the given context, to avoiding two kinds of error that might be made: rejecting the hypothesis under test although it is true, and accepting it although it is false. The importance of this point is particularly clear when acceptance or rejection of the hypothesis is to serve as a basis for practical action. Thus, if the hypothesis concerns the probable effectiveness and safety of a new vaccine, then the decision about its acceptance will have to take into account not only how well the statistical test results accord with the probabilities specified by the hypothesis, but also how serious would be the consequences of accepting the hypothesis and acting on it (e.g., by inoculating children with the vaccine) when in fact it is false, and of rejecting the hypothesis and acting accordingly (e.g., by destroying the vaccine and modifying or discontinuing the process of manufacture) when in fact the hypothesis is true. The complex problems that arise in this context form the subject matter of the theory of statistical tests and decisions, which has been developed in recent decades on the basis of the mathematical theory of probability and statistics.[2]

Many important laws and theoretical principles in the natural sciences are of probabilistic character, though they are often of more complicated form than the simple probability statements we have discussed. For example, according to current physical theory, radioactive decay is a random phenomenon in which the atoms of each radioactive element possess a characteristic probability of disintegrating during a specified period of time. The corresponding probabilistic laws are usually formulated as statements giving the "half-life" of the element concerned. Thus, the statements that the half-life of radium^{226} is 1,620 years and that of polonium^{218} is 3.05 minutes are laws to the effect that the probability for a radium^{226} atom to decay within 1,620 years, and for an atom of polonium^{218} to decay within 3.05 minutes, are both one-half. According to the statistical interpretation cited earlier, these laws imply that of a large number of radium^{226} atoms or of polonium^{218} atoms given at a certain time, very close to one-half will still exist 1,620 years, or 3.05 minutes, later; the others having disintegrated by radioactive decay.

Again, in the kinetic theory various uniformities in the behavior of gases, including the laws of classical thermodynamics, are explained by means of certain assumptions about the constituent molecules; and some of these are probabilistic hypotheses concerning statistical regularities in the motions and collisions of those molecules.

A few additional remarks concerning the notion of a probabilistic law are indicated. It might seem that all scientific laws should be qualified as probabilistic since the supporting evidence we have for them is always a finite and logically inconclusive body of findings, which can confer upon them only a more or less high probability. But this argument misses the point that the distinction between laws of universal form and laws of probabilistic form does not refer to the strength of the evidential support for the two kinds of statements, but to their form, which reflects the logical character of the claim they make. A law of universal form is basically a statement to the effect that in *all* cases where conditions of kind F are realized, conditions of kind G are realized as well; a law of probabilistic form asserts, basically, that under certain conditions, constituting the performance of a random experiment R, a certain kind of outcome will occur in a specified percentage of cases. No matter whether true or false, well supported or poorly supported, these two types of claims are of a logically different character, and it is on this difference that our distinction is based.

As we saw earlier, a law of the universal form "Whenever F then G" is by no means a brief, telescoped equivalent of a report stating for each occurrence of F so far examined that it was associated with an occurrence of G. Rather, it implies assertions also for all unexamined cases of F, past as well as present and future; also, it implies counterfactual and hypothetical conditionals which concern, so to speak "possible occurrences" of F: And it is just this characteristic that gives such laws their explanatory power. Laws of probabilistic form have an analogous status. The law stating that the radioactive decay of $radium^{226}$ is a random process with an associated half-life of 1,620 years is plainly not tantamount to a report about decay rates that have been observed in certain samples of $radium^{226}$. It concerns the decaying process of any body of $radium^{226}$ —past, present, or future; and it implies subjunctive and counterfactual conditionals, such as: If two particular lumps of $radium^{226}$ were to be combined into one, the decay rates would remain the same as if the lumps had remained separate. Again, it is this characteristic that gives probabilistic laws their predictive and their explanatory force.

5.6 THE INDUCTIVE CHARACTER OF PROBABILISTIC EXPLANATION

One of the simplest kinds of probabilistic explanation is illustrated by our earlier example of Jim's catching the measles. The general form of that explanatory argument may be stated thus:

$$\frac{\begin{array}{l} p(O, R) \text{ is close to } 1 \\ i \text{ is a case of } R \end{array}}{i \text{ is a case of } O} \quad \text{[makes highly probable]}$$

Now the high probability which, as indicated in brackets, the explanans confers upon the explanandum is surely not a statistical probability, for it characterizes a relation between sentences, not between (kinds of) events. Using a term introduced in chapter 4, we might say that the probability in question represents the rational credibility of the explanandum, given the information provided by the explanans; and as we noted earlier, in so far as this notion can be construed as a probability, it represents a logical or inductive probability.

In some simple cases, there is a natural and obvious way of expressing that probability in numerical terms. In an argument of the kind just considered, if the numerical value of $p(O, R)$ is specified, then it is reasonable to say that the inductive probability that the explanans confers upon the explanandum has the same numerical value. The resulting probabilistic explanation has the form:

$$\frac{\begin{array}{l} p(O, R) = r \\ i \text{ is a case of } R \end{array}}{i \text{ is a case of } O} \quad \text{[r]}$$

If the explanans is more complex, the determination of corresponding inductive probabilities for the explanandum raises difficult problems, which in part are still unsettled. But whether or not it is possible to assign definite numerical probabilities to all such explanations, the preceding considerations show that when an event is explained by reference to probabilistic laws, the explanans confers upon the explanandum only more or less strong inductive support. Thus, we may distinguish deductive-nomological from probabilistic explanations by saying that the former effect a deductive subsumption under laws of universal form, the latter an inductive subsumption under laws of probabilistic form.

It is sometimes said that precisely because of its inductive character, a probabilistic account does not explain the occurrence of an event, since the explanans does not logically preclude its nonoccurrence. But the important, steadily expanding role that probabilistic laws and theories play in science and its applications makes it preferable to view accounts based on such principles as affording explanations as well, though of a less stringent kind than those of deductive-nomological form. Take, for example, the radioactive decay of a sample of one milligram of polonium^{218}. Suppose that what is left of this initial amount after 3.05 minutes is found to have a mass that falls within the interval from .499 to .501 milligrams. This finding can be explained by the probabilistic law of decay for polonium^{218}; for that law, in combination with the principles of mathematical probability, deductively implies that given the huge number of atoms in a milligram of polonium^{218}, the probability of the specified outcome is overwhelmingly large, so that in a particular case its occurrence may be expected with "practical certainty."

Or consider the explanation offered by the kinetic theory of gases for an empirically established generalization called Graham's law of diffusion. The law states that at fixed temperature and pressure, the rates at which different gases in a container escape, or diffuse, through a thin porous wall are inversely proportional to the square roots of their molecular weights, so that the amount of a gas that diffuses through the wall per second will be the greater, the lighter its molecules. The explanation rests on the consideration that the mass of a given gas that diffuses through the wall per second will be proportional to the average velocity of its molecules, and that Graham's law will therefore have been explained if it can be shown that the average molecular velocities of different pure gases are inversely proportional to the square roots of their molecular weights. To show this, the theory makes certain assumptions broadly to the effect that a gas consists of a very large number of molecules moving in random fashion at different speeds that frequently change as a result of collisions, and that this random behavior shows certain probabilistic uniformities—in particular, that among the molecules of a given gas at specified temperature and pressure, different velocities will occur with definite, and different, probabilities. These assumptions make it possible to compute the probabilistically expected values—or, as we might briefly say, the "most probable" values—that the average velocities of different gases will possess at equal temperatures and pressures. These most probable average values, the theory shows, are indeed inversely proportional to the square roots of the molecular weights of the gases. But the actual diffusion rates, which are measured experimentally and are the subject of Graham's law, will depend on the actual values that the average velocities have in the large but finite swarms of molecules constituting the given bodies of gas. And the actual average values are related to the corresponding probabilistically estimated, or "most probable," values in a manner that is basically analogous to the relation between the proportion of aces occurring in a large but finite series of tossings of a given die and the corresponding probability of rolling an ace with that die. From the theoretically derived conclusion concerning the probabilistic estimates, it follows only that in view of the very large number of molecules involved, it is overwhelmingly *probable* that at any given time the actual average speeds will have values very close to their probability estimates and that, therefore, it is *practically certain* that they will be, like the latter, inversely proportional to the square roots of their molecular masses, thus satisfying Graham's law.[3]

It seems reasonable to say that this account affords an explanation, even though "only" with very high associated probability, of why gases display the uniformity expressed by Graham's law; and in physical texts and treatises, theoretical accounts of this probabilistic kind are indeed very widely referred to as explanations.

NOTES

1. Further details on the concept of statistical probability and on the limit-definition and its shortcomings will be found in E. Nagel's monograph, *Principles of the Theory of Probability* (Chicago: University of Chicago Press, 1939). Our version of the statistical interpretation follows that given by H. Cramér on pp. 148–49 of his book, *Mathematical Methods of Statistics* (Princeton: Princeton University Press, 1946).

2. On this subject, see R. D. Luce and H. Raiffa, *Games and Decisions* (New York: John Wiley & Sons, Inc., 1957).

3. The "average" velocities here referred to are technically defined as root-mean-square velocities. Their values do not differ very much from those of average velocities in the usual sense of the arithmetic mean. A succinct outline of the theoretical explanation of Graham's law can be found in Chap. xxiv of Holton and Roller, *Foundations of Modern Physical Science*. The distinction, not explicitly mentioned in that presentation, between the average value of a quantity for some finite number of cases and the probabilistically estimated or expected value of that quantity is briefly discussed in Chap. vi (especially Sec. 4) of R. P. Feynman, R. B. Leighton, and M. Sands, *The Feynman Lectures on Physics* (Reading, Mass.: Addison-Wesley Publishing Co., 1963).

A. J. Ayer

WHAT IS A LAW OF NATURE?

There is a sense in which we know well enough what is ordinarily meant by a law of nature. We can give examples. Thus it is, or is believed to be, a law of nature that the orbit of a planet around the sun is an ellipse, or that arsenic is poisonous, or that the intensity of a sensation is proportionate to the logarithm of the stimulus, or that there are 303,000,000,000,000,000,000,000 molecules in one gram of hydrogen. It is not a law of nature, though it is necessarily true, that the sum of the angles of a Euclidean triangle is 180 degrees, or that all the presidents of the third French Republic were male, though this is a legal fact in its way, or that all the cigarettes which I now have in my cigarette case are made of Virginian tobacco, though this again is true and, given may tastes, not wholly accidental. But while there are many such cases in which we find no difficulty in telling whether some proposition, which we take to be true, is or is not a law of nature, there are cases where we may be in doubt. For instance, I suppose that most people take the laws of nature to include the first law of thermodynamics, the proposition that in any closed physical system the sum of energy is constant: But there are those who maintain that this principle is a convention, that it is interpreted in such a way that there is no logical possibility of its being falsified, and for this reason they may deny that it is a law of nature at all. There are two questions at issue in a case of this sort: first, whether the principle under discussion is in fact a convention, and secondly whether its being a convention, if it is one, would disqualify it from being a law of nature. In the same way, there may be a dispute whether statistical generalizations are to count as laws of nature, as distinct from the dispute whether

From A. J. Ayer, "What is a Law of Nature" in *Revue Internationale de Philosophie* (Brussels, Belgium), No. 36, fasc. 2 (1956). Reprinted by permission of the author and editor. This article was also printed in A. J. Ayer, *The Concept of a Person*, and is reprinted here also by permission of St. Martin's Press, Inc., The Macmillan Company of Canada Ltd., and Macmillan & Co. Ltd.

certain generalizations, which have been taken to be laws of nature, are in fact statistical. And even if we were always able to tell, in the case of any given proposition, whether or not it had the form of a law of nature, there would still remain the problem of making clear what this implied.

The use of the word *law*, as it occurs in the expression *laws of nature*, is now fairly sharply differentiated from its use in legal and moral contexts: We do not conceive of the laws of nature as imperatives. But this was not always so. For instance, Hobbes in his *Leviathan* lists fifteen "laws of nature" of which two of the most important are that men "seek peace, and follow it" and "that men perform their covenants made": But he does not think that these laws are necessarily respected. On the contrary, he holds that the state of nature is a state of war, and that covenants will not in fact be kept unless there is some power to enforce them. His laws of nature are like civil laws except that they are not the commands of any civil authority. In one place he speaks of them as "dictates of Reason" and adds that men improperly call them by the name of "laws," "for they are but conclusions or theorems concerning what conduceth to the conservation and defence of themselves: Whereas Law, properly, is the word of him, that by right hath command over others." "But yet," he continues, "if you consider the same Theorems, as delivered in the word of God, that by right commandeth all things; then they are properly called Laws."[1]

It might be thought that this usage of Hobbes was so far removed from our own that there was little point in mentioning it, except as a historical curiosity; but I believe that the difference is smaller than it appears to be. I think that our present use of the expression *laws of nature* carries traces of the conception of Nature as subject to command. Whether these commands are conceived to be those of a personal deity or, as by the Greeks, of an impersonal fate, makes no difference here. The point, in either case, is that the sovereign is thought to be so powerful that its dictates are bound to be obeyed. It is not as in Hobbes's usage a question of moral duty or of prudence, where the subject has freedom to err. On the view which I am now considering, the commands which are issued to Nature are delivered with such authority that it is impossible that she should disobey them. I do not claim that this view is still prevalent; at least not that it is explicitly held. But it may well have contributed to the persistence of the feeling that there is some form of necessity attaching to the laws of nature, a necessity which, as we shall see, it is extremely difficult to pin down.

In case anyone is still inclined to think that the laws of nature can be identified with the commands of a superior being, it is worth pointing out that this analysis cannot be correct. It is already an objection to it that it burdens our science with all the uncertainty of our metaphysics, or our theology. If it should turn out that we had no good reason to believe in the existence of such a superior being, or no good reason to believe that he issued any commands, it would follow, on this analysis, that we should not be entitled to believe that there were any laws of nature. But the main argument against this view is independent of any doubt that one may have about the existence of a superior being. Even if we knew that such a one existed, and that he regulated nature, we still could not identify the laws of nature with his commands. For it is only by discovering what were the laws of nature that we could know what form these commands had taken. But this implies that we have some independent criteria for deciding what the laws of nature are. The assumption that they are imposed by a superior being is therefore idle, in the same way as the assumption of providence is idle. It is only if there are independent means of finding out what is going to happen that one is able to say what providence has in store. The same objection applies to the rather more fashionable view that moral laws are the commands of a superior being: But this does not concern us here.

There is, in any case, something strange about the notion of a command which it is impossible to disobey. We may be sure that some command will never in fact be disobeyed. But what is meant by saying that it cannot be? That the sanctions which sustain it are too strong? But might not one be so rash or so foolish as to defy them? I am inclined to say that it is in the nature of commands that it should be possible to disobey them. The necessity which is ascribed to these supposedly irresistible commands belongs in fact to something different: It belongs to the laws of logic. Not that the laws of logic cannot be disregarded; one can make mistakes in deductive reasoning, as in anything else. There is, however, a sense in which it is impossible for anything that happens to contravene the laws of logic. The restriction lies not upon the events themselves but on our method of describing them. If we break the rules according to which our method of description functions, we are not using it to describe anything. This might suggest that the events themselves really were disobeying the laws of logic, only we could not say so. But this would be an error. What is describable as an event obeys the laws of logic: And what is not describable as an event is not an event at all. The chains which logic puts upon nature are purely formal: Being formal they weigh nothing, but for the same reason they are indissoluble.

From thinking of the laws of nature as the commands of a superior being, it is therefore only a short step to crediting them with the necessity that belongs to the laws of logic. And this is in fact a view which many philosophers have held. They have taken it for granted that a proposition could express a law of nature only if it stated that events, or properties, of certain kinds were necessarily connected; and they have interpreted this necessary connection as being identical with, or closely analogous to, the necessity with which the conclusion follows from the premisses of a deductive argument—as being, in short, a logical relation. And this has enabled them to reach the strange conclusion that the laws of nature can, at least in principle, be established independently of experience: For if they are purely logical truths, they must be discoverable by reason alone.

The refutation of this view is very simple. It was decisively set out by Hume. "To convince us," he says, "that all the laws of nature and all the operations of bodies, without exception, are known only by experience, the following reflections may, perhaps, suffice. Were any object presented to us, and were we required to pronounce concerning the effect, which will result from it, without consulting past observation: After what manner, I beseech you, must the mind proceed in this operation? It must invent or imagine some event, which it ascribes to the object as its effect: And it is plain that this invention must be entirely arbitrary. The mind can never find the effect in the supposed cause, by the most accurate scrutiny and examination. For the effect is totally different from the cause, and consequently can never be discovered in it."[2]

Hume's argument is, indeed, so simple that its purport has often been misunderstood. He is represented as maintaining that the inherence of an effect in its cause is something which is not discoverable in nature; that as a matter of fact our observations fail to reveal the existence of any such relation—which would allow for the possibility that our observations might be at fault. But the point of Hume's argument is not that the relation of necessary connection which is supposed to conjoin distinct events is not in fact observable: It is that there could not be any such relation, not as a matter of fact but as a matter of logic. What Hume is pointing out is that if two events are distinct, they are distinct: From a statement which does no more than assert the existence of one of them it is impossible to deduce anything concerning the existence of the other. This is, indeed, a plain tautology. Its importance lies in the fact that Hume's opponents denied it. They wished to maintain both that the events which were coupled by the laws of nature were logically distinct from one another, and that they were united by a logical relation. But this is a manifest contradiction. Philosophers who hold this view are apt to express it in a form which leaves the contradiction latent: It was Hume's achievement to have brought it clearly to light.

In certain passages Hume makes his point by saying that the contradictory of any law of nature is at least conceivable; he intends thereby to show that the truth of the statement which expresses such a law is an empirical matter of fact and not an a priori certainty. But to this it has been objected that the fact that the contradictory of a proposition is conceivable is not a decisive proof that the proposition is not necessary. It may happen, in doing logic or pure mathematics, that one formulates a statement which one is unable either to prove or disprove. Surely in that case both the alternatives of its truth and falsehood are conceivable. Professor W. C. Kneale, who relies on this objection,[3] cites the example of Goldbach's conjecture that every even number greater than two is the sum of two primes. Though this conjecture has been confirmed so far as it has been tested, no one yet knows for certain whether it is true or false: No proof has been discovered either way. All the same, if it is true, it is necessarily true, and if it is false, it is necessarily false. Suppose that it should turn out to be false. We surely should not be prepared to say that what Goldbach had conjectured to be true was actually inconceivable. Yet we should have found it to be the contradictory of a necessary proposition. If we insist that this does prove it to be inconceivable, we find ourselves in the strange position of having to hold that one of two alternatives is inconceivable, without our knowing which.

I think that Professor Kneale makes his case, but I do not think that it is an answer to Hume. For Hume is not primarily concerned with showing that a given set of propositions, which have been taken to be necessary, are not so really. This is only a possible consequence of his fundamental point that "there is no object which implies the existence of any other if we consider these objects in themselves, and never look beyond the idea which we form of them,"[4] in short, that to say that events are distinct is incompatible with saying that they are logically related. And against this Professor Kneale's objection has no force at all. The most that it could prove is that, in the case of the particular examples that he gives, Hume might be mistaken in supposing that the events in question really were distinct. In spite of the appearances to the contrary, an expression which he interpreted as referring to only one of them might really be used in such a way that it included a reference to the other.

But is it not possible that Hume was always so mistaken—that the events, or properties, which are coupled by the laws of nature never are distinct? This question is complicated by the fact that once a generalization is accepted as a law of nature it tends to change its status. The meanings which we attach to our expressions are not completely constant. If we are firmly convinced that every object of a kind which is designated by a certain term has some property which the term does not originally cover, we tend to include the property in the designation; we extend the definition of the object, with or without altering the words which refer to it. Thus, it was an empirical discovery that loadstones attract iron and steel: For someone who uses the word "loadstone" only to refer to an object which has a certain physical appearance and constitution, the fact that it behaves in this way is not formally deducible. But, as the word is now generally used, the proposition that loadstones attract iron and steel is analytically true: An object which did not do this would not properly be called a loadstone. In the same way, it may have become a necessary truth that water has the chemical composition H_2O. But what then of heavy water which has the composition D_2O? Is it not really water? Clearly this question is quite trivial. If it suits us to regard heavy water as a species of water, then we must not make it necessary that water consists of H_2O. Otherwise, we may. We are free to settle the matter whichever way we please.

Not all questions of this sort are so trivial as this. What, for example, is the status in Newtonian physics of the principle that the acceleration of a body is equal to the force which is acting on it divided by its mass? If we go by the textbooks in which "force" is defined as the product of mass and acceleration, we shall conclude that the principle is evidently analytic. But are there not other ways of defining force which allow this principle to be empirical? In fact there are, but as Henri Poincaré has shown,[5] we may then find ourselves obliged to treat some other Newtonian principle as a convention. It would appear that in a system of this kind there is likely to be a conventional element, but that, within limits, we can situate it where we choose. What is put to the test of experience is the system as a whole.

This is to concede that some of the propositions which pass for laws of nature are logically necessary, while implying that it is not true of all of them.

But one might go much further. It is at any rate conceivable that at a certain stage the science of physics should become so unified that it could be wholly axiomatized: It would attain the status of a geometry in which all the generalizations were regarded as necessarily true. It is harder to envisage any such development in the science of biology, let alone the social sciences, but it is not theoretically impossible that it should come about there too. It would be characteristic of such systems that no experience could falsify them, but their security might be sterile; what would take the place of their being falsified would be the discovery that they had no empirical application.

The important point to notice is that, whatever may be the practical or aesthetic advantages of turning scientific laws into logically necessary truths, it does not advance our knowledge, or in any way add to the security of our beliefs: For what we gain in one way, we lose in another. If we make it a matter of definition that there are just so many million molecules in every gram of hydrogen, then we can indeed be certain that every gram of hydrogen will contain that number of molecules; but we must become correspondingly more doubtful, in any given case, whether what we take to be a gram of hydrogen really is so. The more we put into our definitions, the more uncertain it becomes whether anything satisfies them: This is the price that we pay for diminishing the risk of our laws being falsified. And if it ever came to the point where all the "laws" were made completely secure by being treated as logically necessary, the whole weight of doubt would fall upon the statement that our system had application. Having deprived ourselves of the power of expressing empirical generalizations, we should have to make our existential statements do the work instead.

If such a stage were reached, I am inclined to say that we should no longer have a use for the expression *laws of nature*, as it is now understood. In a sense, the tenure of such laws would still be asserted: They would be smuggled into the existential propositions. But there would be nothing in the system that would count as a law of nature, for I take it to be characteristic of a law of nature that the proposition which expresses it is not logically true. In this respect, however, our usage is not entirely clear-cut. In a case where a sentence has originally expressed an empirical generalization, which we reckon to be a law of nature, we are inclined to say that it still expresses a law of nature, even when its meaning has been so modified that it has come to express an analytic truth. And we are encouraged in this by the fact that it is often very difficult to tell whether this modification has taken place or not. Also, in the case where some of the propositions in a scientific system play the rôle of definitions, but we have some freedom in deciding which they are to be, we tend to apply the expression *laws of nature* to any of the constituent propositions of the system, whether or not they are analytically true. But here it is essential that the system as a whole should be empirical. If we allow the analytic propositions to count as laws of nature, it is because they are carried by the rest.

Thus, to object to Hume that he may be wrong in assuming that the events between which his causal relations hold are "distinct existences" is merely to make the point that it is possible for a science to develop in such a way that axiomatic systems take the place of natural laws. But this was not true of the propositions with which Hume was concerned, nor is it true, in the main, of the sciences of today. And, in any case, Hume is right in saying that we cannot have the best of both worlds: If we want our generalizations to have empirical content, they cannot be logically secure; if we make them logically secure, we rob them of their empirical content. The relations which hold between things, or events, or properties, cannot be both factual and logical. Hume himself spoke only of causal relations, but his argument applies to any of the relations that science establishes, indeed to any relations whatsoever.

It should perhaps be remarked that those philosophers who still wish to hold that the laws of nature are "principles of necessitation"[6] would not agree that this came down to saying that the propositions which expressed them were analytic. They would maintain that we are dealing here with relations of objective necessity, which are not to be identified with logical entailments, though the two are in certain respects akin. But what are these relations of objective necessity supposed to be? No explanation is given except that they are just the relations that hold between events, or properties, when they are connected by some natural law. But this is simply to restate the problem, not even to attempt to solve it. It is not as if this talk of objective necessity enabled us to detect any laws of nature. On the contrary it is only *ex post facto*, when the existence of some connection has been empirically tested, that philosophers claim to see that it has this mysterious property of being necessary. And very often what they do "see" to be necessary is shown by further observation to be false. This does not itself prove that the events which are brought together by a law of nature do not stand in some unique relation. If all attempts at its analysis fail, we may be reduced to saying that it is *sui generis*. But why then describe it in a way which leads to its confusion with the relation of logical necessity?

A further attempt to link natural with logical necessity is to be found in the suggestion that two events *E* and *I* are to be regarded as necessarily connected when there is some well-established universal statement *U*, from which, in conjunction with the proposition *i*, affirming the existence of *I*, a proposition *e*, affirming the existence of *E*, is formally deducible.[7] This suggestion has the merit of bringing out the fact that any necessity that there may be in the connection of two distinct events comes only through a law. The proposition which describes "the initial conditions" does not by itself entail the proposition which describes the "effect" It does so only when it is combined with a causal law. But this does not allow us to say that the law itself is necessary. We can give a similar meaning to saying that the law is necessary by stipulating that it follows, either directly or with the help of certain further premisses, from some more general principle. But then what is the status of these more general principles? The question what constitutes a law of nature remains, on this view, without an answer.

II

Once we are rid of the confusion between logical and factual relations, what seems the obvious course is to hold that a proposition expresses a law of nature when it states what invariably happens. Thus, to say that unsupported bodies fall, assuming this to be a law of nature, is to say that there is not, never has been, and never will be a body that being unsupported does not fall. The "necessity" of a law consists, on this view, simply in the fact that there are no exceptions to it.

It will be seen that this interpretation can also be extended to statistical laws. For they too may be represented as stating the existence of certain constancies in nature—only, in their case, what is held to be constant is the proportion of instances in which one property is conjoined with another, or, to put it in a different way, the proportion of the members of one class that are also members of another. Thus it is a statistical law that when there are two genes determining a hereditary property, say the colour of a certain type of flower, the proportion of individuals in the second generation that display the dominant attribute, say the colour white as opposed to the colour red, is three quarters. There is, however, the difficulty that one does not expect the proportion to be maintained in every sample. As Professor R. B. Braithwaite has pointed out, "when we say that the proportion (in a non-literal sense) of the male births among births is 51 percent, we are not saying of any particular class of births that 51 percent are births of males, for the actual proportion might differ very widely from 51 percent in a particular class of births, or in a number of particular classes of births, without our wishing to reject the proposition that the proportion (in the nonliteral sense) is 51 percent."[8] All the same, the "non-literal" use of the word *proportion* is very close to the literal use. If the law holds, the proportion must remain in the neighbourhood of 51 percent, for any sufficiently large class of cases, and the deviations from it which are found in selected sub-classes must be such as the application of the calculus of probability would lead one to expect. Admittedly, the question what constitutes a sufficiently large class of cases is hard to answer. It would seem that the class must be finite, but the choice of any particular finite number for it would seem also to be arbitrary. I shall not, however, attempt to pursue this question here. The only point that I here wish to make is that a statistical law is no less "lawlike" than a causal law. Indeed, if the propositions which express causal laws are simply statements of what invariably happens, they can themselves be taken as expressing statistical laws, with ratios of 100 percent. Since a 100 percent ratio, if it really holds, must hold in every sample, these "limiting cases" of statistical laws escape the difficulty which we have just remarked on. If henceforth we confine our attention to them, it is because the analysis of "normal" statistical laws brings in complications which are foreign to our purpose. They do not affect the question of what makes a proposition lawlike, and it is in this that we are mainly interested.

On the view which we have now to consider, all that is required for there to be laws in nature is the existence of *de facto* constancies. In the most straightforward case, the constancy consists in the fact that events, or properties, or processes of different types are invariably conjoined with one another. The attraction of this view lies in its simplicity, but it may be too simple. There are objections to it which are not easily met.

In the first place, we have to avoid saddling ourselves with vacuous laws.

If we interpret statements of the form "All *S* is *P*" as being equivalent, in Russell's notation, to general implications of the form $(x)\,\Phi x \supset \Psi x$, we face the difficulty that such implications are considered to be true in all cases in which their antecedent is false. Thus we shall have to take it as a universal truth both that all winged horses are spirited and that all winged horses are tame; for assuming, as I think we may, that there never have been or will be any winged horses, it is true both that there never have been or will be any that are not spirited, and that there never have been or will be any that are not tame. And the same will hold for any other property that we care to choose. But surely we do not wish to regard the ascription of any property whatsoever to winged horses as the expression of a law of nature.

The obvious way out of this difficulty is to stipulate that the class to which we are referring should not be empty. If statements of the form "All *S* is *P*" are used to express laws of nature, they must be construed as entailing that there are *S*'s. They are to be treated as the equivalent, in Russell's notation, of the conjunction of the propositions $(x)\,\Phi x \supset \Psi x$ and $(\exists x)\,\Phi x$. But this condition may be too strong. For there are certain cases in which we do wish to take general implications as expressing laws of nature, even though their antecedents are not satisfied. Consider, for example, the Newtonian law that a body on which no forces are acting continues at rest or in uniform motion along a straight line. It might be argued that this proposition was vacuously true, on the ground that there are in fact no bodies on which no forces are acting; but it is not for this reason that it is taken as expressing a law. It is not interpreted as being vacuous. But how then does it fit into the scheme? How can it be held to be descriptive of what actually happens?

What we want to say is that if there *were* any bodies on which no forces were acting then they *would* behave in the way that Newton's law prescribes. But we have not made any provision for such hypothetical cases. According to the view which we are now examining, statements of law cover only what is actual, not what is merely possible. There is, however, a way in which we can still fit in such "non-instantial" laws. As Professor C. D. Broad has suggested,[9] we can treat them as referring not to hypothetical objects, or events, but only to the hypothetical consequences of instantial laws. Our Newtonian law can then be construed as implying that there are instantial laws, in this case laws about the behaviour of bodies on which forces are acting, which are such that when combined with the proposition that there are bodies on which no forces are acting, they entail the conclusion that these bodies continue at rest, or in uniform motion along a straight line. The proposition that there are such bodies is false, and so, if it is interpreted existentially, is the conclusion, but that does not matter. As Broad puts it, "What we are concerned to assert is that this false conclusion is a necessary consequence of the conjunction of a certain false instantial supposition with certain true instantial laws of nature."

This solution of the present difficulty is commendably ingenious, though I am not sure that it would always be possible to find the instantial laws which it requires. But even if we accept it, our troubles are not over. For, as Broad himself points out, there is one important class of cases in which it does not help us. These cases are those in which one measurable quantity is said to depend upon another, cases like that of the law connecting the volume and temperature of a gas under a given pressure, in which there is a mathematical function which enables one to calculate the numerical value of either quantity from the value of the other. Such laws have the form $x = Fy$, where the range of the variable y covers all possible values of the quantity in question. But now it is not to be supposed that all these values are actually to be found in nature. Even if the number of different temperatures which specimens of gases have or will acquire is infinite, there still must be an infinite number missing. How then are we to interpret such a law? As being the compendious assertion of all its actual instances? But the formulation of the law in no way indicates which the actual instances are. It would be absurd to construe a general formula about the functional dependence of one quantity on another as committing us to the assertion that just these values of the quantity are actually realized. As asserting that for a value n of y, which is in fact not realized, the proposition that it is realized, in conjunction with the set of propositions describing all the actual cases, entails the proposition that there is a corresponding value m of x? But this is open to the same objection, with the further drawback that the entailment would not hold. As asserting with regard to any given value n of y that either n is not realized or that there is a corresponding value m of x? This the most plausible alternative, but it makes the law trivial for all the values of y which happen not to be realized. It is hard to escape the conclusion that what we really mean to assert when we formulate such a law is that there is a corresponding value of x to every *possible* value of y.

Another reason for bringing in possibilities is that there seems to be no other way of accounting for the difference between generalizations of law and generalizations of fact. To revert to our earlier examples, it is a generalization of fact that all the Presidents of the Third French Republic are male, or that all the cigarettes that are now in my cigarette case are made of Virginian tobacco. It is a generalization of law that the planets of our solar system move in elliptical orbits, but a generalization of fact that, counting the earth as Terra, they all have Latin names. Some philosophers refer to these generalizations of fact as "accidental generalizations," but this use of the word *accidental* may be misleading. It is not suggested that these generalizations are true by accident, in the sense that there is no causal explanation of their truth, but only that they are not themselves the expression of natural laws.

But how is this distinction to be made? The formula $(x)\,\Phi x \supset \Psi x$ holds equally in both cases. Whether the generalization be one of fact or of law, it will state at least that there is nothing which has the property Φ but lacks the property Ψ. In this sense, the generality is perfect in both cases, so long as the statements are true. Yet there seems to be a sense in which the generality of what we are calling generalizations of fact is less complete. They seem to be restricted in a way that generalizations of law are not. Either they involve some spatio-temporal restriction, as in the example of the cigarettes *now* in my cigarette case, or they refer to particular individuals, as in the example of the presidents of France. When I say that all the planets have Latin names, I am referring definitely to a certain set of individuals—Jupiter, Venus, Mercury, and so on—but when I say that the planets move in elliptical orbits I am referring indefinitely to anything that has the properties that constitute being a planet in this solar system. But it will not do to say that generalizations of fact are simply conjunctions of particular statements, which definitely refer to individuals; for in asserting that the planets have Latin names, I do not individually identify them: I may know that they have Latin names without being able to list them all. Neither can we mark off generalizations of law by insisting that their expression is not to include any reference to specific places or times. For with a little ingenuity, generalizations of fact can always be made to satisfy this condition. Instead of referring to the cigarettes that are now in my cigarette case, I can find out some general property which only these cigarettes happen to possess, say the property of being contained in a cigarette case with such and such markings which is owned at such and such a period of his life by a person of such and such a sort, where the descriptions are so chosen that the description of the person is in fact satisfied only by me and the description of the cigarette case, if I possess more than one of them, only by the one in question. In certain instances these descriptions might have to be rather complicated, but usually they would not; and, anyhow, the question of complexity is not here at issue. But this means that, with the help of these "individuating" predicates, generalizations of fact can be expressed in just as universal a form as generalizations of law. And conversely, as Professor Nelson Goodman has pointed out, generalizations of law can themselves be expressed in such a way that they contain a reference to particular individuals, or to specific places and times. For, as he remarks, "even the hypothesis 'All grass is green' has as an equivalent 'All grass in London or elsewhere is green.' "[10] Admittedly, this assimilation of the two types of statement looks like a dodge; but the fact that the dodge works shows that we cannot found the distinction on a difference in the ways in which the statement can be expressed. Again, what we want to say is that whereas generalizations of fact cover only actual instances, generalizations of law cover possible instances as well. But this notion of possible, as opposed to actual, instances has not yet been made clear.

If generalizations of law do cover possible as well as actual instances, their range must be infinite; for while the number of objects which do throughout the course of time possess a certain property may be finite, there can be no limit to the number of objects which might possibly possess it. For once we enter the realm of possibility we are not confined even to such objects as actually exist. And this shows how far removed these generalizations are from being conjunctions: not simply because their range is infinite, which might be true even if it were confined to actual instances, but because there is something absurd about trying to list all the possible instances. One can imagine an angel's undertaking the task of naming or describing all the men that there ever have been or will be, even if their number were infinite, but how would he set about naming, or describing, all the possible men? This point is developed by F. P. Ramsey who remarks that the variable hypothetical $(x)\ \Phi x$ resembles a conjunction (1) in that it contains all lesser, i.e., here all finite conjunctions, and appears as a sort of infinite product, (2) when we ask what would make it true, we inevitably answer that it is true if and only if every x has Φ, i.e., when we regard it as a proposition capable of the two cases, truth and falsity, we are forced to make it a conjunction which we cannot express for lack of symbolic power.[11] But, he goes on, "What we can't say, we can't say, and we can't whistle it either," and he concludes that the variable hypothetical is not a conjunction and that "If it is not a conjunction, it is not a proposition at all." Similarly, Professor Ryle, without explicitly denying that generalizations of law are propositions, describes them as "seasonal inference warrants,"[12] on the analogy of season railway tickets, which implies that they are not so much propositions as rules. Professor Schlick also held that they were rules, arguing that they could not be propositions because they were not conclusively verifiable; but this is a poor argument, since it is doubtful if any propositions are conclusively verifiable, except possibly those that describe the subject's immediate experiences.

Now to say that generalizations of law are not propositions does have the merit of bringing out their peculiarity. It is one way of emphasizing the difference between them and generalizations of fact. But I think that it emphasizes it too strongly. After all, as Ramsey himself acknowledges, we do want to say that generalizations of law are either true or false. And they are tested in the way that other propositions are, by the examination of actual instances. A contrary instance refutes a generalization of law in the same way as it refutes a generalization of fact. A positive instance confirms them both. Admittedly, there is the difference that if all the actual instances are favourable, their conjunction entails the generalization of fact, whereas it does not entail the generalization of law. But still there is no better way of confirming a generalization of law than by finding favourable instances. To say that lawlike statements function as seasonal inference warrants is indeed illuminating, but what it comes to is that the inferences in question are warranted by the facts. There would be no point in issuing season tickets if the trains did not actually run.

To say that generalizations of law cover possible as well as actual cases is to say that they entail subjunctive conditionals. If it is a law of nature that the planets move in elliptical orbits, then it must not only be true that the actual planets move in elliptical orbits; it must also be true that if anything were a planet it would move in an elliptical orbit—and here "being a planet" must be construed as a matter of having certain properties, not just as being identical with one of the planets that there are. It is not indeed a peculiarity of statements which one takes as expressing laws of nature that they entail subjunctive conditionals, for the same will be true of any statement that contains a dispositional predicate. To say, for example, that this rubber band is elastic is to say not merely that it will resume its normal size when it has been stretched, but that it would do so if ever it were stretched: An object may be elastic without ever in fact being stretched at all. Even the statement that this is a white piece of paper may be taken as implying not only how the piece of paper does look but also how it would look under certain conditions, which may or may not be fulfilled. Thus one cannot say that generalizations of fact do not entail subjunctive conditionals, for they may very well contain dispositional predicates—indeed they are more likely to do so than not—but they will not entail the subjunctive conditionals which are entailed by the corresponding statements of law. To say that all the planets have Latin names may be to make a dispositional statement, in the sense that it implies not so much that people do always call them by such names but that they would so call them if they were speaking correctly. It does not, however, imply with regard to anything whatsoever that if it were a planet it would be called by a Latin name. And for this reason it is not a generalization of law, but only a generalization of fact.

There are many philosophers who are content to leave the matter there. They explain the "necessity" of natural laws as consisting in the fact that they hold for all possible, as well as actual, instances, and they distinguish generalizations of law from generalizations of fact by bringing out the differences in their entailment of subjunctive conditionals. But while this is correct so far as it goes, I doubt if it goes far enough. Neither the notion of possible, as opposed to actual, instances nor that of the subjunctive conditional is so pellucid that these references to them can be regarded as bringing all our difficulties to an end. It will be well to try to take our analysis a little further if we can.

The theory which I am going to sketch will not avoid all talk of dispositions; but it will confine it to people's attitudes. My suggestion is that the difference between our two types of generalization lies not so much on the side of the facts which make them true or false, as in the attitude of those who put them forward. The factual information which is expressed by a statement of the form "for all x, if x has Φ then x has Ψ," is the same whichever way it is interpreted. For if the two interpretations differ only with respect to the possible, as opposed to the actual values of x, they do not differ with respect to anything that actually happens. Now I do not wish to say that a difference in regard to mere possibilities is not a genuine difference, or that it is to be equated with a difference in the attitude of those who do the interpreting. But I do think that it can best be elucidated by referring to such differences of attitude. In short I propose to explain the distinction between generalizations of law and generalizations of fact, and thereby to give some account of what a law of nature is, by the indirect method of analysing the distinction between treating a generalization as a statement of law and treating it as a statement of fact.

If someone accepts a statement of the form $(x)\,\Phi x \supset \Psi x$ as a true generalization of fact, he will not in fact believe that anything which has the property Φ has any other property that leads to its not having Ψ. For since he believes that everything that has Φ has Ψ, he must believe that whatever other properties a given value of x may have, they are not such as to prevent its having Ψ. It may be even that he knows this to be so. But now let us suppose that he believes such a generalization to be true, without knowing it for certain. In that case there will be various properties X, X_1 . . . such that if he were to learn, with respect to any value of α of x, that α had one or more of these properties as well as Φ, it would destroy, or seriously weaken his belief that α had Ψ. Thus I believe that all the cigarettes in my case are made of Virginian tobacco, but this belief would be destroyed if I were informed that I had absent mindedly just filled my case from a box in which I keep only Turkish cigarettes. On the other hand, if I took it to be a law of nature that all the cigarettes in this case were made of Virginian tobacco, say on the ground that the case had some curious physical property which had the effect of changing any other tobacco that was put into it into Virginian, then my belief would not be weakened in this way.

Now if our laws of nature were causally independent of each other, and if, as Mill thought, the propositions which expressed them were always put forward as being unconditionally true, the analysis could proceed quite simply. We could then say that a person A was treating a statement of the form "for all x, if Φx then Ψx" as expressing a law of nature, if and only if there was no property X which was such that the information that a value α of x had X as well as Φ would weaken his belief that α had Ψ. And here we should have to admit the proviso that X did not logically entail not-Ψ, and also, I suppose, that its presence was not regarded as a manifestation of not-Ψ; for we do not wish to make it incompatible with treating a statement as the expression of a law that one should acknowledge a negative instance if it arises. But the actual position is not so simple. For one may believe that a statement of the form "for all x, if Φx then Ψx" expresses a law of nature while also believing, because of one's belief in other laws, that if something were to have the property X as well as Φ it would not have Ψ. Thus one's belief in the proposition that an object which one took to be a loadstone attracted iron might be weakened or destroyed by the information that the physical composition of the supposed loadstone was very different from what one had thought it to be. I think, however, that in all such cases, the information which would impair one's belief that the object in question had the property Ψ would also be such that, independently of other considerations, it would seriously weaken one's belief that the object ever had the property Φ. And if this is so, we can meet the difficulty by stipulating that the range of properties which someone who treats "for all x, if Φx then Ψx" as a law must be willing to conjoin with Φ, without his belief in the consequent being weakened, must not include those the knowledge of whose presence would in itself seriously weaken his belief in the presence of Φ.

There remains the further difficulty that we do not normally regard the propositions which we take to express laws of nature as being unconditionally true. In stating them we imply the presence of certain conditions which we do not actually specify. Perhaps we could specify them if we chose, though we might find it difficult to make the list exhaustive. In this sense a generalization of law may be weaker than a generalization of fact, since it may admit exceptions to the generalization as it is stated. This does not mean, however, that the law allows for exceptions: If the exception is acknowledged to be genuine, the law is held to be refuted. What happens in the other cases is that the exception is regarded as having been tacitly provided for. We lay down a law about the boiling point of water, without bothering to mention that it does not hold for high altitudes. When this is pointed out to us, we say that this qualification was meant to be understood, and so in other instances. The statement that if anything has Φ it has Ψ was a loose formulation of the law: What we really meant was that if anything has Φ but not X, it has Ψ. Even in the case where the existence of the exception was not previously known, we often regard it as qualifying rather than refuting the law. We say, not that the generalization has been falsified, but that it was inexactly stated. Thus, it must be allowed that someone whose belief in the presence of Ψ, in a given instance, is destroyed by the belief that Φ is accompanied by X may still be treating $(x)\ \Phi \supset \Psi x$ as expressing a law of nature if he is prepared to accept $(x)\ \Phi x \sim Xx \supset \Psi x$ as a more exact statement of the law.

Accordingly I suggest that for someone to treat a statement of the form "if anything has Φ it has Ψ" as expressing a law of nature, it is sufficient (1) that subject to a willingness to explain away exceptions he believes that in a non-trivial sense everything which in fact has Φ has Ψ, (2) that his belief that something which has Φ has Ψ is not liable to be weakened by the discovery that the object in question also has some other property X, provided (a) that X does not logically entail not-Ψ, (b) that X is not a manifestation of not-Ψ, (3) that the discovery that something had X would not in itself seriously weaken his belief that it had Φ, (4) that he does not regard the statement "if anything has Φ and not-X it has Ψ" as a more exact statement of the generalization that he was intending to express.

I do not suggest that these conditions are necessary, both because I think it possible that they could be simplified and because they do not cover the whole field. For instance, no provision has been made for functional laws, where the reference to possible instances does not at present seem to me eliminable. Neither am I offering a definition of natural law. I do not claim that to say that some proposition expresses a law of nature entails saying that someone has a certain attitude towards it; for clearly it makes sense to say that there are laws of nature which remain unknown. But this is consistent with holding that the notion is to be explained in terms of people's attitudes. My explanation is indeed sketchy, but I think that the distinctions which I have tried to bring out are relevant and important, and I hope that I have done something towards making them clear.

NOTES

1. *Leviathan*, Part I, Chap. xv.
2. *An Enquiry concerning Human Understanding*, iv, I.25.
3. *Probability and Induction*, pp. 79 ff.
4. *A Treatise of Human Nature*, i, iii, vi.
5. Cf. *La Science et l'hypothèse*, pp. 119–29.
6. Cf. Kneale, *op. cit.*

7. Cf. K. Popper, "What Can Logic Do For Philosophy?" *Supplementary Proceedings of the Aristotelian Society*, Vol. XXII: and papers in the same volume by W. C. Kneale and myself.

8. *Scientific Explanation*, pp. 118–29.

9. "Mechanical and Teleological Causation," *Supplementary Proceedings of the Aristotelian Society*, XIV, 98 ff.

10. *Fact, Fiction and Forecast*, p. 78.

11. *Foundations of Mathematics*, p. 238.

12. "'If,' 'So,' and 'Because'," *Philosophical Analysis* (Essays edited by Max Black), p. 332.

R. B. Braithwaite

LAWS OF NATURE AND CAUSALITY

Scientific hypotheses have been taken to be general empirical propositions whose generality is not restricted to limited regions of space or of time. For the reasons explained in the first chapter, scientific laws, corresponding to true scientific hypotheses, have been taken to assert no more than constant conjunctions of properties, so that the scientific law that everything which is *A* is *B* asserts no more than that all the things which are *A* as a matter of fact are also *B*. It is now time to examine this assumption in order to see whether the Humean analysis of scientific laws is adequate, or whether it is necessary to suppose that the sort of necessity of scientific law ("nomic" necessity) requires some extra element of "necessary connexion" over and above a merely factual uniformity. To express the matter in another way, everyone will agree that everything which is, was or will be *A* is, was or will be *B* is a logical consequence of the scientific law expressed apodeictically as "Every *A* nomically must be *B*"; the question is whether or not it is justifiable to regard the former proposition, not as a consequence, but as an analysis of the meaning of "Every *A* nomically must be *B*."

I have used W. E. Johnson's adjective "nomic" rather than the more usual "causal" to express, without prejudging the analysis, the characteristic sort of necessary connexion with which we are here concerned, because the notion of causality might well be held to involve considerations of temporal precedence and of spatio-temporal continuity which are irrelevant to the present issue. For here we are concerned with the nature of the difference, if any, between "nomic laws" and "mere generalizations"; in Johnson's language, between "universals of law" and "universals of fact."[1]

David Hume maintained that objectively there is no difference, but that a psychological fact about the way in which our minds work causes us to ascribe

---

From R. B. Braithwaite, *Scientific Explanation*. Reprinted by permission of the publishers, Cambridge University Press.

necessity to scientific laws, the "idea of necessary connexion" being derived from our experience of the constant conjunction of properties and not from anything in nature over and above constant conjunction. In common with most of the scientists who have written on the philosophy of science from Ernst Mach and Karl Pearson to Harold Jeffreys, I agree with the principal part of Hume's thesis—the part asserting that universals of law are objectively just universals of fact, and that in nature there is no extra element of necessary connexion. The time has now come to defend this thesis against philosophers who disagree with it.

The discussion has been postponed until this point because the principal argument used by philosophers against a Humean analysis of nomic necessity into constant conjunction is that such an analysis of scientific laws makes justification of induction impossible. But such a criticism thinks along the lines of of assimilating induction to deduction. If we do not attempt such an assimilation and instead propound a Peircean account of induction, as was done in the last chapter, the argument hangs fire. For the Peircean account bases itself upon the inductive policies which are used in making the inferences, and to pursue these policies certainly does not presuppose that the conclusions of the inferences made by pursuing them are anything more than universals of fact. So the positive justification of induction given in the last chapter is my answer to the argument that a Humean view of scientific laws makes their establishment invalid.

If this book were concerned with attacking rival views, cogent criticisms could be made of attempts to justify induction which build on a nomic necessity distinct from constant conjunction. Those philosophers, for example, who wish to identify nomic necessity with logical necessity lay themselves wide open to the charge that, since all the premisses in a valid inference to a logically necessary conclusion must be logically necessary propositions, to treat scientific laws as being logically necessary propositions removes all possibility of basing them upon empirical data. And those philosophers who wish to make nomic necessity a third ultimate category distinct both from logical necessity and from constant conjunction lay themselves open to the charge that their nomic connexion is, as "substance" was for Locke, "something I know not what," and that a philosopher is shirking his duty who uses Butler's maxim "Every thing is what it is, and not another thing," to avoid having to consider whether the difference between universals of law and universals of fact may not lie in the different roles which they play in our thinking, rather than in any difference in their objective contents.

For it cannot be disputed that we do make a distinction of some sort between those empirical general propositions which we dignify with the name of "laws of nature" or "natural laws" and those which we call, sometimes derogatorily, "mere generalizations." A Humean philosopher may well deny that this distinction is one of objective fact; but if he denies that there is any distinction whatever, he runs counter to ordinary usages of language.

SUBJUNCTIVE CONDITIONALS

One of the relevant usages of language to which philosophers have recently given a great deal of attention has been the use of conditional sentences of the form "If a thing is *A*, it is *B*" under circumstances in which nothing is *A* and the use of hypothetical sentences "If *p* then *q*" under circumstances in which *p* is false. These sentences have been called "conditionals" or "hypotheticals" which were "contrary to fact," "counterfactual," or, since the subjunctive mood is one way of expressing them in English, "subjunctive." We shall use the term *subjunctive conditional* for an assertion of the form: "Although there are no *A*'s, if there were to be any *A*'s, all of them would be *B*'s," e.g., "If there were to be a gas whose molecules had zero extension and did not attract one another (although in fact there are no such gases) its pressure and volume would be related by Boyle's law." We shall use the term *subjunctive hypothetical* for an assertion of the form: "Although *p* is false, if *p* were to be true, *q* would be true," e.g., "If the picture-wire had broken (although it didn't), the picture would have fallen to the ground."[2] At present we are concerned with subjunctive conditionals. The problem which they present to a Humean is the following dilemma. The constant-conjunction analysis leaves two choices open for the analysis of «If a thing is *A*, it is *B*». One choice is «Every *A* is *B*» taken, as traditional logic would say, "existentially," i.e., understood in such a way as to assert the existence of at least one thing which is *A*. The other alternative is «Every *A* is *B*» taken non-existentially, i.e., understood as not to assert the existence of an *A*. On the first interpretation, «Every *A* is *B*» is equivalent to the conjunction of «Nothing is both *A* and non-*B*» with «Something is *A*». A subjunctive conditional would combine this conjunctive assertion with the assertion that there is nothing which is *A*, and would thus be self-contradictory. On the second interpretation «Every *A* is *B*» is equivalent to the single proposition «Nothing is both *A* and non-*B*». A subjunctive conditional would conjoin this assertion with the assertion that there is nothing which is *A*. But since, if nothing is *A*, *a fortiori* nothing is both *A* and non-*B*, the conjunction of these two propositions is logically equivalent to the former one alone. Thus a Humean analysis, it is alleged, makes the assertion of a subjunctive conditional either self-contradictory or one which adds nothing to the assertion that nothing is *A*, which is expressed by the subjunctive mood being used. Each horn of the dilemma is equally uncomfortable, since neither horn will account for the fact that we make subjunctive conditional assertions freely and without consciousness of paradox. The opponents of the constant conjunction view conclude that "If a thing is *A*, it is *B*," used to express a nomic connexion, must mean more than «Nothing is both *A* and non-*B*» (with or without the conjunction of «Something is *A*») in order to account for the function played in our thinking by subjunctive conditionals.

This criticism can be met without requiring that the proposition expressed by the sentence "If a thing is *A*, it is *B*," used nomically, should be distinguished from the proposition that nothing is both *A* and non-*B*. What is required is that what is involved in *asserting* the subjunctive conditional expressed by such a sentence as "Although nothing is *A*, yet, if a thing were to be *A*, it would be *B*" is distinct from what is involved in *asserting* a conjunction of «Nothing is *A*» with «Nothing is both *A* and non-*B*». We can make this distinction by taking the assertion of «Nothing is both *A* and non-*B*» involved in asserting the subjunctive conditional as being not simply the assertion of «Nothing is both *A* and non-*B*» as being a true proposition, but also the assertion of «Nothing is both *A* and non-*B*» as being deduced from a higher-level hypothesis in a true and established scientific deductive system. To put the matter metaphorically, the generalization «Nothing is both *A* and non-*B*» enters into an assertion of a subjunctive conditional accompanied by a certificate of origin. Though the generalization itself conjoined with the proposition «Nothing is *A*» is logically equivalent to this latter proposition alone, a belief in the truth of the generalization which is accompanied by a belief about its origin, conjoined with a belief that nothing is *A*, is by no means equivalent to this latter belief alone.

Perhaps it is easiest to think of the matter in terms of the temporal order in which beliefs are acquired. Suppose that a person who has never considered whether or not there are any *A*'s has come to accept a scientific deductive system in which the proposition that nothing is both *A* and non-*B* is deducible from higher-level hypotheses in the system which have been established by induction from evidence which does not include any instances of the generalization «Every *A* is *B*». If the person then makes this deduction in the scientific system, he will have confirmed the proposition «Nothing is both *A* and non-*B*» indirectly; if he regards the higher-level hypotheses as established, he will also regard it as established that nothing is both *A* and non-*B*, and will add this proposition to his body of rational belief. Now suppose that he subsequently discovers that in fact there are no *A*'s. Had he acquired reasonable belief that there are no *A*'s before he had acquired his reasonable belief that nothing is both *A* and non-*B*, he could have deduced this latter proposition from the former, and would not have required to establish it by deducing it from higher-level hypotheses in the scientific deductive system. But he did not do this; he arrived at his reasonable belief in the generalization «Nothing is both *A* and non-*B*» quite independently of his subsequently acquired belief that this generalization was "vacuously" satified. The assertion of a subjunctive conditional may be regarded as a summary statement of this whole situation.

Take, for example, the statement "Although there are no gases whose molecules have zero extension and do not attract one another, yet if there were to be such gases, all of them would obey Boyle's law, $PV =$ a constant." The assertion of this statement envisages a situation in which, before it was known that there were no such gases, a functional law had been established relating the pressure and volume of a gas by examining gases with extended molecules which did attract one another, e.g., van der Waals's equation $(P + a/V^2)(V - b) =$ a constant. From this functional law the special law for gases whose molecules have zero extension and do not attract one another can be deduced by putting $a = b = 0$, i.e., from van der Waals's equation deducing Boyle's law. This special law will then have been established quite independently of any knowledge as to whether or not there are any gases whose molecules have zero extension and do not attract one another. To assert the subjunctive conditional is then to refer to the fact that the proposition that no gases whose molecules have zero extension and do not attract one another fail to obey Boyle's law has been established independently of the fact, which the subjunctive conditional also asserts, that there are no such gases.

Since «Nothing is *A*» is logically equivalent to the conjunction of «Nothing is both *A* and non-*B*» with «Nothing is both *A* and *B*», to establish «Nothing is *A*» after «Nothing is both *A* and non-*B*» has been established is to establish in addition only «Nothing is both *A* and *B*». The evidence for «Nothing is both *A* and non-*B*» provided by the evidence, direct or indirect, for «Nothing is *A*» will, of course, be additional to the evidence for «Nothing is both *A* and non-*B*» provided by the evidence for higher-level hypotheses from which this generalization logically follows; but, since this generalization is supposed to have been already established by a deduction from the established higher-level hypotheses, the additional evidence will not serve to establish it. In Freudian language, its establishment is "over-determined": There are two sets of evidence each sufficient to establish it, and the set which in fact establishes it is the one which gets in first.

Let us now remove the condition that the generalization «Nothing is both *A* and non-*B*» has been established by deducing it from established higher-level hypotheses before it has been considered whether or not it is vacuously true. We then have the situation that there are two ways of establishing the generalization: I can choose which of the two I regard as having got in first and as being the genuine establishment. One way is to deduce the generalization from the proposition, supposed to have been established, that nothing is *A*—call this the "vacuous" establishment; the other way is to deduce the generalization from the supposedly established higher-level hypotheses—call this the "hypothetico-deductive" establishment. The assertion of the subjunctive conditional «Although there are no *A*'s, yet if there were to be any *A*'s all of them would be *B*'s» asserts that there are no *A*'s, that nothing is both *A* and non-*B*, and that the latter of these propositions is establishable hypothetico-deductively without reference to the establishment of the former. The peculiarity of the subjunctive conditional is that to assert it is not only assert two propositions, one of which is a logical consequence of the other, but is also to assert that this former proposition, though vacuously establishable by deduction from the latter, it also hypothetico-deductively establishable in an independent way. The assertion of the subjunctive conditional makes a remark about the relation of two of the propositions asserted in regard to the way that they can be established. This analysis, it seems to me, satisfactorily explains the peculiarity of subjunctive conditionals without our having to suppose that the sentence "Every *A* is *B*," used nomically, need mean any more than that nothing is both *A* and non-*B*.

To consider without asserting a subjunctive conditional is to perform a highly sophisticated activity. It is essentially to consider two propositions which are logically related but to consider them separately within two scientific deductive systems. The proposition «Nothing is both *A* and non-*B*» has to be considered as a deduction from higher-level hypotheses which do not include the proposition «Nothing is *A*», and not as a deduction from this proposition itself. The proposition «Nothing is both *A* and non-*B*» appears as a consequence in two deductive systems; but the fact that it appears in one of these systems has to be attended to, that it appears in the other unattended to. Thus the meaning of a subjunctive conditional sentence cannot be given simply by stating the proposition which is entertained whenever the sentence is used; the function of the sentence in our language is primarily to be used assertively, and in this case, as has been explained, it refers to the origin of the asserter's belief in one of the propositions asserted. Logicians have naturally concentrated their interest upon the meaning of sentences in which consideration of the meaning of the sentence —entertainment of the proposition—can be separated from what is, in addition, involved in using the sentence assertively; subjunctive conditional sentences cannot be treated in this way, since they have a very definite and important use when used assertively, which cannot be broken up into a set of unrelated assertions.

It is important to notice that we have no normal use for subjunctive conditionals which would deny the existence of things specified by a theoretical concept, in the sense of theoretical concept of chapter III. This is because it is self-contradictory to conjoin a statement about a theoretical concept with a statement asserting that there are in fact no instances of the concept, since the truth of a statement about a theoretical concept is sufficient to endow the concept with existence. The only use of subjunctive conditionals in such cases is as a means for insisting that the statement about the theoretical concept, though it appears to be a contingent proposition, is in fact being used as a sterile formula (in the sense of chapter IV) defining the meaning of the symbol for the concept. Consider, for example, the subjunctive conditional statement "Although there are in fact no hydrogen atoms, yet, if there were, all of them would be systems composed of one proton and one electron each." On the supposition that the term "hydrogen atom" in this sentence is to be used to stand for a theoretical concept, I can think of no use for this sentence except as used by someone who was using it as a sterile formula to give a definition of a new theoretical term "hydrogen atom" as a logical construction out of the theoretical concepts proton and electron already in use in his mode of thinking about physics.

LAWS OF NATURE

Our solution of the problem of how our use of subjunctive conditionals is consistent with a constant conjunction analysis of nomic generalizations will enable us to solve a related problem which is posed to Humeans, namely, that of distinguishing between what are laws of nature (or natural laws) and what anti-Humeans contemptuously call "mere generalizations." Surely, they say, this distinction must be admitted: For us it consists in the laws of nature asserting principles of nomically necessary connexion; since you decline to admit such principles, how can you make the distinction? The problem is sometimes put in the form that we all distinguish between uniformities due to natural law and those which are merely accidentally true, "historical accidents on the cosmic scale";[3] if natural laws are just uniformities, how can this distinction be made?

It seems to me foolish to deny (as some Humeans do) that such a distinction is made in common speech; but it also seems perfectly sensible to try to give a rationale for this distinction within the ambit of a constant conjunction view. The distinction will then have to depend upon knowledge or belief in the general proposition rather than in anything intrinsic to the general proposition itself; but this is exactly how we have solved the related problem of subjunctive conditionals. Let us try to use this solution to pick out some among true contingent general propositions to be given the honorific title of "natural law."

Let us tentatively try the following criterion: A true contingent general proposition «Every *A* is *B*» whose generality is not limited to any particular regions of space or of time will be called by a person *C* a "law of nature" or "natural law" if either the corresponding subjunctive conditional «Although there are no *A*'s, yet if there were any *A*'s they would all be *B*'s» is reasonably believed by *C*, or this subjunctive conditional would be reasonably believed by *C* if he were reasonably to believe that there are no *A*'s. In terms of the notion of *C*'s rational corpus of knowledge and reasonable belief, the criterion requires that the corresponding subjunctive conditional should form part of his rational corpus if this includes a belief that there are no *A*'s, or, if it does not include this belief, would form part of his rational corpus if it did include this belief. In terms of the notion of assertion used in the discussion of subjunctive conditionals, an assertion of a natural law together with an assertion that there are no *A*'s would come to the same thing as an assertion of the corresponding subjunctive conditional.

In addition all true hypotheses containing theoretical concepts will be given the title of natural laws.

The condition for an established hypothesis *h* being "lawlike" (i.e., being, if true, a natural law) will then be that the hypothesis either occurs in an established scientific deductive system as a higher-level hypothesis containing theoretical concepts or that it occurs in an established scientific deductive system as a deduction from higher-level hypotheses which are supported by empirical evidence which is not direct evidence for *h* itself. This condition will exclude a hypothesis for which the only evidence is evidence of instances of it, but it will not exclude a hypothesis which is supported partly directly by evidence of its instances and partly indirectly by evidence of instances of same-level hypotheses which, along with it, are subsumed under a higher-level hypothesis. This account of natural law makes the application of the notion dependent upon the way in which the hypothesis is regarded by a particular person at a particular time as having been established: "Lawlike" may be thought of as a honorific epithet which is employed as a mark of origin. If the hypothesis that all men are mortal is regarded as supported solely by the direct evidence that men have died, then it will not be regarded as a law of nature; but if it is regarded as also being supported by being deduced from the higher-level hypothesis that all animals are mortal, the evidence for this being also that horses have died, dogs have died, etc., then it will be accorded the honorific title of "law of nature" which will then indicate that there are other reasons for believing it than evidence of its instances alone.

This criterion for lawlikeness has the paradoxical consequence that the hypothesis that all men are mortal will be regarded as a natural law if it occurs in an established scientific deductive system at a lower-level than a hypothesis (e.g., All animals are mortal) which has other lower-level hypotheses under it which are directly confirmed by experience; whereas the higher-level hypothesis that all animals are mortal, if it appears as the highest-level hypothesis in the established deductive system, will not be regarded as a natural law, since the ground for its establishment is solely the evidence of its instances. However, it seems to me that this corresponds to the way in which, generally speaking, we use the notion of natural law. A hypothesis to be regarded as a natural law must be a general proposition which can be thought to *explain* its instances; if the reason for believing the general proposition is solely direct knowledge of the truth of its instances, it will be felt to be a poor sort of explanation of these instances. If, however, there is evidence for it which is independent of its instances, such as the indirect evidence provided by instances of a same-level general proposition subsumed along with it under the same higher-level hypothesis, then the general proposition will *explain* its instances in the sense that it will provide grounds for believing in their truth independently of any direct knowledge of such truth. And this connexion with a notion of explanation fits in well with the honorific title of "natural law" being ascribable to every hypothesis containing theoretical concepts, whether or not such a hypothesis stands at the highest-level in the established scientific system. For even if the hypothesis with theoretical concepts is not deducible from an established higher-level hypothesis, yet it will not have been established simply by induction by simple enumeration; it will have been obtained by the hypothetico-deductive method of proposing it as a hypothesis and deducing its testable consequences. The case for accepting any particular higher-level hypothesis containing theoretical concepts is exactly that it serves as an explanation of the lower-level generalizations deducible from it, whereas the case for accepting a particular generalization not containing theoretical concepts and not deducible from any higher-level hypothesis is the fact that it covers its known instances rather than that it explains them.

I do not wish to emphasize unduly this relation between explanation and natural law: The marginal uses of both of these concepts are indefinite, and the boundaries of their uses will certainly not agree. Generally speaking, however, a true scientific hypothesis will be regarded as a law of nature if it has an explanatory function with regard to lower-level hypotheses or its instances; vice versa, to the extent that a scientific hypothesis provides an explanation, to that extent will there be an inclination to endow it with the honourable status of natural law.

To consider whether or not a scientific hypothesis would, if true, be a law of nature is to consider the way in which it could enter into an established scientific deductive system. As with the case of subjunctive conditionals, to consider this question is a sophisticated activity; the question cannot be answered by a straight yes or no without reference to how the hypothesis is related to other hypotheses which are used in our scientific thinking.

NOTES

1. W. E. Johnson, *Logic, Part III* (Cambridge, 1924), Chap. i.

2. Many contemporary logicians use the terms "conditional" and "hypothetical" synonymously, and call what I am calling a "conditional" a "general conditional" or "general hypothetical." My distinction between the use of "conditional" and "hypothetical" was suggested by that made by J. N. Keynes, *Studies and Exercises in Formal Logic*, (4th ed.), (London, 1906), pp. 249 ff. In the considerable amount of recent literature dealing with subjunctive conditionals and the related question of natural laws solutions resembling the account to be given here have been published by Hans Reichenbach, *Elements of Symbolic Logic* (New York, 1947), Chap. viii, and by J. R. Weinberg, *Journal of Philosophy*, XLVIII (1951), 17ff. F. P. Ramsey, *The Foundations of Mathematics and other logical essays*, pp. 237ff. and David Pears, *Analysis*, X (1950), 49ff. have also approached the questions along lines similar to mine.

3. William Kneale, *Analysis*, X (1950), 123.

Alternative Models for the Physical Sciences

Sylvain Bromberger

WHY-QUESTIONS

In this paper we seek to pin down the conditions that define correct answers to why-questions. The problem can be stated more precisely. We will mean by a *why-question* a question that can be put in English in the form of an interrogative sentence of which the following is true: (1) the sentence begins with the word *why*; (2) the remainder of the sentence has the (surface) structure of an interrogative sentence designed to ask a whether-question, i.e., a question whose right answer in English, if any, must be either "yes" or "no"; (3) the sentence contains no parenthetical verbs, in Urmson's sense.[1] A why-question put as an English sentence that satisfies (1), (2), and (3) will be said to be in *normal form*. By the *inner question* of a why-question we will mean the question alluded to in (2) above, i.e., the question reached by putting the why-question in normal form, then deleting the initial "why" and uttering the remaining string as a question. By the *presupposition* of a why-question we will mean that which one would be saying is the case if, upon being asked the inner question of the why-question through an affirmative interrogative sentence, one were to reply "yes," or what one would be saying is the case if, upon being asked the inner question through a negative sentence, one were to reply "no." Thus, "Why does copper turn green when exposed to air?" is a why-question in normal form; its inner question is "Does copper turn green when exposed to air?"; and its presupposition is that copper turns green when exposed to air. The presupposition of "Why doesn't iron turn green when exposed to air?" is that iron does not turn green when exposed to air.[2]

This work was supported in part by the Joint Services Electronics Program under Contract DA36–039-AMC-03200(E); in part by the National Science Foundation (Grant GP-2495), the National Institutes of Health (Grant MH-04737-05), the National Aeronautics and Space Administration (Grant NsG-496), and the U.S. Air Force (ESD Contract AF19(628)-2487). This essay is dedicated to Carl G. Hempel and to Peter Hempel as a token of gratitude for the toughmindedness of the one and the gentle-mindedness of the other.

We will not be concerned with every sort of why-question. We will ignore why-questions whose normal forms are not in the indicative. We will ignore why-questions whose presupposition refers to human acts or intentions or mental states. Finally, we will ignore why-questions whose correct answer cannot be put in the form "because p," where *p* indicates a position reserved for declarative sentences. Notice that this last stipulation affects not only why-questions whose correct answer must be put in some such form as "in order to. . ." or "to . . . ,"[3] but also why-questions that one might wish to say have no correct answer, and in particular why-questions with false presupposition and why-questions whose inner question itself has no answer, e.g., "Why doesn't iron form any compounds with oxygen?" and "Why does phlogiston combine with calx?" More may be ruled out, and we shall have to come back to this point.[4]

To simplify matters, we will disregard the fact that correct answers to the why-questions that do concern us can often be put in some other form than "because p" with a declarative sentence at the *p*. Furthermore, we will reserve the term *answer* to refer to what is conveyed by the sentence at *p* abstracted from the "because . . ." environment. Thus, if "because the temperature is rising" is the correct answer to some why-question we will speak of "The temperature is rising" as the answer.

We can now put our problem very simply. Let *a* and *b* be any two true propositions; what necessary and sufficient conditions must they jointly satisfy if *b* is to be a correct answer to a why-question whose presupposition is *a*?[5]

II

So far we have relied on a characterization of why-questions in which features peculiar to the English lexicon and to English grammar play an essential role. We have carefully avoided identifying why-questions as a class of English interrogative sentences, but we have nevertheless defined them as questions that must be expressible in a certain way in English. This may seem to detract from the interest of the problem. Philosophers of science in particular may feel wary of a typology of questions that rests squarely on the availability of certain forms in a specific natural language. There are good grounds for such suspicion. After all, scientific questions are for the most part only accidentally expressible in English. They can also be put in French, German, Russian, Japanese, etc., not to mention artificial languages. Furthermore, some of these questions may not be expressible in English at all, especially so if by "English" we mean contemporary, "ordinary" English. "Why is the emf induced in a coiled conductor a function of the rate of change of magnetic flux through it and of the resistance of the coil?" could probably not have been asked in seventeenth-century English, and a similar situation may hold for questions that have not yet arisen.

One could try to meet such reservations by providing at the outset a language-independent definition of why-questions, or rather of *Why*-questions, a class of questions that would include all why-questions but that would not be limited to questions expressible in English. However, it is not clear how one is to be guided in setting up such a definition. We propose to deal with the matter somewhat differently. We will set as one condition on the solution of our problem that it abstract completely from the peculiarities of English, i.e., that it be stated in terms that transcend linguistic idiosyncracies and are applicable to expressions in any relevantly rich language. Having done this we should be able to give a definition of Why-questions that preserves whatever warrants an interest in the nature of why-questions on the part of philosophers of science.[6]

III

What we have just said commits us to two hypotheses. The first of these hypotheses is that the relation between presupposition and (correct) answer to a why-question can be analyzed in language-independent terms. This hypothesis may be false, in which case we will not be able to solve our problem within the restrictions that we have adopted. However, it should be clear that the hypothesis cannot prevent us from accepting as relevant intuitions about the presence or absence of the relation in specific cases available to us as speakers of English. When we say that the relation is language independent, we do *not* mean that it hinges only on extra-linguistic facts. We mean that insofar as it hinges on linguistic features it hinges only on syntactic and semantic properties that expressions from every language share. Thus, the properties of being true and of being mutually implied are properties that expressions may have whether they belong to English or Chinese or Beulemans. The property of being the result of a do-transformation (the transformation that inserts "do" in, e.g., "He did not eat" or in "Didn't he eat?" but not in "He will not eat" or in "Hasn't he eaten?") is a property shared only by English expressions. Our hypothesis is therefore compatible with the tenet that any speaker of English has the faculty to perceive whether the semantic and syntactic properties of two given English sentences meet (or fail to meet) the conditions that would make one of these sentences express the answer of a why-question whose presupposition is expressed by the other. He must, of course, understand the sentences, and he must also have certain relevant beliefs. On the other hand, to say that he has the faculty to perceive whether this sort of condition is satisfied in specific instances is not to say that he can describe them or analyze them. Nor is it to say that he will never or ought never to hesitate before pronouncing something to be a correct answer to a why-question. Hesitation is to be expected where the case at hand is complex and demands slow and careful scrutiny. It is also to be expected when the truth of the sentences or of the relevant beliefs are themselves objects of hesitation. But there are clear-cut cases and these constitute a corpus for which, as speakers of English, we must account.

IV

The second hypothesis is that there are issues in the philosophy of science that warrant an interest in the nature of why-questions. The most obvious of these issues are whether science (or some branch of science or some specific scientific doctrine or some approach) ought to, can, or does provide answers to why-questions, and if so, to which ones. In other words, when appraising critically the state of scientific knowledge (or of some branch of science or some doctrine or some approach), how much weight should we give to unanswered why-questions? Should we consider that some why-questions are beyond the reach of scientific methodology or rules of evidence? Should we refrain from accepting as final any doctrine that raises why-questions to which no answers are forthcoming? We will have little to say about these very complex issues here, but since they provide much of the motivation for our inquiry, a few words of caution are called for.

These issues are usually discussed in English with the word "explanation" used instead of "why-question" or "answer to why-question." Analogous substitutions occur in other languages. This way of putting things can be innocuous and is possibly justified by the awkwardness of using the more contrived locutions. But it is ambiguous and may be a source of confusion. To become aware of this we need but notice that "explanation" may refer to the answers of a huge variety of questions besides why-questions, the only requirement being that their *oratio obliqua* form fit as grammatical object of the verb "to explain" and its nominalization "explanation of," e.g., how-questions, what is-the-cause-of-questions, what-corresponds-at-the-microscopic-level-questions, etc. Yet, the issues raised by these other types call for considerations peculiar to each type and different from those called for in the case of why-questions. Confusion is therefore likely to ensue and is apt to be further compounded if we allow ourselves to forget that "explanation" may also refer to things not readily specified as answers to a specific class of questions. To remain aware of the range of issues covered by a given analysis we must therefore keep sharp the differences among questions about (1) truth-conditions of sentences generated from "*A* explains *B*" and from "*A* is the explanation of *B*" by substituting any grammatically appropriate phrase for *B*, (2) truth-conditions of sentences obtained by substituting for *B* only *oratio obliqua* forms of grammatically appropriate questions, (3) truth-conditions of sentences obtained by substituting for *B* the *oratio obliqua* form of some more narrowly defined class of questions (e.g., why-questions, how-questions, what-corresponds-at-the-microscopic-level-questions, etc.), (4) conditions that are satisfied by answers and presupposition of all questions whose *oratio obliqua* form can be substituted for *B*, (5) conditions that are satisfied by answers and presupposition of some narrower class of questions whose *oratio obliqua* form can be substituted for *B*.[7] It should be clear that we will limit ourselves to a special case of (5) in this paper, the case of why-questions.

In fact, our limits are even narrower since we have eliminated from consideration certain types of why-questions.

Offhand, it may seem that the above (1) to (5) enumeration is redundant and that we might have stopped after (3). Actually, subtle but important distinctions underlie the difference between "Explanation of *Q*" and "Answer to *Q*." We have discussed these at some length elsewhere[8] and will say just a few words about them here to suggest the sort of further problems involved.

Let us describe someone as in a *p-predicament* (*p* can be thought of as standing for "puzzled" or "perplexed" but for *mnemonic* purposes only) with regard to some question *Q*, if and only if on that person's views, the question *Q* admits of a right answer, yet the person can think of no answer, can make up no answer, can generate from his mental repertoire no answer to which, given that person's views, there are no decisive objections For instance, a physicist committed to classical physics but aware of the photoelectric effect would be in a *p*-predicament with regard to the question "Why does a photoelectric current appear without delay as soon as light of frequency above the threshold frequency impinges on the target, and this no matter how low the frequency of the impinging light?" Let us also describe someone as in a *b-predicament* with regard to a question *Q* if and only if the question admits of a right answer, no matter what the views of the person, but that answer is beyond what that person can think of, can state, can generate from his mental repertoire. Thus, someone unacquainted with the kinetic-molecular theory of matter would be in a *b*-predicament with regard to the question "What is the mechanism by which water evaporates from uncovered dishes left in the open?" Let us say furthermore that a question *Q* is *unanswerable relative to a certain set of propositions and concepts C* if and only if anyone who subscribes to these propositions and limits himself to these concepts must be in either a *p*-predicament or *b*-predicament with regard to the question *Q*. The search for and discovery of scientific *explanations*, we think, is essentially the search for and discovery of answers to questions that are unanswerable relative to prevailing beliefs and concepts. It is not, therefore, merely a question for evidence to settle which available answer is correct, it is a quest for the unthought-of.

The difference between "explanation" and "answer" just sketched transcends the distinction between why-questions and other questions. It should nevertheless be kept in mind when we deal with the issues described at the beginning of this section. These need not be resolved in the same way for why-questions that are unanswerable relative to the set under consideration and for those that are merely unanswered.

V

According to a very familiar theory, explaining a fact (an event, a phenomenon, a natural law) consists in deducing a statement describing the fact from the statement of a true law and additional true premises. Thus, according to this theory, the explanation of a fact is a valid and sound (i.e , all the premises are true) deduction, none of whose premises are superfluous, some of whose premises are empirical laws, and whose conclusion is a description of the fact explained. The premises of such a deduction are called the *explanans* and the conclusion, the *explanadum*. We will refer to such deductions as *deductive nomological explanations* and to the theory itself, whose most famous and competent exponent has been Carl Hempel, as the *Hempelian doctrine*.[9]

As a general characterization of the notion of explanation, i.e., as a description of the truth-conditions of statements of the form "A explains B" or "A is a correct explanation of B," or their non-English equivalents, the Hempelian doctrine obviously will not do, a fact that its proponents have always recognized. The evidence for this also shows that the doctrine does not describe necessary and sufficient conditions on the answers to all the sound questions whose *oratio obliqua* form may be substituted for *B*. Answers to, or explanations of, how cloud chambers work, of what the nature of light is, of what occurs at the molecular level when water freezes, etc., need not be explanans (nor even a pragmatically selected component of explanans). On the other hand, the doctrine no doubt does describe necessary and sufficient conditions on answers to *some* questions whose *oratio obliqua* form can be substituted for *B*. Thus, every deductive nomological explanation is an explanation or at least a sound answer to questions of the form "How could anyone knowing that . . . (here put the conjunction of all the premises in a deductive nomological explanans) . . . but not that . . . (here put the corresponding explanandum) . . . have predicted that . . . (here repeat the explanadum) . . . ?" and obviously the conjunction of the premises also constitutes a correct answer to questions of the form "From what laws and antecedent conditions can the fact that . . . (here put the explanandum) . . . be deduced?" But does the Hempelian doctrine tell us what we want to know about why-questions? Is a proposition p the correct answer of a why-question whose presupposition is q if and only if p is the conjunction of premises (or of some pragmatically selected subset of premises) of a deductive nomological explanation whose conclusion is q? The following counterexamples (and they are easily multiplied) strike us as settling the matter and this quite apart from some technical difficulties connected with the relevant notions of deducibility and law:

(1) There is a point on Fifth Avenue, M feet away from the base of the Empire State Building, at which a ray of light coming from the tip of the building makes and angle of θ degrees with a line to the base of the building. From the laws of geometric optics, together with the "antecedent" condition that the distance is M feet, the angle θ degrees, it is possible to deduce that the Empire State Building has a height of H feet. Any high school student could set up the deduction given actual numerical values. By doing so, he would not, however, have *explained* why the Empire State Building has a height of H feet, nor would he have *answered* the question "Why does the Empire State Building have a height of H feet?" nor would an exposition of the deduction be the explanation of or answer to (either implicitly or explicitly) why the Empire State Building has a height of H feet.

(2) From the Leavitt-Shapley Law, the inverse square law for light, the periods of Cepheid type variable stars in the Andromedan Galaxy, their apparent range of brightness, one can deduce that the Andromedan Galaxy is 1.5×10^6 light years away from the earth. The premises of the deduction, however, do not tell why or explain why the Andromedan Galaxy is 1.5×10^6 light years away from the earth.

(3) Whenever the pointer of the water meter points to 5, and only the bathtub faucet is open, water flows at a rate of five gallons per minute into the bathtub. The pointer has been on 5 for the last three minutes, and no faucet except the bathtub one is open. Therefore, fifteen gallons of water must have flowed into the bathtub. The deduction does not explain or tell or reveal *why* fifteen gallons of water flowed into the bathtub during the last three minutes.

(4) All of Cassandra's predictions always come true. (Cassandra is a computer.) Yesterday Cassandra predicted that it would rain today. But obviously that is not why it is raining today.

(5) Only men who are more than six feet tall leave footprints longer than fourteen inches. The footprints left by Gargantua on the beach are more than fourteen inches long. Therefore Gargantua is more than six feet tall.

Again the reasoning fails to mention *why* Gargantua is more than six feet tall.

These counterexamples are compatible with the thesis that answers and presuppositions of why-questions *must* be premises and conclusions of deductive nomological explanations They do show, however, that this cannot be *sufficient*.

It has been suggested that these counterexamples and others like them are not really binding on philosophers of science, that they ultimately involve an appeal to ordinary usage and that such appeals are not appropriate when we deal with inquiries that are far removed from ordinary concerns. These objections can be construed in a number of ways:

(1) They may mean that our refusal to call the explanans examples of *explanations*, or to look upon them as telling *why* something is the case, merely reflects allegiance to unscientific intellectual practices that scientists qua scientists have or should have abandoned. But this is hardly plausible. In 1885, Balmer devised a formula from which the frequencies represented in the spectrum of a sample of excited hydrogen could be deduced. But any scientist worthy of the name would have refused to accept such a deduction as the answer to why these particular frequencies were represented. The case is far from unique, and we owe the birth of quantum mechanics and of modern astronomy to that sort of refusal.

(2) They may mean that the verb "to explain" and its cognates have a technical meaning in scientific contexts, a status similar to that of "work," "action," "model," etc. But this is false. "To explain" does not belong to any technical jargon (except perhaps that of some philosophers), and anyhow the crucial words in our inquiry are "why" and "because."

(3) They may mean that although we do not say of these inferences that they explain or tell why something is the case, we could, and that only an unscientific tradition prevents us from doing so. This would make sense if "ordinary use"

merely demanded that we *refrain* from saying of the premises of the above inferences that they tell why something is the case, but words meaning what they do, we must also *deny* it. The deduction about Gargantua does *not* tell why Gargantua is more than six feet tall; "because the footprints he left on the beach were more than fourteen inches long" is *not* the answer to "why was Gargantua more than six feet tall?" My typewriter is neither blind nor not blind. That is a state of affairs for which "ordinary language" is partly responsible and a case might be made for extending the meaning of "blind" so that my typewriter can be said to be blind. That horses are warm-blooded, however, is a fact about horses, not language. It would remain true even if "warm-blooded" meant "member of the Ku Klux Klan," although we would then have to put the matter differently. That the premises of the inference about Gargantua do not make up a correct answer to why Gargantua was so big is a fact about these premises. It would remain a fact even if "why" were to become a request-marker for premises of deductive nomological explanations, although we would then have to put the matter differently.

(4) The relation between the explanans and the explanandum of a deductive nomological explanation—let us call it the *H* relation—can be defined in language-independent terms, i.e., in terms applicable to the expressions of any language rich enough for science. On the basis of such a definition it is also possible to define, in *language-independent* terms, a class of questions very much like why-questions, whose answer and presupposition need only be *H* related. Let us call them *H* why-questions. Their definition is a little complicated and we leave it for a footnote,[10] but anyone familiar with Hempel's doctrine will sense this possibility and will recognize it as one of the virtues of the doctrine. Those who reject the above counterexamples may simply doubt that why-questions can also be defined in language-independent terms and may believe that *H* why-questions are the nearest possible language-independent approximation. Accepting the counterexamples as binding would then mean giving up the principle that scientific questions are essentially language independent. However, such qualms are premature if, as we believe, why-questions *can* be defined in language-independent terms.

(5) The objection may finally mean that by insisting on the relevance of these examples we must not only be insisting on the importance of why-questions (which have their own interrogative in English), but must be denying the importance of *H* why-questions (which do not have an interrogative in English). We do not.

VI

What is essential is not always easy to distinguish from what is accidental in the relation between why-questions and their answers. For instance, it is often assumed that besides being true, presuppositions of why-questions that have answers must also be something surprising, something that conflicts with what had been expected, or at least something unusual. Stated a little more precisely, the view amounts to this: We ask questions for all sorts of reasons and with many different purposes in mind, e.g., to test someone's knowledge, to offer someone the opportunity to show his erudition, to kill time, to attract attention; but questions have one basic function, the asking for information not already in our possession. On the view now considered, why-questions can fulfill that basic function only when asked by someone who finds the truth of the presupposition surprising and unexpected.

Why-questions no doubt are often asked by people to whom the presupposition comes as a surprise and the fact that they ask them is often related to their surprise. Furthermore, some why-questions whose presupposition is not surprising or unexpected seem to have no answer. Why does the earth have only one satellite? Why does every gram-molecular weight of matter contain 6×10^{23} molecules? Why can anything not move with a velocity greater than that of light? Why do bodies attract one another with a force that is directly proportional to their mass and inversely proportional to the square of their distance? Why is *chien* the French word for dog? Why has there never been a President of the United States whose first name was Clovis? Why does anything exist at all? Anyone will feel about at least one of these questions that he cannot provide a "because . . ." answer, although not because he does not know or has forgotten but simply because there is no answer. The view is even compatible with the use of "why-should" questions that challenge one to show that a given why-question has an answer, e.g., "Why should there have been a President with the first name Clovis?" "Why shouldn't every gram-molecular weight contain 6×10^{23} molecules?"

If it were true that presuppositions of why-questions must be surprising, we would now have to seek out the relevant criteria for being surprising. Fortunately, it is not true. There is nothing unsound about the question "Why is the train late today?" asked by the harassed New Haven commuter who would be more surprised if the train were on time; nor is there anything unsound about why-questions raised by scientists about very familiar everyday phenomena. The same sort of considerations show that presuppositions need not be departures from regularities.

The view that we have just described is close to another view that is equally tempting and equally false. According to this second view, why-questions have answers only when there exists a plausible argument in behalf of a contrary of their presupposition. This could account for all the things accounted for by the previous view and for further things as well. If true, it would require us to analyze the relevant notion of plausible argument. But it is not true. There is no such plausible argument forthcoming in the case of "Why has there never been any President of the United States with the first name Clovis?" and yet the question is sound and has an answer: "Because no one by that name has ever been elected to the office or been the Vice President when a President died in office." The example is deliberately chosen from the list of questions cited previously as seeming to have no answer. It suggests that one's attitude toward the presupposition and other "pragmatic" considerations play no crucial role.

VII

The solution that we are about to propose requires a few preliminary definitions. These definitions are stated with the help of predicate logic notation.

The use of this notation introduces a number of theoretical problems that we will simply ignore. The problem of lawlikeness is but one of them. There are others that anyone familiar with the discussions of Hempelian doctrine will immediately detect.[11] We use the notation because it strikes us as providing the simplest way of exhibiting at present certain purely formal matters and we hope that our illustrations will bring out the intentions behind the schematisms. All these definitions must eventually be replaced by ones that make use of better representations. We think, however, that the heart of the analysis is essentially sound and that it may therefore be of some interest even in this temporary form. Each definition will be preceded by paradigms. This should make the formulae easier to read; it should, in fact, enable one to skip them altogether.

First Definition. General rule.

Paradigms. The level of a liquid in a cylindrical container on which a melting object is floating always rises. All French nouns form their plural by adding *s*. The velocity of an object never changes.

A general rule is a *lawlike* statement of the form

$$(x)(F_1x \cdot F_2x \ldots F_jx : \supset : S_1x \cdot S_2x \ldots S_kx) \qquad (j \geqslant 1, k \geqslant 1)$$

Note that the definition does *not* require that a general rule be true or even plausible.

Second Definition. General abnormic law.

Paradigms. (1) The level of liquid in a cylindrical container on which a melting object is floating at room temperature will rise unless the object is made of a substance whose density in liquid form is the same or is greater than that of the original liquid at room temperature. If the density is the same, the level will remain the same; if the density is greater, the level will go down.

(2) The level of liquid in a cylindrical container on which a melting object is floating at room temperature will rise unless upon melting completely the floating object undergoes a decrease in volume equal to or greater than the volume originally above the surface of the water. In the former case, the level remains the same; in the latter case, the level goes down.

(3) All French nouns form their plural by adding *s* unless they end in *al* (except *bal*, *cal*, *carnaval*, etc.) or in *eu*, or in *au*, or in *ou* (except *chou*, *genou*, etc.) or *x*, or *z*, or *s*. If and only if they end in *al* (except *bal*, etc.) they form the plural by dropping the last syllable and replacing it with *aux*; if and only if they end in *eu* or *ou* or *au* (except *chou*, etc.) they form their plural by adding *x*; if and only if they end in *x* or *z* or *s* they form their plural by adding nothing.

These are examples only if we are willing to assume that they are true as they stand.

A general abnormic law is *true, lawlike* statement of the form

$$(x)\quad (F_1x \cdot F_2x \ldots F_jx : \supset :. -Ex \equiv . \quad A_1x \vee A_2x \ldots \vee A_nx \vee B_1x \ldots \vee B_mx \ldots \vee R_ex$$

$$: A_1x \vee A_2x \ldots A_nx . \equiv S_Ax$$

$$: B_1x \vee B_2x \ldots B_mx . \equiv S_Bx$$

$$: \ldots\ldots\ldots\ldots\ldots\ldots\ldots$$

$$: R_1x \vee R_2x \ldots R_ex . \equiv S_Rx) \qquad (n \geqslant 1, j \geqslant 1)$$

of which the corresponding following statements are also true:

(a) $(x)(F_1x \cdot F_2x \ldots F_jx : \supset . Ex \vee S_Ax \vee S_Bx \ldots S_Rx) \qquad (R \geqslant 1)$

(b) $(x)(A_1x \supset : -A_2x \cdot -A_3x \ldots A_nx \cdot -B_1x \ldots -B_mx \ldots -R_ex$

$:. A_2x \supset : -A_1x \cdot -A_3x \ldots A_nx \cdot -B_1x \ldots -B_mx \ldots -R_ex$

$:. \quad . \quad . \quad . \quad . \quad . \quad . \quad . \quad . \quad . \quad . \quad . \quad . \quad . \quad . \quad . \quad . \quad .$

$:. R_ex \supset : -A_1x \cdot -A_2x \ldots -R_{e-1}x)$

(c) It does not remain a true, lawlike statement when one or more disjuncts in any of the internal biconditionals is dropped or when one or more of the conjuncts in the initial antecedent is dropped. (These three conditions are redundant, but we are obviously not after elegance in this sketch.)

(d) The closure of the main antecedent is not a logical truth or contradiction.

(e) The closure of none of the internal disjunctions is a logical truth or contradiction.

(We construe the "unless" in the paradigms as the exclusive disjunction.)

Third Definition. Special abnormic law.
Paradigms. (4) The velocity of an object does not change unless the net force on it is not equal to zero.
(5) No sample of gas expands unless its temperature is kept constant but its pressure decreases, or its pressure is kept constant but its temperature increases, or its absolute temperature increases by a larger factor than its pressure, or its pressure decreases by a larger factor than its absolute temperature.

Again we must assume that these are true.
A special abnormic law is a true, lawlike statement of the form

$$(x)(F_1x \cdot F_2x \ldots F_jx : \supset :. -Ex \equiv . A_1x \vee A_2x \vee \ldots \vee A_nx) \qquad (n \geqslant 1, j \geqslant 1)$$

that satisfies conditions (a) to (e) on general abnormic laws. (It is easy to show that every general abnormic law is equivalent to a conjunction of special abnormic laws but we will not make use of this fact.)[12]

Fourth Definition. Antonymic predicates of an abnormic law.
Paradigms. The antonymic predicates of (3) above are "Forms the plural by adding *s*," "Forms the plural by dropping the last syllable and replacing it with *aux*," "Forms the plural by adding *x*," "Forms the plural by adding nothing." Those of (4) are "Has a velocity that is changing," "Does not have a changing velocity."

The antonymic predicates of a general abnormic law are the predicates that appear in the consequent of (a). Those of a special abnormic law are the predicate substituted for *E* in the statement of that law, and the negation of that predicate.

Fifth Definition. The completion of a general rule by an abnormic law.
Paradigms. (1) and (2) are each a completion of the first paradigm of a general rule. (3) and (4) are the completion of the next two paradigms of a a general rule.

An abnormic law is the completion of a general rule if and only if the general rule is false and is obtainable by dropping the "unless" qualifications, i.e., by closing the statement before the first exclusive disjunction. (With our representation of the exclusive disjunction this requires negating the predicate substituted for *E*—or dropping the negation if it is already negated—deleting the biconditional connective, and making the obvious bracketing adjustments.)

We can now describe what we believe to be the relation between presuppositions and answers to why-questions. Before doing so, we will briefly present an example that points out the relevant features. The example and those to follow will only involve monadic predicates and will therefore fit the formulae in the definitions given above. But the predicates of presuppositions and answers of why-questions will not always be monadic and these definitions are thus too narrow as they stand. The shortcoming is readily remedied. We can either replace the references to the various formulae by references to the closure of the formulae obtainable by substitution from those given, or we can replace the formulae by more abstract schemata that allow for polyadic and for "zero-adic" predicates.[13] We shall assume that some such correction has in fact been adopted without actually carrying it out. Doing so would not solve the deeper problems alluded to in the introductory paragraph of this section, and the apparent gain in rigor would only be deceptive.

Why is the plural of the French noun *cheval chevaux*, i.e., formed by dropping the last syllable and replacing it with *aux*? Answer: (Because) *cheval* ends in *al*.

The answer together with abnormic law (3) and the further premise that *cheval* is a French noun form an explanans whose conclusion is the presupposition. The further premise that is not part of the answer together with the general rule completed by the abnormic law constitute a valid (but not sound) deduction whose conclusion is a *contrary* of the presupposition.

Here then is the relation: *b* is the correct answer to the why-question whose presupposition is *a* if and only if (1) there is an abnormic law *L* (general or special) and *a* is an instantiation of one of *L*'s antonymic predicates; (2) *b* is a member of a set of premises that together with *L* constitute a deductive nomological explanation whose conclusion is *a*; (3) the remaining premises together with the general rule completed by *L* constitute a deduction in every respect like a deductive nomological explanation except for a false lawlike premise and false conclusion, whose conclusion is a contrary of *a*; (4) the general rule completed by *L* has the property that if one of the conjuncts in the antecedent is dropped the new general rule cannot be completed by an abnormic law.[14]

More examples may loosen up this jargon.

Why has there never been a President of the United States with the first name Clovis? We get the answer in the following way.

General rule. Every name is the name of some President of the United States.

Abnormic law that completes this general rule. Every name is the name of some President of the United States unless no one by that name has ever been elected to the Presidency and no one by that name has ever been Vice-President when a President died in office.

Premises that together with the law form a deductive nomological explanation whose conclusion is the presupposition. Clovis is a name; no one with the name Clovis has ever been elected to the Presidency of the United States, and no one by that name has ever been Vice President when a President died in office.

Premises that together with the general rule lead to a contrary of the presupposition. Clovis is a name.

Remaining premise. The answer.

Next is an illustration of a why-question that has more than one correct answer. The case is adapted from a paper by Hempel: "In a beaker filled to the brim with water at room temperature there floats a chunk of ice which partly extends above the surface. As the ice gradually melts, one might expect the water in the beaker to overflow. Actually, however, the water level remains unchanged. How is this to be explained?"[15] We construe the last question as simply meaning, "Why did the level of water not rise?" Two relevant abnormic laws, (1) and (2), are available and both are completions of the same general rule, i.e., that given as our first example. The propositions that the contents of the beaker are a liquid on which a melting object is floating, that the liquid is water, that the object is ice, that ice upon melting becomes water, i.e., has the same density in liquid form as water, together with (1) form a deduction whose conclusion is the presupposition. The answer to the question: (Because) ice upon melting has the same density as water. The other premises together with the general rule lead to a contrary of the presupposition. We leave it to the reader to show that (2) leads in the same way to the answer: (Because) the ice undergoes a decrease in volume equal to the volume originally above the surface of the water.

It is instructive to read what Hempel wrote about this example:

> The key to an answer is provided by Archimedes's principle, according to which a solid body floating in a liquid displaces a volume of liquid which is the same weight as the body itself. Hence the chunk of ice has the same weight as the volume of water its submerged portion displaces. Now since melting does not affect the weights involved, the water into which the ice turns has the same weight as the ice itself, the same and hence, weight as the water initially displaced by the submerged portion of ice. Having the same weight, it also has the same volume as the displaced water; hence the melting ice yields a volume of water that suffices exactly to fill the space initially occupied by the submerged part of the ice. Therefore the water level remains unchanged.

Insofar as there is an answer conveyed in all this, it seems to be roughly equivalent to our second one.

Hempel was undoubtedly right in holding that the key to the explanation is provided by Archimedes' principle. However, if we look upon the question as a why-question, the principle is no more crucial than the principle that melting does not affect weight. It is the key in the sense that it provides a clue, also in the sense that anyone in a *p* predicament or *b* predicament with regard to the why-question must in all likelihood be told or be reminded of the principle; it is also an essential piece of knowledge for establishing that the answers are true, but it is not essential to establish that the answers, granted that they are true, are also correct answers to the why-question.

Our last illustration was a why-question that has more than one correct answer. Most why-questions are probably like that, i.e., true presuppositions seldom if ever determine unique answers. According to our analysis, this is to be expected since more than one abnormic law is usually available from which a given presupposition can be derived. Our analysis, then, does not segregate good answers from poor ones, only correct ones from incorrect ones. We could, therefore, expect it even to account for the degenerate cases made famous by Molière: "Why does opium put people to sleep? Because it has dormitive power." One might as well have said, "Because it puts people to sleep." These cases almost go through because of the availablity of such abnormic laws as "No substance puts people to sleep unless it puts people to sleep." Instances of the valid $(x)(Fx \supset . - Ex \veebar Ex)$ and of other schemata obtainable from $p \supset . q \equiv q$ by substitution and generalization are always available. However, these cases do not quite go through insofar as (2) on page 78 together with the definition of "deductive nomological explanation" require that the abnormic law be *empirical*. Thus, we see why, on the one hand, one can assimilate such answers with correct answers and why, on the other hand, one knows that they ought to be rejected.

VIII

Our analysis accounts for some familiar facts about why-questions. In general, a question arises whenever there is reason to believe that it has an answer, although the answer is not known. This will happen in the case of why-questions when one believes that the presupposition is true, views it as a departure from a general rule, and thinks that the conditions under which departures from the general rule occur can be generalized by an abnormic law. One may be mistaken about this. One may, for instance, be mistaken in thinking that the presupposition is true. In that case, no answer (as we have defined the term, i.e., correct reply in the form of "because . . . ") will be forthcoming. There will be, of course, appropriate replies. A statement to the effect that the presupposition is false will provide the relevant information.

One may, on the other hand, be mistaken in thinking that the presupposition represents a departure from a general rule. In that case, again, there will be no answer, although there will be other appropriate replies. "Why does this live oak keep its leaves during the winter?" "All live oaks do!" (Not however, "Because all live oaks do.") This sort of reply, like the previous one, has the force of a correction and entails that the question does not really arise.

One may, finally, be mistaken in thinking that the conditions under which departures from the general rule occur can be generalized. Here once more, no answer will be forthcoming: "Why is Johnny immune to poison ivy?" "Some people are and some people are not." However, an "answer" built from a degenerate abnormic law will also do, e.g., "No one is immune to poison ivy unless he is," and from that, "Because he is."

These everyday situations should not be taken too lightly by philosophers of science. Why-questions must sometimes be countered with a general rule rather than with an answer. This corresponds to the fact that scientific investigations of why something is the case often end not with the discovery of a "because . . ." answer, but with the establishment of a new general rule. And this poses a problem: When is such a substitution merely a *begging* of the why-question? When does our ignorance demand that we not trade a why-question for an H why-question but find the limits of a general rule? Why-questions even in science must sometimes be dealt with by denying that a departure from a general rule can be nontrivially generalized, which also raises problems: What sort of evidence warrants such denials? Can any fact ever be *shown* to be ultimate and unexplainable?

We mentioned in section VI the view that why-questions can fulfill their basic function only if the presupposition is something surprising or if there is at least a plausible argument forthcoming in behalf of one of its contraries. It is easy to see how such a view might come to be accepted: Many instances support it. This is no accident. One often guides one's expectations by general rules, rules that are sometimes explicitly and sometimes only implicitly acknowledged. Reliance on such rules entails belief that they work in most cases. But it also leads one to view certain facts as departures from general rules, a prerequisite for a why-question to arise. This prerequisite then is often satisfied under circumstances that surprise or that at least provide the grounds for a plausible argument for a contrary of a presupposition (the argument whose lawlike premise is the false general rule). As counterexamples show, such circumstances, although frequent, are not essential, and they do not provide the key to the nature of why-questions. Here too, an interesting problem for philosophers of science comes up. A clear mark of scientific genius is the ability to see certain well-known facts as departures from general rules that may have no actual instances, but that *could* have had some, and the germane ability to ask why-questions that occur to no one else. This way of looking at things can sometimes yield important insights, but it is also sometimes simply foolish. Is the difference analyzable in logical categories, or is it fundamentally a matter of psychology or perhaps theology?

Another view frequently held about why-questions—particularly about why-questions with negative presuppositions—is that the answer must describe the absence of a necessary condition for the contrary of the presupposition. This is not far from the truth for many cases to which these notions are easily applied, but it is an oversimplification, even for those cases. In the typical cases, the answer must describe the absence of (or at least something incompatible with) not merely *any* necessary condition for the contrary of the presupposition, but of a necessary condition belonging to a set (1) only one of whose members can be false, (2) each of whose members is necessary, and (3) all of whose members are jointly sufficient for that contrary. This follows from the definition of abnormic law. This is easily seen by looking at the propositional structure of instantiated special abnormic laws. A typical structure is

$$Ya \supset : Xa \equiv .Aa \vee Ba \vee Ca \tag{1}$$

where $Ya \supset -Xa$ is the propositional structure of the instantiated general rule, Xa is the presupposition, and the answer must be one of the disjuncts to the right of the biconditional. Typically, when (1) is true, so is

$$Xa \equiv .Aa \vee Ba \vee Ca \tag{2}$$

(Ya being the premise that together with the general rule leads to a contrary of the presupposition) and so then is

$$-Xa \equiv : -Aa \cdot -Ba \cdot -Ca \tag{3}$$

This shows that the answer (Aa or Ba or Ca) describes the absence of a necessary condition ($-Aa$ or $-Ba$ or $-Ca$) for the contrary of the presupposition. (Throughout we follow the practice of using "contrary" to mean "contrary or contradictory.")

Condition (b) in the definition of an abnormic law requires that the disjuncts in (1) be mutually exclusive, i.e., that if one of the conjuncts in (3) is false, the others must be true. (3) by itself requires that these conjuncts by jointly sufficient for $-Xa$.

We can test this consequence against an idealized, concrete instance. Two switches, *A* and *B*, are in series in a circuit so that current flows if and only if both switches are closed. Current is not flowing and both switches are open. Why is the current not flowing? Because both *A* and *B* are open. It would be misleading to say "Because *A* is open," although it is true and although it mentions the absence of a necessary condition for the contrary of the presupposition, and similarly for "Because *B* is open." Either of these replies *in this context* would imply that the other switch is closed. The possible answers, then, are: "*A* is open although *B* is closed, *B* is open although *A* is closed; *A* and *B* are both open." These are mutually exclusive. The negations are: "either *A* is closed or *B* is open; either *B* is closed or *A* is open; either *A* or *B* is closed." But this is a set of conditions for the contrary of the presupposition (1) only one of whose members can be false, (2) each of whose members is necessary, (3) all of whose members are jointly sufficient.

We can now understand the function and form of the why-should questions mentioned in section VI. "Why is the current flowing?" "Why shouldn't it be flowing?" They are designed not only to bring out grounds for believing that the original why-question has an answer, but also to narrow down the area within which the answer is expected. They do this by asking what necessary conditions for the contrary of the presupposition *are satisfied*, what necessary conditions belong to a set of jointly sufficient conditions only one of which is presumably false.[16] The answer wanted for the original why-question is thereby defined since it must negate the one remaining condition. Why-should questions take on the force of a challenge when there is reason to doubt that only one condition is missing. On the other hand why-should questions need not have an answer when a necessary and sufficient condition for the presupposition of the why-question is expected, i.e., in cases where (1) has only one disjunct or (3) has only one conjunct.

We must now turn to the examples cited in section V against the Hempelian doctrine. How do they fail as answers to why-questions? Let us look at a simple but typical member of the family. The telephone post at the corner of Elm Street is forty feet high. Its top is connected by a taut wire to a point thirty feet from its foot. The length of the wire is fifty feet. Why is the pole forty feet high? According to one interpretation of the Hempelian doctrine, an answer should be available that is made up of the fact about the wire, since the height can be deduced from these facts and laws of physical geometry. There would be an answer made up that way according to our analysis if it were an abnormic law that no pole is forty-feet high unless a taut fifty-foot-long wire connects its top to a point thirty feet from its foot. But there is no such law. Forty-foot-high poles may have no wires attached to them, and they may also have wires attached to them that are of a different length and connect to a different point on the ground. If we extend the clause after "unless" with disjunctions that include the cases with other wires and with no wires, we will still not end up with an abnormic law; some of the disjuncts will not be mutually exclusive and, furthermore, the law will remain a law if all the disjuncts except that pertaining to the case of no wires are dropped.

There would also be an answer made up of the facts about the wire according to our analysis if it were a law that no pole is forty feet high unless, if there is a taut wire connecting the top to a point on the ground and the wire is fifty feet long, then the point on the ground is thirty feet from the foot. But there is no such law. If there were, it would entail that every pole to which no wire is attached must be forty feet high!

However, the following is a law: No pole whose top is connected to a point on the ground by a wire that is fifty feet long is itself forty feet high unless that point on the ground is thirty feet from the foot of the pole. Still, it does not meet the requirements of the analysis. According to (4) on page 73, in the description of the relation, the general rule completed by the abnormic law must not be such that by dropping one or more of the conjuncts in the antecedent a new general rule is obtained that can also be completed by an abnormic law. But the above abnormic law violates that condition. We know enough about poles to be confident that there is an abnormic law of the form "No pole is forty feet high unless. . . . "

All the cases cited against the Hempelian doctrine will fail for similar reasons. Just as we are confident that there are laws according to which poles will be forty feet high regardless of whether wires are attached to them, so there must be laws according to which the Empire State Building will have the height it has even in total darkness, the distance to the Andromedan galaxy would be what it is even if no light traveled to us from it, the rate of flow of water into the bathtub would be what it is whether or not measured, Gargantua would be more than six feet tall even if he had not gone to the beach.[17]

The very same sorts of considerations, it may be worth noting, will account for certain asymmetries that have puzzled some philosophers. From the laws of the simple pendulum and the length of a piece of string at the end of which a bob is hanging and local free-fall acceleration, one can deduce the period with which that bob is oscillating. From the same law and data about local free-fall acceleration and the period with which the bob is oscillating, one can deduce its length. Yet a statement of the length is an answer to "Why does the bob oscillate with such and such a period?" whereas a statement of the period of oscillation is not an answer to "Why is the length of the string at the end of which the bob is hanging so many inches long?" The asymmetry is traceable, in a manner exactly similar to the previous reasoning, to the fact that whereas the period would not have been what it is if the length had not been what it is, the length would have been what it is whether the bob had been oscillating or not.

Condition (4) may seem at first blush somewhat arbitrary. A little reflection will bring out, however, that it corresponds to a generally acknowledged and reasonable norm. It demands on the one hand that the answer be a consequence of the most general abnormic law[18] available, and it demands on the other hand that questions of the form "Why is this *A* a *C*?" not be given answers that are really designed for "Why is this *AB* a *C*?"

It may seem odd that abnormic laws should be associated with a special interrogative. But they are, after all, the form in which many common-sense generalizations that have been qualified through the ages are put. They are also a form of law appropriate to stages of exploratory theoretical developments when general rules are tried, then amended until finally completely replaced. We are always at such a stage.

NOTES

1. J. O. Urmson, "Parenthetical Verbs," reprinted in *Essays in Conceptual Analysis*, ed. A. Flew (London: 1956), p. 192. (3) eliminates from our discussion questions designed to ask for an opinion rather than a fact. Thus, it eliminates, e.g., "Why do you think that nail biting is a symptom of anxiety neurosis?" in the sense of "Why, in your opinion, is nail biting a symptom of anxiety neurosis?" although not in the sense of "Why do you hold the belief that nail biting is a symptom of anxiety neurosis?"

2. A little care is needed in using the notions introduced here. A given why-question can often be put in more than one normal form, some of which will be ambiguous. This is particularly true of those that may be put in interrogative sentences with token reflexive expressions (e.g., "Why is your temperature above normal?" as put to Henry and "Why is Henry's temperature above normal?" as put to his doctor). Whenever this is the case, the inner question can also be ambiguous. We must therefore always think of the inner question as put under circumstances that give ambiguous expressions the same disambiguation given to them in the mother question.

We could have introduced the notion of presupposition by availing ourselves of some grammatical devices, e.g., the presupposition is what one would be saying is the case by asserting the sentence whose underlying structure preceded by a Why morpheme yields the why-question (or at least the interrogative sentence) when subjected to the Question Transformation. But we wish to avoid complicating the exposition beyond necessity or involving ourselves in grammatical issues that are still in flux. See in this connection particularly Sec. 4.2.4 of Jerrold J. Katz and Paul M. Postal, *An Integrated Theory of Linguistic Descriptions* (Cambridge: The Massachusetts Institute of Technology Press, 1964). Note that what we call the presupposition of a why-question does not turn out to be what they call the presupposition of a why-question.

3. It is not clear whether there are such questions, i.e., whether, for example, answers of the form "in order to . . ." cannot always be replaced without loss of meaning by answers of the form "because (subject) wished to . . . ," and similarly for the other types of answer. The issues involved here are extremely interesting but not central for this paper.

4. Cf. Sec. VIII.

5. As should be clear by now, "correct answer" must be understood in a narrow sense. "Correct answer to Q" (where *Q* is a question) covers a possible reply to *Q* if and only if a statement of the form "A told B W" (where *W* indicates a position occupied by the *oratio obliqua* form of *Q*, and *A* and *B* indicate positions occupied by expressions through which persons are mentioned) would be true of any episode in which that reply had been given by *A* to *B* in response to *Q*. "Correct answer," therefore, does not cover such possibly warranted replies as "I don't know" or "The question involves a false presupposition."

6. See n. 14 below.

7. The literature abounds with discussions that are weakened by a failure to see all these possibilities. A classical example will be found in Part I of Pierre Duhem, *La Theorie Physique* (Paris: 1914) in which it is argued that the object of a physical theory is not to explain a set of empirical laws. However, "explain" is construed in effect to mean giving the answers to questions of the form "What fundamental entities involved in what processes and governed by what laws underlie . . . ?" As a consequence, Duhem did not examine a number of other types of explanations that one might plausibly assign to theoretical physics.

The notion of presupposition used in this section is broader than that defined in Sec. I, since it also pertains to questions that are not why-questions. No analysis of this broader notion is needed for this paper. N. 14 may suggest the line that such an analysis might follow since it provides an instance of the schematisms to be generalized. It should be obvious to anyone who bothers to seek out the suggestion that it would be premature to attempt the analysis given the present state of our understanding of other types of questions. In this connection see again J. J. Katz and P. M. Postal, *op. cit.*

8. Some of the ideas in this section have been discussed in greater detail but at too great a length in my "An Approach to Explanation," in *Analytical Philosophy*, 2d Ser., ed. R. J. Butler (Oxford: Basil Blackwell, 1965) and in my "A Theory About the Theory of Theory and About the Theory of Theories" in *Philosophy of Science: The Delaware Seminar*, II, ed. Bernard Baumrin (New York: Interscience Publishers, 1963).

9. A complete bibliography on the subject probably appears in Carl G. Hempel, *Aspects of Scientific Explanation* (New York: The Free Press, 1965), which I have unfortunately not yet seen as this essay is being written. Otherwise, consult, e.g., the bibliography at the end of Hempel's magnificent "Deductive-Nomological vs. Statistical Explanation" in *Minnesota Studies in the Philosophy of Science*, III, eds. H. Feigl and G. Maxwell (Minneapolis: U. of Minnesota Press, 1962); Part I of Israel Scheffler, *The Anatomy of Inquiry* (New York: Knopf, 1963); Chap. ix of Adolf Grunbaum, *Philosophical Problems of Space and Time* (New York: Knopf, 1963); and *Philosophy of Science: The Delaware Seminar*, I, Part II; II, Part I.

10. See n. 14 below.

11. See particularly R. Eberle, D. Kaplan, and P. Montague, "Hempel and Oppenheim on Explanation," *Philosophy of Science*, 28 (1961), 418–28; D. Kaplan, "Explanation Revisited," *Philosophy of Science*, 28, 429–36; and J. Kim, "Discussion on the Logical Conditions of Deductive Explanation," *Philosophy of Science*, 30 (1963), 286–91.

12. Since every special abnormic law is also a general abnormic law, we could have dispensed with one of these two notions but not without complicating the exposition.

13. A "zero-adic" predicate will occur if, for instance, a position indicated in one of our schemata by a predicate letter and variable bound to an initial quantifier is replaced by a sentence with no free variables, i.e., with no variable bound to the initial quantifier. Abnormic laws with occurrences of such internal closed sentences are required for why-questions whose presupposition or answer are expressed by closed sentences, as is the case when they are laws.

14. We asked that this relation abstract from the peculiarities of English and be capable of serving as the basis of a definition of the notion of why-questions, a type of question in every respect like why-questions except that they need not be expressed in English. To satisfy ourselves that it meets these demands we will sketch a more formal analysis that clearly uses only the vocabulary of predicate-cum-identity logic and language-independent predicates, and we will then use the relation to define the notion of why-questions within the same limits. The analysis will be somewhat crude, its only function being to exhibit language independence. It will suffer in at least the following respects: (1) The second half of condition (*1*) and condition (*4*) are not incorporated on the ground that it seems obvious that their incorporation can be accomplished without introducing language-dependent concepts but would complicate matters beyond the point of diminishing returns; (2) we assume without argument that any language rich enough for the purposes of science includes sentences with the logical structure

of abnormic laws. This, we believe, involves no more than the assumption that such a language must possess the equivalent of truth-functional connectives, quantifiers, and lawlikeness; (3) we assume without argument that if a set of sentences implies some conclusion in one language, then any set of sentences that expresses the same thing in another language must imply any sentence that expresses the same conclusion in that language, i.e., that although logic may be reflected by syntax, it is nevertheless independent of it; (4) we assume without argument—although not without qualms—that interrogative sentences of different languages may express the same question, that declarative sentences of different languages may express the same proposition, and that one may use a relational term to speak of a sentence and of what it expresses. It seems, however, that ontologically sounder rephrasings cannot introduce language-dependent elements; (5) we assume without argument that any language rich enough for the purposes of science will contain interrogative as well as declarative sentences; that it will also have methods for transforming declarative sentences into interrogative ones; that furthermore all the answers to all the questions generated by some of these methods must stand in a characteristic relation to the transformed sentence.

We will list a lexicon of language-independent predicates and will then define others in terms of these. Two things ought to be noted. First, we do not assume that being abnormic is a property of laws but assume rather that it is a property of certain sentences that express laws. Thus, certain laws may be expressible as abnormic lawlike statements in some languages but not in other languages, depending on the lexicon of each. Second, let us call the relation between presupposition and answer described above the *W*-relation. We do not assume that *any* question whose presupposition and correct answer, if any, stand in the *W*-relation is a why-question. Instead, we make allowance for the fact that the *W*-relation need not exclude relations characteristic of other questions (see particularly the definition of "TW" below in this connection).

To simplify the reading, we use numerals as free variables.

Initial lexicon:

L1 — *1* is an empirical law.
F1 — *1* is a fact.
E123 — *1* expresses *2* in *3* and *3* is a language.
A12 — *1* is an abnormic lawlike sentence of *2* and *2* is a language.
G12 — *1* is a lawlike general rule in *2* and *2* is a language.
C123 — *1* is a completion of *2*, both being sentences in *3* and *3* is a language.
H1234 — *1* is a deduction whose conclusion is *3* and one of whose premises is *2*, all of whose premises are necessary for the conclusion, *2* is a lawlike sentence, all the premises and conclusion being in *4*, *4* is a language.
P12 — *1* is an argument and *2* is a premise of *1*.
T1234 — *1* is a method of generating (transformation) *2* from *3* in *4* and *2* is an interrogative sentence and *3* is a declarative sentence, *2* and *3* being sentences in *4* and *4* is a language.
R123 — *1* is an interrogative sentence, *2* expresses a correct answer in *3* to the question expressed in *3* by *1* and *3* is a language.
N123 — *1* expresses in *3* a contrary of what is expressed by *2* in *3* and *3* is a language.

Defined terms:

"*LA123*" = "*L1.A23.E213*"
(*1* is an empirical law expressed as an abnormic law by *2* in *3*)
"*GE123*" = "*−F1.G23.E213*"
(*2* is a false general rule expressing *1* in *3*)
"*FA123*" = "*F1.E213*"
(*1* is a fact expressed by *2* in *3*)
"*W123*" = "$(\exists s)(\exists t)(\exists u)(\exists v)(\exists w)(\exists x)(\exists y)$
$[LAst3.GEuv3.Ctv3.Hwvx3.Nx13.Hyt13.(p)(\exists z)(Pwp. p \neq y: \supset :Pyp.Fzp3) . (q)(Pyq. - Pwq . q \neq t: \equiv .q = 2)]$"
(*1* and *2* stand in *3* in the relation in which the presupposition and the answer of a why-question stand in English, i.e., the *W*-relation).

"$TW12$" = "$(x)(y)[\exists z)(T1zx2.Ryz) \equiv Wxy2]$"
(*1* is a method in *2*—*2* being a language—of generating questions whose presupposition and answers are *W*-related.)
"$VY123$" = "$(\exists m)(TWm3.Tm123)$"
(*1* has been generated out of *2* by a method of generating why-questions in *3*—i.e., *1* expresses a Why-question in *3* whose presupposition is expressed by *2* in *3*.)
"$WY12$" = "$(x)(y)(z)(Ey1x.Ez2x: \equiv VYyzx)$"
(*1* is a Why-question whose presupposition is *2*)
"$WHY1$" = "$(x)WY1x$"
(*1* is a Why-question)

We can follow the same procedure to define *H* why-Questions, i.e., questions calling for deductive nomological explanations.

"$HW123$" = "$(\exists x)(z)(\exists y)(Gx3.H2x13:P2z \supset FAyz3)$"
(*1* and *2* stand in the *H*-*W*-relation)
"$THW12$" = "$(x)(y)[\exists z)(T1zx2.Ryz) \equiv HWxy2]$"
"$HVY123$" = "$(\exists m)(THWm3.Tm123)$"
"$HWY12$" = "$(x)(y)(z)(Ey1x.Ez2x: \equiv HVYyzx)$"
"$HWHY1$" = "$(\exists x)HWY1x$"
(*1* calls for a Hempelian explanation.)

15. Carl G. Hempel, "The Logic of Functional Analysis," *Symposium on Sociological Theory*, ed. Llewellyn Gross (Evanston, Ill.: Peterson, 1959), p. 272.

16. Such a request obviously need not be met with an actual listing of conditions. The set can be indicated in many other ways—e.g., by pointing to other cases that seem in all relevant respects like those of the presupposition, but of which the predicate of the presupposition is not true.

17. We can cope with these cases in a different way. Instead of individuating why-questions by their presupposition, as we have done so far, we individuate them by an ordered pair consisting of their presupposition and a false general rule. Thus distinct why-questions can now share the same presupposition. We restate the characteristic relation of why-questions as follows: *b* is the correct answer to a why-question whose presupposition is *a* and whose general rule is *g* if and only if (1) there is an abnormic law *L* that is a completion of *g*, and *a* is an instantiation of one of *L*'s antonymic predicates; (2) *b* is a member of a set of premises that together with *L* constitute a deductive nomological explanation whose conclusion is *a*; (3) the remaining premises together with *g* constitute a deduction in every respect like a deductive nomological explanation except for a false lawlike premise and false conclusion, whose conclusion is a contrary of *a*. We eliminate (4) on p. 78. Instead of appealing to it in analyzing the failure of the counter-examples to the Hempelian doctrine (when these examples are reconstructed to include true abnormic laws), we construe the failure as that of not containing the answer to the why-question most reasonably inferred from context and general background. This is compatible with the possibility that the examples contain the answer to *some* why-question with the given presupposition. The failure is nevertheless fatal. Note that this approach still allows for why-questions—even under this new individuation—that have more than one correct answer.

There is much to be said for this approach. It conforms to many of our practices. It does justice to our intuition that why-questions are governed not only by presuppositions but also by presumptions. It avoids certain difficulties with the argument in the text. However, it introduces certain pragmatic issues that we prefer to delay as long as possible.

18. For a discussion of a similar demand in connection with something the author calls "scientific understanding" (a notion whose relevance to the topic of why-questions, we confess, is not clear to us), cf. p. 310 of A. Grünbaum's *Philosophical Problems of Space and Time* or p. 93 of *Philosophy of Science: The Delaware Seminar*, I.

Michael Scriven

EXPLANATIONS, PREDICTIONS, AND LAWS

4. FUNDAMENTAL ISSUES

4.1. The Distinction between Explanations and the Grounds for Explanations.

It is certainly not the case that our grounds for thinking a plain descriptive statement to be true are part of the statement itself; no one thinks that a more complete analysis of "Gandhi died at an assassin's hand in 1953" would include "I read about Gandhi's death in a somewhat unreliable newspaper" or "I was there at the time and saw it happen, the only time I've been there, and it was my last sabbatical leave so I couldn't be mistaken about the date," etc. Why, then, should one suppose that our grounds for (believing ourselves justified in putting forward)[1] a particular explanation of a bridge collapsing, e.g., the results of our tests on samples of the metal, our knowledge about the behavior of metals, eye-witness accounts, are part of the explanation? They might indeed be produced as part of a *justification* of (the claim that what has been produced is) the explanation. But surely an explanation does not have to contain its own justification any more than a statement about Gandhi's death has to contain the evidence on which it is based. Yet, the deductive model of explanation requires that an explanation include what are often nothing but the grounds for the explanation.

Not only linguistic impropriety but absolute impossibility is involved in the attempt to market the joint package as the "whole explanation" or "complete explanation." The linguistic impropriety is twofold: First, perfectly proper

From Michael Scriven, "Explanations, Predictions, and Laws," *Minnesota Studies in the Philosophy of Science*, III, Herbert Feigl and Grover Maxwell, eds., University of Minnesota Press, Minneapolis.

explanations would be rejected for the quite unjust reason that they did not contain the grounds on which they were asserted; second, the indefinite number of possible grounds for an explanation makes absurd the idea of a single correct explanation since there is, in terms of the model, nothing more or less correct about any one of the wide range of possible sets of deductively adequate true grounds. And clearly these are circumstances in which we do identify a particular account as "The correct explanation." The impossibility derives from the second impropriety. There is no sense in which one could ever provide a complete justification of an explanation, out of context; for a justification is a defense against some specific doubt or complaint, and there is an indefinite number of possible doubts.

The deductive model apparently provides an answer to the latter objection in an interesting way. It prescribes that the only kind of justification required is deduction from general laws and specific antecedent conditions. Once this is given, a complete explanation has been given; until this has been done, only (at best) an "explanation sketch" has been given.

When we say that a perfectly good explanation of one event, e.g., a bridge collapsing, may be no more than an assertion about another event, e.g., a bomb exploding, might it not plausibly be said that this can only *be* an explanation if some laws are assumed to be true, which *connect* the two events? After all, the one is an explanation of the other, not because it came before it, but because it *caused* it. In which case, a full statement of the explanation would make explicit these essential, presupposed laws.

The major weakness in this argument is the last sentence; we can put the difficulty again by saying that, if completeness requires not merely the existence but the quoting of all necessary grounds, there are no complete explanations at all. For just as the statement about the bomb couldn't be an explanation of the bridge collapsing unless there was some connection between the two events, it couldn't be an explanation unless it was true. So, if we must include a statement of the relevant laws to justify our belief in the connection, i.e., in the soundness of the explanation, then we must include a statement of the relevant data to justify our belief in the claim that a bomb burst, on which the soundness of the explanation also depends."[2]

Certainly in putting forward one event as an explanation of another in the usual cause-seeking contexts, we are committed to the view that the first event caused the second, and we are also committed to the view that the first took place. Of course, we may be wrong about either view and then we are wrong in thinking we have given the explanation. But it is a mistake to suppose this error can be eliminated by quoting further evidence (whether laws or data); it is merely that the error may be then located in a more precise way—as due to a mistaken belief in such and such a datum or law. The function of deduction is only to shift the grounds for doubt, though doubts sometimes get tired and give up after a certain amount of this treatment.

Perhaps the most important reason that Hempel and Oppenheim have for insisting on the inclusion of laws in the explanation is what I take to be their belief (at the time of writing the paper in question) that only if one had such laws in mind could one have any rational grounds for putting forward one's explanation. This is simply false as can be seen immediately by considering an example of a simple physical explanation of which we can be quite certain. If you reach for a cigarette and in doing so knock over an ink bottle which then spills onto the floor, you are in an excellent position to explain to your wife how that stain appeared on the carpet, i.e., why the carpet is stained (if you cannot clean it off fast enough). You knocked the ink bottle over. This is the explanation of the state of affairs in question, and there is no nonsense about it being in doubt because you cannot quote the laws that are involved, Newton's and all the others; in fact, it appears one cannot here quote any unambiguous true general statements, such as would meet the requirements of the deductive model.

The fact you cannot quote them does not show they are not somehow "involved," but the catch lies in the term *involved.* Some kind of connection must hold, and if we say this means that laws are involved, then of course the point is won. The suggestion is debatable, but even if true, it does not follow that we will be able to state a law that guarantees the connection. The explanation requires that there be a connection, but not any particular one—just one of a wide range of alternatives. Certainly it would not be the explanation if the world was governed by *anti*gravity. But then it would not be the explanation if you *had not* knocked over the ink bottle—and you have just as good reasons for believing that you did knock it over as you have for believing that knocking it over led to (caused) the stain. Having reasons for causal claims thus does not always mean being able to quote laws. We shall return to this example later. For the moment, it is useful mainly to indicate that (1) there is a reply to the claim that one cannot have good reasons for a causal ascription unless one can quote intersubjectively verifiable general statements and (2) there is an important similarity between the way in which the production of an appropriate law supports the claim that one event explains another, and the way in which the production of further data (plus laws) to confirm the claim that the prior event occurred supports the same claim. They are defenses against two entirely different kinds of error or doubt, indeed, but they are also both support for the same kind of claim, viz., the claim that one event (state, etc.) explains another.

This is perhaps obscured by the fact that when we make an assertion our claim is in full view, so to speak, whereas when we put forward an assertion *as* an explanation, its further role is entirely derived from the context, e.g., that it is produced in answer to a request for an explanation, and so its further obligations seem to require explicit statement. This is a superficial view. All that we actually identify in the linguistic entity of a "declarative statement" is the subject, predicate, tense, etc. We have no reason at all, apart from the context of its utterance, for supposing it to be *asserted*, rather than proposed for consideration, pronounced for a grammatical exercise, mouthed by an actor, produced as an absurdity, etc.[3] *That* it is asserted to be true we infer from the context just as we infer that it is proffered as the explanation of something else;

and for both these tasks it may need support. We may concede that assertion is the *primary* role of indicative sentences without weakening this point.

It is in fact the case that considerations of context, seen to be necessary even at the level of identifying assertions and explanations themselves, not only open up another dimension of error for an explanation, that of pragmatic inappropriateness, but simultaneously offer a possible way of identifying *the* explanation of something, where this notion is applicable.

A particular context—such as a discussion between organic chemists working on the same problem—may make one of many deductively acceptable explanations of a biochemical phenomenon entirely inappropriate, and make another of exactly the right type. (Of course, I also wish to reject the criteria of the deductive model; but even if one accepted it, the consideration of context turns out to be *also* necessary. So its importance is not only apparent in dealing with alternative analyses.)

We may generalize our observations in the following terms. An explanation is sometimes said to be "incorrect," or "incomplete," or "improper." I suggest we pin down these somewhat general terms along with their slightly more specific siblings as follows: If an explanation explicitly contains false propositions, we can call it "incorrect" or "inaccurate." If it fails to explain what it is supposed to explain because it cannot be "brought to bear" on it, e.g., because no causal connection exists between the phenomenon as so far specified and its alleged effect, we can call it "incomplete or "inadequate." If it is satisfactory in the previous respects but is clearly not the explanation required in the given context, either because of its difficulty or its field of reference, we can call it "irrelevant," "improper," or "inappropriate."

Corresponding to these possible failings there are types of defense which may be relevant. Against the charge of inaccuracy, we produce what I shall call *truth-justifying grounds*. Against the charge of inadequacy, we produce *role-justifying grounds*, and against the complaint of inappropriateness, we invoke *type-justifying grounds*. To put forward an explanation is to commit oneself on truth, role, and type, though it is certainly not to have explicitly considered grounds of these kinds in advance, any more than to speak English in England implies language-type consideration for a life-long but polylingual resident Englishman.

The mere production of, for example, truth-justifying grounds does not guarantee their acceptance, of course. They may be questioned, and they may be defended further by appeal to further evidence; we defend our claim that a bomb damaged a bridge by producing witnesses or even photographs taken at the time; and we may defend the accuracy of the latter by producing the unretouched negatives and so on. The second line of defense involves *second-level grounds*, and they may be of the same three kinds. That they can be of these kinds is partly fortuitous (since they are not explanations of anything) and due to the fact that the relation of being-evidence-for is in certain ways logically similar to being-an-explanation-of. In each case, truth, role, and type may be in doubt; in fact, this coincidence of logical character is extremely important.

We notice, however, that there is no similarity of any importance between these two and being-a-prediction-of, where truth is not relevant in the same way, role is wholly determined by time of utterance and syntax, and only the type can be in some sense challenged.

4.2. Completeness in Explanations.

The possibility of indefinitely challenging the successive grounds of an explanation has suggested to some people—not Hempel and Oppenheim—that a complete explanation cannot be given within science. Such people are adopting another use of "complete"—even less satisfactory than Hempel and Oppenheim's—according to which the idea of a complete explanation becomes not only foreign to science but in fact either wholly empty, essentially teleological, or capable of completion by appeal to a self-caused cause. Interesting though this move is in certain respects, it essentially requires saying that we can better understand something in the world by ultimately ascribing its existence and nature to the activities of a mysterious entity whose existence and nature cannot be explained in the same way, than by relating it to its proximate causes or arguing that the world has existed indefinitely. I shall only add that we are supposed to be studying scientific explanations, and if none of them are complete in this sense, we may as well drop this sense while making a note of the point—which is equivalent to the point that the causal relation is irreflexive and hence rather unexciting—for there is an important and standard use of "complete" which does apply to some suggested scientific explanation, and not to others, and is well worth analyzing.

Now, if some scientific explanations are complete—and think how a question in a physics exam may ask for a complete explanation of, for example, the effects noticed by Hertz in his experiments to determine whether electromagnetic waves existed—it cannot be because there is a last step in the process of challenging grounds, for there is no stage at which a request for further proof could not make sense. But *in any given context* such requests eventually become absurd, because in any scientific context certain kinds of data are taken as beyond question, and there is no meaning to the notions of explanation and justification which is not, directly or indirectly, dependent on a context. This situation is of a very familiar kind of logic. It makes perfectly good sense to ask for the spatial location of any physical object; and perfectly proper and complete answers will involve a reference to the location of some other physical objects. Naturally we can, and often do, go on to ask the further question concerning where these other objects are ("Where's Carleton College?" "In Northfield." "But where's Northfield?"). And no question in a series of this kind is meaningless, unless one includes the question "Where is the universe, i.e., everything?" as of this kind. If one does include this question (which is the analogue of "Where did the universe *come from*?"), then the impossibility of answering it only shows something about the notion of position, and nothing about the incompleteness of our knowledge. If one excludes this question, the absence of a last stage in such a series does not show our inability to give anyone complete directions to the public library but only that the notion of completeness of such descriptions involves context criteria.

Any request for directions logically pressupposes that *some* directions can be understood; if no directions can be understood, then the proper request is for an account of the notions of position and directions. A *complete* answer has been given when the particular object has been comprehensibly related to the directions that are understood. Similarly, then, the request for an explanation pressupposes that *something* is understood, and a complete answer is one that relates the object of inquiry to the realm of understanding in some comprehensible and appropriate way. What this way is varies from subject matter to subject matter just as what makes something better than something else varies from the field of automobiles to solutions of chess problems; but the *logical function* of explanation, as of evaluation, is the same in each field. And what counts as complete will vary from context to context within a field; but the logical category of complete explanation can still be characterized in the perfectly general way just given, i.e., the logical function of "complete," as applied to "explanations," can be described. Hence the notion of the proper context for giving or requesting an explanation, which presupposes the existence of a certain level of knowledge and understanding on the part of the audience or inquirer, *automatically entails* the possibility of a complete explanation being given. And it indicates exactly what can be meant by the phrase *the* (*complete*) *explanation*. For levels of understanding and interest define areas of lack of understanding and interest, and the required explanation is the one which relates to these areas and not to those other areas related to the subject of the explanation but perfectly well understood or of no interest (these would be explanations which could be correct and adequate but inappropriate). It is worth mentioning that the same analogy with spatial location (or evaluation) provides a resolution of the "problem of induction," as a limit case of a request for a "complete justification."

It is also clear that calling an explanation into question is not the same as—though it includes—rejecting it as not itself explained. Type justifying involves more than showing relevance of subject matter, i.e., topical and ontological relevance; it involves showing the appropriateness of the intellectual and logical level of the content; a proposed explanation may be inappropriate because it involves the wrong kind of true statements from the right field, e.g., trivial generalizations of the kind of event to be explained, such that they fulfill the deductive model's requirements but succeed only in generalizing the puzzlement. One cannot explain why this bridge failed in this storm by appealing to a law that all bridges of this design in such sites fail in storms of this strength (there having been only two such cases, but there being independent evidence for the law, not quoted). This might have the desirable effect of making the maintenance boss feel responsible, but it surely does not explain *why* this bridge (or any of the other bridges of the same design) fails in such storms. It may be because of excessive transverse wind pressure, because of the waves affecting the foundations or lower members, because of resonance, etc.

So mere deduction from true general statements is again seen to be less than a sufficient condition for explanation; but what interests us here is that our grounds for *rejecting* such an explanation are not suspicions about its *truth* or its *adequacy*, which are the usual grounds for doubting an explanation, but only its failure to *explain*. Certainly it fails to explain if incorrect or inadequate, but then one feels it fails in a genuine attempt, that the slip is then between the cup and the lip; whereas irrelevance of type is a slip between the hand and the cup—the question of it being a *sound* explanation never even arises. One may react to this situation by declaring with Hempel and Oppenheim that the only *logical* criteria for an explanation are correctness and adequacy, the matter of type being psychological; or, as I think preferable, by saying that the concept of explanation is logically dependent on the concept of understanding, just as the concept of discovery is logically dependent on the concept of knowledge-at-a-particular-time. One cannot discover what one already knows, nor what one never knows; nor can one explain what everyone or no one understands. These are tautologies of logical analysis (I hope) and hardly grounds for saying that we are confusing logic with psychology.

Having distinguished the types of difficulty an explanation may encounter, one can more easily see there is no reason for insisting that it is complete only if it is armed against them in advance, since (1) to display in advance one's armor against *all* possible objections is impossible and (2) the value of such a requirement is adequately retained by requiring that scientific explanations be such that scientifically sound defenses of the several kinds indicated be *available* for them though not necessarily *embodied* in them. Since there is no special reason for thinking that true first-level, role-justifying assumptions are any more necessary for the explanation than any others, it seems quite arbitrary to require that they should be included in a complete explanation; and it is quite independently an error to suppose they must take the form of laws.

4.3. The Elements of an Alternative Analysis.

Hempel and Oppenheim make clear on two occasions that they are genuinely concerned with an analysis of the concept scientists and rational men normally refer to as "explanation," not with the development of some vaguely related and prejudicially labeled "explication" of it; for example, they say, in rejecting a possible condition, "This does not appear to accord with sound common usage. . . ."[4] We are coming a little closer to seeing that their proposal for explanations of events represents an analysis of *one kind of deduction of a singular proposition*; it is neither explanation nor the only kind of justification of explanation that they are describing. Our alternative analysis of the *particular kind of scientific explanation* they discuss is also becoming clearer. A *causal* explanation of an event of state X, in circumstances C, is *exemplified* by a set of statements S, such that (a) S asserts, at least, the existence of the phenomenon Y; (b) the existence of Y is comprehensible; (c) Y is the cause of X; (d) it is understood that Y's can cause X's in C; (e) S is understandable; (f) S is true; (g) S is of the correct type for the context. Without getting carried away and suggesting these are necessary or even independent conditions (which they are not), there is perhaps some value in setting out one set of sufficient conditions, to demonstrate the importance of the crucial concept of understanding.

4.4. The Bearing of Understanding and Context on Completeness.

Who is to say whether S and Y are understood? The *primary* case of explanation is the case of explaining X *to someone*; if there were no cases of this kind, there could be no such thing as "an explanation of X" in the abstract, whereas the reverse is not true. For it makes no sense to talk of *an explanation* which nobody understands now, or has understood, or will, i.e., which is not *an explanation for someone*. In the primary case, the level of understanding is that of the person addressed. The notion of "an (or the) explanation of X" in the abstract makes sense just insofar as it makes sense to suppose a standardized context. When we talk of "the explanation of sunspots," we rely on the fact that to get to the point where one understands the term "sunspot," but does not understand what produces them, is to acquire a fairly definite minimum body of knowledge. "The explanation" will be the one appropriate for a person with this knowledge, and this secondary sense of explanation is quite strong enough to survive certain types of failure of comprehension, i.e., a number of cases where "the explanation of X" is produced without the least understanding by those who hear it. In particular, the failure of two-year-old children to understand the explanation of sunspots is no ground for supposing the explanation to be unsatisfactory; it is "the explanation," not because it is understandable to everyone, but because it is understandable to the group that meets the conditions (a) and (b) just mentioned. It is important to remember how limited a use the notion of "the explanation" has; there is nothing that naturally comes to mind as "the explanation" of the nature of light (too general for any single account to be preferable at all points for any substantial body of people) or of enzyme activity (of interest to several fields, in each of which *an* explanation can be given). When more than one explanation is available, certain relations between them can be inferred which we shall discuss later; but that there is more than one is due to the fact that there is more than one way in which a phenomenon may be hard for *someone* to understand.

The analysis of *4.3* does not exclude giving causal laws which connect X and Y, neither does it require it. There are other occasions, not covered by *4.3*, when the explanation of an event *simply* consists in giving some causal laws. Here the role-justifying grounds may be assertions about the occurrence of events. And there are yet other occasions when an explanation is given in physics which does not consist in the production of causes or of causal laws. These include cases in which the explanation consists simply in demonstrating *how* the known laws lead to the unexplained effect; but there are other cases, including those in which the explanation consists simply in *denying* what was believed to be a law or fact, those in which the explanation consists in identification, and of course those already mentioned in which a theory is explained.

Finally, it is perhaps clearer from the specimen conditions given in *4.3* than in the deductive model's conditions that there may be a difference between *doubting an explanation* and *not understanding the phenomena* to which it refers (both being quite different from not understanding the statements in the explanation, or the derivations involved, etc.). The usual doubts of an explanation apply to (c) and (f); whereas, understanding is directly involved in (b) and indirectly in (d) (in the sense mentioned). This clarifies the difference between the infinite regress of requests for further explanation (which is the one often said by theists to limit scientific explanations and comprehension of the world based on them) and the infinite regress of requests for further justification of *this* explanation, i.e., of the claims made in the explanation itself (the phenomenalists' path). The deductive model insists on the inclusion of something that is simply one of the first steps in the justification regress. As far as I can see there is no good reason either for including it or, if it is included, for refusing to insist on others. What is *supposed* to be a reason for rejecting further steps is commonly produced, but would actually apply only if this were the comprehension regress. It is this: Once we have subsumed the event in question under a general law, we have explained it, and any request for an explanation *of that law* is a different question from the one with which we began. Indeed, it is said, we *can* go on and answer this further question, but we have no obligation to do so in order to be said to have explained the matter first raised. This move is quite similar to the correct defense against the comprehension regress, but it is not satisfactory as it stands, and anyway does not apply to the justification regress. It is the doubts which generate the latter which are responsible for the inclusion of a first-level, role-justifying ground (assuming that we are dealing with the explanation of one event by reference to another—some reformulating is necessary for other cases). For it is the doubt "How can you be sure that Y is connected with X (granted that it happens)?" which is used to justify including the causal law, not the doubt "How does *any of this explain* X (granted that Y happens and Y's cause X's)?" Hence the usual deductivist answer to the theist's complaint that the deductive model provides incomplete explanations is not only too strong, because the objection is often valid, but also leaves them open to similar pressures to include justification of truth or type, or the various second-level grounds. Let us set out the proper answers with some care.

The proper answer to the complaint of "incompleteness" is twofold. With respect to complaints which are based on *failure* to understand part of what has been put forward, we reply in one way; with respect to complaints based on *doubts* about the contents of the explanation, we reply in another. In terms of the distinctions introduced earlier, the former involves type-justifying grounds, the latter truth- and particularly role-justifying grounds. The first complaint sometimes reflects a fault of the inquirer but by no means always so: Sometimes the "explanation" produced is based on a misapprehension of the type of difficulty involved and this may be the fault of the explainer. It then does not explain the matter to this person at this time; if it can nonetheless be called an "explanation," this is because for someone else or for other circumstances (to which it *would* be related by contextual propriety), it would explain the matter. What cannot be dogmatically said is that we have explained the matter, on this occasion, merely because we have gone through the motions appropriate on some other occasion. The decision will depend on what legitimate inferences were possible from the context. For example, it requires the most profound searching of the imagination to discover any occasion on which the "mere generalization" type of "explanation" ("Bridges always do that") could qualify. There are certainly occasions on which the error lies with the inquirer, because he has asked a question inappropriate to his actual problem, suggesting either too much, too little, or the wrong kind of knowledge or interest on his part. In such cases we can rightly claim to have produced the correct explanation or an adequate reply to his request for an explanation (though from charity we may undertake to go ahead and deal with his real worry).

Hempel and Oppenheim's[5] first mistake, then, lies in the supposition that by subsumption under a generalization one has automatically explained *something*, and that queries about this "explanation" represent a request for *further* and *different* explanation. Sometimes these queries merely echo the original puzzlement, and it is wholly illicit to argue that the original matter has been explained. It is as if I asked the way to the town hall from the post office and you replied, "It's in the same relation to the post office as town halls always are"; and upon my indicating dissatisfaction you were to say, "Ah! What you *really* wanted to know (what you now want to know) is the geographic relation of all town halls to their post offices." This was not at all what I wanted to know, though in the light of the facts an answer to this will provide me with the answer I seek. You have produced an interesting regularity related to my question, but you have not, in this case, adequately done what I asked you to do. "Explaining the way" is so called just because it is not successfully accomplished until understanding is conveyed; and there are various ways to do this, none of them identifiable on purely syntactical grounds. It depends on what I know; and only when there are strong contextual reasons for supposing that I know the general relation of post offices to town halls could you continue to sustain any kind of claim to have explained the way to me "though it wasn't understood."

In the second place, the above debate concerns the comprehension regress, not the justificatory regress; and hence Hempel and Oppenheim's defense in terms of the "That's a further question" routine, which requires modification before it can be accepted even in the former case, is not an answer to the series of doubts about justification. In connection with *those* doubts, we are accusing them of inconsistency. The inconsistency, once more, is between (1) insisting that laws connecting X and Y be included in a complete explanation and (2) denying the necessity of including evidence for the truth of these or other assertions in it, or the type of explanation given. Each of these is necessary to the same extent (i.e., for the explanation to be a good explanation), yet they legislate for the inclusion of only one of them in the explanation.

4.5. The Relation between Deduction from True Premises and Sound Inference.

By requiring deduction from true premises, Hempel and Oppenheim impose a pair of conditions which exclude their own examples and most scientific explanations of events. The reasons for this are, I think, clear and interesting. Deduction looks as if it is the only watertight connection between the premises and the conclusion, and it appears obvious that a watertight connection must be insisted on. But the matter is much more complicated, and eventually we must abandon even the idea that they are proposing a useful *ideal* for explanation.

I want to begin with the explanation of events and only then go on to the explanation of laws. Hempel and Oppenheim never deal with the explanation of a particular event instance, but only with events of a certain *kind*; naturally success in this task will enable one to explain particular events of this kind, but it is worth remembering that a degree of generality is already present which is absent in the case of many scientific explanations, e.g., of the formation of the earth, the emergence of Homo sapiens, and the extinction of the dinosaurs. This degree of generality in their examples is one of the factors which lends what I take to be spurious plausibility to their explanation-prediction correlation. For an explanation of events of *kind* X is necessarily couched timelessly and hence may more reasonably be thought applicable prospectively as well as retrospectively.

Turning to the example with which they begin, and which they give in more detail than any other, we read, "A mercury thermometer is rapidly immersed in hot water; there occurs a temporary drop of the mercury column, which is then followed by a swift rise. How is this phenomenon to be explained? The increase in temperature affects at first only the glass tube of the thermometer; it expands and thus provides a larger space for the mercury inside, whose surface therefore drops. As soon as by heat conduction the rise in temperature reaches the mercury, however, the latter expands, and as its coefficient of expansion is considerably larger than that of glass, a rise of the mercury level results."[6] What can be said about this example in the terms of their analysis?

In the first place, it undoubtedly does represent a certain kind of scientific explanation. Not only is what is given, to the best of my knowledge, *an* explanation of the phenomenon described, but it could often perfectly well be called *the* explanation of it, since the appropriate contextual conditions for that description are usually met. Hence, an analysis of it will have some claim to be an analysis of scientific explanation. However, it is equally indubitable that it contains false statements, some of them being the alleged laws, and that the statements given do not entail the desired consequence. Their example therefore violates all three of their own criteria (one trivially, since general statements are "essentially" involved even if they are not true and hence not laws). This rather extraordinary situation was not altogether unnoticed by the authors. Referring to the deduction requirement, they say of the explanation as they give it that "if adequately and completely formulated," it would "entail the consequence that the mercury will first drop, then rise," which implies that the formulation given will *not* do this.

I shall not elaborate in detail the various physical errors in this and their other examples.[7] An example or two will suffice, since I do not take these errors to invalidate the explanation, only the deductivist analysis of it. There are two kinds of error in their explanation. First, they say: "The increase in temperature affects at first only the glass tube of the thermometer; it expands and thus. . . ." No physicist would be willing to accept this as literally true. Radiation effects reach the mercury at the speed of light and are causing it to expand while the slower conduction effects are expanding the glass; but even these reach the mercury long before the glass has expanded enough to produce a visible drop in the mercury column. Second, they do not allow for the fact that glass itself is a physically unique substance, with highly anisotropic *multiple* coefficients of expansion, the relationship between them being dependent upon the minutiae of its chemical composition and details of the annealing process, which vary even from batch to batch of the closely controlled thermometer glass.

Surely these complaints are not serious; a little research or rewording would take care of them, would it not? They are not serious for the explanation: It is correct. But they are serious for the analysis because it requires literal truth. The only way in which literal truth could be salvaged here would be by invoking an extremely complicated inequality, involving all the coefficients mentioned, the specific heats, the radiation rates, energy distribution factors, upper limits obtained from the heat-transfer equations, and so on. And with every complication at the theoretical level—as we try to employ laws that are exactly correct—there is a corresponding complication at the experimental end, since new measurements are required in order to apply the more complex theory.

The upshot of this is not merely that Hempel and Oppenheim's example is "incomplete" according to their own criteria of completeness; it is that it is not an explanation at all, on their analysis. Now could the situation be redeemed by turning to some textbook and using the more complex relationships referred to above? A serious research effort would be required to obtain the material, and it has certainly never been done. That is, there *is* as yet no "complete" explanation of this simple phenomenon, on their criterion of completeness. All we have are "explanation sketches," which they think are characteristic of the *social* sciences, by *contrast* with physics. But in physics, just as in economics, virtually all the explanations given are of this kind, and they are usually just as "complete" as they should be, i.e., as complete as is necessary to attain the requisite degree of reliability. Hempel and Oppenheim have an over-simple logical model: An explanation that fails to measure up to its standards may be a great deal more complete than one that does, i.e., it may identify the relevant effective variables and ignore the ineffective ones, as does their example in this case. Other processes are going on here, but nobody asked for a complete description of the thermodynamic process; they asked for the explanation of a particular effect. Giving an explanation requires *selecting* from among the variables that are involved those whose activities are unknown to the inquirer and crucial for the phenomenon; it does not involve dragging in or constructing *all* the relations of the variables to the others involved. The explanation won't be an explanation unless the variables selected *are* crucial. To *show* they are would thus be appropriate only if the explanation were challenged in this respect.

But what *harm* can it do to say that an explanation is complete only when its role-justifying grounds are included? Apart from the answers of the earlier sections, it must now be stressed that there are ways of supporting an explanation that do not involve anything as simple as quoting physical laws of the kind Hempel and Oppenheim have in mind (i.e., justification is not the same as deduction from true premises). In the first place, when we do quote laws, they are usually not literally true. In the second place, we often support the explanation indirectly by what I shall call elimination analysis or—to use a shorter term—*detection*. The important difference between the social sciences and the physical sciences is not that one has and one does not have complete explanations, but that one has more quantitative laws than the other. Naturally, these will often be used in giving and defending explanations. But they confer no benefit on the explanations that cannot be obtained in other ways, and in particular they do not convey the blessing of deduced truth, since they are usually only approximations.

4.6. The Nature of Physical Laws.

I have discussed this topic in greater detail elsewhere[8] and wish to stress here only the positive side of that treatment. The examples of physical laws with which we are all familiar are distinguished by one feature of particular interest for the traditional analyses—they are virtually all known to be in error. Nor is the error trifling, nor is an amended law available which corrects for all the error. The important feature of laws cannot be their literal truth, since this rarely exists. It is not their closeness to the truth which replaces this, since far better approximations are readily constructed. Their virtue lies in a compound out of the qualities of generality, formal simplicity, approximation to the truth, and theoretical tractability.[9]

These qualities, not truth unvarnished, are extremely important for the plausibility of the Hempel and Oppenheim model, which unfortunately excludes the laws which exemplify them, if carefully examined. Deduction from a "mere empirical generalization" is very rarely explanatory, and it is only because laws usually involve more than this (as well as less) that they carry explanatory force. (The fact that they commonly reflect some underlying processes, albeit imprecisely, accounts for much of the inductive reliability we ascribe to them, and hence for much of our willingness to allow contrary-to-fact inferences from them.) Were it not for this fact, that what we call laws are usually thought to reflect the "inner workings" of the world, the deductive model would be singularly implausible. This fact manifests itself in another equally helpful way. Laws provide a framework for events which we use as a convenient grid for plotting phenomena that may need explanation. When we are trying to locate events with respect to what we know and understand, we often look to see whether they represent departures from patterns we know and understand, and these patterns are the laws. Their importance lies not in the *precision* with which they trace the characteristics of events or substances but in the fact that they provide a readily identifiable pattern. The event in question either conforms to a known pattern or it does not. If it does not, it (probably) needs explaining; if it does, then either it is not puzzling or the puzzle involves the origin or relation of the patterns. This visual metaphor survives the discovery that descriptions of actual physical events and scientific laws are usually not related deductively, and it gives a different twist to the role of laws; they often serve as the *starting point* from which we survey the events, looking for the nonconformists, not only as the *rules under which we try to bring them.* Thus they often have a crucial double role in the process of explanation, but not an ultimate role; the only ultimate element in the logic of explanation is understanding itself, and that comes in many ways. Its exact relationship to the perception of regularities I shall briefly discuss later (5).

When we turn to puzzles about the patterns themselves, i.e., to the explanation of laws, we find the irrelevance of inaccuracy further demonstrated. Can we explain why bodies near the earth's surface fall according to the law $s = kt^2$? According to Hempel and Oppenheim, who use this example, this is done "by deducing it from a more comprehensive set of laws, namely Newton's laws of motion and his law of gravitation, together with some statements about particular facts namely the mass and radius of the earth."[10] There would be circumstances under which this would be a satisfactory explanation, though there are others where it would be hopeless (e.g., where the explanation is "Atmospheric density varies less than 10% over the earth's surface, and $s = 16t^2$ when measured from the top of the campanile at Pisa"). But the points to be made here are that: (1) The Newtonian laws are known to be in error, and (2) even if true, they would not entail $s = kt^2$, since (a) the actual relationship varies from point to point and height to height, so no such formula is derivable, and (b) the premises quoted are inadequate for the proposed deduction anyway (the earth has no single radius, air resistance is not considered,[11] etc.).

I conclude that, where it is appropriate to invoke Newton's laws for this explanation, the point is not *at all* that deduction from these as true premises is possible, but that these relations are the crucial factors for this inquiry, in this context, i.e., the only *important* ones (with respect to the degree of accuracy judged appropriate from the context, and the level of knowledge of the inquirers). We know this to be true, and we could even go out and do the experiments to show it is true. It does not follow from the claim that we know it and that we have to be able to show it true by deduction from true premises now in our possession.

4.7. Reduction.

These comments have obvious consequences for the analysis of reduction, coinciding to an interesting extent with those of Paul Feyerabend[12] in his essay in this volume. We are in agreement against Nagel, Hempel, and Oppenheim about the errors of the "principle of deducibility." However, on the principle of meaning invariance, his position seems to me overstated; I would certainly agree (and I think Nagel might also) that *important changes* in the meaning of certain words and sentences come about with the adoption of new scientific theories. It is equally important and this is what I think Nagel is stressing in the passage Feyerabend cites . . . that important elements in the meaning of terms, given by its primary "unreduced" relationship to observation claims, remain *unchanged.* This is necessary for the claim of reduction to make sense.[13]

Reduction is simply one kind of explanation, namely unilateral interfield explanation, and hence not essentially or even usually deductive. It is a requirement of adequacy for a microtheory that it be capable of "reducing" (explaining) all macrophenomena. In no stronger sense is reduction prossible.

. . .

5. THE ALTERNATIVE ANALYSIS

What is a scientific explanation? It is a topically unified communication, the content of which imparts understanding of some scientific phenomenon. And the better it is, the more efficiently and reliably it does this, i.e., with less redundancy and a higher *over-all* probability. What is understanding? Understanding is, roughly, organized knowledge, i.e., knowledge of the relations between various facts and/or laws. These relations are of many kinds—deductive, inductive, analogical, etc. (Understanding is deeper, more thorough, the greater the span of this relational knowledge.) It is for the most part a perfectly objective matter to test understanding, just as it is to test knowledge,[14] and it is absurd to identify it with a subjective feeling, as have some critics of this kind of view. So long as we give examinations to our students, we think we can test understanding objectively. (On the other hand, it is to be hoped and expected that the subjective feeling of understanding is fairly well *correlated* with real understanding as a result of education.)

Explanation is not "reduction to the familiar," partly because the familiar is often not understood (rainbows, memory, the appeal of music) and partly because we may understand the unfamiliar perfectly well (pure elements, the ideal gas, absolute zero). On the other hand, (1) we do understand much of what is familiar and so much explanation is reduction to the familiar. And (2) "familiar" can be taken as synonymous with "explained," in which case this slogan comes nearer the truth, though by using the tautology route. Finally, (3) there is the great truth implicit in this view that at a particular stage in explanations of a certain kind, there is very little more to understanding besides familiarity. We do come to accept facts and relations which we at first viewed as wholly incomprehensible; we come to accept gravity and to reject Newton's vituperative condemnation of the view that it is inherent in matter and can act through a vacuum without mediation. The stage at which we do this is the last stage in an explanation, and the explanations which lead us to this kind of last stage are those which take us back to the most fundamental features of our knowledge.

Although it is an illusion to suppose that scientific explanation is anything as simple as deductive subsumption under a true generalization, there sometimes comes a point in scientific explanation, especially in physics, cosmology, and psychophysiology, when assimilation into some kind of over-all regularity is about all we have to offer. The weakness of the deductivist's thesis, as a general account of scientific explanation, is well demonstrated by the profound unease which affects scientists when they find themselves in this situation. Why does the distribution of matter determine the curvature of space? Why is mass energy conserved? Why are protons or electrons continually appearing throughout the universe (in the steady-state cosmologies)? Why does a particular electron impinge on the screen at *this* point? Why does this neural configuration correspond to pain? Et cetera. What can we do with these questions except answer by showing *how* these variables are related and distributed? We can do no more at the moment, but (1) we know very well what it would be like to have a more comprehensive theory within which these questions could be answered; hence there is always an incentive for and a possibility of going further, though there are no good reasons for supposing there must *always* be something further.[15] (2) That we finally reach this stage does not show that at earlier stages nothing more is required than reiterating the puzzling regularities. Understanding requires knowing the relations that exist; it is not confined to cases where one of these relations is subsumption under a *higher* level generalization; it may only involve subsuming (i.e., colligation), not subsumption, but it will be the less for that. (3) There is an alternative to vertical subsumption in these areas and that is "horizontal" assimilation, by analogy or formal principle. (4) Even this would leave at least one fundamental principle to be explained in something more than the colligation sense, and of this all that can be said is that if the world exists at all, it is logically necessary that it have some properties and we shall be very fortunate if we can reduce them all to one. (5) Why the physical world exists, or has *this* fundamental property rather than some other, *may* have an answer in terms of some antecedent circumstances, but this only postpones the final question and that final question is probably illegitimate in just the way that the final question in the series, Where is Oxford? Where is England? . . . is illegitimate, i.e., senseless; unaskable not unanswerable.

. . .

NOTES

1. I shall abbreviate some more precise formulations by omitting the words in parentheses where I think they are not essential.

2. Their model requires the truth of the asserted explanation, but it doesn't require the inclusion of evidence for this. Instead of similarly requiring a causal connection, it actually requires the inclusion of *one special kind* of evidence for this. If it treated both requirements equitably, the model would be either trivial (causal explanations must be true and causally relevant) or deviously arbitrary (. . . must include *deductively* adequate grounds for the truth of any assertions and for the causal connection).

3. See Max Black's "Definition, Presupposition, and Assertion," in his *Problems of Analysis* (London: Routledge and Kegan Paul, 1954).

4. Carl G. Hempel and Paul Oppenheim, "Studies in the Logic of Explanation," *Philosophy of Science*, XV, p. 322.

5. I attribute this to the joint authors on the basis of exegetical discussion by Hempel alone; the point is not made in their article.

6. Hempel and Oppenheim, p. 320.

7. See pp. 255–350 of my "Explanations," D. Phil. thesis, Oxford, 1956, microfilmed by the University of Illinois library.

8. "The Key Property of Physical Laws—Inaccuracy," in *Current Issues in the Philosophy of Science*, H. Feigl and G. Maxwell, eds. (New York: Holt, Rinehart, and Winston, 1961).

9. This does not mean I require their actual incorporation into a theory, but only considerable prospective cases of incorporation, something which requires good metatheoretical judgment for identification.

10. Hempel and Oppenheim,p. 321.

11. Hempel and Oppenheim refer to "Galileo's Law for the free fall of bodies," not $s = kt^2$. In the *Dialogue Concerning Two New Sciences* (Third Day), it is made clear that $s = kt^2$ is intended, and that it is held to apply to motion in air. (Details are in my "Explanations," p. 346.)

12. The interest lies not least in his belief that his thesis somehow controverts those of linguistic analysts. I am not sure exactly who is in this ignorant (p. 85, fn. 102), misguided (p. 60, fn. 70) group, but if it is indeed the unhappy disciples from Oxford (as fn. 102 suggests, but see p. 60, fn. 69), it is clear that we share with him almost everything except the recognition of our similarity. (Smart, Presley, Toulmin, and Waisman would be the obvious sources from which to support this claim, along with more study of Wittgenstein and Hanson, whom he does cite; but I certainly do not wish to deny the use of a straw man to a good friend!)

13. This point is elaborated in my "Definitions, Explanations, and Theories," in *Minnesota Studies in the Philosophy of Science*, Vol. II, H. Feigl, M. Scriven, G. Maxwell, eds. (Minneapolis: University of Minnesota Press, 1958).

14. Knowledge stands in relation to understanding rather as explanations do to theories. We know the date of our birthday, but we understand the calendrical system: we know the items and the relations which, combined, make up understanding.

15. There are *always* good reasons for trying to go further when first we come to a limiting principle of this kind. But in the end we may grant autonomy, and abandon the attempt to reduce gravity to mechanics or magnetism, ESP to radio waves, etc., and say that we understand the phenomena all the same.

Functional Explanations in the Biological and Social Sciences

Ernest Nagel

TELEOLOGICAL EXPLANATIONS AND TELEOLOGICAL SYSTEMS

. . .

II

Almost any biological treatise or monograph yields conclusive evidence that biologists are concerned with the functions of vital processes and organs in maintaining characteristic activities of living things. In consequence, if *teleological analysis* is understood to be an inquiry into such functions, and into processes which are directed toward attaining certain end-products, then undoubtedly teleological explanations are pervasive in biology. In this respect, certainly, there appears to be a marked difference between the latter and the physical sciences. It would surely be an oddity on the part of a modern physicist were he to declare, for example, that atoms have outer shells of electrons in order to make chemical unions between themselves and other atoms possible. In ancient Aristotelian science categories of explanation suggested by the study of living things (and in particular by human art) were made canonical for all inquiry. Since non-living as well as living phenomena were thus analyzed in teleological terms—an analysis which made the notion of final cause focal—Greek science did not assume a fundamental cleavage between biology and other natural sciences. Modern science, on the other hand, regards final causes to be vestal virgins which bear no fruit in the study of physical and chemical phenomena; and because of the association of teleological explanations with the

---

From E. Nagel, *Vision and Action*, ed. by Sidney Ratner. Rutgers University Press.

doctrine that goals or ends of activity are dynamic agents in their own realization, it tends to view such explanations as a species of obscurantism. But does the procedure of teleological explanations in biology and the apparent absence of such explanations from the physical sciences entail the absolute autonomy of the former? We shall try to show that it does not.

Quite apart from their association with the doctrine of final causes, teleological explanations are sometimes suspect in modern natural science because they are assumed to invoke purposes or ends-in-view as causal factors in natural processes. Purposes and deliberate goals admittedly play important roles in human activities; but there is no basis whatever for assuming them in the study of physico-chemical and most biological phenomena. However, as has already been noted, there are a great many explanations that are counted as teleological which do not postulate any purposes or ends-in-view; for explanations are often said to be "teleological" only in the sense that they specify the *functions* which things or processes possess. Most contemporary biologists certainly do not impute purposes to the organic parts of living things whose functions are investigated; and most of them would probably also deny that the means-ends relationships discovered in the organization of living creatures are the products of some deliberate plan on the part of a purposeful agent. To be sure, there are biologists who postulate psychic states as concomitants and even as directive forces of all organic behavior. But they are in a minority; and they usually support their view by special considerations which can be distinguished from the facts of functional or teleological dependencies that most biologists do not hesitate to accept. Since the word *teleology* is ambiguous, it would doubtless prevent confusions and misunderstandings were it eliminated from the vocabulary of biologists. But as it is, biologists do use it, and say they are giving a "teleological explanation" when, for example, they explain that the function of the alimentary canal in vertebrates is to prepare ingested materials for absorption into the blood-stream. The crucial point is that when biologists do employ teleological language they are not necessarily committing the pathetic fallacy or lapsing into anthropomorphism.

We must now show, however, that teleological (or functional) explanations are equivalent to non-teleological ones, so that the former can be replaced by the latter without loss in asserted content. Consider some typical teleological statement, for example. "The function of chlorophyll in plants is to enable plants to perform photo-synthesis." But this statement appears to assert nothing which is not asserted by "Plants perform photosynthesis only if they contain chlorophyll," or alternatively by "A necessary condition for the occurrence of photosynthesis in plants is the presence of chlorophyll." These latter statements, however, do not explicitly ascribe a function to chlorophyll, and in that sense are therefore not teleological formulations. If this example is taken as a paradigm, it seems that when a function is ascribed to a constituent of some organism, the content of the teleological statement is fully conveyed by another statement which simply asserts a necessary (or possibly a necessary and sufficient) condition for a certain trait or activity of that organism. On this assumption, therefore, a teleological explanation states the *consequences* for a given biological system of one of the latter's constituent parts or processes; the equivalent non-teleological explanation states some of the *conditions* (though not necessarily in physico-chemical terms) under which the system persists in its characteristic organization and activities. The difference between teleological and non-teleological explanations is thus comparable to the difference between saying that *B* is an effect of *A*, and saying that *A* is a cause or condition of *B*. In brief, the difference is one of selective attention, rather than of asserted content.

This point can be reinforced by another consideration. If a teleological explanation had an asserted content which is different from the content of every non-teleological statement, it would be possible to cite procedures and evidence for establishing the former which are different from the procedures and evidence required for confirming the latter. But in point of fact, there appear to be no such procedures and evidence. Thus, consider the teleological statement: "The function of the leucocytes in human blood is to defend the body against foreign micro-organisms." Now whatever may be the evidence which warrants this statement, it also warrants the statement that, "Unless human blood contains a sufficient number of leucocytes, certain normal activities of the body are injured," and conversely. Accordingly, there is a strong presumption that the two statements do not differ in *factual* content. More generally, if as seems to be the case the conceivable empirical evidence for a teleological explanation is identical with the conceivable evidence for a certain non-teleological one, the conclusion appears inescapable that these statements cannot be distinguished with respect to what they *assert*, even though they may differ in other ways.

However, this proposed equation of teleological and non-teleological explanations must face a fundamental objection. Most biologists would perhaps be prepared to admit that a teleological explanation *implies* a non-teleological one; but some of them, at any rate, would maintain that the latter does not in general imply the former, so that the suggested equivalence does not in fact hold. This latter claim can be forcefully stated as follows. If there were such an equivalence, it may be said, not only could a teleological explanation be replaced by a non-teleological one, but a converse replacement would also be possible. In consequence, the customary statements of laws and theories in the physical sciences must be translatable into teleological formulations. In point of fact, however, modern physical science does not appear to sanction such formulations, and physical scientists would doubtless resist their introduction into their disciplines as an unfortunate attempt to reinstate the point of view of Greek and medieval science. Thus the statement, "The volume of a gas at constant temperature varies inversely with its pressure" is a typical physical law, which is entirely free of teleological connotations. If it were equivalent to a teleological statement, its presumed equivalent would be, "The function of a varying pressure in a gas at constant temperature is to produce an inversely varying pressure," or, perhaps, "Gases at constant temperature under variable pressure alter their volumes in order to keep the product of pressure and volume constant." But most physicists would regard these latter formulations as preposterous, and at best as misleading. There must therefore be some important differences between teleological and nonteleological statements which the discussion has thus far failed to make explicit.

The attitude of physical scientists toward teleological formulations in their own disciplines is doubtless as alleged in this objection. Nevertheless, the objection is not completely decisive on the point at issue. Two general comments are in order which will at least weaken its force.

In the first place, it is not entirely accurate to maintain that the physical sciences never employ formulations that have at least the appearance of teleological statements. As is well known, some physical laws and theories are often expressed in so-called "isoperimetric" or "variational" form, rather than in the more familiar form of numerical or differential equations. When laws and principles are so expressed, they undoubtedly seem to be akin to teleological formulations. For example, the elementary law of optics that the angle of incidence of a light ray with a surface is equal to the angle of reflection, can also be rendered by the statement that a light ray travels in such a manner that when it is reflected from a surface the length of its actual path is the minimum of all possible paths. More generally, a considerable part of classical as well as contemporary physical theory can be stated in the form of "extremal" principles. What these principles assert is that the actual development of a system is such as to minimize or maximize some magnitude which represents the possible configurations of the system.[1]

The discovery that the principles of mechanics can be given such extremal formulations was once considered as evidence (especially by Maupertuis in the eighteenth century) for the operation of a divine plan throughout nature. Such theological interpretations of extremal principles is now recognized almost universally to be entirely gratuitous; and no competent physicist today supposes that extremal principles entail the assumption of purposes animating physical processes. The use of such principles in physical science nevertheless does show that it can formulate the dynamical structure of physical systems so as to bring into focus the incidence of constituent elements and processes upon certain properties of a system taken as a whole. If physical scientists dislike teleological language in their own disciplines, it is not because they regard teleological notions in this sense as foreign to their task. Their dislike stems in some measure from the fear that, except when such teleological language is made rigorously precise through the use of quantitative formulations, it is apt to be misunderstood as connoting the operation of purposes.

In the second place, the physical sciences unlike biology are in general not concerned with a relatively special class of organized bodies, and they do not investigate the conditions making for the persistence of some selected physical system rather than of others. When a biologist ascribes a function to the kidney, he tacitly assumes that it is the kidney's contribution to the maintenance of the living animal which is under discussion; and he ignores as irrelevant to his primary interest the kidney's contribution to the maintenance of any other system of which it may also be a constitutent. On the other hand, a physicist generally attempts to discuss the effects of solar radiation upon a wide variety of things; and he is reluctant to ascribe a "function" to the sun's radiation, because there is no one physical system of which the sun is a part that is of greater interest to him than any other such system. And similarly for the law connecting the pressure and volume of a gas. If a physicist views with suspicion the formulation of this law in functional or teleological language, it is because (in addition to the reasons which have been or will be discussed) he does not regard it his business to assign special importance (even if only by vague suggestion) to one rather than another consequence of varying pressures in a gas.

III

However, the discussion thus far can be accused with some justice of naïveté if not of irrelevance, on the ground that it has ignored completely the fundamental point—namely, the "goal-directed" character of organic systems. It is because living things exhibit in varying degrees adaptive and regulative structures and activities, while the systems studied in the physical sciences do not—so it is frequently claimed—that teleological explanations are peculiarly appropriate for the former but not for the latter. Thus, it is because the solar system, or any other system of which the sun is a part, does not tend to persist in the face of environmental changes in some integrated pattern of activities, and because the constituents of the system do not undergo mutual adjustments so as to maintain this pattern in relative independence from the environment, that it is preposterous to ascribe any function to the sun or to the solar radiation. Nor does the fact that physics can formulate some of its theories in the form of extremal principles—so the objection continues—minimize the differences between biological and purely physical systems. It is true that a physical system develops in such a way as to minimize or maximize a certain magnitude which represents a property of the system as a whole. But physical systems are not organized to *maintain* extremal values of such magnitudes, or to develop under widely varying conditions in the direction of realizing some particular values of such magnitudes.

Biological systems, on the other hand, do possess such organization, as a single example (which could be matched by an indefinite number of others) makes clear. There are complicated but coordinated physiological processes in the human body, which maintain many of its characteristics in a relatively steady state (or homeostasis). Thus, the internal temperature of the body must remain fairly constant if it is not to be fatally injured. In point of fact, the temperature of the normal human being varies during a day only from about 97.3°F. to 99.1°F., and cannot fall much below 75°F. or rise much above 110°F. without permanent injury to the body. However, the temperature of the external environment can fluctuate much more widely than this; and it is clear from elementary physical considerations that the body's characteristic activities would be profoundly curtailed unless it were capable of compensating for such environmental changes. But the body is indeed capable of doing just this; and, in consequence, its normal activities can continue, in relative independence of the temperature of the environment—provided, of course, that the environmental temperature does not fall outside a certain interval of magnitudes. The body achieves this homeostasis by means of a number of mechanisms, which serve as a series of defenses against shifts in the internal temperature. Thus, the thyroid gland is one of several that control the body's basal metabolic rate; the heat radiated or conducted through the skin depends on the quantity of blood flowing through peripheral vessels, a quantity which is regulated by dilation or contraction of these vessels; sweating and the respiration rate determine the quantity of moisture that is evaporated, and so affect the internal temperature; adrenalin in the blood also stimulates internal combustion, and its secretion is affected by changes in the external temperature; and automatic muscular contractions involved in shivering are an additional source of internal heat. There are thus physiological mechanisms in the body such that its internal temperature is automatically preserved, despite disturbing conditions in the body's internal and external environment.[2]

Three separate questions that are frequently confounded are raised by such facts of biological organization. Is it possible to formulate in general but fairly precise terms the distinguishing structure of "goal-directed" systems, but in such a way that the analysis is neutral with respect to assumptions concerning the existence of purposes or the dynamic operation of goals as instruments in their own realization? Is the fact, if it is a fact, that teleological explanations are customarily employed only in connection with "goal-directed" systems, decisive on the issue whether a teleological explanation is equivalent to some non-teleological one? Is it possible to explain in purely physico-chemical terms—that is, exclusively in terms of the laws and theories of current physics and chemistry—the operations of biological systems? This third question will not concern us in this paper; but the other two require our attention.

IV

There have been many attempts since antiquity at constructing machines and physical systems which simulate the behavior of living organisms in one respect or another. None of these attempts has been entirely successful, for it has not been possible thus far to manufacture in the workshop and out of inorganic materials any device which acts fully like a living being. Nevertheless, it has been possible to construct physical systems which are self-maintaining and self-regulating up to a point, and which therefore resemble living organisms in one important respect. In an age in which servo-mechanisms no longer excite wonder, and in which the language of cybernetics and "negative feedbacks" has become widely fashionable, the imputation of "goal-directed" behavior to purely physical systems certainly cannot be rejected as an absurdity. Whether "purposes" can also be imputed to such physical systems, as some expounders of cybernetics claim[3] is perhaps doubtful, though the question is in large measure a semantic one; in any event, the issue is not relevant in the present context of discussion. Moreover, the possibility of constructing self-regulating physical systems does not constitute proof that the activities of living organisms can be explained in exclusively physico-chemical terms. However, the occurrence of such systems does suggest that there is no sharp division between the teleological organization which is often assumed to be distinctive of living things, and the goal-directed organization of many physical systems; and it does offer strong support for the presumption that the structure of such organizations can be formulated without the postulation of purposes or of goals as dynamic agents.

With the homeostasis of the temperature of the human body before us as an exemplar, let us now state in general terms the structure of systems which have a goal-directed organization.[4] The characteristic feature of such systems is that they continue to manifest a certain state or property G, or to develop "in the direction" of attaining G, in the face of a relatively extensive class of changes in their external environments or in some of their internal parts—changes which, if not compensated by internal modifications in the system, would result in the vanishing of G or in an altered direction of development. This feature can be formulated more precisely though schematically as follows.

Let S be some system, E its external environment, and G some state or property which S possesses or is capable of possessing under suitable conditions. Assume for the moment—this assumption will be presently relaxed—that E remains constant in all relevant respects, so that its influence upon S can be ignored. Suppose also that S is analyzable into a structure of parts, such that the activities of a certain number of them are causally relevant for the occurrence of G. For the sake of simplicity, assume that there are just three such parts, the state of each of which at any time can be specified by a determinate form of the complex predicates "A," "B," and "C," respectively; numerical subscripts will serve as indicators of such determinate forms. Accordingly, the state of S at any time causally relevant to G will be expressed by specializations of the matrix "$(A_x B_y C_z)$." One further general assumption must now be made explicit. Each of these state-variables (they are not necessarily numerical variables) can be assigned any determinate values that are compatible with the known character of the part of S whose state it specifies. In effect, therefore, the states which can be values for "A_x" must fall into a certain class K_A; and there are corresponding classes K_B and K_C for the other two state variables. The reason for this restriction will be clear from an example. If S is the human body, and "A_x" states the degree of dilation of peripheral blood vessels, it is obvious that this degree cannot exceed some maximum value; for it would be absurd to suppose that a blood-vessel could acquire a mean diameter of, say, five feet. On the other hand, the possible values of one state-variable at a given time will be assumed to be independent of the possible values of the other state variables at that same time. Accordingly, any combination of values of the state variables will be a permissible specialization of the matrix "$(A_x B_y C_z)$," provided that the values of each variable belong to the classes K_A, K_B, and K_C rspectively. This is tantamount to saying that the state variables which are stipulated to be causally relevant to G are also postulated to be capable of having values at a given time which are mutually independent of one another.

Suppose now that if S is in the state $(A_0B_0C_0)$ at some given time, then S either has the property G, or else a sequence of changes will take place in S in consequence of which S will possess G at some subsequent time. Call such an initial state of S a "causally effective state with respect to G," or a "G-state" for short. Not every possible state of S need be a G-state; for one of the causally relevant parts of S may be in such a state at a given time, that no combination of possible states of the other parts will yield a G-state for S. Thus, suppose that S is the human body, G the property of having an internal temperature lying in the range 97° to 99°F., A_x again the state of peripheral blood-vessels, and B_y the state of the thyroid glands; it may happen that B_y assumes a value (e.g., corresponding to acute hyperactivity) such that for no possible value of A_x will G be realized. It is also conceivable that no possible state of S is a G-state, so that in fact G is never realized in S. For example, if S is the human body and G the property of having an internal temperature lying in the range 150° to 160°F., then there is no G-state for S. On the other hand, more than one possible state of S may be a G-state, though only one of them will be actual at a given time; but if there is more than one possible G-state, we shall assume that the one which is realized at a given time is uniquely determined by the actual state of S at some previous time. In short, we are assuming that S is a deterministic system with respect to the occurrence of G-states. The case in which there is more than one possible G-state for S is of particular relevance to the present discussion, and we must now consider it more closely.

Assume again that at some initial time t_0, S is in the G-state $(A_0B_0C_0)$. But suppose now that a change occurs in S so that in consequence A_0 is caused to vary, and that at time t_1 subsequent to t_0 the state variable "A_x" has some other value. Which value it will have at t_1 will depend on the particular changes that have occurred in S. We shall assume, however, that there is a range of possible changes, and that the values which "A_x" may have at time t_1 fall into some class K'_A (a sub-class of K_A) which contains more than one member. To fix our ideas, suppose that A_1 and A_2 are the members of K'_A; and assume further that neither $(A_1B_0C_0)$ nor $(A_2B_0C_0)$ is a G-state—that is, a variation in A_0 alone would take S out of a G-state. Accordingly, if the changes mentioned thus far were the only changes in the state of S, S would no longer be in a G-state at time t_1. Let us, however, make the contrary assumption. Assume S to be so constituted that if A_0 is caused to vary so that the value of "A_x" at time t_1 falls into K'_A, there will also be further compensatory changes in the values of some or all of the other state variables. More specifically, these further changes are stipulated to be of the following kind: If K'_{BC} is the class of sets of values which "B_y" and "C_z" have at time t_1, then for each value of "A_x" in K'_A there is a unique set in K'_{BC} such that S continues to be in a G-state at time t_1; but these further changes unaccompanied by the first-mentioned ones would take S out of a G-state—

that is, if at time t_1 the state variables of S have a set of values such that two of them belong to a set in K'_{BC} while the remaining one is not the corresponding member in K'_A, then S is not a G-state. For example, suppose that if A_0 is changed into A_1, the initial G-state $(A_0B_0C_0)$ is changed into the G-state $(A_1B_1C_1)$ with $(A_0B_1C_1)$ not a G-state; and if A_0 is changed into A_2 the initial G-state is changed into the G-state $(A_2B_1C_0)$, with $(A_0B_1C_0)$ not a G-state. In this example, K'_A is the class $\{A_1, A_2\}$, and K'_{BC} the class of sets $\{[B_1, C_1], [B_1, C_0]\}$, with A_1 corresponding to $[B_1, C_1]$, and A_2 to $[B_1, C_0]$.

We now introduce some definitions, based upon the above discussion. Assume S to be a system satisfying the following conditions: (1) S can be analyzed into a structure of parts, a certain number of which (say three) are causally relevant to the occurrence in S of some property or feature G; and the causally relevant state of S at any time can be specified by means of a set of state-variables. These state-variables at any given time can be assigned values independently of each other, though the possible values of each variable are restricted to some class of values. (2) If S is in a G-state at some time t_0 during period T, and a variation occurs in one of the state parameters (say "A") such that this variation alone would take S out of its G-state, then the possible values of this parameter at time t_1 subsequent to t_0 but still in T fall into a certain class K'_A. Call this variation a "primary variation" in S. (3) If the state parameter "A" varies in the indicated manner, then the remaining parameters also vary so that their variation alone would take S out of its G-state, and so that their possible values at time t_1 constitute sets belonging to a class K'_{BC}. (4) The elements of K'_A and K'_{BC} correspond to each other in a uniquely reciprocal fashion, such that when the state of S is specified by these corresponding values S is in a G-state at time t_1. Call the variations in S which are represented by the members of K'_{BC} the "adaptive" variations in relation to the variations represented by members of K'_A. When these assumptions hold for S, the parts of S that are causally relevant to G will be said to be "directively organized during the period T with respect to G"—or more shortly "directively organized," if the reference to T and G can be taken for granted. This definition can be easily generalized for a larger number of state-variables, and for the primary variation of more than one state-variable; but the present incompletely general definition will suffice for our purposes.

It will be clear from this account that if S is directively organized, the persistence of G is in a certain sense independent of the variations (up to a point) in any one of the causally relevant parts of S. For although it is the state of these parts which by hypothesis determines the occurrence of G, an altered state in one of them may be compensated by altered states in the other parts of S so as to preserve S in its G-state. The structure or character of so-called "teleological" systems is therefore expressed by the indicated conditions for a directively organized system; and these conditions can be stated, as we have seen, in a manner not requiring the adoption of teleology as a fundamental or unanalyzable category. What may be called the "degree of directive organization" of a system, or perhaps the "degree of persistence" of some trait of a system, can also be made explicit in terms of the above analysis. For the property G is maintained in S (or S persists in its development which eventuates in G) to the extent that the range of K'_A of the possible primary variations is associated with the range of induced compensatory changes K'_{BC} such that S is preserved in its G-state. The more inclusive the range K'_A that is associated with such compensatory changes, the more is the persistence of G independent of variations in the state of S. Accordingly, on the assumption that it is possible to specify a measure for the range K'_{BC} the "degree of directive organization" of S with respect to variations in the state-parameter A can be defined as the measure of this range.

We may now relax the assumption that the external environment E has no influence upon S. But in dropping this assumption, we merely complicate the analysis, without introducing anything novel into it. For suppose that there is some factor in E which is causally relevant to the occurrence of G in S, and whose state at any time can be specified by some determinate form of the state-variable "F_w." Then the state of the system S' (which includes both S and E) that is causally relevant to the occurrence of G in S is specified by some determinate form of the matrix "$(A_x B_y C_z F_w)$"; and the discussion proceeds as before. However, it is generally not the case that a variation in any of the internal parts of S produces any significant variation in the environmental factors. What usually is the case is that the latter vary quite independently of the former; that they do not undergo changes which compensate for changes in the state of S; and that while a limited range of changes in them may be compensated by changes in S so as to preserve S in some G-state, most of the states which environmental factors are capable of assuming cannot be so compensated by changes in S. It is customary, therefore, to talk of the "degree of plasticity" or the "degree of adaptability" of organic systems in relation to their environment, and not conversely. However, it is possible to define these notions without reference to organic systems in particular, in a manner analogous to the definition of "degree of directive organization" already suggested. Thus suppose that the variations in the environmental state F, compensated by changes in S so as to preserve S in some G-state, all fall into the class K'_F; then if a measure for this class is available, the "degree of plasticity" of S with respect to G in relation to F can be defined as the measure of K'_F.

V

This must suffice as an account of the structure of "teleological" or "goal-directed" systems. The account is intended to formulate only the gross pervasive features of such systems, and undoubtedly suffers from neglect of many important complications. Moreover, it does not pretend to indicate what the detailed mechanisms may be which are involved in the occurrence of such systems. It is therefore deliberately neutral with respect to such issues as whether these mechanisms are explicable entirely in physico-chemical terms, or whether the notion of "feedback" is required in analyzing them. But if the account is at least approximately adequate, it implies a positive answer to the question whether the distinguishing features of "goal-directed" systems can be formulated without invoking purposes and goals as dynamic agents.

However, there is one matter that must be briefly discussed. The definition of directively organized systems has been so stated that it may apply both to biological as well as to non-vital systems. It is in fact easy to find illustrations for it from either domain. The human body with respect to the homeostasis of its temperature is an example from biology; a building equipped wiih a furnace and thermostat is an example from physico-chemistry. But though the definition is not intended to distinguish between vital and non-vital systems—the difference between such systems must be stated in terms of the *specific* properties and activities they manifest—it *is* intended to set off systems which have a prima facie "goal-directed" character, from systems which are usually not so characterized. The question therefore remains whether the definition does achieve this aim, or whether on the contrary it is so inclusive that almost *any* system (whether it is ordinarily judged to be goal-directed or not) satisfies it.

Now there certainly are many physico-chemical systems which are ordinarily *not* regarded as being "goal-directed," but which appear to conform to the definition of directively organized systems proposed above. Thus, a pendulum at rest, an elastic solid, a steady electric current flowing through a conductor, a chemical system in thermodynamic equilibrium, are obvious examples of such systems. It seems therefore that the definition of directive organization —and in consequence the proposed analysis of "goal-directed" or "teleological" systems—fails to attain its intended objective. However, two comments are in order on the point at issue. In the first place, though we admittedly do distinguish between systems that are goal-directed and those which are not, the distinction is highly vague, and there are many systems which cannot be classified definitely as being of one kind rather than another. Thus, is the child's toy sometimes known as "the walking beetle"—which turns aside when it reaches the edge of a table and fails to fall off, because an idle wheel is then brought into play through the action of an "antenna"—a goal-directed system or not? Is a virus such a system? Is the system consisting of members of some species which has undergone certain lines of evolutionary development, a goal-directed one? Moreover, some systems have been classified as "teleological" at one time and in relation to one body of knowledge, only to be reclassified as "nonteleological" at a later time when knowledge concerning the physics of mechanisms had improved. "Nature does nothing in vain" was a maxim commonly accepted in pre-Newtonian physics, and on the basis of the doctrine of "natural places" even the descent of bodies and the ascent of smoke were regarded as goal-directed. Accordingly, it is at least an open question whether the current distinction between systems that are goal-directed and those which are not has an identifiable objective basis (i.e., in terms of differences between the actual organization of such systems), and whether the *same* system may not be identified in alternative ways depending on the perspective from which it is viewed and on the antecedent assumptions that are adopted for analyzing its structure.

In the second place, it is by no means clear that physical systems such as the pendulum at rest, which is not usually regarded as goal-directed, really do conform to the definition of "directively organized" systems proposed above. Consider a simple pendulum which is initially at rest, and is then given a small impulse (say by a sudden gust of wind) and assume that apart from the constraints of the system and the force of gravitation the only force that acts on the bob is the friction of the air. Then on the usual physical assumptions, the pendulum will perform harmonic oscillations with decreasing amplitudes, and will finally assume its initial position of rest. The system here consists of the pendulum and the various forces acting on it, while the property G is the state of the pendulum when it is at rest at the lowest point of its path of oscillation. By hypothesis, its length and the mass of the bob are fixed, and so is the force of gravitation acting on it, as well as the coefficient of damping. What is variable is the impulsive force of the gust of wind and the restoring force which operates on the bob as a consequence of the constraints of the system and of the presence of the gravitational field. However, and this is the crucial point, these two forces are *not* independent of one another. Thus, if the effective component of the former has a certain magnitude, the restoring force will have an equal magnitude with an opposite direction. Accordingly, if the state of the system at a given time were specified in terms of state-variables which take these forces as values, these state variables would not satisfy one of the stipulated conditions for state-variables of directively organized systems: for the value of one of them at a given time is uniquely determined by the value of the other at that same time. In short, the values at any specified time of these proposed state-variables are not independent. It therefore follows that the simple pendulum is *not* a directively organized system in the sense of the definition given. And it is possible to show in a similar manner that a number of other physical systems, currently classified as non-teleological, fail to satisfy this definition. Whether one could show this for all such systems is admittedly an open question. But there is at least some ground for holding that the definition does achieve what it is intended to achieve, and that it states the distinctive features of systems commonly characterized as "teleological."

VI

We can now settle quite briefly the second question we undertook to discuss —namely, whether the supposed fact that teleological explanations are usually reserved for "goal-directed" systems casts doubt on the claim that teleological and non-teleological explanations are equivalent in asserted content. But if such systems are always analyzable as directively organized ones, in the sense of the above definition, the answer to this question is clearly in the negative. For the defining characteristics of such systems can be formulated entirely in nonteleological language and, in consequence, every teleological explanation (that is, every explanation which contains a teleological expression) must be translatable into an equivalent statement (or set of statements) which is nonteleological.

Why, then, does it seem odd to render physical statements such as Boyle's law in teleological form? The obvious answer is that we do not usually employ teleological statements except in the context of discussing systems which are assumed to be directively organized. A teleological version of Boyle's law appears strange and unacceptable, because such a formulation is usually taken to imply that any gas enclosed in a volume is a directively organized system, in contradiction to the tacit assumption that it is not such a system. In a sense, therefore, a teleological explanation does assert more than its prima facie equivalent nonteleological translation does. For the former tacitly assumes, while the latter often does not, that the system under consideration is directively organized. But if the above discussion is sound in principle, this "excess" meaning of teleological statements can always be expressed in nonteleological language.

On the assumption that a teleological explanation can always be equated to a nonteleological one with respect to what each asserts, let us now make more explicit in what respects they do differ. The difference appears to be as follows: Teleological explanations focus attention on the culminations and products of specific processes, and upon the contributions of parts of a system to its maintenance. They view the operations of things from the perspective of certain selected wholes to which the things belong; and they are therefore concerned with properties of parts of such wholes only in so far as these properties are relevant to some complex features or activities assumed as characteristic for those wholes. Nonteleological explanations, on the other hand, place chief emphasis on certain conditions under which specified processes are initiated and persist, and on the factors upon which the continued operation of given systems are contingent. They represent the inclusive behavior of a thing as the the operation of certain selected constituents into which the thing is analyzable, and they are therefore concerned with features of complex wholes only to the extent that these features are related to the assumed characteristics of those constituents. The difference between teleological and nonteleological explanations, as has already been suggested, is one of emphasis and of perspective in formulation.

It is sometimes objected, however, that teleological explanations are fallaciously parochial; for they tacitly assume a privileged status for a special set of complex systems, and so make focal the role of things and processes in maintaining just those systems and no others. Processes have no inherent termini, so it is argued, and cannot rightly be assumed to contribute exclusively to the maintenance of some unique set of wholes. It is therefore misleading to say that *the* function of the white cells in the human blood is to defend the human body against foreign micro-organisms. This is admittedly *a* function of the leucocytes; and it may even be said to be *the* function of these cells from the perspective of the human body. But leucocytes are elements in other systems as well—for example, in the system of the blood stream considered in isolation from the rest of the body, in the system composed of some virus colony as well as these white cells, or in the more inclusive and complex solar system. These other systems are also capable of persisting in their "normal" organization and activities only under definite conditions and from the perspective of *their* maintenance the leucocytes possess other functions.

On obvious reply to this objection is a *tu quoque*. It is as legitimate to focus attention on consequences, culminations, and uses, as it is on antecedents and conditions. Processes do not have inherent termini, but neither do they have absolute beginnings; things and processes are not in general exclusively involved in maintaining some unique whole, but neither are wholes analyzable into a unique set of constituents. It is nevertheless intellectually profitable in causal inquiries to focus attention on certain earlier stages in the development of a process rather than on later ones, and on one set of constituents of a system rather than another set. And similarly, it is illuminating to select as the point of departure for the investigation of some problems certain complex wholes rather than others. Moreover, as we have seen, some things are parts of directively organized systems, but do not appear to be parts of more than one such system. The study of the unique function of such parts in such unique teleological systems is therefore not a preoccupation that assigns without warrant a special importance to certain systems. On the contrary, it is an inquiry which is sensitive to fundamental and objectively identifiable differences in nature.

There is nevertheless a point to the objection. For the operation of human interest in the construction of teleological explanations is perhaps more often overlooked than in the case of nonteleological analyses. In consequence, certain end-products of processes and certain directions of changes are frequently assumed to be inherently "natural," "essential," or "proper," while all others are then labelled as "unnatural," "accidental," or even "monstrous." Thus, the development of corn seeds into corn plants is sometimes said to be natural, while their transformation into the flesh of birds or men is asserted to be accidental. In a given context of inquiry, and in the light of the problems which initiate it, there may be ample justification for ignoring all but one direction of possible changes, and all but one system of activities to whose maintenance things and processes contribute. But such disregard of other wholes and of other functions which their constituents may have, does not warrant the conclusion that what is ignored is less genuine or natural than what receives selective attention.

. . .

NOTES

1. It can in fact be shown that when certain very general conditions are satisfied, all quantitative laws and principles can be given an "extremal" formulation.

2. Cf. Walter B. Cannon, *The Wisdom of the Body* (New York, 1932), Chap. xii.

3. Cf. Arturo Rosenblueth, Norbert Wiener, Julian Bigelow, "Behavior, Purpose and Teleology," *Philosophy of Science*, Vol. X (1943); Norbert Wiener, *Cybernetics* (New York, 1948); A. M. Turing, "Computing Machines and Intelligence," *Mind*, Vol. LIX (1950); Richard Taylor, "Comments on a Mechanistic Conception of Purposefulness," *Philosophy of Science*, Vol. XVII (1950), and the reply by Rosenblueth and Wiener with a rejoinder by Taylor in the same volume.

4. The following discussion is heavily indebted to R. B. Braithwaite, "Teleological Explanation," *Proc. of the Aristotelian Society*, Vol. XLVII (1947), and G. Sommerhoff, *Analytical Biology* (London, 1950). Cf. also Alfred J. Lotka, *Elements of Physical Biology* (New York, 1926) Chap xxv.

Carl G. Hempel

THE LOGIC OF FUNCTIONAL ANALYSIS

Empirical science, in all its major branches, seeks not only to *describe* the phenomena in the world of our experience, but also to *explain* or *understand* their occurrence: It is concerned not just with the "What?", "When?", and "Where?", but definitely, and often predominantly, with the "Why?" of the phenomena it investigates.

That explanation and understanding constitute a common objective of the various scientific disciplines is widely recognized today. However, it is often held that there exist fundamental differences between the explanatory *methods* appropriate to the different fields of empirical science, and especially between those of the "exact" natural sciences and those required for an adequate understanding of the behavior of humans or other organisms, taken individually or in groups. In the exact natural sciences, according to this view, all explanation is achieved ultimately by reference to causal or correlational antecedents; whereas in psychology and the social and historical disciplines—and, according to some, even in biology—the establishment of causal or correlational connections, while desirable and important, is not sufficient. Proper understanding of the phenomena studied in these fields is held to require other types of explanation.

Perhaps the most important of the alternative methods that have been developed for this purpose is the method of functional analysis, which has found extensive use in biology, psychology, sociology, and anthropology. This procedure raises problems of considerable interest for the comparative methodology of empirical science. This essay is an attempt to clarify some of these problems; its object is to examine the logical structure of functional analysis

---

and its explanatory and predictive significance by means of an explicit confrontation with the principal characteristics of the explanatory procedures used in the physical sciences. We begin by a brief examination of the latter.

. . .

2. THE BASIC PATTERN OF FUNCTIONAL ANALYSIS

Historically speaking, functional analysis is a modification of teleological explanation, i.e., of explanation not by reference to causes which "bring about" the event in question, but by reference to ends which determine its course. Intuitively, it seems quite plausible that a teleological approach might be required for an adequate understanding of purposive and other goal-directed behavior; and teleological explanation has always had its advocates in this context. The trouble with the idea is that in its more traditional forms, it fails to meet the minimum scientific requirement of empirical testability. The neovitalistic idea of entelechy or of vital force is a case in point. It is meant to provide an explanation for various characteristically biological phenomena, such as regeneration and regulation, which according to neovitalism cannot be explained by physical and chemical laws alone. Entelechies are conceived as goal-directed nonphysical agents which affect the course of physiological events in such a way as to restore an organism to a more or less normal state after a disturbance has occurred. However, this conception is stated in essentially metaphorical terms: No testable set of statements is provided (1) to specify the kinds of circumstances in which an entelechy will supervene as an agent directing the course of events otherwise governed by physical and chemical laws and (2) to indicate precisely what observable effects the action of an entelechy will have in such a case. And since neovitalism thus fails to state general laws as to when and how entelechies act, it cannot explain any biological phenomena; it can give us no grounds to expect a given phenomenon, no reasons to say: "Now we see that the phenomenon had to occur." It yields neither predictions nor retrodictions: The attribution of a biological phenomenon to the supervenience of an entelechy has no testable implications at all. This theoretical defect can be thrown into relief by contrasting the idea of entelechy with that of a magnetic field generated by an electric current, which may be invoked to explain the deflection of a magnetic needle. A magnetic field is not directly observable any more than an entelechy; but the concept is governed by strictly specifiable laws concerning the strength and direction, at any point, of the magnetic field produced by a current flowing through a given wire and by other laws determining the effect of such a field upon a magnetic needle in the magnetic field on the earth. And it is these laws which, by their predictive and retrodictive import, confer explanatory power upon the concept of magnetic field. Teleological accounts referring to entelechies are thus seen to be pseudoexplanations. Functional analysis, as will be seen, though often worded in teleological phraseology, need not appeal to such problematic entities and has a definitely empirical core.

The kind of phenomenon that a functional analysis[1] is invoked to explain is typically some recurrent activity or some behavior pattern in an individual or a group; it may be a physiological mechanism, a neurotic trait, a culture pattern, or a social institution, for example. And the principal objective of the analysis is to exhibit the contribution which the behavior pattern makes to the preservation or the development of the individual or the group in which it occurs. Thus, functional analysis seeks to understand a behavior pattern or a sociocultural institution in terms of the role it plays in keeping the given system in proper working order and thus maintaining it as a going concern.

By way of a simple and schematized illustration, consider first the statement:

(3.1) The heartbeat in vertebrates has the function of circulating blood through the organism.

Before asking whether and how this statement might be used for explanatory purposes, we have to consider the preliminary question: What does the statement *mean*? What is being asserted by this attribution of function? It might be held that all the information conveyed by a sentence such as (3.1) can be expressed just as well by substituting the word "effect" for the word "function." But this construal would oblige us to assent also to the statement:

(3.2) The heartbeat has the function of producing heart sounds; for the heartbeat has that effect.

Yet a proponent of functional analysis would refuse to assert (3.2), on the ground that heart sounds are an effect of the heartbeat which is of no importance to the functioning of the organism; whereas the circulation of the blood effects the transportation of nutriment to, and the removal of waste from, various parts of the organism—a process that is indispensable if the organism is to remain in proper working order, and indeed if it is to stay alive. Thus understood, the import of the functional statement (3.1) might be summarized as follows:

(3.3) The heartbeat has the effect of circulating the blood, and this ensures the satisfaction of certain conditions (supply of nutriment and removal of waste) which are necessary for the proper working of the organism.

We should notice next that the heart will perform the task here attributed to it only if certain conditions are met by the organism and by its environment. For example, circulation will fail if there is a rupture of the aorta; the blood can carry oxygen only if the environment affords an adequate supply of available oxygen and the lungs are in proper condition; it will remove certain kinds of waste only if the kidneys are reasonably healthy; and so forth. Most of the conditions that would have to be specified here are usually left unmentioned, partly no doubt because they are assumed to be satisfied as a matter of course in situations in which the organism normally finds itself. But, in part, the omission reflects lack of relevant knowledge, for an explicit specification of the conditions in question would require a theory in which (a) the possible states of organisms and of their environments could be characterized by the values of certain physicochemical or perhaps biological "variables of state," and in which (b) the fundamental theoretical principles would permit the determination of that range of internal and external conditions within which the pulsations of the heart would perform the function referred to above.[2] At present, a general theory of this kind, or even one that could deal in this fashion with some particular kind of organism, is unavailable, of course.

Also, a full restatement of (3.1) in the manner of (3.3) calls for criteria of what constitutes "proper working," "normal functioning," and the like, of the organism at hand; for the function of a given trait is here construed in terms of its causal relevance to the satisfaction of certain necessary conditions of proper working or survival of the organism. Here again, the requisite criteria are often left unspecified—an aspect of functional analysis whose serious implications will be considered later (in section 5).

The considerations here outlined suggest the following schematic characterization of a functional analysis:

(3.4) *Basic pattern of a functional analysis.* The object of the analysis is some "item" *i*, which is a relatively persistent trait or disposition (e.g., the beating of the heart) occurring in a system *s* (e.g., the body of a living vertebrate); and the analysis aims to show that *s* is in a state, or internal condition, c_i and in an environment presenting certain external conditions c_e such that under conditions c_i and c_e (jointly to be referred to as *c*) the trait *i* has effects which satisfy some "need" or "functional requirement" of *s*, i.e., a condition *n* which is necessary for the system's remaining in adequate, or effective, or proper, working order.

Let us briefly consider some examples of this type of analysis in psychology and in sociological and anthropological studies. In psychology, it is especially psychoanalysis which shows a strong functional orientation. One clear instance is Freud's functional characterization of the role of symptom formation. In *The Problem of Anxiety*, Freud expresses himself as favoring a conception according to which "all symptom formation would be brought about solely in order to avoid anxiety; the symptoms bind the psychic energy which otherwise would be discharged as anxiety."[3] In support of this view, Freud points out that if an agoraphobic who has usually been accompanied when going out is left alone in the street, he will suffer an attack of anxiety, as will the compulsion neurotic who, having touched something, is prevented from washing his hands. "It is clear, therefore, that the stipulation of being accompanied and the compulsion to wash has as their purpose, and also their result, the averting of an outbreak of anxiety."[4] In this account, which is put in strongly teleological terms, the system *s* is the individual under consideration; *i* his agoraphobic or compulsive behavior pattern; *n* the binding of anxiety, which is necessary to avert a serious psychological crisis that would make it impossible for the individual to function adequately.

In anthropology and sociology the object of functional analysis is, in Merton's words, "a standardized (i.e., patterned and repetitive) item, such as social roles, institutional patterns, social processes, cultural pattern, culturally patterned emotions, social norms, group organization, social structure, devices for social control, etc."[5] Here, as in psychology and biology, the function, i.e., the stabilizing or adjusting effect, of the item under study, may be one not consciously sought (and indeed, it might not even be consciously recognized) by the agents; in this case, Merton speaks of *latent* functions—in contradistinction to *manifest* functions, i.e., those stabilizing objective effects which are intended by participants in the system.[6] Thus, e.g., the rain-making ceremonials of the Hopi fail to achieve their manifest meteorological objective, but they "may fulfill the latent function of reinforcing the group identity by providing a periodic occasion on which the scattered members of a group assemble to engage in a common activity."[7]

Radcliffe-Brown's functional analysis of the totemic rites of certain Australian tribes illustrates the same point: "To discover the social function of the totemic rites we have to consider the whole body of cosmological ideas of which each rite is a partial expression. I believe that it is possible to show that the social structure of an Australian tribe is connected in a very special way with these cosmological ideas and that the maintenance of its continuity depends on keeping them alive, by their regular expression in myth and rite.

"Thus, any satisfactory study of the totemic rites of Australia must be based not simply on the consideration of their ostensible purpose. . . , but on the discovery of their meaning and of their social function."[8]

Malinowski attributes important latent functions to religion and to magic: He argues that religious faith establishes and enhances mental attitudes such as reverence for tradition, harmony with environment, and confidence and courage in critical situations and at the prospect of death—attitudes which, embodied and maintained by cult and ceremonial, have "an immense biological value." He points out that magic, by providing man with certain ready-made rituals, techniques, and beliefs, enables him "to maintain his poise and his mental integrity in fits anger, in the throes of hate, of unrequited love, of despair and anxiety. The function of magic is to ritualize man's optimism, to enhance his faith in the victory of hope over fear."[9]

There will soon be occasion to add to the preceding examples from psychoanalysis and anthropology some instances of functional analysis in sociology. To illustrate the general character of the procedure, however, the cases mentioned so far will suffice: They all exhibit the basic pattern outlined in (3.4). We now turn from our examination of the form of functional analysis to a scrutiny of its significance as a mode of explanation.

3. THE EXPLANATORY IMPORT OF FUNCTIONAL ANALYSIS

Functional analysis is widely considered as achieving an *explanation* of the "items" whose functions it studies. Malinowski, for example, says of the functional analysis of culture that it "aims at the explanation of anthropological facts at all levels of development by their function. . ."[10] and he adds, in the same context: "To explain any item of culture, material or moral, means to indicate its functional place within an institution, . . . "[11] At another place, Malinowski speaks of the "functional explanation of art, recreation, and public ceremonials."[12]

Radcliffe-Brown, too, considers functional analysis as an explanatory method, though not as the only one suited for the social sciences: "Similarly one 'explanation' of a social system will be its history, where we know it—the detailed account of how it came to be, what it is, and where it is. Another 'explanation' of the same system is obtained by showing (as the functionalists attempt to do) that it is a special exemplification of laws of social physiology or social functioning. The two kinds of explanation do not conflict, but supplement one another."[13]

Apart from illustrating the attribution of explanatory import to functional analysis, this passage raises two points which bear on the general question as to the nature of explanation in empirical science. We will therefore digress briefly to comment on these points.

First, as Radcliffe-Brown stresses, a functional analysis has to refer to general laws. This is shown also in our schematic characterization (3.4): The statements that *i*, in the specified setting *c*, has effects that satisfy *n*, and that *n* is a necessary condition for the proper functioning of the system, both involve general laws. For a statement of causal connection this is well known; and the assertion that a condition *n* constitutes a functional prerequisite for a state of some specified kind (such as proper functioning) is tantamount to the statement of a law to the effect that whenever condition *n* fails to be satisfied, the state in question fails to occur. Thus, explanation by functional analysis requires reference to laws.[14]

The second point relates to a concept invoked by Radcliffe-Brown, of a historic-genetic explanation, which accounts for an item such as a social system or institution by tracing its origins. Clearly, the mere listing of a series of events preceding the given item cannot qualify as an explanation; temporal precedence does not in itself make an event relevant to the genesis of the item under consideration. Thus, a criterion of relevance is needed for the characterization of a sound historic-genetic explanation. As brief reflection shows, relevance here consists in causal or probabilistic determination. A historic-genetic explanation will normally proceed in stages, beginning with some initial set of circumstances which are said to have "brought about," or "led to," certain events at a later time; of these it is next argued that by virtue of, or in conjunction with, certain further conditions prevailing at that later time, they led to a specified further set of events in the historical development; these are in turn combined with additional factors then prevailing and lead to a still later stage, and so forth, until the final explanandum is reached. In a genetic account of this kind, the assertion that a given set of circumstances brought about certain specified subsequent conditions clearly has to be construed as claiming a nomological connection of causal, or more likely, of probabilistic, character. Thus, there is tacit reference to general laws of strictly universal or of statistical form; and a historic-genetic explanation can be construed schematically as a sequence of steps each of which has the character of a nomological explanation. However, while in each step but the first, some of the particular facts mentioned in the explanans will have been accounted for by preceding explanatory steps, the other particular facts invoked will be brought in simply by way of supplementary information. Thus, even in a highly schematic construal, a historic-genetic explanation cannot be viewed as proceeding from information about circumstances at some initial time, via certain statistical or causal laws alone, to the final explanandum: It is essential that, as the argument goes on, additional information is fed into it, concerning certain events which supervene "from the outside," as it were, at various stages of the process under study. Let us note that exactly the same procedure would be required in the case of the melting ice if, during the period of time under consideration, the system were subject to certain outside influences, such as someone's pushing the beaker and spilling some of the water, or salt being added to the water. Basically, then, historic-genetic explanation is nomological explanation.

Returning now to the main issue of the present section, we have to ask what explanatory import may properly be attributed to functional analysis. Suppose, then, that we are interested in explaining the occurrence of a trait *i* in a system *s* (at a certain time *t*), and that the following functional analysis is offered:

(4.1) (a) At *t*, *s* functions adequately in a setting of kind *c* (characterized by specific internal and external conditions).

(b) *s* functions adequately in a setting of kind *c* only if a certain necessary condition, *n*, is satisfied.

(c) If trait *i* were present in *s* then, as an effect, condition *n* would be satisfied.

(d) (Hence,) at *t*, trait *i* is present in *s*.

For the moment, we will leave aside the question as to what precisely is meant by statements of the types (a) and (b), and especially by the phrase "*s* functions adequately"; these matters will be examined in section 5. Right now, we will concern ourselves only with the *logic* of the argument, i.e., we will ask whether (d) formally follows from (a), (b), (c), just as in a deductive nomological explanation the explanandum follows from the explanans. The answer is obviously in the negative, for, to put it pedantically, the argument (4.1) involves the fallacy of affirming the consequent in regard to premise (c). More explicitly, the statement (d) could be validly inferred if (c) asserted that *only* the presence of trait *i* could effect satisfaction of condition *n*. As it is, we can infer merely that condition *n* must be satisfied in some way or other at time *t*; for otherwise, by reason of (b), the system *s* could not be functioning adequately in its setting, in contradiction to what (a) asserts. But it might well be that the occurrence of any one of a number of alternative items would suffice no less than the occurrence of *i* to satisfy requirement *n*, in which case the account provided by the premises of (4.1) simply fails to explain why the trait *i* rather than one of its alternatives is present in *s* at *t*.

As has just been noted, this objection would not apply if premise (c) could be replaced by the statement that requirement *n* can be met *only* by the presence of trait *i*. And indeed, some instances of functional analysis seem to involve the claim that the specific item under analysis is, in this sense, functionally indispensable for the satisfaction of *n*. For example, Malinowski makes this claim for magic when he asserts that "magic fulfills an indispensable function within culture. It satisfies a definite need which cannot be satisfied by any other factors of primitive civilization,"[15] and again when he says about magic that "without its power and guidance early man could not have mastered his practical difficulties as he has done, nor could man have advanced to the higher stages of culture. Hence the universal occurrence of magic in primitive societies and its enormous sway. Hence we do find magic an invariable adjunct of all important activities."[16]

However, the assumption of functional indispensability for a given item is highly questionable on empirical grounds: In all concrete cases of application, there do seem to exist alternatives. For example, the binding of anxiety in a given subject might be effected by an alternative symptom, as the experience of psychiatrists seems to confirm. Similarly, the function of the rain dance might be subserved by some other group ceremonial. And interestingly, Malinowski himself, in another context, invokes "the principle of limited possibilities, first laid down by Goldenweiser. Given a definite cultural need, the means of its satisfaction are small in number, and therefore the cultural arrangement which comes into being in response to the need is determined within narrow limits."[17] This principle obviously involves at least a moderate liberalization of the conception that every cultural item is functionally indispensable. But even so, it may still be too restrictive. At any rate, sociologists such as Parsons and Merton have assumed the existence of "functional equivalents" for certain cultural items; and Merton, in his general analysis of functionalism, has insisted that the conception of the functional indispensability of cultural items be replaced quite explicitly by the assumption of "functional alternatives, or functional equivalents, or functional substitutes."[18] This idea, incidentally, has an interesting parallel in the "principle of multiple solutions" for adaptational problems in evolution. This principle, which has been emphasized by functionally oriented biologists, states that for a given functional problem (such as that of perception of light) there are usually a variety of possible solutions, and many of these are actually used by different—and often closely related—groups of organisms.[19]

It should be noted here that, in any case of functional analysis, the question whether there are functional equivalents to a given item *i* has a definite meaning only if the internal and external conditions *c* in (4.1) are clearly specified. Otherwise, any proposed alternative to *i*, say *i*′, could be denied the status of a functional equivalent on the ground that, being different from *i*, the item *i*′ would have certain effects on the internal state and the environment of *s* which would not be brought about by *i*; and that therefore, if *i*′ rather than *i* were realized, *s* would not be functioning in the same internal and external situation.

Suppose, for example, that the system of magic of a given primitive group were replaced by an extension of its rational technology plus some modification of its religion, and that the group were to continue as a going concern. Would this establish the existence of a functional equivalent to the original system of magic? A negative answer might be defended on the grounds that as a result of adopting the modified pattern the group had changed so strongly in regard to some of its basic characteristics (i.e., its internal state, as characterized by c_i, had been so strongly modified) that it was not the original kind of primitive group any more; and that there simply was no functional equivalent to magic which would leave all the "essential" features of the group unimpaired. Consistent use of this type of argument would safeguard the postulate of the functional indispensability of every cultural item against any conceivable empirical disconfirmation, by turning it into a covert tautology.

Let i be the class of those items, $i, i', i'', \ldots$, any one of which, if present in s under conditions c, would effect satisfaction of condition n. Then those items are functional equivalents in Merton's sense, and what the premises of (4.1) entitle us to infer is only:

(4.2) Some one of the items in class I is present in s at t. But the premises give us no grounds to expect i rather than one of its functional alternatives.

So far, we have viewed functional analysis only as a presumptive deductive explanation. Might it not be construed instead as an inductive argument which shows that the occurrence of i is highly probable in the circumstances described by the premises? Might it not be possible, for example, to add to the premises of (4.1) a further statement to the effect that the functional prerequisite n can be met only by i and by a few specifiable functional alternatives? And might not these premises make the presence of i highly probable? This course is hardly promising, for in most, if not all, concrete cases it would be impossible to specify with any precision the range of alternative behavior patterns, institutions, customs, or the like that would suffice to meet a given functional prerequisite or need. And even if that range could be characterized, there is no satisfactory method in sight for dividing it into some finite number of cases and assigning a probability to each of these.

Assume, for example, that Malinowski's general view of the function of magic is correct: How are we to determine, when trying to explain the system of magic of a given group, all the different systems of magic and alternative cultural patterns any one of which would satisfy the same functional requirements for the group as does the actually existing system of magic? And how are we to ascribe probabilities of occurrence to each of these potential functional equivalents? Clearly, there is no satisfactory way of answering these questions, and practitioners of functional analysis do not claim to achieve their explanation in this extremely problematic fashion.

Nor is it any help to construe the general laws implicit in the statements (b) and (c) in (4.1) as statistical rather than strictly universal in form, i.e., as expressing connections that are very probable, but do not hold universally; for the premises thus obtained would still allow for functional alternatives of i (each of which would make satisfaction of n highly probable), and thus the basic difficulty would remain: The premises taken jointly could still not be said to make the presence just of i highly probable.

In sum then, the information typically provided by a functional analysis of an item *i* affords neither deductively nor inductively adequate grounds for expecting *i* rather than one of its alternatives. The impression that a functional analysis does provide such grounds, and thus explains the occurrence of *i*, is no doubt at least partly due to the benefit of hind-sight: When we seek to explain an item *i*, we presumably know already that *i* has occurred.

But, as was briefly noted earlier, a functional analysis provides, in principle, the basis for an explanation with a weaker explanandum; for the premises (a) and (b) of (4.1) imply the consequence that the necessary condition *n* must be fulfilled in some way or other. This much more modest kind of functional explanation may be schematized as follows:

(4.3) (a) At time *t*, system *s* functions adequately in a setting of kind *c*.
(b) *s* functions adequately in a setting of kind *c* only if condition *n* is satisfied.

(e) Some one of the items in class I is present in *s* at *t*.

This kind of inference, while sound, is rather trivial, however, except in cases where we have additional knowledge about the items contained in class I. Suppose, for example, that at time *t*, a certain dog (system *s*) is in good health in a "normal" kind of setting *c* which precludes the use of such devices as artificial hearts, lungs, and kidneys. Suppose further that in a setting of kind *c*, the dog can be in good health only if his blood circulates properly (condition *n*). Then schema (4.3) leads only to the conclusion that in some way or other, the blood must be kept circulating properly in the dog at *t*—hardly a very illuminating result. If, however, we have additional knowledge of the ways in which the blood may be kept circulating under the circumstances and if we know, for example, that the only feature that would ensure proper circulation (the only item in class I) is a properly working heart, then we may draw the much more specific conclusion that at *t* the dog has a properly working heart. But if we make explicit the additional knowledge here used by expressing it as a third premise, then our argument assumes a form considered earlier, namely that of a functional analysis which is of the type (4.1), except that premise (c) has been replaced by the statement that *i* is the *only* trait by which *n* can be satisfied in setting *c*; and, as was pointed out above, the conclusion (d) of (4.1) does follow in this case; in our case, (d) is the sentence stating that the dog has a properly working heart at *t*.

In general, however, additional knowledge of the kind here referred to is not available, and the explanatory import of functional analysis is then limited to the precarious role schematized in (4.3).

4. THE PREDICTIVE IMPORT OF FUNCTIONAL ANALYSIS

We noted earlier the predictive significance of nomological explanation; now we will ask whether functional analysis can be put to predictive use.

First of all, the preceding discussion shows that the information which is typically provided by a functional analysis yields at best premises of the forms (a), (b), (c) in (4.1); and these afford no adequate basis for the deductive or inductive prediction of a sentence of the form (d) in (4.1). Thus, functional analysis no more enables us to predict than it enables us to explain the occurrence of a particular one of the items by which a given functional requirement can be met.

Second, even the much less ambitious explanatory schema (4.3) cannot readily be put to predictive use; for the derivation of the weak conclusion (e) relies on the premise (a); and if we wish to infer (e) with respect to some future time t, that premise is not available, for we do not know whether s will or will not be functioning adequately at that future time. For example, consider a person developing increasingly severe anxieties, and suppose that a necessary condition for his adequate functioning is that his anxiety be bound by neurotic symptoms, or be overcome by other means. Can we predict that some one of the modes of "adjustment" in the class I thus roughly characterized will actually come to pass? Clearly not, for we do not know whether the person in question will in fact continue to function adequately or will suffer some more or less serious breakdown, perhaps to the point of self-destruction.

It is of interest to note here that a somewhat similar limitation exists also for the predictive use of nomological explanations, even in the most advanced branches of science. For example, if we are to predict, by means of the laws of classical mechanics, the state in which a given mechanical system will be at a specified future time t, it does not suffice to know the state of the system at some earlier time t_o, say the present; we also need information about the boundary conditions during the time interval from t_o to t, i.e., about the external influences affecting the system during that time. Similarly, the "prediction," in our first example, that the water level in the beaker will remain unchanged as the ice melts assumes that the temperature of the surrounding air will remain constant, let us say, and that there will be no disturbing influences such as an earthquake or a person upsetting the beaker. Again when we predict for an object dropped from the top of the Empire State Building that it will strike the ground about eight seconds later, we assume that during the period of its fall, the object is acted upon by no forces other than the gravitational attraction of the earth. In a full and explicit formulation then, nomological predictions such as these would have to include among their premises statements specifying the boundary conditions obtaining from t_o up to the time t to which the prediction refers. This shows that even the laws and theories of the physical sciences do not actually enable us to predict certain aspects of the future exclusively on the basis of certain aspects of the present: The prediction also requires certain assumptions about the future. But, in many cases of nomological prediction, there are good inductive grounds, available at t_o, for the assumption that during the time interval in question, the system under study will be practically "closed," i.e., not subject to significant outside interference (this case is illustrated, for example, by the prediction of eclipses) or that the boundary conditions will be of a certain specified kind—a situation illustrated by predictions of events occurring under experimentally controlled conditions.

Now, the predictive use of (4.3) likewise requires a premise concerning the future, namely (a); but there is often considerable uncertainty as to whether (a) will in fact hold true at the future time *t*. Furthermore, if in a particular instance there should be good inductive grounds for considering (a) as true, the forecast yielded by (4.3) is still rather weak; for the argument then leads from the inductively warranted assumption that the system will be properly functioning at *t* to the "prediction" that a certain condition *n*, which is necessary for such functioning, will be satisfied at *t* in some way or other.

The need to include assumptions about the future among the premises of predictive arguments can be avoided, in nomological predictions as well as in those based on functional analysis, if we are satisfied with predictive conclusions which are not categorical, but only conditional, or hypothetical, in character. For example, (4.3) may be replaced by the following argument, in which premise (a) is avoided at the price of conditionalizing the conclusion:

(5.1) (b) System *s* functions adequately in a setting of kind *c* only if condition *n* is satisfied.

(f) If *s* functions adequately in a setting of kind *c* at time *t*, then some one of the items in class I is present in *s* at *t*.

This possibility deserves mention because it seems that at least some of the claims made by advocates of functional analysis may be construed as asserting no more than that functional analysis permits conditional predictions of the kind schematically represented by (5.1). This might be the intent, for example, of Malinowski's claim: "If such [a functional] analysis discloses to us that, taking an individual culture as a coherent whole, we can state a number of general determinants to which it has to conform, we shall be able to produce a number of predictive statements as guides for field-research, as yardsticks for comparative treatement, and as common measures in the process of cultural adaptation and change."[20] The statements specifying the determinants in question would presumably take the form of premises of type (b); and the "predictive statements" would then be of a hypothetical character.

Many of the predictions and generalizations made in the context of functional analysis, however, eschew the cautious conditional form just considered. They proceed from a statement of a functional prerequisite or need to the categorical assertion of the occurrence of some trait, institution, or other item suited to meet the requirement in question. Consider, for example, Sait's functional explanation of the emergence of the political boss: "Leadership is necessary; and *since* it does not develop readily within the constitutional framework, the boss provides it in a crude and irresponsible form from the outside";[21] or take Merton's characterization of one function of the political machine: Referring to various specific ways in which the political machine can serve the interests of business, he concludes, "These 'needs' of business, as presently constituted, are not adequately provided for by conventional and culturally approved social structures; *consequently*, the extra-legal but more-or-less efficient organization of the political machine comes to provide these services."[22] Each of these arguments, which are rather typical of the functionalist approach, is an inference from the existence of a certain functional prerequisite to the categorical assertion that the prerequisite will be satisfied in some way. What is the basis of these inferences, which are marked by the words, "since" and "consequently" in the passages just quoted? When we say that *since* the ice cube was put in warm water it melted; or that the current was turned on, and *consequently*, the ammeter in the circuit responded, these inferences can be explicated and justified by reference to certain general laws of which the particular cases at hand are simply special instances; and the logic of the inferences can be exhibited by putting them into the form of the schema (2.1). Similarly, each of the two functionalist arguments under consideration clearly seems to presuppose a general law to the effect that, within certain limits of tolerance or adaptability, a system of the kind under analysis will—either invariably or with high probability—satisfy, by developing appropriate traits, the various functional requirements (necessary conditions for its continued adequate operation) that may arise from changes in its internal state or in its environment. Any assertion of this kind, no matter whether of strictly universal or of statistical form, will be called a (*general*) *hypothesis of self-regulation.*

Unless functional analyses of the kind just illustrated are construed as implicitly proposing or invoking suitable hypotheses of self-regulation, it remains quite unclear what connections the expressions "since," "consequently," and others of the same character are meant to indicate, and how the existence of those connections in a given case is to be objectively established.

Conversely, if a precise hypothesis of self-regulation for systems of a specified kind is set forth, then it becomes possible to explain, and to predict categorically, the satisfaction of certain functional requirements simply on the basis of information concerning antecedent needs; and the hypothesis can then be objectively tested by an empirical check of its predictions. Take, for example, the statement that if a hydra is cut into several pieces, most of these will grow into complete hydras again. This statement may be considered as a hypothesis concerning a specific kind of self-regulation in a particular kind of biological system. It can clearly be used for explanatory and predictive purposes, and indeed the success of the predictions it yields confirms it to a high degree.

We see, then, that wherever functional analysis is to serve as a basis for categorical prediction or for generalizations of the type illustrated by the passages from Sait and from Merton, it is of crucial importance to establish appropriate hypotheses of self-regulation in an objectively testable form.

The functionalist literature does contain some explicitly formulated generalizations of the kind here referred to. Merton, for example, after citing the passage from Sait quoted above, comments thus: "Put in more generalized terms, *the functional deficiencies of the official structure generate an alternative* (*unofficial*) *structure to fulfill existing needs somewhat more effectively.*"[23] This statement seems clearly intended to make explicit a hypothesis of self-regulation that might be said to underlie Sait's specific analysis and to provide the rationale for his "since." Another hypothesis of this kind is suggested by Radcliffe-Brown: "It may be that we should say that. . . a society that is thrown into a condition of functional disunity or inconsistency. . . will not die, except in such comparatively rare instances as an Australian tribe overwhelmed by the white man's destructive force, but will continue to struggle toward. . . some kind of social health. . . ."[24]

But, as was briefly suggested above, a formulation proposed as a hypothesis of self-regulation can serve as a basis for explanation or prediction only if it is a reasonably definite statement that permits of objective empirical test. And indeed many of the leading representatives of functional analysis have expressed very clearly their concern to develop hypotheses and theories which meet this requirement. Malinowski, for example, in his essay significantly entitled "A scientific theory of culture," insists that "each scientific theory must start from and lead to observation. It must be inductive and it must be verifiable by experience. In other words, it must refer to human experiences which can be defined, which are public, that is, accessible to any and every observer, and which are recurrent, hence fraught with inductive generalizations, that is, predictive."[25] Similarly, Murray and Kluckhohn have this to say about the basic objective of their functionally oriented theory, and indeed about any scientific "formulation," of personality: "The general purposes of formulation are three: (1) to *explain* past and present events; (2) to *predict* future events (the conditions being specified); and (3) to serve, if required, as a basis for the selection of effective measures of *control*."[26]

Unfortunately, however, the formulations offered in the context of concrete functional analyses quite often fall short of these general standards. Among the various ways in which those conditions may be violated, two call for special consideration because of their pervasiveness and central importance in functional analysis. They will be referred to as (i) *inadequate specification of scope* and (ii) *nonempirical use of functionalist key terms* (such as "need," "functional requirement," "adaptation," and others). We will consider these two defects in turn: the former in the balance of the present section, the latter in the next.

Inadequate specification of scope consists in failure to indicate clearly the kind of system to which the hypothesis refers, or the range of situations (the limits of tolerance) within which those systems are claimed to develop traits that will satisfy their functional requirements. Merton's formulation, for example, does not specify the class of social systems and of situations to which the proposed generalization is meant to apply; as it stands, therefore, it cannot be put to an empirical test or to any predictive use.

The generalization tentatively set forth by Radcliffe-Brown has a similar shortcoming: Ostensibly, it refers to any society whatever, but the conditions under which social survival is claimed to occur are qualified by a highly indefinite "except" clause, which precludes the possibility of any reasonably clear-cut test. The clause might even be used to protect the proposed generalization against any conceivable disconfirmation: If a particular social group should "die," this very fact might be held to show that the disruptive forces were as overwhelming as in the case of the Australian tribe mentioned by Radcliffe-Brown. Systematic use of this methodological strategy would, of course, turn the hypothesis into a covert tautology. This would ensure its truth, but at the price of depriving it of empirical content: Thus construed, the hypothesis can yield no explanation or prediction whatever.

A similar comment is applicable to the following pronouncement by Malinowski, in which we italicize the dubious qualifying clause: "When we consider any culture *which is not on the point of breaking down or completely disrupted, but which is a normal going concern*, we find that need and response are directly related and tuned up to each other."[27]

To be sure, Radcliffe-Brown's and Malinowski's formulations do not *have to* be construed as covert tautologies, and their authors no doubt intended them as empirical assertions; but, in this case, the vagueness of the qualifying clauses still deprives them of the status of definite empirical hypotheses that might be used for explanation or prediction.

5. THE EMPIRICAL IMPORT OF FUNCTIONALIST TERMS AND HYPOTHESES

In the preceding section, we mentioned a second flaw that may vitiate the scientific role of a proposed hypothesis of self-regulation. It consists in using key terms of functional analysis, such as "need" and "adequate (proper) functioning"[28] in a nonempirical manner, i.e., without giving them a clear "operational definition," or more generally, without specifying objective criteria of application for them.[29] If functionalist terms are used in this manner, then the sentences containing them have no clear empirical meaning; they lead to no specific predictions and thus cannot be put to an objective test; nor, of course, can they be used for explanatory purposes.

A consideration of this point is all the more important here because the functionalist key terms occur not only in hypotheses of self-regulation, but also in functionalist sentences of various other kinds, such as those of the types (a), (b), and (f) in our schematizations (4.1), (4.3), and (5.1) of functionalist explanation and prediction. Nonempirical use of functionalist key terms may, therefore, bar sentences of these various kinds from the status of scientific hypotheses. We turn now to some examples.

Consider first the terms "functional prerequisite" and "need," which are used as more or less synonymous in the functionalist literature, and which serve to define the term "function" itself. "Embedded in every functional analysis is some conception, tacit or expressed, of the functional requirements of the system under observation,"[30] and, indeed, "a definition [of function] is provided by showing that human institutions, as well as partial activities within these, are related to primary, that is, biological, or derived, that is, cultural needs. Function means, therefore, always the satisfaction of a need. . . . "[31]

How is this concept of need defined? Malinowski gives a very explicit answer: "By need, then, I understand the system of conditions in the human organism, in the cultural setting, and in the relation of both to the natural environment, which are sufficient and necessary for the survival of group and organism."[32] This definition sounds clear and straightforward; yet it is not even quite in accord with Malinowski's own use of the concept of need. For he distinguishes, very plausibly, a considerable number of different needs, which fall into two major groups: primary biological needs and derivative cultural ones; the latter include "technological, economic, legal, and even magical, religious, or ethical"[33] needs. But if every single one of these needs did actually represent not only a necessary condition of survival but also a sufficient one, then clearly the satisfaction of just one need would suffice to ensure survival, and the other needs could not constitute necessary conditions of survival at all. It seems reasonable to assume, therefore, that what Malinowski intended was to construe the needs of a group as a set of conditions which are individually necessary and jointly sufficient for its survival.[34]

However, this correction of a minor logical flaw does not remedy a more serious defect of Malinowski's definition, which lies in the deceptive appearance of clarity of the phrase "survival of group and organism." In reference to a biological organism, the term "survival" has a fairly clear meaning, though even here, there is need for further clarification. For when we speak of biological needs or requirements e.g., the minimum daily requirements, for human adults, of various vitamins and minerals, we construe these, not as conditions of just the barest survival but as conditions of persistence in, or return to, a "normal," or "healthy" state, or to a state in which the system is a "properly functioning whole." For the sake of objective testability of functionalist hypotheses, it is essential, therefore, that definitions of needs or functional prerequisites be supplemented by reasonably clear and objectively applicable criteria of what is to be considered a healthy state or a normal working order of the systems under consideration; and that the vague and sweeping notion of survival then be construed in the relativized sense of survival in a healthy state as specified. Otherwise, there is definite danger that different investigators will use the concept of functional prerequisite—and hence also that of function—in different ways, and with valuational overtones corresponding to their diverse conceptions of what are the most "essential" characteristics of "genuine" survival for a system of the kind under consideration.

Functional analyses in psychology, sociology, and anthropology are even more urgently in need of objective empirical criteria of the kind here referred to; for the characterization of needs as necessary conditions of psychological or emotional survival for an individual, or of survival of a group is so vague as to permit, and indeed invite, quite diverse subjective interpretations.

Some authors characterize the concept of functional prerequisite or the concept of function without making use of the term "survival" with its misleading appearance of clarity. Merton, for example, states: "*Functions* are those observed consequences which make for the adaptation or adjustment of a given system; and *dysfunctions*, those observed consequences which lessen the adaptation or adjustment of the system."[35] And Radcliffe-Brown characterizes the function of an item as its contribution to the maintenance of a certain kind of unity of a social system, "which we may speak of as a functional unity. We may define it as a condition in which all parts of the social system work together with a sufficient degree of harmony or internal consistency, i.e., without producing persistent conflicts which can neither be resolved nor regulated."[36] But like the definitions in terms of survival, these alternative characterizations, though suggestive, are far from giving clear empirical meanings to the key terms of functional analysis. The concepts of adjustment and adaptation, for example, require specification of some standard; otherwise, they have no definite meaning and are in danger of being used tautologically or else subjectively, with valuational overtones.

Tautological use could be based on construing *any* response of a given system as an adjustment, in which case it becomes a trivial truth that any system will adjust itself to any set of circumstances. Some instances of functional analysis seem to come dangerously close to this procedure, as is illustrated by the following assertion: "Thus we are provided with an explanation of suicide and of numerous other apparently antibiological effects as so many forms of relief from intolerable suffering. Suicide does not have *adaptive* (survival) value but it does have *adjustive* value for the organism. Suicide is *functional* because it abolishes painful tension."[37]

Or consider Merton's formulation of one of the assumptions of functional analysis: ". . . when *the net balance of the aggregate of consequences* of an existing social structure is clearly dysfunctional, there develops a strong and insistent pressure for change."[38] In the absence of clear empirical criteria of adaptation and thus of dysfunction, it is possible to treat this formulation as a covert tautology and thus to render it immune to empirical disconfirmation. Merton is quite aware of such danger: In another context he remarks that the notion of functional requirements of a given system "remains one of the cloudiest and empirically most debatable concepts in functional theory. As utilized by sociologists, the concept of functional requirement tends to be tautological or *ex post facto*."[39] Similar warnings against tautological use and against *ad hoc* generalizations about functional prerequisites have been voiced by other writers, such as Malinowski[40] and Parsons.[41]

On the other hand, in the absence of empirical criteria of adjustment or adaptation, there is also the danger of each investigator's projecting into those concepts (and thus also into the concept of function) his own ethical standards of what would constitute a "proper" or "good" adjustment of a given system—a danger which has been pointed out very clearly by Levy.[42] This procedure would obviously deprive functionalist hypotheses of the status of precise objectively testable scientific assertions. And, as Merton notes, "If theory is to be productive, it must be sufficiently *precise* to be *determinate*. Precision is an integral element of the criterion of *testability*."[43]

It is essential, then, for functional analysis as a scientific procedure that its key concepts be explicitly construed as relative to some standard of survival or adjustment. This standard has to be specified for each functional analysis, and it will usually vary from case to case. In the functional study of a given system s, the standard would be indicated by specifying a certain class or range R of possible states of s, with the understanding that s was to be considered as "surviving in proper working order," or as "adjusting properly under changing conditions" just in case s remained in, or upon disturbance returned to, some state within the range R. A need, or functional requirement, of system s relative to R is then a necessary condition for the system's remaining in, or returning to, a state in R; and the function, relative to R, of an item i in s consists in i's effecting the satisfaction of some such functional requirement.

In the field of biology, Sommerhoff's analysis of adaptation, appropriateness, and related concepts, is an excellent illustration of a formal study in which the relativization of the central functionalist concepts is entirely explicit.[44] The need of such relativization is made clear also by Nagel, who points out that "the claim that a given change is functional or dysfunctional must be understood as being relative to a specified G (or sets of G's)"[45] where the G's are traits whose preservation serves as the defining standard of adjustment or survival for the system under study. In sociology, Levy's analysis of the structure of society[46] clearly construes the functionalist key concepts as relative in the sense just outlined.

Only if the key concepts of functional analysis are thus relativized can hypotheses involving them have the status of determinate and objectively testable assumptions or assertions; only then can those hypotheses enter significantly into arguments such as those schematized in (4.1), (4.3), and (5.1).

But, although such relativization may give definite empirical content to the functionalist hypotheses that serve as premises or conclusions in those arguments, it leaves the explanatory and predictive import of the latter as limited as we found it in sections 4 and 5; for our verdict on the logical force of those arguments depended solely on their formal structure and not on the meanings of their premises and conclusions.

It remains true, therefore, even for a properly relativized version of functional analysis, that its explanatory force is rather limited; in particular, it does not provide an explanation of why a particular item i rather than some functional equivalent of it occurs in system s. And the predictive significance of functional analysis is practically nil—except in those cases where suitable hypotheses of self-regulation can be established. Such a hypothesis would be to the effect that within a specified range C of circumstances, a given system s (or: any system of a certain kind S, of which s is an instance) is self-regulating relative to a specified range R of states, i.e., after a disturbance which moves s into a state outside R, but which does not shift the internal and external circumstances of s out of the specified range C, the system s will return to a state in R. A system satisfying a hypothesis of this kind might be called *self-regulating with respect to R*.

Biological systems offer many illustrations of such self-regulation. For example, we mentioned earlier the regenerative ability of a hydra. Consider the case, then, where a more or less large segment of the animal is removed and the rest grows into a complete hydra again. The class R here consists of those states in which the hydra is complete; the characterization of range C would have to include (1) a specification of the temperature and the chemical composition of the water in which a hydra will perform its regenerative feat (clearly, this will not be just one unique composition, but a class of different ones: The concentrations of various salts, for example, will each be allowed to take some value within a specified, and perhaps narrow, range; the same will hold of the temperature of the water) and (2) a statement as to the kind and size of segment that may be removed without preventing regeneration.

It will no doubt be one of the most important tasks of functional analysis in psychology and the social sciences to ascertain to what extent such phenomena of self-regulation can be found, and clearly represented by laws of self-regulation, in these fields.

6. FUNCTIONAL ANALYSIS AND TELEOLOGY

Whatever specific laws might be discovered by research along these lines, the kind of explanation and prediction made possible by them does not differ in its logical character from that of the physical sciences.

It is quite true that hypotheses of self-regulation, which would be characteristic results of successful functionalist research, seem to have a teleological character, asserting, as they do, that within specified conditions systems of some particular kind will tend toward a state within the class R, which thus assumes the appearance of a final cause determining the behavior of the system.

But, first of all, it would be simply untenable to say of a system s which is self-regulating with respect to R that the future event of its return to (a state in) R is a "final cause" which determines its present behavior. For even if s is self-regulating with respect to R and if it has been shifted into a state outside R, the future event of its return to R may never come about: In the process of its return toward R, s may be exposed to further disturbances, which may fall outside the permissible range C and lead to the destruction of s. For example, in a hydra that has just had a tentacle removed, certain regenerative processes will promptly set in; but these cannot be explained teleologically by reference to a final cause consisting in the future event of the hydra being complete again. For that event may never actually come about since in the process of regeneration, and before its completion, the hydra may suffer new, and irreparably severe, damage, and may die. Thus, what accounts for the present changes of a self-regulating system s in not the "future event" of s being in R, but rather the *present disposition* of s to return to R; and it is this disposition that is expressed by the hypothesis of self-regulation governing the system s.

Whatever teleological character may be attributed to a functionalist explanation or prediction invoking (properly relativized) hypotheses of self-regulation lies merely in the circumstance that such hypotheses assert a tendency of certain systems to maintain, or return to, a certain kind of state. But such laws attributing, as it were, a characteristic goal-directed behavior to systems of specified kinds are by no means alien to physics and chemistry. On the contrary, it is these latter fields which provide the most adequately understood instances of self-regulating systems and corresponding laws. For example, a liquid in a vessel will return to a state of equilibrium, with its surface horizontal, after a mechanical disturbance; an elastic band, after being stretched (within certain limits), will return to its original shape when it is released. Various systems controlled by negative feedback devices, such as a steam engine whose speed is regulated by a governor, or a homing torpedo, or a plane guided by an automatic pilot, show, within specifiable limits, self-regulation with respect to some particular class of states.

In all of these cases, the laws of self-regulation exhibited by the systems in question are capable of explanation by subsumption under general laws of a more obviously causal form. But this is not even essential, for the laws of self-regulation themselves are causal in the broad sense of asserting, essentially, that for systems of a specified kind, any one of a class of different "initial states" (any one of the permissible states of disturbance) will lead to the same kind of final state. Indeed, as our earlier formulations show, functionalist hypotheses, including those of self-regulation, can be expressed without the use of any teleological phraseology at all.[47]

There are, then, no systematic grounds for attributing to functional analysis a character *sui generis* not found in the hypotheses and theories of the natural sciences and in the explanations and predictions based on them. Yet, psychologically, the idea of function often remains closely associated with that of purpose, and some functionalist writing has no doubt encouraged this association, by using a phraseology which attributes to the self-regulatory behavior of a given system practically the character of a purposeful action. For example, Freud, in stating his theory of the relation of neurotic symptoms to anxiety, uses strongly teleological language, as when he says that "the symptoms are created in order to remove or rescue the ego from the situation of danger"[48]; the quotations given in section 2 provide further illustrations. Some instructive examples of sociological and anthropological writings which confound the concepts of function and purpose are listed by Merton, who is very explicit and emphatic in rejecting this practice.[49]

It seems likely that precisely this psychological association of the concept of function with that of purpose, though systematically unwarranted, accounts to a large extent for the appeal and the apparent plausibility of functional analysis as a mode of explanation; for it seems to enable us to "understand" self-regulatory phenomena of all kinds in terms of purposes or motives, in much the same way in which we "understand" our own purposive behavior and that of others. Now, explanation by reference to motives, objectives, or the like may be perfectly legitimate in the case of purposive behavior and its effects. An explanation of this kind would be causal in character, listing among the causal antecedents of the given action, or of its outcome, certain purposes or motives on the part of the agent, as well as his beliefs as to the best means available to him for attaining his objectives. This kind of information about purposes and beliefs might even serve as a starting point in explaining a self-regulatory feature in a human artifact. For example, in an attempt to account for the presence of the governor in a steam engine, it may be quite reasonable to refer to the purpose its inventor intended it to serve, to his beliefs concerning matters of physics, and to the technological facilities available to him. Such an account, it should be noted, might conceivably give a probabilistic explanation for the presence of the governor, but it would not explain why it functioned as a speed-regulating safety device: To explain this latter fact, we would have to refer to the construction of the machine and to the laws of physics, not to the intentions and beliefs of the designer. (An explanation by reference to motives and beliefs can be given as well for certain items which do not, in fact, function as intended, e.g., some superstitious practices, unsuccessful flying machines, or ineffective economic policies, etc.) Furthermore—and this is the crucial point in our context—for most of the self-regulatory phenomena that come within the purview of functional analysis, the attribution of purposes is an illegitimate transfer of the concept of purpose from its domain of significant applicability to a much wider domain, where it is devoid of objective empirical import. In the context of purposive behavior of individuals or groups, there are various methods of testing whether the assumed motives or purposes are indeed present in a given situation; interviewing the agent in question might be one rather direct way, and there are various alternative "operational" procedures of a more indirect character. Hence, explanatory hypotheses in terms of purposes are here capable of reasonably objective tests. But such empirical criteria for purposes and motives are lacking in other cases of self-regulating systems, and the attribution of purposes to them has therefore no scientific meaning. Yet, it tends to encourage the illusion that a profound type of understanding is achieved, that we gain an insight into the nature of these processes by likening them to a type of behavior with which we are thoroughly familiar from daily experience. Consider, for example, the law of "adaptation to an obvious end" set forth by the sociologist L. Gumplowicz with the claim that it holds both in the natural and the social domains. For the latter, it asserts that "every social growth, every social entity, serves a definite end, however much its worth and morality may be questioned. For the universal law of adaptation signifies simply that no expenditure of effort, no change of condition, is purposeless on any domain of phenomena. Hence, the inherent reasonableness of all social facts and conditions must be conceded."[50] The suggestion is rather strong here that the alleged law enables us to understand social dynamics in close analogy to purposive behavior aimed at the achievement of some end. Yet the purported law is completely devoid of empirical meaning since no empirical interpretation has been given to such key terms as "end," "purposeless," and "inherent reasonableness" for the contexts to which it is applied. The "law" asserts nothing whatever, therefore, and cannot possibly explain any social—or other—phenomena.

Gumplowicz's book antedates the writings of Malinowski and other leading functionalists by several decades, and certainly these more recent writers have been more cautious and sophisticated in stating their ideas. Yet, there are certain quite central assertions in the newer functionalist literature which are definitely reminiscent of Gumplowicz's formulation in that they suggest an understanding of functional phenomena in the image of deliberate purposive behavior or of systems working in accordance with a preconceived design. The following statements might illustrate this point: "[Culture] is a system of objects, activities, and attitudes in which every part exists as a means to an end,"[51] and "The functional view of culture insists therefore upon the principle that in every type of civilization, every custom, material object, idea and belief fulfills some vital function, has some task to accomplish, represents an indispensable part within a working whole."[52] These statements express what Merton, in a critical discussion, calls the postulate of universal functionalism.[53] Merton qualifies this postulate as premature;[54] the discussion presented in the previous section shows that, in the absence of a clear empirical interpretation of the functionalist key terms, it is even less than that, namely, empirically vacuous. Yet, formulations of this kind may evoke a sense of insight and understanding by likening sociocultural developments to purposive behavior and in this sense reducing them to phenomena with which we feel thoroughly familiar. But scientific explanation and understanding are not simply a reduction to the familiar: Otherwise, science would not seek to explain familiar phenomena at all; besides, the most significant advances in our scientific understanding of the world are often achieved by means of new theories which, like quantum theory, assume some quite unfamiliar kinds of objects or processes which cannot be directly observed, and which sometimes are endowed with strange and even seemingly paradoxical characteristics. A class of phenomena has been scientifically understood to the extent that they can be fitted into a testable, and adequately confirmed, theory or a system of laws; and the merits of functional analysis will eventually have to be judged by its ability to lead to this kind of understanding.

7. THE HEURISTIC ROLE OF FUNCTIONAL ANALYSIS

The preceding considerations suggest that what is often called "functionalism" is best viewed, not as a body of doctrine or theory advancing tremendously general principles such as the principle of universal functionalism, but rather as a program for research guided by certain heuristic maxims or "working hypotheses." The idea of universal functionalism, for example, which becomes untenable when formulated as a sweeping empirical law or theoretical principle, might more profitably be construed as expressing a directive for research, namely to search for specific self-regulatory aspects of social and other systems and to examine the ways in which various traits of a system might contribute to its particular mode of self-regulation. (A similar construal as heuristic maxims for empirical research might be put upon all the "general axioms of functionalism" suggested by Malinowski, and considered by him as demonstrated by all the pertinent empirical evidence.[55])

In biology, for example, the contribution of the functionalist approach does not consist in the sweeping assertion that all traits of any organism satisfy some need and thus serve some function; in this generality, the claim is apt to be either meaningless or covertly tautologous or empirically false (depending on whether the concept of need is given no clear empirical interpretation at all, or is handled in a tautologizing fashion, or is given one definitive empirical interpretation). Instead, functional studies in biology have been aimed at showing, for example, how in different species, specific homeostatic and regenerative processes contribute to the maintenance and development of the living organism; and they have gone on (1) to examine more and more precisely the nature and limits of those processes (this amounts basically to establishing various specific empirical hypotheses or laws of self-regulation) and (2) to explore the underlying physiological or physicochemical mechanisms, and the laws governing them, in an effort to achieve a more thorough theoretical understanding of the phenomena at hand.[56] Similar trends exist in the study of functional aspects of psychological processes, including, for example, symptom formation in neurosis.[57]

Functional analysis in psychology and in the social sciences no less than in biology may thus be conceived, at least ideally, as a program of inquiry aimed at determining the respects and the degrees in which various systems are self-regulating in the sense here indicated. This conception clearly underlies, for example, Nagel's essay, "A Formalization of Functionalism,"[58] a study which develops an analytic scheme inspired by, and similar to, Sommerhoff's formal analysis of self-regulation in biology[59] and uses it to exhibit and clarify the structure of functional analysis, especially in sociology and anthropology.

The functionalist mode of approach has proved highly illuminating, suggestive, and fruitful in many contexts. If the advantages it has to offer are to be reaped in full, it seems desirable and indeed necessary to pursue the investigation of specific functional relationships to the point where they can be expressed in terms of reasonably precise and objectively testable hypotheses. At least initially, these hypotheses will likely be of quite limited scope. But this would simply parallel the present situation in biology, where the kinds of self-regulation, and the uniformities they exhibit, vary from species to species. Eventually, such "empirical generalizations" of limited scope might provide a basis for a more general theory of self-regulating systems. To what extent these objectives can be reached cannot be decided in a priori fashion by locial analysis or philosophical reflection: The answer has to be found by intensive and rigorous scientific research.

NOTES

1. In developing the characterization of functional analysis presented in this section, I have obtained much stimulation and information from the illuminating and richly documented essay "Manifest and Latent Functions" in R. K. Merton's book, *Social Theory and Social Structure* (Glencoe, Ill.: Free Press; revised and enlarged edition, 1957), pp. 19–84. Each of the passages from this work which is referred to in the present essay may also be found in the first edition (1949), on a page with approximately the same number.

2. For a fuller statement and further development of this point, see Part I of the essay "A Formalization of Functionalism" in E. Nagel, *Logic Without Metaphysics* (Glencoe, Ill.: Free Press, 1957), pp. 247–83. Part I of this essay is a detailed analytical study of Merton's essay mentioned in n. 1, and thus is of special significance for the methodology of the social sciences.

3. S. Freud, *The Problem of Anxiety*, trans. H. A. Bunker (New York: Psychoanalytic Quarterly Press, and W. W. Norton & Company, Inc., 1936), p. 111.

4. *Ibid.*, p. 112.

5. Merton, *op. cit.*, p. 50.

6. *Ibid.*, p. 51. Merton defines manifest functions as those which are both intended and recognized, and latent functions as those which are neither intended nor recognized. But this characterization allows for functions which are neither manifest nor latent, e.g., those which are recognized though not intended. It would seem to be more in keeping with Merton's intentions, therefore, to base the distinction simply on whether or not the stabilizing effect of the given item was deliberately sought.

7. *Ibid.*, pp. 64–65.

8. A. R. Radcliffe-Brown, *Structure and Function in Primitive Society* (London: Cohen and West Ltd., 1952), p. 145.

9. B. Malinowski, *Magic, Science and Religion, and Other Essays* (Garden City, N.Y.: Doubleday Anchor Books, 1954), p. 90. For an illuminating comparison of Malinowski's views on the functions of magic and religion with those advanced by Radcliffe-Brown, see G. C. Homans, *The Human Group* (New York: Harcourt, Brace & Company, Inc., 1950), pp. 321 ff. (Note also Homan's general comments on "the functional theory," *ibid.*, pp. 268–72.) This issue and other aspects of functional analysis in anthropology are critically examined in the following article, which confronts some specific applications of the method with programmatic declarations by its proponents: Leon J. Goldstein, "The Logic of Explanation in Malinowskian Anthropology," *Philosophy of Science*, 24 (1957), 156–66.

10. B. Malinowski, "Anthropology," *Encyclopaedia Britannica*, First Supplementary volume (London and New York: The Encyclopaedia Britannica, Inc., 1926), p. 132.

11. *Ibid.*, p. 139.

12. B. Malinowski, *A Scientific Theory of Culture, and Other Essays* (Chapel Hill: University of North Carolina Press, 1944), p. 174.

13. Radcliffe-Brown, *op. cit.*, p. 186.

14. Malinowski, at one place in his writings, endorses a pronouncement which might appear to be at variance with this conclusion: "Description cannot be separated from explanation, since in the words of a great physicist, 'explanation is nothing but condensed description.'" (Malinowski, "Anthropology," *op. cit.*, p. 132.) He seems to be referring here to the views of Ernst Mach or of Pierre Duhem, who took a similar position on this point. Mach conceived the basic objective of science as the brief and economic description of recurrent phenomena and considered laws as a highly efficient way of compressing, as it were, the description of an infinitude of potential particular occurrences into a simple and compact formula. But, thus understood, the statement approvingly quoted by Malinowski is, of course, entirely compatible with our point about the relevance of laws for functional explanation.

Besides, a law can be called a description only in a Pickwickian sense. For even so simple a generalization as "All vertebrates have hearts" does not describe any particular individual, such as Rin-Tin-Tin, as being a vertebrate and having a heart; rather, it asserts of Rin-Tin-Tin —and of any other object, whether vertebrate or not—that *if* it is a vertebrate *then* it has a heart. Thus, the generalization has the import of an indefinite set of conditional statements about particular objects. In addition, a law might be said to imply statements about "potential events" which never actually take place. The gas law, for example, implies that if a given body of gas were to be heated under constant pressure at time *t*, its volume would increase. But if in fact the gas is not heated at *t* this statement can hardly be said to be a description of any particular event.

15. Malinowski, "Anthropology," *op. cit.*, p. 136.

16. Malinowski, *Magic, Science and Religion, and Other Essays*, *op. cit.*, p. 90. (Note the explanatory claim implicit in the use of the word "hence.")

17. B. Malinowski, "Culture," *Encyclopedia of the Social Sciences*, IV (New York: The Macmillan Company, 1931), 626.

18. Merton, *op. cit.*, p. 34. Cf. also T. Parsons, *Essays in Sociological Theory, Pure and Applied* (Glencoe, Ill.: Free Press, 1949), p. 58. For an interesting recent attempt to establish the existence of functional alternatives in a specific case, see R. D. Schwartz, "Functional alternatives to inequality," *American Sociological Review*, 20 (1955), 424–30.

19. See G. G. Simpson, *The Meaning of Evolution* (New Haven: Yale University Press, 1949), pp. 164 ff., 190, 342–43; and G. G. Simpson, C. S. Pittendrigh, L. H. Tiffany, *Life* (New York: Harcourt, Brace & Company, Inc., 1957), p. 437.

20. Malinowski, *A Scientific Theory of Culture, and Other Essays*, *op. cit.*, p. 38.

21. E. M. Sait, "Machine, Political," *Encyclopedia of the Social Sciences*, IX (New York: The Macmillan Company, 1933), 659. (Italics supplied.)

22. Merton, *op. cit.*, p. 76. (Italics supplied.)

23. Merton, *op. cit.*, p. 73. (Italics the author's.)

24. Radcliffe-Brown, *op. cit.*, p. 183.

25. Malinowski, *A Scientific Theory of Culture, and Other Essays*, *op. cit.*, p. 67.

26. Henry A. Murray and Clyde Kluckhohn, "Outline of a Conception of Personality," in Clyde Kluckhohn and Henry A. Murray, eds., *Personality in Nature, Society, and Culture* (New York: Alfred A. Knopf, Inc., 1950), pp. 3–32; quotation from p. 7; italics the authors.

27. Malinowski, *A Scientific Theory of Culture, and Other Essays*, *op. cit.*, p. 94.

28. In accordance with a practice followed widely in contemporary logic, we will understand by terms certain kinds of words or other linguistic expressions, and we will say that a term expresses or signifies a concept. For example, we will say that the term "need" signifies the concept of need. As this illustration shows, we refer to, or mention, a linguistic expression by using a name for it which is formed by simply enclosing the expression in single quotes.

29. A general discussion of the nature and significance of "operational" criteria of application for the terms used in empirical science, and references to further literature on the subject, may be found in C. G. Hempel, *Fundamentals of Concept Formation in Empirical Science* (University of Chicago Press, 1952), Secs. 5–8; and in the symposium papers on the present state of operationalism by G. Bergmann, P. W. Bridgman, A. Grunbaum, C. G. Hempel, R. B. Lindsay, H. Margenau, and R. J. Seeger, which form Chap. ii of Philipp G. Frank, ed., *The Validation of Scientific Theories* (Boston: The Beacon Press, 1956).

30. Merton, *op. cit.*, p. 52.

31. Malinowski, *A Scientific of Culture, and Other Essays*, *op. cit.*, p. 159.

32. Malinowski, *ibid.*, p. 90.

33. Malinowski, *ibid.*, p. 172; see also *ibid.*, pp. 91 ff.

34. In some of his statements Malinowski discards, by implication, even the notion of function as satisfaction of a condition that is at least *necessary* for the survival of group or organism. For example, in the same essay containing the two passages just quoted in the text, Malinowski comments as follows on the function of some complex cultural achievements:

"Take the airplane, the submarine, or the steam engine. Obviously, man does not need to fly, nor yet to keep company with fishes, and move about within a medium for which he is neither anatomically adjusted nor physiologically prepared. In defining, therefore, the function of any of those contrivances, we can not predicate the true course of their appearance in any terms of metaphysical necessity." (*Ibid.*, pp. 118–19.)

35. Merton, *op. cit.*, p. 51. (Italics the author's.)

36. Radcliffe-Brown, *op. cit.*, p. 811.

37. Murray and Kluckhohn, *op. cit.*, p. 15. (Italics the authors'.)

38. Merton, *op. cit.*, p. 40.

39. Merton, *op. cit.*, p. 52.

40. See, for example, Malinowski, *A Scientific Theory of Culture, and Other Essays, op. cit.*, pp. 169–70; but also compare this with pp. 118–19 of the same work.

41. See, for example, T. Parsons, *The Social System* (Glencoe, Ill.: Free Press, 1951), p. 29, fn. 4.

42. Marion J. Levy, Jr., *The Structure of Society* (Princeton: Princeton University Press, 1952), pp. 76 ff.

43. R. K. Merton, "The Bearing of Sociological Theory on Empirical Research" in Merton, *Social Theory and Social Structure, op. cit.*, pp. 85–101; quotation from p. 98. (Italics the author's.)

44. See G. Sommerhoff, *Analytical Biology* (New York: Oxford University Press, 1950).

45. Nagel, "A Formalization of Functionalism," *op. cit.*, p. 269. See also the concluding paragraph of the same essay (pp. 282–83).

46. Levy speaks of eufunction and dysfunction of a unit (i.e., a system) and characterizes these concepts as relative to "the unit as defined." He points out that this relativization is necessary "because it is to the definition of the unit that one must turn to determine whether or not 'adaptation of adjustment' making for the persistence or lack of persistence of the unit is taking place." (Levy, *ibid.*, pp. 77–78.)

47. For illuminating discussions of further issues concerning "teleological explanation," especially with respect to self-regulating systems, see R. B. Braithwaite, *Scientific Explanation* (Cambridge: Cambridge University Press, 1953), Chap. x; and E. Nagel, "Teleological Explanation and Teleological Systems" in S. Ratner, ed., *Vision and Action: Essays in Honor of Horace Kallen on His Seventieth Birthday* (New Brunswick, N.J.: Rutgers University Press, 1953); reprinted in H. Feigl and M. Brodbeck, eds., *Readings in the Philosophy of Science* (New York: Appleton-Century-Crofts, Inc., 1953).

48. Freud, *op. cit.*, p. 112.

49. Merton, "Manifest and Latent Functions," *op. cit.*, pp. 23–25, 60 ff.

50. L. Gumplowicz, *The Outlines of Sociology*, trans. F. W. Moore (Philadelphia: American Academy of Political and Social Science, 1899), pp. 79–80.

51. Malinowski, *A Scientific Theory of Culture, and Other Essays, op. cit.*, p. 150.

52. Malinowski, "Anthropology," *op. cit.*, p. 133.

53. Merton, "Manifest and Latent Functions," *op. cit.*, pp. 30 ff.

54. *Ibid.*, p. 31.

55. Malinowski, *A Scientific Theory of Culture, and Other Essays, op. cit.*, p. 150.

56. An illuminating general account of this kind of approach to homeostatic processes in the human body will be found in Walter B. Cannon, *The Wisdom of the Body* (New York: W. W. Norton & Company, Inc.; rev. ed. 1939).

57. See, for example, J. Dollard and N. E. Miller, *Personality and Psychotherapy* (New York: McGraw-Hill Book Company, Inc., 1950), Chap. xi, "How symptoms are learned," and note particularly pp. 165–66.

58. Nagel, "A Formalization of Functionalism," *op. cit.* See also the more general discussion of functional analysis included in Nagel's paper, "Concept and Theory Formation in the Social Sciences," in *Science, Language, and Human Rights;* American Philosophical Association, Eastern Division, I (Philadelphia: University of Pennsylvania Press, 1952), 43–64. Reprinted in J. L. Jarrett and S. M. McMurrin, eds., *Contemporary Philosophy* (New York: Henry Holt & Co., Inc., 1954).

59. Sommerhoff, *op. cit.*

The Logic of Historical Explanation

W. B. Gallie

EXPLANATIONS IN HISTORY AND THE GENETIC SCIENCES

I

The claim that historical study gives rise to a unique kind of understanding, or at least to understanding of a very different kind from that afforded by the natural sciences, is sometimes advanced on grounds and with intentions which no one will wish to dispute. Thus no one denies that history is concerned with particular facts, with what actually happened on this or that particular occasion, whereas the natural sciences are concerned with specimen or sample facts and with what is sure or likely to happen on *any* occasion of given definite description; or, in case this statement should seem to neglect the historian's concern with generality in one sense of that term, we might say that his interest is in virtually any of the conditions and implications of some particular event that fall within that event's actual historical context, whereas the scientist's interest is in implications and conditions that will hold in any conceivable context in which an event of definite description may occur. This contrast in direction of interest can quite properly be described as resulting in different kinds of understanding in the two cases. At other times, however, the claim that historical understanding is unique is urged on grounds that, far from being platitudinous, call for careful philosophical scrutiny; and this is true in particular when it is argued that the general concept of explanation, as ordinarily elucidated by philosophers, is not adequate to the kind—or at least to a certain kind—of explanation which we find exemplified in historical writings. Now this claim seems to me justified; but the form in which I wish to defend it is so moderate that few of those who urge the uniqueness of historical understanding are likely to thank me for my pains. What I have to say affords no support to so-called "intuitionist" theories of historical explanation and understanding: It is much

Reprinted by permission of the editor of *Mind*, where this article first appeared in 1955.

nearer to the view, which has been expounded by Professor Popper and others,[1] that any causal argument can be regarded as historical in so far as it is applied to some particular event; only, where that view emphasizes that, theoretically, *any* kind of causal argument can serve to explain historical events, I wish to show that one kind of causal argument is peculiarly characteristic of historical explanation.

Historians, I shall argue, sometimes explain events in a perfectly good sense of "explain" by referring us to one or a number of their temporally prior *necessary* conditions; they tell us how a particular event happened by pointing out hitherto unnoticed, or at least undervalued, antecedent events, but for which, they claim on broadly inductive grounds, the event in question would not or could hardly have happened. In such cases explanation commences from our recognition of the event to be explained as being of such a kind that *some one* of a disjunction of describable conditions is necessary to its occurrence; and the explanation consists in elucidating *which one* of this disjunctive set is applicable, in the sense of being necessary, to the event in question. Explanations of this broad pattern are, of course, found in other branches of inquiry, e.g., as a first step towards functional explanations; but in so far as conclusions drawn from the assertion or supposal of *necessary conditions only* are found in the developed sciences, they are usually regarded as partial or interim results, to be replaced ultimately by conclusions of arguments of more complete, and by prevailing standards, more satisfactory kinds. Thus we often come to see in a provisional way how the working of one part of a machine is necessary to the working of the whole, whose general character we thus begin to appreciate before we have fully grasped the mechanically relevant properties of all its working parts; ultimately or ideally, however, we should come to appreciate the function of any one such part as a necessary deducible consequence of the actual movements of all the parts considered together. Roughly similar remarks would seem to apply to the use of functional explanations in the biological sciences. In history on the other hand, I wish to claim, explanations in terms of temporally prior necessary conditions are commonly put forward when there is no good ground for accepting—and when indeed there is no consideration of—further explanation of a more complete, and in particular of a predictive, character.

Suppose a historian is asked to explain how a certain statement came to be made, or deliberate action to be taken, or coherent policy pursued. It would be perfectly natural to say that he has explained, e.g., the statement, when he has discovered or inferred the kind of question—or comment or threat or taunt—that evoked it. But in offering this kind of explanation the historian would not necessarily be claiming, and indeed is not likely to be claiming, that some ideally clever person could have predicted the making of the statement given the occurrence of the question; his claim would more naturally be taken to mean that but for the question's having been put—or but for some other of a disjunction of describable conditions—the statement would remain unintelligible in the sense of lacking an appropriate historical context. The predictive explanation in this case, although it would support the same conclusion, would be a further and quite different inference or explanation, of a different logical pattern, and resting upon partially different evidence. It is, to be sure, no part of my thesis to deny that explanations of predictive pattern are ever found in historical narratives. Of course they are found there; of course for the historian, as for the rest of us, to explain an event very often means to show how it could have been predicted, or how it exemplifies some, perhaps very vague, predictive rule, e.g., "Power corrupts" or "You can do everything with bayonets except sit on them." What I am urging is that, side by side with such predictive explanations, supporting and supported by them, we also find in histories explanations of the kind which I have just outlined, explanations which in their own way do explain, and which for convenience I propose to call "characteristically historical explanations." My main question in what follows is, why are such characteristically historical explanations in their own way perfectly satisfactory?

I shall try to show that there are at least two main grounds for this, of which the second can be regarded as a highly specific form of the first. Summarized very briefly, my thesis is as follows: (1) In pointing back to certain temporally prior conditions, alleged to be necessary to the event to be explained, characteristically historical explanations emphasize either a continuity in direction of development or else a persistence of certain elements, within a particular succession of events. (2) But, in the second place, the continuity or persistence of elements which a characteristically historical explanation emphasizes may be of a kind which serves to render the *explicandum*—when it is some human action or sequence of actions—intelligible or justifiable. Now where the first of these grounds holds but not the second, explanations of the type we are considering are not confined to history proper: They are equally characteristic of the genetic sciences, e.g., of much of biology, geology, and of the social and psychological sciences. From one point of view, therefore, the problem or puzzle about characteristically historical explanations is simply one instance of the wider problem of genetic explanations. It is in terms of such wider questions as, "Is there such a thing as a characteristically genetic explanation?", "If there is, what distinguishes such an explanation from other forms of causal explanation?", and "What makes us perfectly satisfied with a 'good' genetic explanation when we get one?" that I advance a large part of my argument (section II below). But this is by no means the whole story. The existence, the puzzle, the problem, the interest of characteristically historical explanations are due also to the way in which they make use of such notions as purpose, motive, belief, understanding, communication, etc., notions which are relevant presumably to human actions alone and about whose applications there are notorious philosophical difficulties; and I shall try to show that my account of characteristically historical explanations throws some much needed light on the "logic" of these notions, and on some of the difficulties which philosphical discussion encounters in connexion with them. Roughly, my claim is that these difficulties are due in large measure to the fact that notions such as purpose, motive, belief, understanding, communication, etc., figure *both* within characteristically historical *and* within predictive explanations, and that the logical force of any one of these notions inevitably differs according as it is taken as occurring in one sort of explanation or the other (section III).

II

Is there such a thing as a characteristically genetic explanation? We may usefully approach this question by recalling the account of genetic explanation which has usually been favoured by philosophically-minded biologists during the last ninety-odd years. Serious discussion of genetic explanation was stimulated by the gradual recognition that Darwin's account of evolutionary change through the agency of natural selection was very far from affording a complete causal explanation of how species had developed from a presumed common ancestor; the most that Darwin's argument had established was one very general necessary condition to which changes (or for that matter absences of change) in any given species must conform if that species is to survive. An ideally complete causal explanation of any evolutionary change, it came to be realized, would involve, first, a genetic statement—a detailed account of the ancestry and ancestral environments of any particular form of life, and secondly, a causal statement, i.e., a statement of certain universal laws, conceived after the model of known physical and chemical laws, in terms of which passage from one definitely described form of life to another might always and everywhere be inferred. As against the Darwinians it was necessary to emphasize the importance of the hoped-for causal or physico-chemical statement; but, as the prospect of a developed biochemistry grew brighter, emphasis came to be placed rather on the indispensable positive role of the genetic statement. Thus, supposing that, through the dispensation of a God who loved chemists, biochemistry was able to produce under artificial conditions all the forms of life that are known to have existed at any time on or near the earth's surface: Then, in a sense, "the riddle of life" would have been explained; the dream of a complete causal explanation of all known forms of life would have come true. But this triumph or miracle of causal explanation would not necessarily answer the question how *historically* some given form of life first arose, or how *historically* it came to undergo this or that change in view of this or that pressure or opportunity within its successive environments. Indeed, not only the incidents of that history, but the particular combinations of causal laws which were relevant to those incidents, might still remain to be disclosed.

Unfortunately, this latter point has never, to my knowledge, been developed by philosophers of biology, who have persisted in regarding genetic explanations (which would go back, ideally, to "primordial causes") as purely *descriptive*. This way of thinking is illustrated in the following passage from Professor Woodger's *Biological Principles*. "The tortuous course taken by a nerve or artery may strike us as strange and we seek an explanation . . . and we find one if it can be shown that in the course of individual development this state of affairs has been brought about by the shifting of neighbouring parts. This seems to be what is meant by an embryological explanation which is 'merely descriptive' .... If now we ask why such shifting of parts takes place we can investigate the racial forerunners of the organism in question. From the comparative anatomical and palaeontological data there may be good reason for supposing that the condition in the animal or plant in question has been reached through an evolutionary shifting of the parts. In this way we should reach a phylogenetic explanation: an historical explanation which *could not be generalized because it would describe a unique series of changes characterising an evolutionary succession*" (my itals.).[2] Professor Woodger then proceeds to contrast this kind of explanation with the kind of causal explanation which physiology might here contribute; and he concludes, "What distinguishes the physiologist's procedure is the fact that he does not record changes normally observable, but the changes which are observed when the organism or its environment is systematically interfered with. He is thus able to discover more than the descriptionist; he is able to investigate the mutual internal dependence of the parts and the role they fulfil in pervasive types of change, i.e., types of change *which can be generalized*" (my itals.).

From this account, we might easily be led to believe that the former "genetic" explanation has nothing causal about it. But that there is a causal reference latent in this explanation is at least very strongly suggested by the sentence, "From comparative anatomical and palaeontological data there may be good reason for supposing that the condition . . . has been reached through an evolutionary shifting of the parts." Quite clearly this statement amounts to an application of an observed general regularity to the particular case of change which the biologist is seeking to explain; and as such it has—anyhow on the currently prevalent view of causation—every right to be accounted a causal explanation, albeit an evidently incomplete one. The same point could be urged, perhaps more forcibly, as follows. Suppose the presumed earlier phases of the process of shifting had had *no* causal relevance whatsoever to the later result; then reference to these phases would be as logically redundant as a reference to, e.g., the earlier positions of a shadow whose present position is being explained in terms of the sun's movement and the fixed position of the object which is casting the shadow. But does Professor Woodger want us to believe that there really is such a parallel between the two cases?

The truth seems to be that there is an important contrast between the two kinds of explanation which Professor Woodger, like a number of earlier philosophers of biology, has distinguished; but this contrast is not between explanations that are in causal and explanations that are in strictly non-causal terms. How then should it be expressed? It is to answer this question that we require the notion of a "characteristically genetic explanation," and the answer I wish to urge is that such an explanation, as exemplified in an embryological or phylogenetic explanation, functions in a fashion that is logically parallel to that of the characteristically historical explanations which we discussed provisionally in section I above. The first prerequisite of a characteristically genetic explanation is that we shall recognize the *explicandum* as a temporal whole whose structure either contains certain persistent factors or else shows a certain definite direction of change or development. Thereupon we look for an antecedent event, the *explicans,* which can be accounted a necessary condition of the explicandum, on ordinary inductive grounds (observations of analogous cases), but more specifically on the ground of a probable continuity—in respect either of persistent factors or of direction of change—between explicans and explicandum. It just happens to be the case that evidence of the kind required for explanations in this sense is repeatedly found in certain fields, e.g., in rock-strata, fossil-series, the recorded observations of embryologists, natural historians, anthropologists, genetic psychologists and others. In the simplest cases, e.g., many rock-strata or the observed and reported sequence of development of a fertilized ovum, a characteristic order is repeatedly presented to observation; but in many other cases—those in which the need for characteristically genetic explanation is most obvious—the order has to be inferred: e.g., the strata have been inverted or certain aspects of an ovum's development are such that they cannot all be observed in any one particular instance. In general, causal explanation of the characteristically genetic kind passes insensibly in certain cases into a mere assertion of the persistence of some factor or of the existence of some, as yet unexplained, trend in an observed or recorded sequence of events. More commonly, however, the assertion that a later phase of a genetic sequence requires an earlier phase as one of its necessary conditions is explicitly inferential, in the sense both that we are led to assert it through a consideration of the available relevant evidence, and, more important, that it is this evidence which justifies our assertion of it.

The function of characteristic genetic explanations, it should be noted, is by no means confined to the tracing out of single causal lines. It takes only a little reflection to see that the process of relative dating of different series of living forms, rock strata, etc., depends for the most part on our knowledge simply of antecedent necessary conditions of the occurrence of certain observed or reported events. The geologist's knowledge that certain rocks in Scotland and others in Wales date from approximately the same period does not depend on any theories he may entertain as to the sufficient conditions of their formation, and similarly as regards the dating of human affairs. Too often in their discussions of historical and genetic explanations postivistically-minded philosophers appear to assume that the broad chronological pattern of past events is something that is given to the historian or genetic scientist ready-made, or else that it is something that is deduced from certain, miraculously known, initial conditions with the aid of established predictive laws.

We can now bring together our findings as to the need, functions, and limitations of characteristically genetic explanations as follows:

(1) A characteristically genetic explanation seeks to establish, or at least helps to indicate, some kind of continuity, between one or a number of temporally prior conditions and a subsequent result.

(2) On the other hand, a characteristically genetic explanation does not pretend to predictive power: The prior event is not taken, in conjunction with certain universal laws, to constitute a sufficient condition of the occurrence of the subsequent event.
(3) Moreover, a characteristically genetic explanation emphasizes the one-way passage of time—what came earlier explains, in the genetic sense, what came later, and not vice versa. In other words the prior event is not taken, in conjunction with certain universal laws, to constitute both a sufficient and a necessary condition of the occurrence of the subsequent event.

Let us assume that satisfaction of these three conditions defines a characteristically genetic explanation: Then two important consequences follow:

(a) A characteristically genetic explanation cannot, in all cases, be considered as simply the converse of a retrodiction, i.e., any inference from effect to prior cause. In any such inference, to be sure, since the effect is taken as sufficient to establish the occurrence of the cause, the latter must be considered—as in characteristically genetic explanations—as a necessary temporal prior condition of the former. But we must remember that the term "retrodiction" is commonly and properly taken to include the kind of case (e.g., calculation of earlier positions and accelerations of a freely swinging pendulum) in which the "cause" or prior event to which we infer is seen to be a sufficient as well as a necessary condition of the effect or later event as described in our premiss. In other words, "retrodictions," as the term is commonly used by logicians, include the kind of case in which we can equally well deduce cause from effect and effect from cause; and this is the kind of case which our condition (3) above expressly excludes. But further, in the second place, the term "retrodiction" is commonly used to cover, *inter alia*, inferences from effects to causes, in which there is no claim to establish continuity of any kind between cause and effect; and such inferences are, of course, excluded by our condition (1) above.

(b) Still more obviously, characteristically genetic explanations do not conform to that conception of explanation which insists—or simply assumes—that an earlier event can be said to explain a later event if and only if the former, taken in conjunction with certain universal laws, provides us with sufficient grounds for deducing the occurrence of the later. This assertion[3] is indeed tantamount to a denial that characteristically genetic explanations explain in any genuine sense at all. But we have seen that in their own way, and in a perfectly normal and reasonable sense of the word, characteristically genetic explanations do explain—do establish conclusions which it is very important to know, if only because they commonly provide the premisses—the instantial or historical premisses—of further explanations of predictive pattern.

To assert the existence of characteristically genetic explanations, or, in other words, to attempt to describe their peculiar features and functions with some care, is therefore highly pertinent if only in the interests of logical theory. But in fact, our arguments to this end have a much more directly practical value in that they help us to understand the prevalence in genetic studies of a brand of fallacy which, had not the phrase already been appropriated for an important purpose, might well be termed the "genetic fallacy" par excellence. Broadly, this kind of fallacy arises because genetic scientists, adhering consciously or unconsciously to the view that only predictive or "sufficiency" explanations genuinely explain, are temped to believe, when they hit on a perfectly good characteristically genetic explanation, that it either must be, or ought to be, or anyhow—if valid and useful—can easily be translated into, an explanation of predictive or quasi-predictive pattern. To cite one well-known instance:

We have all been told that the species giraffe has arisen through a succession of longer and longer necked species because, at every stage in this sequence, possession of unusually long necks gave the species in question certain advantages, e.g., as regards food-getting, and thus enabled it to survive and leave descendants whose inherent variability allowed for further changes in the same general direction. But what is the force—or the legitimate interpretation—of the "because" in this explanation? If it means simply that the giraffe and the intermediate species would probably not have survived unless they had possesed certain advantages connected with the possession of long necks, then it is an extremely plausible conjecture. There can be no doubt but that this is a conjecture which really explains: It enables us to pass from the relatively platitudinous assertion that the species giraffe could not have arisen except from the ancestry which, on grounds of continuity, it would appear to have had, to the assertion that unless its ancestry had displayed certain characteristics which are specifically mentioned, or unless this ancestry had conformed in certain specified ways to the general requirement of competition for survival, the species giraffe would not exist today. But if, on the other hand, it is intended—as it often has been in, anyhow, the popular literature of the subject—that possession of progressively longer and longer necks by the giraffe-ancestors suffices to explain or even contributes in a notable way to some quasi-predictive explanation of the ultimate emergence of the species giraffe, then the argument is utterly fallacious. For one thing, as Professor Woodger has pointed out in discussing this very example, no one in fact knows the detailed characters of the environments within which possession of progressively longer and longer necks is supposed to have been advantageous (*op. cit.* p. 401). But perhaps more pertinent is the fact that variations in this respect must in all cases have called for a vast number of minute adjustments in other organs, without which this variation—or rather this more or less single direction of a great number of successive variations—would have proved, and perhaps in many cases did prove, disastrous rather than advantageous. The truth is that in this kind of spurious "sufficiency" explanation *everything* is assumed that needs to be assumed to get our alleged explanation to do its appointed work.

The same fallacy arises whenever economic or sociological generalizations, which enable us to make valid predictions within a known institutional framework, are applied to past civilizations whose institutional framework remains largely unknown. It is thus, to take one extreme case, that Marxists have exploited our sheer ignorance of "pre-history," and have succeeded in "making sense" for us of remote ages for which the sum of our historical evidence lies, say, in the traces of successive irrigation systems or the survival of a few buried metal tools.

To put the responsibility for reasonings such as these upon the shortcomings or one-sidedness of our inherited inductive logic would, of course, be unjust. What is true, however, is that this one-sidedness reflects, and has unfortunately appeared to endorse a habit of thought which is highly characteristic of our age, viz., that which looks to the physical, to the almost total neglect of the genetic sciences, for examples and models of successful explanation.

III

Characteristically historical explanations can be described in the first instance as such explanations of the deeds of men (individuals, groups, nations, etc.) as conform to the logical pattern and evidential conditions which define characteristically genetic explanations. As with characteristically genetic, so with characteristically historical explanations: There are some in which the antecedent is among the facts already known to or accepted by the historian, and others in which it has to be inferred. Cases of the latter sort are no doubt more characteristic of history, since they include the great majority of those in which the emphasized necessary antecedent is a motive or a belief or a decision or a communication received or a principle or policy or precept adhered to by some agent; and certainly such explanations involve important features and difficulties of their own. But for the purpose of establishing and articulating the general function of characteristically historical explanations, it will be best to concentrate at the outset upon cases of the former sort, i.e., those in which the emphasized antecedent necessary condition is a fact already known. A live example will serve to underline my central thesis here.

Suppose we want to explain the two already known and presumably closely connected facts, first, that the personal expectations of the Christian disciples after the death of Christ were turned in a remarkably short time into a proselytizing religion, and second, that this religion was spread in a remarkably short time over the length and breadth of the Mediterranean world. M. Alfred Loisy has suggested that one and the same explanation largely accounts for both of these facts, viz., that the Christian "good news" was voiced over the Mediterranean world largely on a platform already provided by the Jewish synagogues, which were themselves the vehicles of a fully established and actively, if not very successfully, proselytizing religion. Thus it turned out that the first Christian emissaries found themselves with audiences ready made and presumably well suited to receive their news—in particular of the fulfilment of the Messianic promise—with joy and acclamation. Now, without scrutinizing the many subordinate points which this explanation serves to clarify, and without wishing to claim that it is altogether historically acceptable, we can at least affirm that all that Loisy has here done is to point out one very important and necessary condition of the result to be explained; certainly he has not established—and indeed would not claim to have established—that the connection of facts to which he has called attention *suffices* to explain in the predictive sense how, given the antecedent conditions which he has emphasized, the known result could have been predicted or deduced.

With this example in mind we can see more easily and can express in homelier terms what is lacking in the account of historical explanation which Professor Popper and others have favoured. Because they identify the field of explained history—or reasoned narrative—with the field of application of any number of scientific and commonsense generalizations of predictive character, they fail to notice the way in which many historical explanations help to "thicken up" the historian's narrative by bringing out the continuity between its successive phases, and at the same time to "tighten up" the narrative by emphasizing the dependence—though not the predictable sequence—of its later upon its earlier phases. As a corrective to this mistake, it is reasonable enough to insist, in a currently popular phrase, that the main job of the historian is simply to tell his story: But to this we must add that to tell or to follow even the simplest story is not just to assert or to accept "one dammed thing after another"—a brute sequence of temporally and spatially continuous or overlapping events. To follow a story—or a conversation, or a game, or the development and execution of a policy—involves for one thing some vague appreciation of its drift or direction, a vague sense of its alternative possible outcomes: But much more important for our purpose, it involves a relatively clear appreciation of certain relations of dependence of the sort that characteristically historical explanations serve to articulate, i.e., an appreciation of how what comes later depends upon what came earlier, in the sense that but for the latter the former could not have, or could hardly have, occurred in the way it did occur. Consider, e.g., what we do when a child complains that he cannot follow the story we read aloud to him, or when we ourselves are unable to follow a conversation which we have broken in upon half way through. We re-read to the child the earlier stages of the story, or re-tell them in simpler language so as to emphasize those incidents which give sense or context to the present, puzzling episode. But in doing this we do not try to show that the present episode was a predictable consequence of earlier events, else the story would have been not unfollowable, but just unbearably dull as a story. Similarly in our second example we ask our friends how the conversation started, what considerations, examples, etc., led to its present juncture; but in doing this we do not claim that the present state of the discussion could have been predicted from its origins, since the flow of conversation of intelligent adults is never, or anyhow very seldom, wholly predictable.

This account of the logic of *following* a narrative or a discussion can equally well be applied to the case of *understanding a particular statement*—in contrast to our understanding of a grammarian's specimen sentence—or for that matter to our understanding of a single intelligent action. An essential part of what we mean by our understanding of a statement is that we appreciate the kind of prior statement or question, whether actually articulated or not, which is required to give it a sense or context but which is most unlikely to afford a sufficient explanation of the original statement's having been made. Similarly, to understand an intelligent action involves the ability, not indeed to have predicted its occurrence from certain earlier events, but to recognize among these earlier events a disjunction of conditions such that some or one of them was required if the action in question was to be of the sort, of the intelligent and intelligible sort, we claim it to have been.

We are now in a position to consider those characteristically historical explanations in which the explicans refers to some motive or purpose or belief or information received, that is taken to be a necessary condition of some recorded overt event or deed. Notoriously, explanations of physical results in terms of mental antecedents have given rise to prolonged philosophical perplexity and debate; and our account of characteristically historical explanations might seem particularly questionable on this score since it emphasizes the need of some kind of continuity between antecedents—which in the cases now to be considered will be mental antecedents—and the physical results which they explain. But what kind of continuity, unless in respect of sheer temporal succession, can be observed or inferred between, say, the (mental) acceptance of some proposition and the performance of some appropriately related physical action? Can we always observe or infer a temporal continuity in cases of this kind? Is there not very often an observed temporal gap between, e.g., decision and deed?

It is useful to notice, as a first step in dealing with these difficulties, that they are specifically philosophical difficulties: The historian is virtually insensitive to them. Let us imagine ourselves, as philosophers, approaching a historian with the questions which we have just raised, and pressing him to tell us precisely what he means by the continuity which he seeks to disclose between this or that physical happening and its presumed necessary mental antecedents. Our historian, if so approached, will almost certainly hover between two lines of reply; one modest and negative, the other positive and self-confident. He will confess that of course he does not know precisely what he means by the continuity which he habitually seeks to establish: After all, it is not his business to distinguish carefully between the different possible uses of the word "continuity," still less to pronounce on very abstract questions concerning the continuity or lack of continuity between the mental and physical aspects of human action. But in another mood he will rally and protest that he, like everyone else, knows perfectly well how to distinguish a sequence of events in which there is a continuous passage between, say, appreciation and action, from a sequence which is at some point interrupted, frustrated, rendered discontinuous. He will tell us that in his search for continuity in some sequence of events, he is simply working on the hypothesis that these events make up one intelligible whole, recognized by either the persistence or the continuous development of certain of its elements and within which at every stage what happens could not or could hardly have happened but for what had just gone before.

To bring out the substance of this answer by an example, let us suppose that a General has received intelligence, the contents of which help us to some extent to make sense of his subsequent decisions and orders. How does the idea of continuity enter into this characteristically historical explanation? Well, the following things seem either undeniable or very reasonably inferred. The General's physical receiving of the message was a necessary condition of, and in some sense continuous with, his physical process of reading it; this process was a necessary condition of and in some sense continuous with his understanding of it; this process—if the word *process* can here be allowed—was a necessary condition of and in some sense continuous with his acceptance of the veracity and practical relevance of its contents; this process—again if the word *process* be allowed—can reasonably be inferred to have been a necessary condition of and in some sense continuous with his decision to give and his actual physical giving of the spoken order; this in turn was almost certainly a necessary condition of, and in some sense continuous with, his army's deployment, the attack, etc. But in treating any such episode as a continuous and intelligible whole the historian does not indulge in any philosophical theorizing as to the nature or grounds of the continuity which he assumes to hold between, for instance, reading and understanding, understanding and assenting, assenting and deciding, deciding and the final issuing of a command; nor does he—or should he—as a historian indulge in any metaphysical speculations as to the General's intervening state of mind assuming that there was some temporal gap between, e.g., his acceptance of the veracity of the report and his decision to act in the light of it.

To be sure, the continuity which is essential to characteristically historical explanations is not always of the direct and obvious kind which we have here examplified. Thus an action may perfectly well be explained in terms of the *mis*understanding of a message, or as the result of someone's *dis*obeying an order, or through the influence of a *conflict of* motives or through a partial *failure* of nerve or concentration in the carrying out of a decision. But such complicated cases, it would seem, do not necessitate any revision of our general account of characteristically historical explanation. The motive, no matter how complex or confused, the understanding or decision, no matter how imperfect, which explains in a characteristically historical explanation, is always conceived as a necessary antecedent condition of the action which follows, and as in certain specifiable ways continuous with it.

But even if it be granted that the explicans in a characteristically historical explanation is always suggested in the first instance on the kind of grounds that I have just described, it may still be objected that if our explanation is to be accepted it must enable us to predict certain results which can then be verified independently: For instance, any plausible characteristically historical explanation of some recorded fact must enable us to make predictions of a kind that other documents or archaeological evidence can confirm. But this is by no means the only way in which a characteristically historical explanation can receive confirmation: It can be said quite properly to be confirmed if and in so far as it helps us to explain *in characteristically historical fashion* facts other than those which it was first advanced to explain. To give an extremely simplified example, suppose I hear someone shout down a telephone and infer from this that the speaker at the other end has complained that he cannot hear: This explanation would clearly be strengthened if I were to notice other facts which could be explained in characteristically historical fashion on the basis of it, e.g., if I noticed that the speaker proceeds to spell out letter by letter certain words, to use very simple familiar words, and so on. In general, verification of a characteristically historical explanation is very commonly of the kind which defenders of the coherence theory of truth have laboured to describe—and have claimed quite unjustifiably to be the necessary and sufficient condition of the acceptability of *any* explanation; and in this connexion it is worth recalling what has often been remarked, that defenders of that theory seem to have elaborated their views with the problems of historical rather than of scientific explanation in mind.

There remains for consideration the important fact that the notions of purpose, motive, belief, etc., figure not only in characteristically historical explanations but in explanations of historic events that are essentially predictive. A historian, like a jury, may conclude that given certain circumstances the existence of a certain motive or the declaration of a certain purpose or even the receipt of certain information suffices to explain some action, which is therefore conceived as having been at least theoretically predictable. The question naturally arises: Do not such predictive explanations provide the model or ideal of all historical explanation in terms of motive, purpose, belief, etc., so that, anyhow in theory, the historian is committed to maintaining the general predictability of all intelligent or justifiable human conduct? This ideal, it may be admitted, is not even suggested in our uses of the notions of motive, purpose, etc., in characteristically historical explanations; but, it may be urged, this is simply because of the palpable incompleteness of these explanations, which the historian is indeed forced to make use of, but only *fautre de mieux*, because of the paucity of his evidence. The natural hope and aspiration of all forward-looking historians must be that this situation will gradually be altered, not only by further accumulation of historical evidence, but as a result of the findings of psychologists, sociologists, and others. The old-fashioned historian, explaining events for the most part in terms of their at least necessary conditions, had perhaps little reason for affirming the general predictability of human conduct; but the modern historian, who comes to his task armed with the predictive explanations of psychology, sociology, economics, and so on, is in a quite different case.

It would be tedious to reapply in detail here the arguments by which we countered this claim, made with regard to the theoretical replaceability of all characteristically genetic explanations, in section II above. Predictive explanations for the most part neither affirm the necessity of their antecedents to their consequents nor emphasize any continuity of direction or development between them; whilst those predictive explanations which satisfy these conditions satisfy them at too high a cost—they present antecedent and consequent as mutually determining, i.e., either is deducible from the other, given the postulates and definitions of the system to which they belong; and there is no good reason for thinking, or indeed for hoping, that all the generalizations of the social sciences will eventually be expressible in this way, as theorems of deductive systems. Finally, there are two lessons in practical logic which need to be emphasized in connection with the claim which we are here rebutting. (1) Whilst explanations of predictive pattern are naturally attractive to the historian on the score of explanatory completeness, they are just as naturally suspect to him on the score of accuracy and reliability. The reflective historian knows—in his bones even if he lacks the terminology needed to express it—that any motive-explanation of characteristically historical pattern which he offers for some action is likely to be far more reliable than the related motive-explanation which claims predictive power: e.g., to inter that a wounding word would not have been spoken had not the speaker been jealous of another is almost always easier—and safer—than to claim that the utterance of it could have been predicted from the speaker's known jealous state of mind. (2) There is precisely the same danger in history proper as within the genetic sciences of the kind of "sham sufficiency" explanation which we exposed on page 157 above. The fact that a certain kind of motive provides us in a large number of instances with explanations of a characteristically historical pattern is by *itself* no warrant for the conclusion that by generalizirg these instances we can obtain, or even come within reasonable hope of obtaining, an explanation possessing predictive power.

The upshot would therefore appear to be that historians will continue to employ characteristically historical explanations side by side with others of predictive pattern so long as history continues to be written. Yet, it would be unsatisfactory to leave the matter on this somewhat negative note; for it seems to me that the situation in which the reflective historian finds himself—the fact that he is at once drawn to and mistrustful of the ideal of explanatory completeness which the predictive generalizations of the social sciences seem to offer—may well help to illuminate a number of the most central and stubborn problems of philosophy. In particular the historian's situation has a very marked relevance to the age-old problem of the accountability and/or predictability of intelligent human conduct, since it enables us to regard this problem as a necessary consequence of the fact that there are *two* distinct ways—the predictive and the characteristically historical—in which we can be said to explain or to offer a reasoned account of any human action, and that the second of these ways is perfectly compatible with the continued use of the vulgar notion of accountability. Moreover, this way of regarding the predictablity/accountability problem has the merit of rebutting the familiar criticism of the commonsense libertarian thesis, viz., that this thesis leaves "real choices" entirely unexplained in any acceptable sense of "explain"; for if our account of characteristically historical explanations be accepted it is perfectly possible for the intelligibility or rationality of an action to be explained, without any claim being made as to its actual or theoretical predictability.

To generalize from this example, I would urge that the disinction I have drawn between predictive and characteristically historical explanations provides philosophers with a valuable guide in their researches into the inner logic of "motive-," "purpose-," "belief-," and other mental words. The line of research which I have in mind can perhaps be indicated by the following questions: (1) Can it be plausibly maintained that we learn to use motive- and other mental words in the first instance not to predict but to explain in something like characteristically historical fashion certain observed physical actions? (2) If so, when motive- and other mental words come to be used for elementary predictive purposes also, how are their original meanings affected?—in the way of extended connotation, of greater precision, or of greater vagueness? (3) Assuming that any one clear answer can be given to this last question, does this answer hold good of our familiar mental words when they come to be employed for the relatively stringent predictive purposes of psychology—when they are subordinated to the overriding quest of "the sufficient stimulus"? Researches conducted along this line of questioning might, I think, give some much-needed shape to those discussions of the logic of mental words which bulk so large in contemporary philosophy.

IV

In conclusion I should like to indicate very briefly the bearings of my account of characteristically historical explanations on two long-vexed problems in the philosophy of history: (1) the nature of the historian's interpretation of his documents, and (2) the kind of practical wisdom that can justly be attributed to the study of history.

(1) When a historian or epigraphist interprets a document he always looks, *inter alia*, for a characteristically historical explanation of it. This search may proceed at any of the following stages: (a) The historian has to satisfy himself that what he has before him is a meaningful document. This involves not only that the shapes he is presented with are formed and ordered in accordance with the conventions of some known language, and not only that the words and sentences he reads have sense in that language, but also—and this is what distinguishes the historian's from the linguist's interpretation—that this sense could not have belonged to the words unless they had been inscribed by someone who intended them to convey a recognizable meaning. (b) He must satisfy himself, on either internal or external evidence, that this meaningful document is in part at least veridical in intention, i.e., that it could not have been produced except by someone wishing to record some actual (or believed to be actual) happenings. (c) He must satisfy himself, usually on a combination of internal and external evidence, that his document is also, in some degree, trustworthy, i.e., that it could not have been produced except by someone relatively well informed about the happenings that are described. Other stages or aspects of the historian's interpretation involve explanations of predictive type; e.g., he must satisfy himself that the material inscribed is of such a kind that it *could* date from the alleged time of inscription, and so on.

In at least these respect the document from which a historian usually, though by no means always, begins his reconstruction of the past, calls for characteristically historical explanation in precisely the same sense as do the events which that document is presumed to describe. Hence, it may, on certain occasions, be difficult to decide whether a historian is seeking to explain in characteristically historical fashion some statement in his document or some past event which his document claims to have taken place. But this is by no means the typical case. Usually, the historian knows perfectly well when he is pondering over the character—as revealed in the characteristically historical explanation—of his document, and when he is pondering over the character—as revealed in the characteristically historical explanation—of some past event which is the primary object of his interest and study. There is, therefore, some slight excuse for the mystification which idealist philosophers of history have spread around the historian's task of interpretation. But there is no excuse for the way in which they have apparently delighted to jumble together things that are for common sense perfectly distinct—the known past and the historian's present activity, the reasonableness or justification of some past deed, and the reasonableness or justification of the historian's assessment of it.

(2) Certainly historical knowledge contributes indirectly to practical wisdom in as much as many—and in a wide sense all—of our predictive generalizations are based upon it. But many historians would claim that, irrespective of the practical guidance that can be obtained from the predictive generalizations of the social sciences, a "sense of history" is indispensable to the wise conduct of all human affairs; and I think my account of characteristically historical explanation helps us to see what is valid and important in this claim. Consider the kind of case in which a long series of actions is explained in characteristically historical fashion by reference to a continuously developing policy or to some complex movement of intellectual or religious ideas. Clearly such an explanation may contribute to the way in which the historian characterizes the whole period within which the actions in question fall. Equally clearly, where such an extension is justified, the policy or movement referred to will have contributed to the ways in which men actually thought and acted during the period: Their practical appreciation of its value and indeed of its existence will have been a necessary condition of the ways they regarded their particular problems, and took and kept their bearings in meeting these problems. Here, I think, we have a clue to the kind of practical wisdom for which some knowledge and sense of history may rightly be claimed to be indispensable. This wisdom is naturally described as "conservative," since it serves as a check to the radical experimentalist spirit (the spirit of Professor Popper's "Social Engineering"), which where given a free hand may indeed supply an effective solution to this or that practical problem, but may well do this at the cost of utterly transforming the general pattern of living within which these particular problems arose. Looked at from this point of view, therefore, "conservative wisdom" can be regarded as a generalization of the familiar moralist's warning against the dangers of shortsighted opportunism—against the slick move that saves the day

but ruins the cause. There can, I think, be no doubt about the need for this kind of wisdom in all practical affairs, or about the dependence of this kind of wisdom upon a knowledge and sense of history.

This account of the matter, however, may easily be misconstrued; certainly it does not provide us with a full explanation or justification of the kind of wisdom we are discussing. If it did, then even the wisest conservatism would be, as its radical detractors too often accuse it of being, an altogether backward-looking attitude. But conservatism is serious only in so far as it manifests its loyalty to certain inherited methods, principles, policies, etc., in its determination to embody these in its treatment of current and future problems; and it is positively wise only in so far as this determination stems from a reasoned judgment as to the applicability of inherited methods, policies, etc., to the changing political scene. Now judgment of this latter kind can seldom be equated with predictive knowledge: e.g., no one today can predict for any assignable future date the most probable consequences of maintaining free political and intellectual institutions in this country, yet determination to maintain such institutions can be accounted reasonable (a) in so far as they are seen to be necessary conditions of many other activities which we prize, and (b) in so far as none of the predictive knowledge we possess appears to preclude the possibility of their being maintained in their present form. On the other hand, we have here no reason for a blind or "anti-experimentalist" loyalty to inherited principles, policies, or institutions. A great policy or institution may well founder because opportunist radicals fail to appreciate its essential contribution to some general pattern of living; but it is just as likely to founder because unresourceful minds fail to adapt it to at least roughly predictable changes in the world for which it was framed. It would, therefore, be a complete mistake to regard the kind of conservative wisdom which we have been discussing as the preserve of backward-looking, anti-experimentalist, or even Fabian spirits.

NOTES

1. See in particular Popper's "The Poverty of Historicism, III," *Economica*, May, 1945.

2. See Woodger, *op. cit.*, pp. 394–95.

3. In so far as it agrees, or anyhow seems to agree with this assertion, Professor Popper's otherwise admirable account of historical explantion is liable to be misleading. See "The Poverty of Historicism, III," esp. p. 76 para. 3 and p. 79 paras. 1 and 2.

W. H. Dray

HISTORICAL UNDERSTANDING AS RE-THINKING

The theory of historical understanding set forth in R. G. Collingwood's *Idea of History*[1] has produced some sharp divisions of opinion among philosophers of history. While the "nays" have usually been written off by their opponents as insensitive positivists, obsessed by a model of inquiry derived from the natural sciences, the "ayes" have been considered, in their turn, as woolly-minded idealists, unable to distinguish between the essential logical structure of an enquiry and its psychological and methodological frills. Even those who have accepted Collingwood's doctrine, however, have often found it difficult to say exactly what they think it commits them to. For *The Idea of History* is an irritating, as well as an exciting book. It is full of paradoxes and apparent contradictions, which in some cases appear to be not entirely unconnected with a certain contempt which Collingwood from time to time displayed towards his philosophical opponents (indeed, his readers do not always escape): a contempt well exemplified by his remark, at one point, in response to the objections of an imaginary opponent: "I am not arguing; I am telling him" (p. 263). In many cases, no doubt, a kinder explanation might be sought; for the book was left unfinished at its author's death, and some parts of it apparently record the results of fairly raw, if vigorous and stimulating, reflection.

Since some of the papers incorporated into *The Idea of History* were published by Collingwood during his lifetime, however, it seems reasonable to regard them as especially authoritative on points in his general thesis which have been disputed. The best short summary of his theory of understanding is, in fact, to be found in one of these: a lecture entitled "Human Nature and Human His-

---

From *University of Toronto Quarterly*, 1958. Reprinted by permission of the University of Toronto Press and the author.

tory," which Collingwood delivered before the British Academy in 1936. The theory which he there elaborates, in the space of a few pages (pp. 213–15), could be reduced to the following three propositions: first, that human action, which is the proper concern of history, cannot be described *as* "action" at all, without mentioning the thought which it expresses—it has, in Collingwood's terms, a "thought-side"; second, that once the thought in question has been grasped by the historian, the action is understood in the sense appropriate to actions, so that it is unnecessary to go on to ask for the cause which produced it, or the law which it instantiates; third, that the understanding of action in terms of thought requires the re-thinking of the thought in question by the historian, so that, in essence, all history is "the re-enactment of past thought in the historian's own mind." No doubt there is much more to Collingwood's account than can be stated in such a condensed, schematic way. But the propositions stated do seem to contain the core of the theory.

In what follows, I propose to discuss each of Collingwood's three propositions in turn, clarifying it where that is possible, amending it where that seems necessary. The discussion cannot, of course, hope to deal with more than a few of the many objections which have been advanced against his theory in recent years. But it may perhaps serve to clear the ground a little for a sympathetic consideration of what he had to say.

I

That history is concerned with human actions, perhaps few would want to dispute; but that actions necessarily have what Collingwood calls a "thought-side" may not pass quite so easily without challenge. For exactly what is meant here by "thought"? And how is the relation between such thought and the action itself to be conceived?

It must be admitted that Collingwood's treatment of these questions is often puzzling. In the paper entitled "The Subject Matter of History," for example, it seems to be his view that the thought-side of an action is an activity of reflection[2]—as if, in order to act, an agent must first consider what to do, and then act in accordance with his reflection. If history is said to be concerned exclusively with human actions, its field of study, on such a view of the nature of action, must appear ridiculously narrow. Even a sympathetic commentator like W. H. Walsh has compared Collingwood unfavourably with the German philosopher, Willhelm Dilthey, in his apparent attempt to confine the historian's attention to "intellectual operations."[3] And Arnold Toynbee, in one of his latest volumes, visibly smarting under Collingwood's opinion of his own vast enterprise as a kind of "pigeon-holing," has expressed great amusement at Collingwood's supposed ignorance of the fact that politicians, whose activities he had represented as especially suitable to historical investigation, behave in precisely the "intellectually horrifying way" which seems, for him, to place a subject matter out of the reach of the historian's techniques.[4] Such naïvete, Toynbee maliciously suggests, could be found only in a philosopher's "theorizing."

If the paper on subject matter contained Collingwood's only discussion of the thought-side of actions, it would be difficult to escape the conclusion that, as a philosopher, he sets up a definition of history so restrictive and arbitrary that it would rule out much of what he himself wrote as an historian. But the paper in question, which is one of the least finished ones, must be weighed carefully against what is said in other parts of *The Idea of History*. If this is done, I believe that a view of the relation between thought and action will emerge which is not open to the charge of intellectualism in any damaging sense, although it safeguards, nevertheless, the distinction which Collingwood was so anxious to maintain, between historical and scientific modes of understanding.

A hint of the rather innocuous character of Collingwood's supposed intellectualism might be gleaned from a remark he makes in discussing the weaknesses of Graeco-Roman historiography. If we look back "over our actions, or over any stretch of past history," he tells us, it becomes obvious that "to a very great extent people do not know what they are doing until they have done it, if then" (p. 42). That we may nevertheless *discover*, by careful inquiry, what they were doing—including its thought-side—is clearly implied in the British Academy Lecture already referred to, in which Collingwood argues that the historical understanding which we achieve of the actions of people remote in time from ourselves is identical in kind with that which we obtain every day of the actions of our friends and neighbours. We discover their thoughts by considering their actions. The difference between an action and a mere physical movement, according to Collingwood, is that the former is the "outward expression" of a thought.[5] We discover the thought by interpreting the action *as* an expression.

It might perhaps be argued that such an account of the relation of thought to action is quite compatible with the claim that when we recognize an action as expressing a certain thought, this is equivalent to treating it as evidence that some prior activity called "reflection" must have gone on, although, since the agent's "stream of consciousness" is not open to us, we are unable to verify this directly. But when Collingwood is found to extend his analysis to knowledge of our *own* thoughts, such an interpretation of his meaning becomes very unplausible. In "Human Nature and Human History" Collingwood begins his discussion with the claim that historical inquiry can provide us with a kind of self-knowledge which a natural science of human nature fails to afford. The conclusion which his argument eventually reaches is that *all* knowledge of the human mind is historical. "It is only by historical thinking," he declares, "that I can discover what I thought . . . five minutes ago, by reflecting on an action that I then did, which surprised me when I realized what I had done" (p. 219). Elsewhere he makes the same point by distinguishing between memory knowledge of myself and autobiography—the latter being the discovery, by the techniques of the historian, of the thoughts actually expressed in my past actions, by contrast with what, at the time of acting, I (perhaps mistakenly) assumed my thoughts to be (pp. 295–96). As his equally celebrated successor in the chair of metaphysical philosophy at Oxford University, Professor Gilbert Ryle, was to put it: Self-knowledge is obtained, not by *introspection*, but by *retrospection.*[6]

Why does Collingwood insist that knowledge of even my own thoughts must come by a process of interpreting "expressions," after the fact? It is surely because he would agree with Ryle and Wittgenstein that having a certain thought is not, in essence, a matter of reciting certain propositions to oneself, or focusing certain images on one's internal cinema screen, or "going over" in any such way what one is about to do.[7] The being of a thought is in its expression; and thoughts are expressed in actions as well as in those internal monologues and private screenings which are associated with reflecting and planning. Covert activities of this sort cannot, on Collingwood's account, be regarded as thinking *by contrast with* overt ones. The fact that thought has *inward* expressions as well as outward ones does not give the former any privileged status as "the thought." If this is so, there is no reason to assume that reflection—however brief—must precede, or even accompany, an action for the latter to be properly regarded as an expression of thought, and hence as having a thought-side. For if this were Collingwood's view, he could scarcely claim that our knowledge of the thought-side of our own actions is by historical analysis, after the fact, since we have privileged access to the stream of consciousness in which the activity of reflection presumably goes on.

In claiming that *all* knowledge of mind is historical, however, Collingwood appears to have overstated his case. For if we allowed this claim, the following difficulty would arise. The doctrine, as he states it, is that if, in fact, I think a certain thought, *x*, at a certain time, *t*, I cannot know that I think *x* at *t*, but only at, say, *t* plus five minutes, when I can retrospect what I did and experienced at *t*. But if I say at *t plus 5* I can know, by historical inquiry, that at *t* I thought *x*, then I am claiming to know what I think, not only at *t*, but also at *t plus 5*; and on Collingwood's theory this is something I cannot justifiably do. To know what I think at *t plus 5*, I must make still another historical investigation at a later time, and so on ad infinitum. The regress which opens up would, of course, make it impossible to claim that I could *ever* know what I think at any time whatever. To stop the regress, it is necessary to claim that at some time or other I can know *at the time* what I think. Collingwood appears to have recognized this in the paper entitled "History as the Re-enactment of Past Experience," when, having asked himself whether "a person who performs an act of knowing" knows "that he 'is performing or has performed' that act," concludes that he can know both (p. 292).

But although Collingwood apparently abandons, in the end, his claim that all knowledge of mind is historical, what he says about self-knowledge in the passage quoted from the British Academy Lecture makes it clear that he believes some to be. And if it is, it will be unplausible to contend that what we discover by interpreting "expressions" is an activity of reflection which we engaged in at some earlier time without being aware of it. The claim that Collingwood did not confine the thought-side of actions to "intellectual operations" gains additional support from what he says at many other points. In discussing, for example, the historian's investigation of the mind "of a community or an age"—a case where deliberate planning or reflecting would be out of the question—we are told once again that the problem is essentially one of the historical interpretation of "expressions" (p. 219). And elsewhere, in attacking what he calls "scissors and paste" conceptions of the use of documents, Collingwood claims that it is possible for the historian to discover not only what has been completely forgotten—that is, is nowhere recorded in any extant document—but also "what, until he discovered it, no one ever knew to have happened at all" (p. 238). Since, for Collingwood, historical facts always concern actions, and actions always have a thought-side, the conclusion would appear to follow that historians can discover *thoughts* which were unknown, not only to any contemporary eyewitness of the actions concerned, but even to the agents themselves. And it seems to me that this claim is perfectly justified.

It may be worth adding that, on the interpretation of Collingwood's meaning which I have argued for here, even thought*less* actions may express a thought, and hence be properly regarded as having a thought-side. For to act thoughtlessly is not necessarily to act to no purpose; it is to act, rather, without taking into account certain considerations which ought to have been taken into account. Similarly, impulsive actions are in no way extruded from the historian's concerns on the ground that they lack a thought-side; for although to do a thing on impulse is incompatible with doing it as a matter of policy, or in order to implement a plan, it is not necessarily to do it for no reason. W. H. Walsh has distinguished usefully, in this connection, between saying that a person acted with something *before his mind*, and saying that he had something *in mind*; and he points out that the historian is as interested in cases falling into the second category as into the first.[8] In arguing, however, that actions in the second category may also be said to express a thought, Walsh offers what he regards as an amendment required to bring Collingwood's theory to terms with historical inquiry as it now exists. What I have tried to show is that such cases are already provided for by a theory of mind and thought which there are grounds for believing that Collingwood actually held. In arguing thus, I am, of course, appealing without much scruple from Collingwood drunk to Collingwood sober; from "The Subject Matter of History" to "Human Nature and Human History," on the assumption that the latter can be considered as the more authoritative. That the criticisms of Walsh and Toynbee find justification in the contents of the former paper, I should not attempt to deny.

II

If we grant the foregoing analysis of what it is that the historian, in order to know another's thought, must set himself to discover, what are we to say of Collingwood's second proposition: that once the thought has been discovered, the action can be understood without our having also to know the law under which it falls, or the cause which brought it about? Collingwood expresses his claim in characteristically provocative fashion when he asserts that, in history, once we know *what* happened, we already know *why* it happened (p. 214). He says this, presumably, because what happened is an action, and this, on his view, includes an explanatory thought-side. But does the thought of the agent really have this explanatory force? Can it provide us by itself with a *complete* explanation of what was done?

In asking ourselves whether an action can be completely or satisfactorily explained in terms of the agent's thought, it will be helpful, at the outset, to remind ourselves of a peculiar feature of the question "Why?"—namely, that a person can go on asking it as long as he pleases. Any theory of the nature of explanation, therefore, which does not rule out in advance objections arising from this possibility, can never hope to gain acceptance by a really determined opponent. Thus, if a historian explains Caesar's crossing of the Rubicon by referring to his determination to oust Pompey from the capital, it cannot be taken as a defect of Collingwood's theory that this explanation is incomplete—if by this we mean only that we can still ask why Caesar wanted to get rid of Pompey. It is not a defect of Collingwood's theory because, if a defect at all, it would be a defect of *any* theory of what counts as an answer to the question "Why?" Explanations can be regarded as given at successive levels of inquiry; and the farther we carry the questioning process, the deeper the explanation may be said to be. What Collingwood has to say, however, can be fairly assessed only as a theory of what counts as an answer to the question "Why?" at a single level.

What we must ask, therefore, is whether a thought has full explanatory force—by contrast with great explanatory depth—even at one level of inquiry. And it will be seen at once that if we merely cited Caesar's goal or purpose in crossing the Rubicon, the "thought" in question would not achieve such explanatory completeness by itself. To achieve this we should have to fill in many other things which, in the course of his narrative, a historian would ordinarily be content to leave implicit or understood. We should have to take account of such other thoughts as, for example, Caesar's belief that Pompey was in Rome and would remain there; and before we were through, we should have collected a sizable and complex group of what might be called the considerations which moved Caesar to action. But although Collingwood would certainly admit that the original explanation, if criticized on the ground of its incompleteness, would have to be added to in some such way, he would insist that such addition need not refer beyond the action itself to laws or causes. The explanation is completed by rounding out the thought-side, not by adding something of an altogether different kind.

But what is the criterion of completeness, in this second, and relevant sense of the term? Collingwood himself speaks of the explanatory thought as what "moves" or "determines" the agent to act.[9] And I think it would do no violence to his meaning if we said that until the thought-side is rounded out to the point where we can see that it *necessitated* the action, no complete explanation will have been given. Perhaps this will appear, at first sight, to be too rigorous a demand. But upon reflection, I think we should have to agree that a set of considerations which shows only that a certain course of action was one of a number which *might* have been pursued by the agent, yet falls short of showing that he *had* to choose the one he did, would not explain his doing what he did. No doubt such considerations would explain something. They might explain, perhaps, how it was possible for the agent to have done what he did. But they could scarcely be regarded as answering the question "Why?"

Unfortunately, Collingwood was not as careful as he should have been to draw such distinctions between kinds of explanatory problems. In the passage of the British Academy Lecture already referred to, he gives as an example of an explanatory thought, Caesar's "defiance of Republican law"; and although this does indeed appear to be a thought which the crossing of the Rubicon could be said to express, it is difficult to see how it could be an answer, or even part of an answer, to the question: "Why did Caesar cross the Rubicon?" Caesar *could*, of course, have crossed it in order to defy Republican law; but if his purpose had been only to carry out a demonstration of this sort, there would have been no need for him to march on Rome. The thought elicited in this case seems to me to be much more plausibly represented as an answer to an entirely different sort of question: a question about the *significance* of Caesar's action.[10] Our concern here, however, is with "explanation why." The fact that, in Collingwood's example, "defiance of Republican law" does not appear to have been one of the considerations which "moved" Caesar to action, is therefore irrelevant to our assessment of the claim that a complete set of explanatory "thoughts" would be one which represented what was done as necessarily done.

If it is still not clear that this is what Collingwood is saying, the fault may lie in a certain lack of precision which should be noticed in the claim that once the thought-side of an action has been discovered, the conditions of understanding have been satisfied. For, strictly speaking, an action cannot be explained in terms of its own thought-side. As Collingwood originally states his theory, the thought-side of an action is something that must be included in the very *specification* of the action (p. 213). The action is not a "doing" at all, but simply a "movement," unless it is an expression of thought. The thought which is required to make it the action it is, therefore, cannot be considered as something logically distinct from it, by reference to which the action itself can be explained. The thought-side of an action might, of course, be called upon to explain the mere movement which is its own other "side": We might say, for instance, that Caesar's wanting to get to the other shore explains his physical progress across the river. But it is clear enough that in history, the problem will almost always be the explanation of actions, not mere movements; and in order to explain an action in terms of thought it will be necessary to refer beyond the action to a thought which is not itself part of the action as specified.

It would always be possible, of course, to incorporate this *further* thought into the specification of the action, thus expanding its thought-side. If we explained Caesar's crossing of the Rubicon, for example, in terms of his wanting to oust Pompey from the capital, we might say that his crossing expressed the latter thought as well as his merely wanting to get to the other shore. But in assessing the applicability of a theory of explanation or understanding, it is important to keep the questions, as well as the answers, straight; and it should be clear that if we allow such incorporation, we can no longer claim that Caesar's wanting to oust Pompey explains *the action*; for the action to be explained is now, not the crossing of the Rubicon, but the crossing of the Rubicon to oust Pompey. To insist on such logical niceties may appear unduly pedantic. Yet unless they are recognized, Collingwood's theory of explanation in terms of thought lies open to the charge—commonly made—that he believes human actions, by contrast with natural events, to be *self*-explanatory. And in spite of the misleading way in which he sometimes states the "inside-outside" theory, it is difficult to believe, in the light of his general view of historical thinking, that he intended to make any such claim.

If a complete explanation of an action is said to be achieved when a set of considerations, or "thoughts," not themselves included in the specification of the action, can be seen to have necessitated it, the question arises whether Collingwood can maintain the distinction between his own account and that of his positivist opponents. For it is a central doctrine of the positivists whom he attacks—a doctrine derived ultimately from Hume's classical discussion of causation—that necessities asserted of the real world are read into it on the basis of a belief in general laws. On such a view, the only way to vindicate my claim that, given x, y necessarily happened, is to argue: "Whenever x then y"; so that to say that the historian's explanation represents the action as necessitated by the thought which explains it, would commit the historian to the view that it does so by virtue of some law. Now Collingwood, at this point, is indeed questioning this positivist assumption quite directly: But it is important to see exactly how he is doing it. For his frequent remarks about the limited applicability of the generalizations which might be derived from a study of human activities in any historical period, and his denial that man's reason is itself subject to laws at all, may make it appear that his point is simply that human actions cannot be explained on the model of the natural sciences because they are not necessitated.[11] His point would, in fact, appear to be quite otherwise. For, as a study of "Human Nature and Human History" will confirm, his claim is rather that, whether "positive" laws of human nature are discoverable or not, they are not *required* for historical understanding (pp. 214, 223). They are not required because human actions can be understood when they are seen to have been necessitated in an entirely different way.

The issue between Collingwood and his opponents thus turns on the question: "How can an action be necessitated except in terms of natural law?" Collingwood's answer would be that it can be necessitated in the sense of its being rationally required. A set of antecedent conditions which explains a consequent event by virtue of a law of nature, shows the event to have been necessary in the sense of being "the thing to have expected, the laws of nature being what they are." The thoughts or considerations which explain an action in

Collingwood's context of discussion show the action to have been necessary in the sense of being "the thing to have done, the principles of reason being what they are."[12] We could put the point by saying that the necessity which is required for the explanation of action in history, according to Collingwood, is a *rational* rather than a *natural* necessity. If something happens in spite of natural necessity, we call it a miracle. If an action is done in spite of rational necessity, we call it a stupidity, a mistake, an irrationality. It is Collingwood's claim that if, and only if, rational necessity can be shown, then we understand what the agent did. And he adds that such understanding does not require the further demonstration, by the methods of natural science, that what happened was a natural necessity as well.

The exact rôle which the concept of rational necessity plays in Collingwood's theory will appear more clearly when we go on to investigate the notion of "re-thinking." But a word should perhaps be added first about the concept of causation; for it may seem, once again, that Collingwood is simply flying in the face of the facts when he says that historians are not, or should not be, interested in the causes of actions. But his claim turns out, on closer examination, to be only that historians—their concept of understanding being what it is—do not need to discover *natural* causes for the actions they claim to understand. That historians commonly and legitimately use causal language, he does not attempt to deny. But he regards the historian's use of the term as a special one; for when it is said that it was his wanting to get rid of Pompey that caused Caesar to march on Rome, the word *cause* here implies a rational connection of the sort already examined, between what Caesar did and the considerations said to explain it (pp. 214–15). Collingwood himself speaks at various points of a thought not only causing an action but also determining or inducing it. But I think it is clear that he does not regard these locutions as reintroducing the concept of natural necessitation in terms of general laws of human mind of behaviour, which his general theory of historical understanding has ruled out as irrelevant.

III

We have still to examine Collingwood's third proposition: that discovering the explanatory thought requires its re-thinking in the historian's mind. From some critics, this feature of the theory has encountered the objection that such re-thinking is impossible; from others the complaint that it is unnecessary. It seems to me that neither objection is valid if Collingwood's point is taken in its proper sense. Let me therefore attempt, once again, to separate what I think is the basic doctrine from the misconceptions which may perhaps arise out of unfortunate forms of expression.

At least one of the reasons for questioning the possibility of re-thinking disappears in the light of our discussion of Collingwood's general conception of thought. For if thinking is logically distinguished from private imaginings and recitations, and if it is allowed that a person may sometimes not know what he thinks, there is no need to boggle at the notion of the historian *re*-thinking thoughts which may not have been fully articulated or explicitly recognized by the agent himself—even when those thoughts can be expressed by the historian as an argument. Since there are at least as many ways of thinking the same thought as there are ways of expressing it, there is nothing in Collingwood's theory which requires the historian to go through the same overt actions, or undergo the same private experiences, as the subject of his inquiry.[13] The fact that the original thought may not have been thought propositionally would therefore be no barrier to its being re-thought propositionally—the way we should naturally expect the historian to re-think it. It does not seem to be absolutely necessary, however, that the historian's re-thinking should take this form. He might, for example, re-think Nelson's thoughts with models of his ships on a table of naval operations, for a start—although he would, of course, have to express the thoughts propositionally when writing his monograph on "Nelson at Trafalgar."

The demand that the agent's thoughts be re-thought has also been seen to raise a different sort of problem. For, as Collingwood himself declares, "the mere re-enactment of another's thought does not make historical knowledge; we must also know that we are re-enacting it" (p. 289). How can the historian know that the thought he thinks is identical with the agent's? Collingwood unfortunately does not give this question the kind of consideration we should have liked it to get from a philosopher with his experience in historical research. Apart from a denial that historical arguments are either deductive or inductive in the ordinary sense,[14] and a lively dissection of the reasoning implicit in crime detection—illustrated by a detective yarn of his own invention (pp. 266 ff.)—we are left with little more than the rather dogmatic assurance that, in so far as there is knowledge at all, there are occasions on which the historian can be sure of his conclusions. Perhaps this contention would have seemed more acceptable if Collingwood had not, in an apparently unguarded moment, claimed that historical conclusions can, in some cases, be known as certainly as a proof in mathematics (p. 262). Since history proceeds, on his own showing, by the intepretation of evidence, not by formal deduction, the precise meaning of this curious claim is not obvious. But that a "weight of evidence" may, from time to time, justify a non-mathematical kind of certainty in history, would appear to be deniable only by someone whose scepticism embraced a good deal more than historical arguments.

If we grant that the historian *can* re-think his subject's thoughts, must we go on to agree that historical understanding *requires* this? Collingwood is most insistent that it does. "To know 'what someone is thinking' (or 'has thought')," he writes, "involves thinking it for oneself" (p. 288). And it is clear that he means this quite literally. In order to grasp the agent's thought, and see that it really does explain his action, the historian must do more than merely reproduce the agent's argument, whether implicit or explicit; he must also *draw his conclusion*. It is not enough merely to examine a report of the agent's "thought-process"; the historian must, on inspecting the thoughts, and treating them as premises of practical deliberation, actually *think that* the conclusion follows. The historian's "seeing" the connection between the agent's "considerations" and his action entails his *certification* of the connection between them—this entailment being a logical one. If the attempt to re-think, and thus to certify, the agent's thought-action complex breaks down—as Collingwood admits it has done in the case of certain early Roman emperors—then we have a dark spot, an unintelligibility, a failure to understand.[15]

Collingwood sometimes puts the point in terms of a distinction between knowing the agent's thoughts as "object" and as "act." In order to understand an action, he declares, "it is necessary to know 'what someone else is thinking,' not only in the sense of knowing the same object that he knows, but in the further sense of knowing the act by which he knows it" (p. 288). And again: "The act of thinking can be studied only as an act" (p. 293). To put the point in these terms is perhaps to be unnecessarily obscure. But to represent re-thinking as a sharing in an activity does serve to bring out the requirement that the connection between thought and action be *tested* by the historian, if he wishes to understand what was done. The doctrine may appear less formidable when Collingwood goes on to consider examples. In order to undersand the Theodosian Code, we are told, the historian "must envisage the situation with which the emperor was trying to deal, and he must envisage it just as the emperor envisaged it. Then he must see for himself, just as if the emperor's situation were his own, how such a situation might be dealt with; he must see the possible alternatives, and the reasons for choosing one rather than another; and thus he must go through the process which the emperor went through in deciding on this particular course" (p. 283). To see that the emperor had reasons for doing what he did is thus to understand his doing it. And to see that he had reasons is, at the same time, to certify the reasons *as* reasons.

The explanation might thus be said to succeed to the extent to which it reveals the rationality of the agent. An action is said to be understood, on Collingwood's view, when it is seen to have been rationally necessary. Does this way of putting it impute to Collingwood a rationalism as divorced from the realities of historical inquiry as the intellectualism we have already denied? Some of Collingwood's dicta might seem to support such a conclusion; others point the other way. The strategy of Admiral Villeneuve at Trafalgar, he tells us, cannot possibly be understood; for the fact that he lost the battle shows that Villeneuve's strategy was wrong—as if an action must be rationally appropriate to the *actual* situation or be unintelligible.[16] On the other hand, in disusssing Hegel's philosophy of history. Collingwood welcomes the "fertile and valuable principle" asserted there that "every historical character in every historical situation thinks and acts rationally as that person in that situation *can* think and act."[17] What was said above about the understanding of the Theodosian Code should be sufficient to show that in Collingwood's hands, such a principle attributes no absolute rationality to the agents whose actions historians claim to understand. For what renders an action "rational," in the sense of "understandable," is its being required by the situation as it was envisaged, not by the situation as it actually was. The historian's certification might thus be said to be a "hypothetical" one. Its implication is simply that *if* the situation had been as the agent envisaged it, then what he did would have been the rational thing to do.

But even if we interpret Hegel's "fertile principle" in terms of such a "hypothetical" rationality, the principle may possibly lead us astray. For it apparently suggests to Collingwood that when understanding fails, we can only attribute this failure to a breakdown of historical analysis—"It is the historian himself who stands at the bar of judgment . . ." (p. 219). Historical understanding being the sort of thing it is, it seems to me that a number of other possibilities would have to be taken into account.

We must allow, for example, for cases where the agent's reasoning about his situation is itself mistaken. For an agent may be in error, not only in the way he envisages his situation, but also in concluding that what he did was required by the situation as he envisaged it. Such an agent would have to be regarded by the historian as "irrational" in a stronger sense than the agent who, although he misconceived his situation, nevertheless acted in the way required by the situation as he conceived it. The thought of the latter agent can be re-thought by the historian, albeit only hypothetically; but the thought of the former cannot be re-thought at all. In so far as a person makes a mistake in the reasoning which represents what he did as rationally required, his action can only be explained *as* mistaken—a word whose function here is to rule out the possibility of explaining what was done in terms of reasons. Such an action could therefore not be "understood," in Collingwood's sense of the word.

We must allow also for cases where what is done was not intended by the agent: cases of accidental or inadvertent action, illustrated, for example, by a cabinet minister's subversion of his own government's position by an unwise speech made in its defence. It would appear to be cases of this kind which Collingwood has in mind when he denies that "every agent is wholly and directly responsible for everything that he does" (p. 41). It might be argued, I suppose, that in such cases, although there is doubtless an action (giving the speech) which is explicable in the ordinary way, what is specified as requiring explanation (bringing down the government) is not really the agent's *action* at all. Yet what was done *was*, in this case, something which the agent could be held responsible for; and his colleagues would not be slow to say that he *did* it. If for these reasons we called it an action, then it must be admitted that this, too, is a kind of action which could not be understood in Collingwood's sense.

Still another sort of case which Hegel's principle must allow for is that of arbitrary or capricious action: cases where an agent may have known certain things about his situation, and may have realized that a certain response was, in reason, demanded, yet deliberately acted otherwise. The possibility of such action has sometimes been denied. It has been alleged that what we do, we always do for some reason—even if it takes a psychoanalyst to discover it. I cannot undertake to argue the contrary here in any detail. But I think we might perhaps be a trifle wary of a theory of action which, on some occasion, might condemn a historical agent to the fate of Buridan's ass, who, on finding himself stationed equidistant between two equally succulent bundles of hay, followed reason and starved to death. The ass, admittedly, had every reason to eat one bundle before the other; and if he had eaten them successively, *something* could doubtless have been explained—for example, that he ate any hay at all, or even that he ate one bundle before the other. But it would clearly have been impossible, *ex hypothesi,* to explain in Collingwood's sense why he chose to eat the one he did before eating the other. I do not, of course, suggest that the historian will very often be confronted with an agent placed in the unhappy position of such a philosophical ass. But it is hard to see, if the ass's philosophy be not refuted, how Hegel's principle, even in its attenuated Collingwoodian form, can be accepted as universally valid.

NOTES

1. Oxford, 1946.

2. *Idea of History*, pp. 305–15.

3. *Introduction to Philosophy of History* (London, 1951), p. 50.

4. *A Study of History* (London, 1954), IX, 722.

5. *Idea of History*, p. 217. See also pp. 212, 214, 220.

6. *The Concept of Mind* (London, 1949), Chap. vi. Rylian language is employed in the two paragraphs following.

7. *Idea of History*, pp. 286, 301, 303.

8. *Philosophy of History*, p. 54.

9. E.g., *Idea of History*, pp. 216–17.

10. For a different solution, see Toynbee, *Study of History*, pp. 720–21. According to him, Caesar's action expresses "will," not "thought."

11. See *Idea of History*, pp. 220, 224.

12. I am discussing here the general *concept* of rational necessity. The qualifications requiring to be made when it is employed in historical understanding are introduced in Sec. III.

13. Collingwood's term "re-enactment" may appear more troublesome in this connection than "re-thinking"; but for Collingwood, thinking is an activity.

14. *Idea of History*, pp. 261–62. Mr. Alan Donagan has discussed this aspect of Collingwood's theory in an illuminating way in "The Verification of Historical Theses," *Philosophical Quarterly* (July 1956).

15. *Idea of History*, p. 310.

16. R. G. Collingwood, *An Autobiography* (Oxford, 1941), p. 70. T. M. Knox, in the "Editor's Preface" to *The Idea of History*, assumes that this is Collingwood's meaning. I think it possible that what he had in mind was rather the extent to which the evidence, upon which conclusions about Villeneuve's actual strategy during the battle would have been based, was destroyed by Nelson's victory.

17. *Idea of History*, p. 116.

part 2

THE STRUCTURE AND FUNCTION OF SCIENTIFIC THEORIES

INTRODUCTION

We have so far discussed the functions of empirical laws and scientific theories, but we have not yet made any attempt to distinguish them. In this section we will be concerned with the differences between them and with the philosophical problems that arise because of these differences.

Let us begin by contrasting Boyle's Law, an empirical law stating that when the temperature of a gas is constant the pressure of the gas is inversely proportional to its volume, with the kinetic theory of gases, a scientific theory stating that gases are composed of many small molecules in constant motion whose impact on the containing vessel produces the pressure of the gas and whose mean kinetic energy determines the temperature of the gas. The most obvious difference between Boyle's Law and the kinetic theory is that all the descriptive terms in the former refer to observable objects and their properties, whereas several of the descriptive terms in the latter, e.g., "molecule," refer to nonobservable objects and their nonobservable properties.

Following this distinction, we can divide the descriptive vocabulary of science into two parts, the *observable* vocabulary and the *theoretical* vocabulary; we can then distinguish laws from theories on the grounds that the latter, but not the former, contain terms which are part of the theoretical vocabulary of science. This is not, as we shall see below, the only way in which philosophers draw the distinction, but it will do for now.

Many scientists and philosophers have challenged the use of theories, but most scientists have continued to theorize, and theories now play a central role in the scientific enterprise. Philosophers have therefore tried to understand, rather than to criticize, the role of theories, and have discussed in particular: (1) The structure of a scientific theory and how the scientist uses it; (2) the problem of understanding terms that refer to the unobservable; and (3) the question of whether these theories necessarily presuppose the existence of the theoretical entities, and if so, the question of how the scientist knows that these entities do exist. The logical empiricists discussed all of these issues, and their views are described in the selections by Professors Carnap, Hempel, and Nagel which begin this section.

Professor Carnap presents the standard logical empiricist conception of the structure of a theory: A theory is a partially interpreted formal system. Its precisely formulated axioms contain both observational and theoretical primitive terms, and from these axioms one can deduce the consequences of the theory. Not all of the primitive terms are interpreted; in particular, the theoretical primitive terms, which refer to nonobservable entities, properties, or processes, cannot be directly interpreted. Nevertheless, the scientist understands the whole theory in the sense that the knows how to use it to derive non-theoretical consequences that are completely interpretable.

Consider, as an example, the kinetic theory of gases whose axioms might be:

(1) All gases are composed of many small molecules that are in a state of perpetual rapid motion in straight lines.

(2) At normal pressures, the total volume of the molecules is very small compared to the total volume of the gas.

(3) The molecules are spherical, elastic, and smooth.

(4) The pressure exerted by the gas is the force per unit area due to the impact of the individual molecules on the wall of the containing vessel.

(5) Two gases are at the same temperature when the mean kinetic energy of the individual molecules of the two gases are the same.

From these axioms one can deduce experimental laws like Boyle's Law which are stated in purely observational terms ("temperature of the gas," "pressure of the gas") which can all be directly interpreted. But some of the primitive terms of the theory cannot be so interpreted because they refer to the nonobservable. Nevertheless, the physicist understands the whole theory because he knows how to use it to derive empirical consequences.

Given this picture of the structure of a scientific theory, one can easily see its function. As in the case of an ordinary experimental law, one can deduce empirical consequences from a theory, and one can therefore use the theory to explain data that has been observed and to predict the occurrence of future events.

Professor Hempel points out that some of the axioms play a central role in the interpretation of the theory. These are the axioms [like (4) and (5)] that relate the observational and theoretical terms, thus enabling the axioms containing theoretical terms to have empirical consequences. For this reason, Hempel calls these axioms "interpretative sentences" (others have called them "correspondence rules").

As Hempel points out, this approach to the meaning of scientific terms is far less demanding than Professor P. W. Bridgman's "operationalist" approach according to which every descriptive term in science should be defined by a rule stating that the term applies in a given case if and only if the performance of a given operation yields a specific result. Operationalism runs into serious difficulties with theoretical terms (and also with dispositional terms) because the appropriate operations do not seem to exist. The logical empiricists, therefore, have modified the operationalist position to merely require that each term in science have some experimental import, even the weak one afforded by appearing in a partially interpreted formal system.

Although most logical empiricists and many other philosophers agree with Carnap's picture of the structure and meaning of a theory, there is far less agreement about the scientist's commitment to the actual existence of theoretical entities. Does acceptance of the kinetic theory, for example, imply a commitment to the existence of unobservable molecules with the properties postulated by the theory? The view that it does is based upon the assumption that a correct scientific theory is a true description of some unobservable reality. This position is known as *realism*. Professor Scheffler, in his article, discusses a variety of *fictionalist* positions that maintain that the acceptance of theoretical statements when properly understood does not commit us to the existence of theoretical entities.

In recent years, many philosophers have objected to part or all of the logical empiricist conception of theories, and the rest of this section will be concerned with these objections. Perhaps the most fundamental of these is the very possibility of drawing an observational-theoretical distinction. Professor Grover Maxwell, for example, argues that the particular point at which one draws the distinction between the observational and the theoretical is a combinatory function of our physiological makeup, our current state of knowledge, and the instruments available to us. Since these latter two factors vary, the distinction will be drawn at different places at different times. Consequently, Maxwell argues, there can be no ontological significance to the observational-theoretical distinction, and the controversy over the reality of theoretical entities is therefore a pseudo-controversy.

Professor Peter Achinstein, another critic of the distinction, challenges two existing ways of drawing it. One is the standard approach based upon observability. The other is a recent attempt to draw the distinction according to the extent to which the term is theory-laden, i.e., the extent to which one's understanding of the term depends upon one's knowledge of a particular scientific theory. Achinstein argues that either way of drawing the distinction, rather than neatly dividing the vocabularly of science into two parts, generates many different distinctions in different contexts. He concludes from this that there is no general account of the relation between theories and observation, that different types of theoretical terms are related to observations in different ways, and that the philosopher of science should give us an account of these different relations.

A second group of critics claims that the logical empiricists have failed to give a proper account of the role of models in the understanding and development of a scientific theory. A model of a theory is a complete interpretation of that theory, other than the intended interpretation, which makes all the statements of the theory true. Thus, one can construct a model for the kinetic theory by interpreting its statements so that they are about a large number of perfectly elastic and spherical physical objects moving in straight lines through a much larger container.

Professor N. R. Campbell argues that such models are necessary for the understanding of a theory. Campbell feels that the correspondence rules (which he calls the "dictionary of the theory") merely make possible the derivation of empirical consequences from the theory, but it is the model or analogy that gives the theory its meaning.

Professor Braithwaite, in defending the logical empiricist position on this point, argues that since the meaning of any claim is its empirically testable consequences, and since correspondence rules give the theory these consequences, they are all that is needed in order for the theory to have meaning. There is no need of, nor no possibility for, the theory having additional meaning via the model.

Braithwaite also describes the role that a model may play in the further development of a scientific theory: that of suggesting extensions and developments of such a theory. In these cases, the theory must then be developed and tested independently of the model. Consequently, according to Braithwaite, the role of the model is limited to suggesting the extensions of the theory.

Several philosophers have recently argued that Braithwaite's position does not do justice to the role of models and that, at least in some cases, Campbell's view is correct. Professor Marshall Spector, for one, argues that some models, like the model for the kinetic theory, eventually become the standard interpretation of the theory. This occurs when we discover an identity between the designata of the observable terms of the theory under the intended interpretation and under the model. We then conclude that the designata of the theoretical terms are identical to their designata under the model. The whole theory is then directly interpreted via its model and independently of its correspondence rules.

A third group of critics has been concerned with the logical empiricist conception of the relation between observational generalizations and theories. This account is carefully articulated by Professor Nagel in his account of the reduction of one science to another, in the subsumption of the results of one science under the results of another science. Nagel claims that the goal of the process of reduction is the deduction of the laws of the one science (which is usually purely observational) from the theories of the other science. He points out that this process is possible only because of the existence of correspondence rules (which, he points out, should be treated as additional assumptions of the theory). This view is clearly based upon many of the standard positions of logical empiricism. One law is explained in terms of another (in this case, a theory) only when one can deduce the former from the latter, and theories are seen as having empirical consequences only because of their correspondence rules.

Professors Kemeny and Oppenheim, while generally in agreement with this view, insist that two modifications must be made in it. To begin with, they require that the reducing theory be at least as well systematized as the theory reduced. Second, and somewhat more radically, they claim that, while reducing one theory to another, one can sometimes only deduce from the reducing theory the empirical consequences of the reduced theory (or some modification of it) and not the reduced theory itself. Nagel's picture of reduction, in which we can actually deduce the reduced theory, is only a special case of the general process of reduction.

Professor Paul Feyerabend has been one of the leading critics of the logical empiricist view that observational laws are deduced from theories. He claims that this presupposes that the laws and the theories never have conflicting empirical consequences and that the meaning of the terms in the laws does not change when the new theory is formulated. Feyerabend calls these two conditions the "consistency condition" and "the condition of meaning invariance," and he argues that neither is actually satisfied by scientific practice. He also argues that it would be unreasonable for the scientist to confine himself to the formulation of theories that satisfy these conditions. Feyerabend concludes, therefore, that the logical empiricists were mistaken and that we should view new theories, not as covering laws for earlier laws and theories, but as attempts to replace these earlier laws with which they conflict.

Professor Sellars has also argued that theories are meant as replacements for older theories and laws. But his reasons for holding this view differ from the ones offered by Feyerabend. Sellars is concerned about the problem of "fitting together" observational and theoretical entities. Are there, to use a standard problem first raised by Professor Eddington, two tables, the one we observe and the one we learn about in physics, or is there only one? Sellars believes that there is only one, the latter table, and that the correspondence rules state that the observational entities and properties do not exist and that the real entities and properties are the theoretical ones referred to by the theoretical terms in the correspondence rules.

A great deal hinges upon this dispute. We normally think of science as a cumulative enterprise in which scientists build upon the laws and theories that earlier scientists have established. It is just this picture of science that is captured by the logical empiricist account. But if Feyerabend and/or Sellars are right, then we must consider a newly formulated theory as an attempt to do without earlier laws and theories.

The reader must decide for himself the degree to which these critics have weakened the case for the logical empiricist view of scientific theories. But there is no doubt that they have at least heightened our awareness of many features of scientific theories.

There is one additional question about scientific theories that has attracted much interest in recent years. It has been claimed by some philosophers, most notably Professor T. Kuhn, that some theories play a special role in science that has not been adequately described by previous philosophers. These are the paradigm-theories. As a result of his historical and sociological investigations, Kuhn believes that normal scientific activity in a field of study is paradigm-governed, i.e., governed by a fundamental theory which is accepted as true, and which has been applied to many, but not all, of the problems of the field. And the goal of normal scientific activity in that field is the successful use of that theory to solve the remaining problems in the field. In contrast to normal activity, there are two types of scientific activity that are not paradigm-governed. One occurs at the beginning of the development of a science, before the fundamental theory has been articulated, whereas the other occurs in revolutionary periods when the fundamental theory breaks down.

Kuhn concedes that paradigm-governed scientific activity is in a certain sense dogmatic since the scientist is committed to this theory, and all his activities are therefore devoted to articulating the paradigm and to interpreting new data according to it, and he is not particularly concerned with testing it. But in contrast to many who say that the scientist should never accept a theory in such an unequivocal fashion, Kuhn claims that the dogmatic acceptance of a paradigm in normal scientific activity has many important advantages. He does distinguish, however, between normal and revolutionary scientific activity, and in the latter there is no place for this type of dogmatism. He even claims that the existence of the paradigm aids in the recognition of the need for revolutionary change.

Kuhn's remarks raise many interesting conceptual problems. Some are concerned with the clarification of the notion of a paradigm and of its role in normal and revolutionary scientific activity. Others are concerned with an evaluation of the advantages of, and the problems raised by, the presence of this dogmatism in normal scientific activity. Kuhn's ideas have also led many to question the validity of the concepts and models that we have discussed in this book. If he is right, then perhaps the relation between explanans and explanandum, theory and observation, and hypothesis and evidence, differs from revolutionary science to normal science and from normal science under one paradigm to normal science under the other. This is an exciting question, one which will have to be carefully considered by all who are interested in the philosophy of science. Aside from these considerations, however, there can be no doubt that Kuhn's ideas have illuminated the role some theories play in the development of a science.

The Classical View

Rudolf Carnap

THEORIES AS PARTIALLY INTERPRETED FORMAL SYSTEMS

. . .

23. PHYSICAL CALCULI AND THEIR INTERPRETATIONS

The method described with respect to geometry can be applied likewise to any other part of physics: We can first construct a calculus and then lay down the interpretation intended in the form of semantical rules, yielding a physical theory as an interpreted system with factual content. The customary formulation of a physical calculus is such that it presupposes a logico-mathematical calculus as its basis, e.g., a calculus of real numbers in any of the forms discussed above (§18). To this basic calculus are added the specific primitive signs and the axioms, i.e., specific primitive sentences, of the physical calculus in question.

Thus, for instance, a calculus of mechanics of mass points can be constructed. Some predicates and functors (i.e., signs for functions) are taken as specific primitive signs, and the fundamental laws of mechanics as axioms. Then semantical rules are laid down stating that the primitive signs designate, say, the class of material particles, the three spatial coordinates of a particle x at the time t, the mass of a particle x, the class of forces acting on a particle x or at a space point s at the time t. (As we shall see later [§ 24], the interpretation can also be given indirectly, i.e., by semantical rules, not for the primitive signs, but for

---

certain defined signs of the calculus. This procedure must be chosen if the semantical rules are to refer only to observable properties.) By the interpretation, the theorems of the calculus of mechanics become physical laws, i.e., universal statements describing certain features of events; they constitute physical mechanics as a theory with factual content which can be tested by observations. The relation of this theory to the calculus of mechanics is entirely analogous to the relation of physical to mathematical geometry. The customary division into theoretical and experimental physics corresponds roughly to the distinction between calculus and interpreted system. The work in theoretical physics consists mainly in constructing calculi and carrying out deductions within them; this is essentially mathematical work. In experimental physics interpretations are made and theories are tested by experiments.

In order to show by an example how a deduction is carried out with the help of a physical calculus, we will discuss a calculus which can be interpreted as a theory of thermic expansion. To the primitive signs may belong the predicates "Sol" and "Fe", and the functors "lg", "te", and "th." Among the axioms may be A1 and A2. (Here, x, β and the letter with subscripts are real number variables; the parentheses do not contain explanations as in former examples, but are used as in algebra and for the arguments of functors.)

A1. For every $x, t_1, t_2, l_1, l_2, T_1, T_2, \beta$ [if [x is a Sol and lg $(x, t_1) = l_1$ and lg$(x, t_2) = l_2$ and te$(x, t_1) = T_1$ and te$(x, t_2) = T_2$ and th$(x) = \beta$] then $l_2 = l_1 \times (1 + \beta \times (T_2 - T_1))$].

A2. For every x, if [x is a Sol and x is a Fe] then th$(x) = 0.000012$.

The *customary interpretation*, i.e., that for whose sake the calculus is constructed, is given by the following semantical rules. lg(x, t) designates the length in centimeters of the body x at the time t (defined by the statement of a method of measurement); te(x, t) designates the absolute temperature in centigrades of x at the time t (likewise defined by a method of measurement); th(x) designates the coefficient of thermic expansion for the body x; Sol designates the class of solid bodies; Fe the class of iron bodies. By this interpretation, A1 and A2 become physical laws. A1 is the law of thermic expansion in quantitative form, A2 the statement of the coefficient of thermic expansion for iron. As A2 shows, a statement of a physical constant for a certain substance is also a universal sentence. Further, we add semantical rules for two signs occuring in the subsequent example: The name c designates the thing at such and such a place in our laboratory; the numerical variable t as time coordinate designates the time-point t seconds after August 17, 1938, 10:00 A.M.

Now we will analyze an example of a derivation within the calculus indicated. This derivation D_2 is, when interpreted by the rules mentioned, the deduction of a prediction from premisses giving the results of observations. The construction of the derivation D_2, however, is entirely independent of any interpretation. It makes use only of the rules of the calculus, namely, the physical calculus indicated together with a calculus of real numbers as basic calculus. We have discussed, but not written down, a similar derivation D_1 (§ 19), which, however, made use only of the mathematical calculus. Therefore the physical laws used had to be taken in D_1 as premisses. But here in D_2 they belong to the axioms of the calculus (A1 and A2, occurring as [6] and [10]). Any axiom or theorem proved in a physical calculus may be used within any derivation in that calculus without belonging to the premisses of the derivation, in exactly the same way in which a proved theorem is used within a derivation in a logical or mathematical calculus, e.g., in the first example of a derivation in §19 sentence (7), and in D_1 (§19) the sentences which in D_2 are called (7) and (13). Therefore only singular sentences (not containing variables) occur as premisses in D_2. (For the distinction between premisses and axioms see the remark at the end of §19.)

Derivation D_2:

Premisses	1. c is a Sol. 2. c is a Fe. 3. te(c, 0) = 300. 4. te(c, 600) = 350. 5. lg(c, 0) = 1,000.
Axiom A1	6. For every $x, t_1, t_2, l_1, l_2, T_1, T_2, \beta$ [if [x is a Sol and lg(x, t_1) = l_1 and lg(x, t_2) = l_2 and te(x, t_1) = T_1 and te(x, t_2) = T_2 and th(x) = β] then $l_2 = l_1 \times (1 + \beta \times (T_2 - T_1))$].
Proved mathem. theorem:	7. For every $l_1, l_2, T_1, T_2, \beta$ [$l_2 - l_1 = l_1 \times \beta \times (T_2 - T_1)$ if and only if $l_2 = l_1 \times (1 + \beta \times (T_2 - T_1))$].
(6) (7)	8. For every $x, t_1, \ldots$ (as in [6]) . . . [if [- - -] then $l_2 - l_1 = l_1 \times \beta \times (T_2 - T_1)$].
(1)(3)(4)(8)	9. For every l_1, l_2, β [if [th(c) = β and lg(c, 0) = l_1 and lg(c, 600) = l_2] then $l_2 - l_1 = l_1 \times \beta \times (350 - 300)$].
Axiom A2	10. For every x, if [x is a Sol and x is a Fe] then th(x) = 0.000012.
(1)(2)(10)	11. th(c) = 0.000012.
(9)(11)(5)	12. For every l_1, l_2 [if [lg(c, 0) = l_1 and lg(c, 600) = l_2] then $l_2 - l_1 = 1{,}000 \times 0.000012 \times (350 - 300)$].
Proved mathem. theorem:	13. $1{,}000 \times 0.000012 \times (350 - 300) = 0.6$.
(12)(13) *Conclusion:*	14. lg(c, 600) − lg(c, 0) = 0.6.

On the basis of the interpretation given before, the premisses are singular sentences concerning the body c. They say that c is a solid body made of iron, that the temperature of c was at 10:00 A.M. 300° abs., and at 10:10 A.M. 350° abs., and that the length of c at 10:00 A.M. was 1,000 cm. The conclusion says that the increase in the length of c from 10:00 to 10:10 A.M. is 0.6 cm. Let us suppose that our measurements have confirmed the premisses. Then the derivation yields the conclusion as a prediction which may be tested by another measurement.

Any physical theory, and likewise the whole of physics, can in this way be presented in the form of an interpreted system, consisting of a specific calculus (axiom system) and a system of semantical rules for its interpretation; the axiom system is, tacitly or explicitly, based upon a logico-mathematical calculus with customary interpretation. It is, of course, logically possible to apply the same method to any other branch of science as well. But practically the situation is such that most of them seem at the present time to be not yet developed to a degree which would suggest this strict form of presentation. There is an interesting and successful attempt of an axiomatization of certain parts of biology, especially genetics, by Woodger (Vol. I, No. 10). Other scientific fields which we may expect to be soon accessible to this method are perhaps chemistry, economics, and some elementary parts of psychology and social science.

Within a physical calculus the mathematical and the physical theorems, i.e., *C*-true formulas, are treated on a par. But there is a fundamental difference between the corresponding *mathematical* and the *physical propositions* of the physical theory, i.e., the system with customary interpretation. This difference is often overlooked. That physical theorems are sometimes mistaken to be of the same nature as mathematical theorems is perhaps due to several factors, among them the fact that they contain mathematical symbols and numerical expressions and that they are often formulated incompletely in the form of a mathematical equation (e.g., A1 simply in the form of the last equation occurring in it). A mathematical proposition may contain only logical signs, e.g., "for every m, n, $m + n = n + m$," or descriptive signs also, if the mathematical calculus is applied in a descriptive system. In the latter case the proposition, although it contains signs not belonging to the mathematical calculus, may still be provable in this calculus, e.g., $\mathrm{lg}(c) + \mathrm{lg}(d) = \mathrm{lg}(d) + \mathrm{lg}(x)$ (lg designates length as before). A physical propostion always contains descriptive signs because otherwise it could not have factual content; in addition, it usually contains also logical signs. Thus the difference between mathematical theorems and physical theorems in the interpreted system does not depend upon the kinds of signs occurring but rather on the kind of truth of the theorems. The truth of a mathematical theorem, even if it contains descriptive signs, is not dependent upon any facts concerning the designata of these signs. We can determine its truth if we know only the semantical rules: Hence it is L-true. (In the example of the theorem just mentioned, we need not know the length of the body c). The truth of a physical theorem, on the other hand, depends upon the properties of the designata of the descriptive signs occuring. In order to determine its truth, we have to make observations concerning these designata; the knowledge of the semantical rules is not sufficient. (In the case of A2, e.g., we have to carry out experiments with solid iron bodies.) Therefore, a physical theorem, in contradistinction to a mathematical theorem, has factual content.

24. ELEMENTARY AND ABSTRACT TERMS

We find among the concepts of physics—and likewise among those of the whole of empirical science—differences of abstractness. Some are more elementary than others, in the sense that we can apply them in concrete cases on the basis of observations in a more direct way than others. The others are more abstract; in order to find out whether they hold in a certain case, we have to carry out a more complex procedure, which, however, also finally rests on observations. Between quite elementary concepts and those of high abstraction there are many intermediate levels. We shall not try to give an exact definition for "degree of abstractness"; what is meant will become sufficiently clear by the following series of sets of concepts, proceeding from elementary to abstract concepts: bright, dark, red, blue, warm, cold, sour, sweet, hard, soft (all concepts of this first set are meant as properties of things, not as sense-data); coincidence; length; length of time; mass, velocity, acceleration, density, pressure; temperature, quantity of heat; electric charge, electric current, electric field; electric potential, electric resistance, coefficient of induction, frequency of oscillation; wave function.

Suppose that we intend to construct an interpreted system of physics—or of the whole of science. We shall first lay down a calculus. Then we have to state semantical rules of the kind SD for the specific signs, i.e., for the physical terms. (The SL-rules are presupposed as giving the customary interpretation of the logico-mathematical basic calculus.) Since the physical terms form a system, i.e., are connected with one another, obviously we need not state a semantical rule for each of them. For which terms, then, must we give rules, for the elementary or for the abstract ones? We can, of course, state a rule for any term, no matter what its degree of abstractness, in a form like this: "The term te designates temperature," provided the meta-language used contains a corresponding expression (here the word *temperature*) to specify the designatum of the term in question. But suppose we have in mind the following purpose for our syntactical and semantical description of the system of physics: The description of the system shall teach a layman to understand it, i.e., to enable him to apply it to his observations in order to arrive at explanations and predictions. A layman is meant as one who does not know physics but has normal senses and understands a language in which observable properties of things can be described (e.g., a suitable part of everyday nonscientific English). A rule like "the sign *P* designates the property of being blue" will do for the purpose indicated; but a rule like "the sign *Q* designates the property of being electrically charged" will not do. In order to fulfil the purpose, we have to give semantical rules for elementary terms only, connecting them with observable properties of things. For our further discussion we suppose the system to consist of rules of this kind, as indicated in the diagram on the next page.

Now let us go back to the construction of the calculus. We have first to decide at which end of the series of terms to start the construction. Should we take elementary terms as primitive signs, or abstract terms? Our decision to lay down the semantical rules for the elementary terms does not decide this question. Either procedure is still possible and seems to have some reasons in its favor, depending on the point of view taken. The *first method* consists in taking elementary terms as primitive and then introducing on their basis further terms step

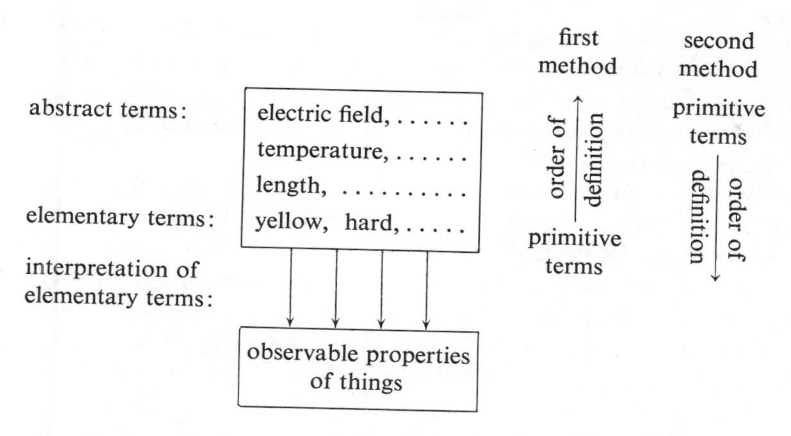

by step, up to those of highest abstraction. In carrying out this procedure, we find that the introduction of further terms cannot always take the form of explicit definitions; conditional definitions must also be used (so-called reduction sentences [see Vol. I, No. 1, p. 50]). They describe a method of testing for a more abstract term, i.e., a procedure for finding out whether the term is applicable in particular cases, by referring to less abstract terms. The first method has the advantage of exhibiting clearly the connection between the system and observation and of making it easier to examine whether and how a given term is empirically founded. However, when we shift our attention from the terms of the system and the methods of empirical confirmation to the laws, i.e., the universal theorems, of the system, we get a different perspective. Would it be possible to formulate all laws of physics in elementary terms, admitting more abstract terms only as abbreviations? If so, we would have that ideal of a science in sensationalistic form which Goethe in his polemic against Newton, as well as some positivists, seems to have had in mind. But it turns out—this is an empirical fact, not a logical necessity—that it is not possible to arrive in this way at a powerful and efficacious system of laws. To be sure, historically, science started with laws formulated in terms of a low level of abstractness. But for any law of this kind, one nearly always later found some exceptions and thus had to confine it to a narrower realm of validity. The higher the physicists went in the scale of terms, the better did they succeed in formulating laws applying to a wide range of phenomena. Hence we understand that they are inclined to choose the *second method.* This method begins at the top of the system, so to speak, and then goes down to lower and lower levels. It consists in taking a few abstract terms as primitive signs and a few fundamental laws of great generality as axioms. Then further terms, less and less abstract, and finally elementary ones, are to be introduced by definitions; and here, so it seems at present, explicit definitions will do. More special laws, containing less abstract terms, are to be proved on the basis of the axioms. At least, this is the direction in which physicists have been striving with remarkable success, especially in the past few decades. But at the present time, the method cannot yet be carried through in the pure form indicated. For many less abstract terms no definition on the basis of abstract terms alone is as yet known; hence those terms must also be taken as primitive. And many more special laws, especially in biological fields, cannot yet be proved on the basis of laws in abstract terms only; hence those laws must also be taken as axioms.

Now let us examine the result of the interpretation if the first or the second method for the construction of the calculus is chosen. In both cases the semantical rules concern the elementary signs. In the first method these signs are taken as primitive. Hence, the semantical rules give a complete interpretation for these signs and those explicitly defined on their basis. There are, however, many signs, especially on the higher levels of abstraction, which can be introduced not by an explicit definition but only by a conditional one. The interpretation which the rules give for these signs is in a certain sense incomplete. This is due not to a defect in the semantical rules but to the method by which these signs are introduced; and this method is not arbitrary but corresponds to the way in which we really obtain knowledge about physical states by our observations.

If, on the other hand, abstract terms are taken as primitive—according to the second method, the one used in scientific physics—then the semantical rules have no direct relation to the primitive terms of the system but refer to terms introduced by long chains of definitions. The calculus is first constructed floating in the air, so to speak; the construction begins at the top and then adds lower and lower levels. Finally, by the semantical rules, the lowest level is anchored at the solid ground of the observable facts. The laws, whether general or special, are not directly interpreted, but only the singular sentences. For the more abstract terms, the rules determine only an *indirect interpretation*, which is—here as well as in the first method—incomplete in a certain sense. Suppose *B* is defined on the basis of *A*; then, if *A* is directly interpreted, *B* is, although indirectly, also interpreted completely; if, however, *B* is directly interpreted, *A* is not necessarily also interpreted completely (but only if *A* is also definable by *B*).

To give an example, let us imagine a calculus of physics constructed, according to the second method, on the basis of primitive specific signs like "electromagnetic field," "gravitational field," "electron," "proton," etc. The system of definitions will then lead to elementary terms, e.g., to Fe, defined as a class of regions in which the configuration of particles fulfils certain conditions, and Na-yellow as a class of space-time regions in which the temporal distribution of the electromagnetic field fulfils certain conditions. Then semantical rules are laid down stating that Fe designates iron and Na-yellow designates a specified yellow color. (If "iron" is not accepted as sufficiently elementary, the rules can be stated for more elementary terms.) In this way the connection between the calculus and the realm of nature, to which it is to be applied, is made for terms of the calculus which are far remote from the primitive terms.

Let us examine, on the basis of these discussions, the example of a derivation D_2 (§23). The premisses and the conclusion of D_2 are singular sentences, but most of the other sentences are not. Hence the premisses and the conclusion of this as of all other derivations of the same type can be directly interpreted, understood, and confronted with the results of observations. More of an interpretation is not necessary for a practical application of a derivation. If, in confronting the interpreted premisses with our observations, we find them confirmed as true, then we accept the conclusion as a prediction and we may base a decision upon it. The sentences cocurring in the derivation between premisses and conclusion are also interpreted, at least indirectly. But we need not make their interpretation explicit in order to be able to construct the derivation and to apply it. All that is necessary for its construction are the formal rules of the calculus. This is the advantage of the method of formalization, i.e., of the separation of the calculus as a formal system from the interpretation. If some persons want to come to an agreement about the formal correctness of a given derivation, they may leave aside all differences of opinion on material questions or questions of interpretation. They simply have to examine whether or not the given series of formulas fulfils the formal rules of the calculus. Here again, the function of calculi in empirical science becomes clear as instruments for transforming the expression of what we know or assume.

Against the view that for the application of a physical calculus we need an interpretation only for singular sentences, the following objection will perhaps be raised. Before we accept a derivation and believe its conclusion we must have accepted the physical caluclus which furnishes the derivation; and how can we decide whether or not to accept a physical calculus for application without interpreting and understanding its axioms? To be sure, in order to pass judgment about the applicability of a given physical calculus we have to confront it in some way or other with observation, and for this purpose an interpretation is necessary. But we need no explicit interpretation of the axioms, nor even of any theorems. The empirical examination of a physical theory given in the form of a calculus with rules of interpretation is not made by interpreting and understanding the axioms and then considering whether they are true on the basis of our factual knowledge. Rather, the examination is carried out by the same procedure as that explained before for obtaining a prediction. We construct derivations in the calculus with premisses which are singular sentences describing the results of our observations, and with singular sentences which we can test by observations as conclusions. The physical theory is indirectly confirmed to a higher and higher degree if more and more of these predictions are confirmed and none of them is disconfirmed by observations. Only singular sentences with elementary terms can be directly tested; therefore, we need an explicit interpretation only for these sentences.

25. "UNDERSTANDING" IN PHYSICS

The development of physics in recent centuries, and especially in the past few decades, has more and more led to that method in the construction, testing, and application of physical theories which we call *formalization*, i.e., the construction of a calculus supplemented by an interpretation. It was the progress of knowledge and the particular structure of the subject matter that suggested and made practically possible this increasing formalization. In consequence it became more and more possible to forego an "intuitive understanding" of the abstract terms and axioms and theorems formulated with their help. The possibility and even necessity of abandoning the search for an understanding of that kind was not realized for a long time. When abstract, nonintuitive formulas, as, e.g., Maxwell's equations of electromagnetism, were proposed as new axioms, physicists endeavored to make them "intuitive" by constructing a "model," i.e., a way of representing electromagnetic micro-processes by an analogy to known macro-processes, e.g., movements of visible things. Many attempts have been made in this direction, but without satisfactory results. It is important to realize that the discovery of a model has no more than an aesthetic or didactic or at best a heuristic value, but is not at all essential for a successful application of the physical theory. The demand for an intuitive understanding of the axioms was less and less fulfilled when the development led to the general theory of relativity and then to quantum mechanics, involving the wave function. Many people, including physicists, have a feeling of regret and disappointment about this. Some, especially philosophers, go so far as even to contend that these modern theories, since they are not intuitively understandable, are not at all theories about nature but "mere formalistic constructions," "mere calculi." But this is a fundamental misunderstanding of the function of a physical theory. It is true a theory must not be a "mere calculus" but possess an interpretation, on the basis of which it can be applied to facts of nature. But it is sufficient, as we have seen, to make this interpretation explicit for elementary terms; the interpretation of the other terms is then indirectly determined by the formulas of the calculus, either definitions or laws, connecting them with the elementary terms. If we demand from the modern physicist an answer to the question what he means by the symbol ψ of his calculus, and are astonished that he cannot give an answer, we ought to realize that the situation was already the same in classical physics. There the physicist could not tell us what he meant by the symbol E in Maxwell's equations. Perhaps, in order not to refuse an answer, he would tell us that E designates the electric field vector. To be sure, this statement has the form of a semantical rule, but it would not help us a bit to understand the theory. It simply refers from a symbol in a symbolic calculus to a corresponding word expression in a calculus of words. We are right in demanding an interpretation for E but that will be given indirectly by semantical rules referring to elementary signs together with the formulas connecting them with E. This interpretation enables us to use the laws containing E for the derivation of predictions. Thus we understand E, if "understanding" of an expression, a sentence, or a theory means capability of its use for the description of known facts or the prediction of new facts. An "intuitive understanding" or a direct translation of E into terms referring to observable properties is neither necessary nor possible. The situation of the modern physicist is not essentially different. He knows how to use the symbol ψ in the calculus in order to derive predictions which we can test by observations. (If they have the form of probability statements, they are tested by statistical results of observations.) Thus the physicist, although he cannot give us a translation into everyday language, understands the symbol ψ and the laws of quantum mechanics. He possesses that kind of understanding which alone is essential in the field of knowledge and science.

Carl G. Hempel

A LOGICAL APPRAISAL OF OPERATIONISM

Operationism, in its fundamental tenets, is closely akin to logical empiricism. Both schools of thought have put much emphasis on definite experiential meaning or import as a necessary condition of objectively significant discourse, and both have made strong efforts to establish explicit criterions of experiential significance. But logical empiricism has treated experiential import as a characteristic of statements—namely, as their susceptibility to test by experiment or observation—whereas operationsism has tended to construe experiential meaning as a characteristic of concepts or of the terms representing them—namely, as their susceptibility to operational definition.

BASIC IDEAS OF OPERATIONAL ANALYSIS

An operational definition of a term is conceived as a rule to the effect that the term is to apply to a particular case if the performance of specified operations in that case yields a certain characteristic result. For example, the term *harder than* might be operationally defined by the rule that a piece of mineral, x, is to be called harder than another piece of mineral, y, if the operation of drawing a sharp point of x across the surface of y results in a scratch mark on the latter. Similarly, the different numerical values of a quantity such as length are thought of as operationally definable by reference to the outcomes of specified measuring operations. To safeguard the objectivity of science, all operations invoked in this kind of definition are required to be intersubjective in the sense that different observers must be able to perform "the same operation" with reasonable agreement in their results.[1]

Reprinted from *Scientific Monthly*, I, No. 79 (October 19, 1954), 215–20 by permission of the editor.

P. W. Bridgman, the originator of operational analysis, distinguishes several kinds of operation that may be invoked in specifying the meanings of scientific terms.[2] The principal ones are (1) what he calls *instrumental operations*—these consist in the use of various devices of observation and measurement—and (2) paper-and-pencil operations, verbal operations, mental experiments, and the like—this group is meant to include, among other things, the techniques of mathematical and logical inference as well as the use of experiments in imagination. For brevity, but also by way of suggesting a fundamental similarity among the procedures of the second kind, I shall refer to them as *symbolic operations.*

The concepts of operation and of operational definition serve to state the basic principles of operational analysis, of which the following are of special importance:

(1) "Meanings are operational." To understand the meaning of a term, we must know the operational criterions of its application,[3] and every meaningful scientific term must therefore permit of an operational definition. Such definition may refer to certain symbolic operations and it always must ultimately make reference to some instrumental operation.[4]

(2) To avoid ambiguity, every scientific term should be defined by means of one unique operational criterion. Even when two different operational procedures (for instance, the optical and the tactual ways of measuring length) have been found to yield the same results, they still must be considered as defining different concepts (for example, optical and tactual length), and these should be distinguished terminologically because the presumed coincidence of the results is inferred from experimental evidence, and it is "not safe" to forget that the presumption may be shown to be spurious by new, and perhaps more precise, experimental data.[5]

(3) The insistence that scientific terms should have unambiguously specifiable operational meanings serves to insure the possibility of an objective test for the hypotheses formulated by means of those terms.[6] Hypotheses incapable of operational test or, rather, questions involving untestable formulations, are rejected as meaningless: "If a specific question has meaning, it must be possible to find operations by which an answer may be given to it. It will be found in many cases that the operations cannot exist, and the question therefore has no meaning."[7]

The emphasis on "operational meaning" in scientifically significant discourse has unquestionably afforded a salutary critique of certain types of procedure in philosophy and in empirical science and has provided a strong stimulus for methodological thinking. Yet, the central ideas of operational analysis as stated by their proponents are so vague that they constitute not a theory concerning the nature of scientific concepts but rather a program for the development of such a theory. They share this characteristic with the insistence of logical empiricism that all significant scientific statements must have experiential import, that the latter consists in testability by suitable data of direct observation, and that sentences which are entirely incapable of any test must be ruled out as meaningless "pseudo hypotheses." These ideas, too, constitute not so much a thesis or a theory as a program for a theory that needs to be formulated and amplified in precise terms.

An attempt to develop an operationist theory of scientific concepts will have to deal with a least two major issues: The problem of giving a more precise explication of the concept of operational definition; and the question whether operational definition in the explicated sense is indeed necessary for, and adequate to, the introduction of all nonobservational terms in empirical science.

I wish to present here in brief outline some considerations that bear on these problems. The discussion will be limited to the descriptive, or extralogical, vocabulary of empirical science and will not deal, therefore, with Bridgman's ideas on the status of logic and mathematics.

A BROADENED CONCEPTION OF OPERATIONAL DEFINITION AND OF THE PROGRAM OF OPERATIONAL ANALYSIS

The terms "operational meaning" and "operational definition," as well as many of the pronouncements made in operationist writings, convey the suggestion that the criterions of application for any scientific term must ultimately refer to the outcome of some specified type of manipulation of the subject matter under investigation. Such emphasis would evidently be overly restrictive. An operational definition gives experiential meaning to the term it introduces because it enables us to decide on the applicability of that term to a given case by observing the response the case shows under specifiable test conditions. Whether these conditions can be brought about at will by "instrumental operations" or whether we have to wait for their occurrence is of great interest for the practice of scientific research, but it is inessential in securing experiential import for the defined term; what matters for this latter purpose is simply that the relevant test conditions and the requisite response be of such kind that different investigators can ascertain, by direct observation and with reasonably good agreement, whether, in a given case, the test conditions are realized and whether the characteristic response does occur.

Thus, an operational definition of the simplest kind—one that, roughly speaking, refers to instrumental operations only—will have to be construed more broadly as introducing a term by the stipulation that it is to apply to all and only those cases which, under specified observable conditions *S*, show a characteristic observable response *R*.

However, an operational definition cannot be conceived as specifying that the term in question is to apply to a given case only if *S* and *R* actually occur in that case. Physical bodies, for example, are asserted to have masses, temperatures, charges, and so on, even at times when these magnitudes are not being measured. Hence, an operational definition of a concept—such as a property or a relationship, for example—will have to be understood as ascribing the concept to all those cases that *would* exhibit the characteristic response if the test conditions *should* be realized. A concept thus characterized is clearly not "synonymous with the corresponding set of operations."[8] It constitutes not a manifest but a potential character, namely, a disposition to exhibit a certain characteristic response under specified test conditions.

But to attribute a disposition of this kind to a case in which the specified test condition is not realized (for example, to attribute solubility-in-water to a lump of sugar that is not actually put into water) is to make a generalization, and this involves an inductive risk. Thus, the application of an operationally defined term to an instance of the kind here considered would have to be adjudged "not safe" in precisely the same sense in which Bridgman insists it is "not safe" to assume that two procedures of measurement that have yielded the same results in the past will continue to do so in the future. It is now clear that if we were to reject any procedure that involves an inductive risk, we would be prevented not only from using more than one operational criterion in introducing a given term but also from ever applying a disposition term to any case in which the characteristic manifest conditions of application are not realized; thus, the use of dispositional concepts would, in effect, be prohibited.

A few remarks might be added here concerning the non-instrumental operations countenanced for the introduction especially of theoretical terms. In operationist writings, those symbolic procedures have been characterized so vaguely as to permit the introduction, by a suitable choice of "verbal" or "mental" operations, of virtually all those ideas that operational analysis was to prohibit as devoid of meaning. To meet this difficulty, Bridgman has suggested a distinction between "good" and "bad" operations;[9] but he has not provided a clear criterion for this distinction. Consequently, this idea fails to plug the hole in the operationist dike.

If the principles of operationism are to admit the theoretical constructs of science but to rule out certain other kinds of terms as lacking experiential, or operational, meaning, then the vague requirement of definability by reference to instrumental and "good" symbolic operations must be replaced by a precise characterization of the kinds of sentences that may be used to introduce, or specify the meanings of, "meaningful" nonobservational terms on the basis of the observational vocabulary of science. Such a characterization would eliminate the psychologistic notion of mental operations in favor of a specification of the logicomathematical concepts and procedures to be permitted in the context of operational definition.

The reference just made to the observational vocabulary of science is essential to the idea of operational definition; for it is in terms of this vocabulary that the test conditions and the characteristic response specified in an operational definition are described and by means of which, therefore, the meanings of operationally defined terms are ultimately characterized. Hence, the intent of the original operationist insistence on intersubjective repeatability of the defining operations will be respected if we require that the terms included in the observational vocabulary must refer to attributes (properties and relationships) that are directly and publicly observable—that is, whose presence or absence can be ascertained, under suitable conditions, by direct observation, and with good agreement among different observers.[10]

In sum, then, a precise statement and elaboration of the basic tenets of operationism require an explication of the logical relationships between theoretical and observational terms, just as a precise statement and elaboration of the basic tenets of empiricism require an explication of the logical relationships connecting theoretical sentences with observation sentences describing potential data of direct observation.

SPECIFICATION OF MEANING BY EXPLICIT DEFINITION AND BY REDUCTION

Initially, it may appear plausible to assume that all theoretical terms used in science can be fully defined by means of the observational vocabulary. There are various reasons, however, to doubt this assumption.

First of all, there exists a difficulty concerning the definition of the scientific terms that refer to dispositions—and, as is noted in a foregoing paragraph, all the terms introduced by operational definition have to be viewed as dispositional in character. Recent logical studies strongly suggest that dispositions can be defined by reference to manifest characteristics, such as those presented by the observational vocabulary, only with the help of some "nomological modality" such as the concept of nomological truth, that is, truth by virtue of general laws of nature.[11] But a concept of this kind is presumably inadmissible under operationist standards, since it is neither a directly observable characteristic nor definable in terms of such characteristics.

Another difficulty arises when we attempt to give full definitions, in terms of observables, for quantitative terms such as "length in centimeters," "duration in seconds," "temperature in degrees Celsius." Within scientific theory, each of these is allowed to assume any real-number value within a certain interval; and the question therefore arises whether each of the infinitely many permissible values, say of length, is capable of an operational specification of meaning. It can be shown that it is impossible to characterize every one of the permissible numerical values by some truth-functional combination of observable characteristics, since the existence of a threshold of discrimination in all areas of observation allows for only a finite number of nonequivalent combinations of this kind.[12]

Difficulties such as these suggest the question whether it is not possible to conceive of methods more general and more flexible than definition for the introduction of scientific terms on the basis of the observational vocabulary. One such method has been developed in considerable detail by Carnap. It makes use of so-called reduction sentences, which constitute a considerably generalized version of definition sentences and are especially well suited for a precise reformulation of the intent of operational definitions. As we noted earlier, an operational definition of the simplest kind stipulates that the concept it introduces, say C, is to apply to those and only those cases which, under specified test conditions S, show a certain characteristic response R. In Carnap's treatment, this stipulation is replaced by the sentence

$$Sx \rightarrow (Cx \equiv Rx) \quad (1)$$

or, in words: If a case x satisfies the test condition S, then x is an instance of C if and only if x shows the response R. Formula (1), called a bilateral reduction sentence, is not a full definition (which would have to be of the form $Cx \equiv$ . . . , with Cx constituting the definiendum); it specifies the meaning of Cx, not for all cases, but only for those that satisfy the condition S. In this sense, it constitutes only a partial, or conditional, definition for C.[13] If S and R belong to the observational vocabulary of science, formula (1) schematizes the simplest type of operational definition, which invokes (almost) exclusively instrumental operations or, better, experiential findings. Operational definitions that also utilize symbolic operations would be represented by chains of reduction sentences containing logical or mathematical symbols. Some such symbols occur even in formula (1), however; and clearly, there can be no operational definition that makes use of no logical concepts at all.

INTERPRETATIVE SYSTEMS

Once the idea of a partial specification of meaning is granted, it appears unnecessarily restrictive, however, to limit the sentences effecting such partial interpretation to reduction sentences in Carnap's sense. A partial specification of the meanings of a set of nonobservational terms might be expressed, more generally, by one or more sentences that connect those terms with the observational vocabularly but do not have the form of reduction sentences. And it seems well to countenance, for the same purpose, even stipulations expressed by sentences containing only nonobservational terms; for example, the stipulation that two theoretical terms are to be mutually exclusive may be regarded as a limitation and, in this sense, a partial specification of their meanings.

Generally, then, a set of one or more theoretical terms, $t_1, t_2, \ldots, t_n$, might be introduced by any set M of sentences such that (i) M contains no extra-logical terms other than $t_1, t_2, \ldots, t_n$, and observation terms, (ii) M is logically consistent, and (iii) M is not equivalent to a truth of formal logic. The last two of these conditions serve merely to exclude trivial extreme cases. A set M of this kind will be referred to briefly as an *interpretative system*, its elements as *interpretative sentences.*

Explicit definitions and reduction sentences are special types of interpretative sentences, and so are the meaning postulates recently suggested by Kemeny and Carnap.[14]

The interpretative sentences used in a given theory may be viewed simply as postulates of that theory,[15] with all the observation terms, as well as the terms introduced by the interpretative system, being treated as primitives. Thus construed, the specification of the meanings of nonobservational terms in science resembles what has sometimes been called the implicit definition of the primitives of an axiomatized theory by its postulates. In this latter procedure, the primitives are all uninterpreted, and the postulates then impose restrictions on any interpretation of the primitives that is to turn the postulates into true sentences. Such restrictions may be viewed as partial specifications of meaning. The use of interpretative systems as here envisaged has this distinctive peculiarity, however: The primitives include a set of terms—the observation terms—which are antecedently understood and thus not in need of any interpretation, and by reference to which the postulates effect a partial specification of meaning for the remaining, nonobservational, primitives. This partial specification again consists in limiting those interpretations of the nonobservational terms that will render the postulates true.

IMPLICATIONS FOR THE IDEA OF EXPERIENTIAL MEANING AND FOR THE DISTINCTION OF ANALYTIC AND SYNTHETIC SENTENCES IN SCIENCE

If the introduction of nonobservational terms is conceived in this broader fashion, which appears to accord with the needs of a formal reconstruction of the language of empirical science, then it becomes pointless to ask for the operational definition or the experiential import of any one theoretical term. Explicit definition by means of observables is no longer generally available, and experiential—or operational—meaning can be attributed only to the set of all the nonobservational terms functioning in a given theory.

Furthermore, there remains no satisfactory general way of dividing all conceivable systems of theoretical terms into two classes: Those that are scientifically significant and those that are not; those that have experiential import and those that lack it. Rather, experiential, or operational, significance appears as capable of gradations. To begin with one extreme possibility: The interpretative system *M* introducing the given terms may simply be a set of sentences in the form of explicit definitions that provide an observational equivalent for each of those terms. In this case, the terms introduced by *M* have maximal experiential significance, as it were. In another case, *M* might consist of reduction sentences for the theoretical terms; these will enable us to formulate, in terms of observables, a necessary and a (different) sufficient condition of application for each of the introduced terms. Again *M* might contain sentences in the form of definitions or reduction sentences for only some of the nonobservational terms it introduces. And finally, none of the sentences in *M* might have the form of a definition or of a reduction sentence; and yet, a theory whose terms are introduced by an interpretative system of this kind may well permit of test by observational findings, and in this sense, the system of its nonobservational terms may possess experiential import.[16]

Thus, experiential significance presents itself as capable of degrees, and any attempt to set up a dichotomy allowing only experientially meaningful and experientially meaningless concept systems appears as too crude to be adequate for a logical analysis of scientific concepts and theories.

The use of interpretative systems is a more inclusive method of introducing theoretical terms than the method of meaning postulates developed by Carnap and Kemeny. For although meaning postulates are conceived as analytic and, hence, as implying only analytic consequences, an interpretative system may imply certain sentences that contain observation terms but no theoretical terms and are neither formal truths of logic nor analytic in the customary sense. Consider, for example, the following two interpretative sentences, which form what Carnap calls a reduction pair, and which interpret C by means of observation predicates, R_1, S_1, R_2, S_2:

$$S_1x \rightarrow (R_1x \rightarrow Cx) \tag{2.1}$$

$$S_2x \rightarrow (R_2x \rightarrow -Cx). \tag{2.2}$$

Since in no case the sufficient conditions for C and for $-C$ (non-C) can be satisfied jointly, the two sentences imply the consequence[17] for every case x,

$$-(S_1x \cdot R_1x \cdot S_2x \cdot R_2x), \tag{3}$$

that is, no case x exhibits the attributes S_1, R_1, S_2, R_2 jointly. Now an assertion of this kind is not a truth of formal logic, nor can it generally be viewed as true solely by virtue of the meanings of its constituent terms. Carnap therefore treats this consequence of formulas (2.1) and (2.2) as empirical and as expressing the factual content of the reduction pair from which it was derived. Occurrences of this kind are by no means limited to reduction sentences, and we see that in the use of interpretative systems, specification of meaning and statement of empirical fact—two functions of language often considered as completely distinct—become so intimately bound up with each other as to raise serious doubt about the advisability or even the possibility of preserving that distinction in a logical reconstruction of science. This consideration suggests that we dispense with the distinction, so far maintained for expository purposes, between the interpretative sentences, included in M, and the balance of the sentences constituting a scientific theory: We may simply conceive of the two sets of sentences as constituting one "intepreted theory."

The results obtained in this brief analysis of the operationist view of significant scientific concepts are closely analogous to those obtainable by a similar study of the logical empiricist view of significant scientific statements, or hypotheses.[18] In the latter case, the original requirement of full verifiability or full falsifiability by experiential data has to give way to the more liberal demand for confirmability—that is, partial vertifiability. This demand can be shown to be properly applicable to entire theoretical systems rather than to individual hypotheses—a point emphasized, in effect, already by Pierre Duhem. Experiential significance is then seen to be a matter of degree, so that the originally intended sharp distinction between cognitively meaningful and cognitively meaningless hypotheses (or systems of such) has to be abandoned; and it even appears doubtful whether the distinction between analytic and synthetic sentences can be effectively maintained in a formal model of the language of empirical science.

NOTES

1. P.W. Bridgman, "Some general principles of operational analysis" and "Rejoinders and second thoughts," *Psychol. Rev.*, LII (1945), 246; "The nature of some of our physical concepts," *Brit. J. Phil. Sci.*, I (1951), 258.

2. ———, "Operational analysis," *Phil. Sci.*, V (1938), 123; *Brit. J. Phil. Sci.*, I, (1951), 258.

3. ———, *Phil. Sci.*, V (1938), 116.

4. ———, *Brit. J. Phil. Sci.*, I (1951), 260.

5. ———, *The Logic of Modern Physics* (New York: Macmillan, 1927), pp. 6, 23–24; *Phil. Sci.*, V (1938), 121; *Psychol. Rev.*, LII (1945), 247; "The operational aspect of meaning," *Synthése*, VIII (1950–51), 255.

6. ———, *Psychol. Rev.*, LII (1945), 246.

7. ———, *The Logic of Modern Physics*, p. 28.

8. ———, *ibid.*, p. 5; qualified by Bridgman's reply [*Phil. Sci.*, V (1938), 117] to R. B. Lindsay, "A critique of operationalism in physics," *Phil. Sci.*, IV (1937), a qualification that was essentially on the ground, quite different from that given in the present paper, that operational meaning is only a necessary, but presumably not a sufficient characteristic of scientific concepts.

9. ———, *Phil. Sci.*, V (1938), 126; "Some implications of recent points of view in physics," *Rev. intern. phil.*, III (1949), 484. The intended distinction between good and bad operations is further obscured by the fact that in Bridgman's discussion the meaning of "good operation" shifts from what might be described as "operation whose use in operational definition insures experiential meaning and testability" to "scientific procedure—in some very broad sense—which leads us to correct predictions."

10. The condition thus imposed upon the observational vocabulary of science is of a pragmatic character; it demands that each term included in that vocabulary be of such a kind that under suitable conditions, different observers can, by means of direct observation, arrive at a high degree of agreement on whether the term applies to a given situation. The expression *coincides with* as applicable to instrument needles and marks on scales of instruments is an example of a term meeting this condition. That human beings are capable of developing observational vocabularies that satisfy the given requirement is a fortunate circumstance: without it, science as an intersubjective enterprise would be impossible.

11. To illustrate briefly, it seems reasonable, prima facie, to define "x is soluble in water" by "if x is put in water then x dissolves." But if the phrase *if . . . then . . .* is here construed as

the truth-functional, or "material," conditional, then the objects qualified as soluble by the definition include, among others, all those things that are never put in water—no matter whether or not they are actually soluble in water. This consequence—one aspect of the "paradoxes of material implication"—can be avoided only if the aforementioned definiens is construed in a more restrictive fashion. The idea suggests itself to construe "x is soluble in water" as short for "by virtue of some general laws of nature, x dissolves if x is put in water," or briefly, "it is nomologically true that if x is put in water then x dissolves." The phrase *if . . . then . . .* may now be understood in the truth-functional sense again. However, the acceptability of this analysis depends, of course, upon whether nomological truth can be considered as a sufficiently clear concept. For a fuller discussion of this problem complex, see especially R. Carnap, "Testability and meaning," *Phil. Sci.*, III (1936) and IV (1937) and N. Goodman, "The problem of counterfactual conditionals," *J. Phil.*, XLIV (1947).

12. In other words, it is not possible to provide, for every theoretically permissible value r of the length $l(x)$ of a rod x, a definition of the form

$$[l(x) = r] =_{df} C(P_1x, P_2x, \ldots, P_nx),$$

where $P_1, P_2, \ldots, P_n$ are observable characteristics, and the definiens is an expression formed from $P_1x, P_2x, \ldots, P_nx$ with help of the connective words *and or*, and *not* alone.

It is worth noting, however, that if the logical constants allowed in the definiens include, in addition to truth-functional connectives, also quantifiers and the identity sign, then a finite observational vocabulary may permit the explicit definition of a denumerable infinity of further terms. For instance, if "x spatially contains y" and "y is an apple" are included in the observational vocabulary, then it is possible to define the expressions "x contains 0 apples," "x contains exactly 1 apple," "x contains exactly 2 apples," and so forth, in a manner familiar from the Frege-Russell construction of arithmetic out of logic. Yet even if definitions of this type are countenanced—and no doubt they are in accord with the intent of operationist analysis—there remain serious obstacles for an operationist account of the totality of real numbers which are permitted as theoretical values of length, mass, and so forth. On this point, see C. G. Hempel, *Fundamentals of Concept Formation in Empirical Science* (Univ. of Chicago Press, Chicago, 1952), Sec. 7. Gustav Bergman, in his contribution to the present symposium, deplores this argument—although he agrees with its point—on the ground that it focuses attention on a characteristic shared by all quantitative concepts instead of bringing out the differences between, say, length and the psi-function. He thinks this regretable because, after all, as he puts it, "the real numbers are merely a part of the logical apparatus; concept formation is a matter of the descriptive vocabulary." I cannot accept the suggestion conveyed by this statement. To be sure, the theory of real numbers can be developed as a branch (or as an extension) of logic; however, my argument concerns not the definability of real numbers in logical terms, but the possibility of formulating an observational equivalent for each of the infinitely many permissible real-number values of length, temperature, and so forth. And this is clearly a question concerning the descriptive vocabulary rather than merely the logical apparatus of empirical science. I quite agree with Bergmann, however, that it would be of considerable interest to explicate whatever logical differences may obtain between quantitative concepts which, intuitively speaking, exhibit different degrees of theoretical abstractness, such as length on the one hand and the psi-function on the other.

13. The use of reduction sentences circumvents one of the difficulties encountered in the attempt to give explicit and, thus, complete definitions of disposition terms: The conditional and biconditional signs occurring in formula (1) may be construed truth-functionally without giving rise to underirable consequences of the kind characterized in n. 11. For details, see R. Carnap, "Testability and meaning," *Phil. Sci.* (1936–37), Part II; also C. G. Hempel, *Fundamentals of Concept Formation in Empirical Science*, Secs. 6 and 8. Incidentally, the use of nomological concepts is not entirely avoided in Carnap's procedure; the reduction sentences that are permitted for the introduction of new terms are required to satisfy certain conditions of logical or of nomological validity. See R. Carnap, *Phil. Sci.*, III and IV (1936–37), 442–43.

14. J. G. Kemeny, "Extension of the methods of inductive logic," *Philosophical Studies*, III (1952); R. Carnap, "Meaning postulates," *ibid.*, III (1952).

15. For the case of Carnap's reduction sentences, the postulational interpretation was suggested to me by N. Goodman and by A. Church.

16. This is illustrated by the following simple model case: The theory T consists of the sentence $(x)((C_1x \cdot C_2x) \rightarrow C_3x)$ and its logical consequences; the three "theoretical" terms occurring in it are introduced by the interpretative set M consisting of the sentences $O_1x \rightarrow (C_1x \cdot C_2x)$ and $(C_1x \cdot C_3x) \rightarrow (O_2x \vee O_3x)$, where O_1, O_2, O_3, belong to the observational vocabulary. As is readily seen, T permits, by virtue of M, the "prediction" that if an object has the observable property O_1 but lacks the observable property O_2, then it will have the observable property O_3. Thus T is susceptible to experiential test, although M provides for none of its constituent terms both a necessary and a sufficient observational, or operational, criterion of application.

17. Carnap calls it the representative sentence of the pair of formulas (2.1) and (2.2). See R. Carnap, *Phil. Sci.*, III and IV (1936–37), pp. 444 and 451. Generally, when a term is introduced by several reduction sentences representing different operational criterions of application, then the agreement among the results of the corresponding procedures, which must be presupposed if the reduction sentences are all to be compatible with one another, is expressed by the representative sentence associated with the given set of reduction sentences. The representative sentence reflects, therefore, the inductive risk which, as Bridgman has stressed, is incurred by using more than one operational criterion for a given term.

18. C. G. Hempel, "Problems and changes in the empiricist criterion of meaning," *Rev. intern. phil.*, IV (1951), and "The concept of cognitive significance: a reconsideration," *Proc. Am. Acad. Arts Sci.*, LXXX (1951); W. V. Quine, "Two dogmas of empiricism," *Phil. Rev.* XL (1951).

Israel Scheffler

THE FICTIONALIST VIEW OF SCIENTIFIC THEORIES

13. THE PROBLEM OF THEORETICAL OR TRANSCENDENTAL TERMS

If disposition terms required for descriptive adequacy may be accommodated in *E* without marring its observational character, as we have above argued, we still face a major obstacle to *E*'s adequacy, i.e., the case of so-called abstract, theoretical, or transcendental terms. These terms, unlike dispositional predicates, do not generally purport to apply to entities within the range of application of our clearly observational terms. They are typically nonobservational and beyond the reach even of reduction sentences. It is generally claimed that they are required not because of their usefulness in expressing available observational evidence, but rather because, by their introduction in the context of certain developed theories, comprehensive relationships become expressible in desirable ways on the observable level.

Such theories seem to commit us to a new range of entities as values of variables to which their transcendental terms may be attached in existentially quantified statements, e.g., "There is an electron"; "Something is a position." For the entities here required are not such as are qualifiable by our hitherto accepted observational predicates. We have, for example, no clearly true sentence such as "*x* is an electron and *x* is red," or "*x* is an electron and *x* is non-red," for any value of *x*. Historically, this is perhaps related to the distinction between primary and secondary qualities, a distinction which, in some form it is dangerous to overlook in the interpretation of modern scientific theories. Thus, it is by now well known that confusion results both from the popularization of advanced theories through pictorial description in the common language of

---

From "Prospects of a Modest Empiricism," *Review of Metaphysics* (1957). Reprinted by permission of the author and the editor.

observation, and from the ascription of exclusive reality to the entities presupposed by such of these theories as we deem true. But to maintain the relevant distinction among ranges seems clearly to mean the abandonment of even our modest, revised empiricism. For the predicates appropriate to one of these ranges are nonobservational, and to admit the necessity of theories couched in such terms in order for E to be adequate is to admit that no adequate, purely observational E exists which is even a sufficient condition of cognitive significance. (Thus those who insist on some such distinction by saying that "theoretical entities" have only such properties as are attributed to them by their respective theoretical contexts are, if they let the matter rest here, abandoning even our modest version of empiricism.) Furthermore, aside from empiricism, the perpetuation of a distinction among ranges seems to have generally puzzling aspects, which have troubled philosophers of science recurrently: (a) If adequacy requires that clearly nonobservational terms be eligible for admission into the language of science, then, since these include terms ordinarily deemed meaningless, *what* are we believing in committing ourselves to science? (b) If science explains by providing *true* premises from which the relevant problematic data may be derived, how can theories that are *meaningless* in the ordinary sense be said to explain? In short, *even if we are not interested in relating cognitive significance to observationality*, but are concerned with constructing some weakest language adequate for formulating some specified segment of our scientific beliefs, we may be troubled to find ourselves explicitly allowing clearly meaningless units to be built into our language structure, in some ordinary sense of "meaningless."

14. PRAGMATISM

The qualification embodied in the final phrase of the last sentence is a clue to one widespread attempt to cope with the troubles discussed: The view which takes the *system* to be the unit of significance, and adopts a wider notion of meaningfulness in deference to scientific practice. This view I shall label "pragmatism." The alternative reaction, which I shall discuss in later sections, I call "fictionalism."[1] Both views may be considered independently of the issue of empiricism, it seems to me, in view of the final set of remarks in the previous section. For however we delimit the terms initially admissible (whether by reference to observation or not), we face the problem of interpreting all the others which seem to be required in increasing numbers with the theoretical development of science. Nevertheless, a consideration of this problem in abstraction from empiricism will bear rather directly on it as embodied in our revised form: Pragmatism will negate this empiricism, while fictionalism will render it at least possible.

Pragmatism, then (in my terminology), accepts as fact that no initial listing of admissible terms is sufficient for the formulation of our scientific beliefs, and that the admission of any term is conceivable (hence legitimate) on the grounds of its utility in prediction and theoretical simplification. It admits, furthermore, that some such terms are meaningless, *in one usual sense*. But here it takes the bull by the horns, claiming that this sense is irrelevant for analyzing our scientific beliefs and practices. For any sense of "meaningless" which renders what is predictively useful meaningless is inadequate for philosophy of science, however relevant it may be in other contexts. If science finds, e.g., transcendental theories fruitful within whole systematic contexts, our notions of cognitive significance must reflect this fact. If science introduces terms not only by reference to pre-scientific usage or explicit test-methods, but within the network of whole theoretical frameworks justified by their predictive utility in subsequent inquiry, this is a bona fide fact about cognitive significance, not a problem. We must, accordingly, for the pragmatists, admit the *significance* of whole systems with unrestricted vocabularies, provided that they are, at some points, functionally tied to our initially specified language. Since, however, this proviso excludes no term at all, and virtually no system at all (every system meets this requirement by addition of one conjunct in the initial language, and every term is part of some system meeting this requirement), pragmatism supplements it by stressing some simplicity factor which presumably is to eliminate certain systems, but which is not to be so stringent as to eliminate every system which overflows the bounds of the initially specified language. In no account is the treatment of simplicity very precise, but in some accounts it is intended as a matter of degree so that cognitive significance is broadened correspondingly. It is, further, not very clear how considerations of simplicity are to be applied in determination of *confirmedness* or *truth* as distinct from *significance*. Nevertheless, the pragmatist reaction to the problem of interpreting transcendental terms and their theoretical contexts is clearly to accept their fruitfulness and to defend, in consequence, a broader notion of significance applicable to whole systems. This systematic emphasis is supported also by reference to well-known analyses of testing which show the theoretical revisability of every segment of a system when any segment is ostensibly under review. A corollary of this pragmatist treatment is its insistence that questions of ontology are scientific questions, since it takes the range of significant ontological assertion to be solely a function of scientific utility in practice and denies all independent language restrictions based on intuitive clarity or observationality: And rejecting such independent restrictions, it solves the two initial difficulties noted above (at the end of section 13) by (a) denying that we believe meaningless assertions in committing ourselves to science and by (b) affirming the possible truth and explanatory power of transcendental theories.

To what extent is the pragmatist position in favor of a broader notion of significance positively supported by the arguments it presents? Its strong point is obviously it congruence with the de facto scientific use of transcendental theories and with the interdependence of parts of a scientific system undergoing test. These facts are, however, not in themselves *conclusive* evidence for significance, inasmuch as many kinds of things are used in science with no implication of cognitive significance, i.e., truth-or-falsity; and many things are interdependent under scientific test without our feeling that they are therefore included within the cognitive system of our assertions. Clearly "is useful," "is fruitful," "is subject to modification under test," etc., are applicable also to non-linguistic entities, e.g., telescopes and electronic computers. On the other hand, even linguistic units judged useful and controllable via empirical test may conceivably be construed as nonsignificant machinery, and such construction is not definitely ruled out by pragmatist arguments. This, even if we accept pragmatism's positive grounds, we *need* not broaden our original notion of literal significance. And it further follows that our revised empiricism is not refuted by pragmatism.

15. PRAGMATISM AND FICTIONALISM

But if not refuted, our empiricism remains beset with the problem of interpreting transcendental terms and theories. If pragmatism's positive grounds do not, that is, *establish* the literal significance of transcendental theories, it is not thereby demonstrated that they are eliminable or otherwise interpretable as nonbeliefs, i.e., mere instruments. Any view which takes them to be either I call "fictionalism." Clearly if fictionalism can show how transcendental terms are eliminable from our corpus of scientific beliefs, it will have removed transcendental theories from the domain of beliefs which need to be encompassed in *E*, and it will have destroyed a major obstacle to our revised empiricism. Short of showing eliminability, if fictionalism can plausibly construe transcendental theories as mere machinery without literal meaning, it will avoid the need for expressing such theories in *E*, and again make way for our revised empiricism.

16. INSTRUMENTALISTIC FICTIONALISM

Perhaps the easiest, and by far the most popular type of fictionalism is one which simply disavows the belief-character of transcendental theories without claiming their eliminability from scientific discourse. Indeed such a fictionalism often goes with a positive indifference to the question of their eliminability, or even champions their ineliminability; we might aptly label this type "instrumentalism," and note in passing that some writers have vacillated between pragmatism and instrumentalism (in our present terminology) or have confused the two. Instrumentalistic fictionalism, then, holds that some scientific theories are not significant, but that they are moreover not intended as formulations of belief or as truths, being employed simply as mechanical devices for coordinating or generating bona fide assertions. Hence, again, transcendental theories are said to pose no problem; since they do not represent *beliefs*, we need not worry about including them within any deliberate statement of our beliefs in some restricted language. Our problem, it will be recalled, was that clearly meaningless terms seem required for adequate expression of our scientific beliefs. Whereas pragmatism's answer is to deny that any terms usefully employed in science are meaningless in the relevant sense, fictionalism's answer is to deny that our objectionable terms are required for expression of *beliefs*, though they may be otherwise required. And our instrumentalistic variant supports this denial not by showing how to eliminate such terms from scientific language, but rather by stipulating how "belief" is to be understood. Correspondingly, certain further stipulations are generally accepted as corollaries, e.g., that transcendental theories be said to *hold* or *fail* rather than to be *true* or *false*, that they are adopted or abandoned rather than believed or denied, etc. Thus instrumentalism takes care of the difficulties mentioned (a) by insisting that we do not strictly *believe* but *hold* or *employ* some statements in science and (b) by generalizing the concept of explanation to allow such held theories to serve as explanatory grounds.

If pragmatism's positive grounds seemed to us unconvicing, instrumentalism's positive grounds seem to consist just in the intuitive meaningless of transcendental theories. But the point at issue is whether science requires us to believe such theories, and this point is not met but begged by arguing that the answer is negative since the theories are intuitively meaningless. We can, however, be more generous to both pragmatism and instrumentalism by taking them not as arguments but as decisions or resolutions: Pragmatism's denial of the meaningless of transcendental theories represents a decision to apply to them the ordinary language of truth and falsity, and this, coupled with denial of the need for further interpretation, involves a rejection of even a modified empiricism, as we have above formulated it. Instrumentalism's denial of the belief-character of transcendental theories represents a decision to talk about such theories in different and special ways without any further changes. Taken as basic decisions, there would seem to be no way of refuting either position, and to this extent at least, ontology is independent of science. There is no way to refute the instrumentalist's denial of the belief-character of various theories which he continues to employ. We may charge his implicit conception of the nature of belief with being tenuous and merely verbal, and we may declare his disavowals of belief to be rather hollow unless he gives up using the sentences which he claims intellectually to disavow. Yet, if he sticks to his guns, and continues to remind us that we all *use* all kinds of objects which we hold meaningless, and feel no guilt upon reflection in continuing to use them, then he is secure. Ontology, then, is relative to the person, and independent of the used language. Just as our common use of available technology does not commit us all equally to the same beliefs, so our common use of scientific language does not dictate that we should all draw the same line between literal sense and nonsense therein.

Coming back to our modified empiricism now, it appears that if pragmatism, in choosing to deny it, does not thereby refute it, instrumentalism renders it trivial. For if the range of our *beliefs* is freely specifiable by intellectual decision independently of the *content of our discourse*, we can always guarantee *E*'s adequacy by simply deciding to exclude recalcitrant sentences from this range.

Our judgments of recalcitrance will, of course, vary; but one consistent with our modified empiricism is trivially always possible.

If, however, we interpret our modified empiricist problem more stringently and more objectively, i.e., as not allowing for such a trivial answer, we must require of the empiricist fictionalist not simply that he appropriately adjust his terminology of belief but that he provide a method for eliminating transcendental terms and theories from scientific *discourse*, or of treating them within his discourse otherwise than as significant.

17. SYNTACTIC FICTIONALISM

One such course open to the fictionalist is to provide a syntax for transcendental theories. Goodman and Quine[2] have in part thus dealt with the problem of treating mathematics nominalistically. Unable, at the time of their study, to translate all of mathematics into a nominalistic language, they developed a nominalistic syntax language enabling them to talk *about* and deal with the untranslated residue, thus *independently* supporting the claim that this residue could be treated as mere machinery without literal significance. Though, in one sense, they did not eliminate this residue, they did go considerably beyond a mere statement that it might be considered as machinery only. For they provided an alternative language without the (to them at that time) objectionable features of the original, such that it was capable of doing much the same job. As they put their view, "Our position is that the formulas of platonistic mathematics are, like the beads of an abacus convenient computational aids which need involve no question of truth. What is meaningful and true in the case of platonistic mathematics as in the case of the abacus is not the apparatus itself, but only the description of it: The rules by which it is constructed and run. These rules we do understand, in the strict sense that we can express them in purely nominalistic language. The idea that classical mathematics can be regarded as mere apparatus is not a novel one among nominalistically minded thinkers; but it can be maintained only if one can produce, as we have attempted to above, a syntax which is itself free from platonistic commitments. At the same time, every advance we can make in finding direct translations for familiar strings of marks will increase the range of the meaningful language at our command (p. 122)."

Such a syntactical approach has relevance far beyond the question of platonistic mathematics. It is in general open to the fictionalist who wishes to disavow the belief-character of some segment of received scientific discourse in more than the trivial sense discussed above in connection with instrumentalism. In the special case of nominalism, it was by no means initially obvious that a syntax could be constructed without platonistic features. Having such a syntax for mathematics, it seems possible to extend it to specified parts of empirical science by addition of predicates applicable to the extra-logical notion contained therein. In particular, such syntax could be developed for transcendental theories which the fictionalist cannot eliminate through translation but which he finds it objectionable to take as significant. For the non-nominalist who objects to taking as significant some particular transcendental theory, the task is, of course, much easier, for he has available all the tools of platonistic syntax. In a less trivial sense than that of instrumentalism, then, our modified empiricism may be feasible through syntactic construction. Note again, incidentally, that ontology turns out independent of our received scientific discourse, through the possibility of variable syntactic reinterpretation, and that only such elements as are needed for the applicability of syntactic predictes may be sufficient in the extreme case.

18. ELIMINATIVE FICTIONALISM

We may, finally, require our modified empiricism to show how transcendental terms may be eliminated from scientific discourse in favor of some other object-language discourse which is equivalent in some appropriate sense. Here it is well to recall that transcendental theories are justified generally as making possible the statement of comprehensive relationships (in desirable ways) on the observable level. Thus, if a way could be shown of appropriately stating these observational relationships in some theory, *S*, which otherwise differed from its transcendental counterpart only by lacking sentences with any transcendental term, *S* would be, in a reasonable sense, equivalent to that counterpart.

One such method is that of Craig, who states as one of his results, ". . . if *K* is any recursive set of nonlogical (individual, function, predicate) constants, containing at least one predicate constant, then there exists a system whose theorems are exactly those theorems of *T* in which no constants other than those of *K* occur. In particular, suppose that *T* expresses a portion of a natural science, that the constants of *K* refer to things or events regarded as 'observable,' and that the other constants do not refer to 'observables' and hence may be regarded as 'theoretical' or 'auxiliary.' Then there exists a system which does not employ 'theoretical' or 'auxiliary' constants and whose theorems are the theorems of *T* concerning 'observables.'"[3]

Professor Hempel, discussing Craig's method, states concisely what is involved and what sense of "equivalence" is here relevant: "Craig's result shows that no matter how we select from the total vocabulary V_T' of an interpreted theory *T*' a subset V_B of experiential or observational terms, the balance of V_T', constituting the 'theoretical terms,' can always be avoided in sense (c)." This sense, which Hempel distinguishes from definability and translatability, he calls "functional replaceability" and describes as follows, "The terms of *T* might be said to be avoidable if there exists another theory T_B, conched in terms of V_B, which is 'functionally equivalent' to *T* in the sense of establishing exactly the same deductive connections between V_B sentences as *T*."

Professor Hempel offers, however, two reasons against the scientific use of

Craig's method: "No matter how welcome the possibility of such replacement may be to the epistemologist." One reason is that the functionally equivalent replacing system constructed by Craig's method "always has an infinite set of postulates, irrespective of whether the postulate set of the original theory is finite or infinite, and that his result cannot be essentially improved in this respect. . . . This means that the scientist would be able to avoid theoretical terms only at the price of forsaking the comparative simplicity of a theoretical system with a finite postulational basis, and of giving up a system of theoretical concepts and hypotheses which are heuristically fruitful and suggestive—in return for a practically unmanageable system based upon an infinite, though effectively specified, set of postulates in observational terms."[4]

It should be obvious that any proposal like Craig's for meeting our present demand for elimination of transcendental terms will be judged in various ways in accordance with varying approval of its tools and subsidiary concepts. Moreover, such variation may be independent of the question of modified empiricism as such. In particular, a dissatisfaction with systems containing infinite, effectively specified sets of postulates may or may not be justified, but is at any rate independent of modified empiricism as we have formulated it. Further, though the relevant notions of heuristic fruitfulness and suggestiveness, simplicity, and practicality are not very precise, suppose it granted that Craig's functionally equivalent system is indeed inferior to its counterpart in all these respects. This is irrelevant to our modified empiricism. If Craig's replacing system renders such empiricism possible, this represents an intellectual gain no worse for the fact that the system is unwieldy and not likely to be used by the practicing scientist. The case is analogous to ordinary definition, where we try to minimize the complexity of our primitive basis at the cost of replacing short and handy definienda by cumbersome definientia in terms of a simple few primitives. Obviously, no one intends these definientia to be used in practice in place of their definienda, but neither does anyone seriously maintain that their formulation therefore represents less of an intellectual gain.

One may however, with Goodman,[5] suggest the infinity of postulates in Craig's replacing system not as representing a practical difficulty, but rather as indicating that the deductive character of the original system is not sufficiently reflected by its replacement. That is to say, if transcendental theories serve to enable finite postulation, no replacement is equivalent *deductively* in every relevant sense if it fails to serve thus also, even though it does accurately reflect the whole class of relevant postulates-or-theorems (assertions) of the original. If specific empiricist programs are to be interpreted in accord with this point of view, then, even granted Craig's result, they are not proven generally achievable, and continue to represent non-trivial problems in individual cases. It seems to me, however, that if we take these programs as requiring simply the reflection of non-transcendental assertions into replacing systems without transcendental terms, then we do not distort traditional notions of empiricism, and we have to acknowledge that Craig's result does the trick; the further cited problems remain but they are independent of empiricism as above formulated.

Professor Hempel's second reason against the scientific use of Craig's method is that "The application of scientific theories in the prediction and explanation of empirical findings involves not only deductive inference, i.e., the exploitation of whatever deductive connections the theory establishes among statements representing potential empirical data, but it also requires procedures of an inductive character, and some of these would become impossible if the theoretical terms were avoided." He illustrates in terms of the following four sentences, where "magnet" is taken to be a theoretical, i.e., nonobservational term:

(5.1) The parts obtained by breaking a rod-shaped magnet in two are again magnets.

(5.2) If x is a magnet, then whenever a small piece y of iron filing is brought into contact with x, then y clings to x. In symbols:

$$Mx \supset (y)(Fxy \supset Cxy)$$

(5.3) Objects b and c were obtained by breaking object a in two, and a was a magnet and rod-shaped.

(5.4) If d is a piece of iron filing that is brought into contact with b, then d will cling to b.

Now, says Hempel, given (5.3) (and assuming (5.2)), we are able to deduce, with the help of (5.1), such sentences as (5.4). But (5.3) is nonobservational, containing Ma, itself not deducible from observational sentences via (5.2) which states only a necessary, but not a sufficient condition for it. Thus, if (5.4) is to be connected by our theory here with other observational sentences, an *inductive* step is necessary, leading to (5.3), i.e., to Ma specifically, from observational sentences, e.g., Ma might be *inductively* based on a number of instances of $Fay \supset Cay$, assuming that we have no instance of $Fay \cdot \sim Cay$. This is so, since such instances confirm $(y)\ (Fay \supset Cay)$, which, by (5.2), partially supports Ma. Thus, our hypothesis (5.1) takes us, in virtue of (5.2), from some observational sentences, i.e., instances of $Fay \supset Cay$, to observational sentences such as (5.4), but the transition requires certain inductive steps along the way, But though Craig's functionally equivalent system retains all the deductive connections among observational sentences of the original system, it does not, in general, retain the inductive connections among such sentences. Hempel concludes, "the transition, by means of the theory, from strictly observational to strictly observational sentences usually requires inductive steps, namely the transition from some set of observational sentences to some non-observational sentence which they support inductively, and which in turn can serve as a premise in the strictly deductive application of the given theory."

With respect to this argument, we might question by what theory of confirmation $(y)\ (Fay \supset Cay)$ supports Ma; it surely is not Hempel's satisfaction criterion of confirmation.[6] But this is irrelevant to the important point brought out by Hempel's argument, viz., that since functionally equivalent systems (of Craig's type) are not logically equivalent to their originals, they need not (on *any* likely view of confirmation) sustain the same confirmation relations as these originals, even among purely observational sentences. And this despite the fact that they do preserve the same deductive relations among such sentences by retaining all original theorems couched in purely observational terms. Thus, if we do not attempt an observational reduction of the *whole* of our theoretical discourse in given scientific domains via definition and translation or syntactic construction, but aim merely to isolate the *observational part* of such discourse, we must be careful to construe this part adequately, i.e., as comprising not only a deductive network but also a wider confirmational range. Specific empiricist programs would then seem to be not achievable generally by means of Craig's result, in the light of Hempel's argument. One further line of attack might be to clarify the inductive relation sufficiently to enable agreement on just which sentences confirm which, relative to a given theoretical system, and then to strive for an independent, observational specification of such confirmation-pairs, to supplement the appropriate Craigian equivalent.[7] A second approach would be to try to meet each specific case by strengthening the replacing functional equivalent with hypotheses designed to yield just those inductive relations borne by the original, in which we are particularly interested.

In the case of Hempel's above example, for instance, the inductive relation in question is between instances of $Fay \supset Cay$ plus the observational parts of (5.3), (objects *b* and *c* were obtained by breaking object *a* in two [*Bbca*] and *a* was rod-shaped [*Ra*]) and (5.4). Such an iductive relation might also be expressed without theoretical terms by the following statement, which, owing to the universal quantification, can at best establish only an inductive relationship between its instances and the observational sentence derived:

$$(x)\{[(y)\ (Fxy \supset Cxy) \cdot Rx] \supset (z)\ [(\exists w)(Bzwx) \supset (u)(Fzu \supset Czu)]\}$$[8]

19. SUMMARY AND CONCLUSION

If our journey has yielded no single, easy solution as a climax, its difficulty has nevertheless earned for us the right to stop and get our bearings. For we have traveled a long way from the conception of empiricism as a shiny, new philosophical doctrine for weeding out obscurantism and cutting down nonsense wherever they crop up. We have, furthermore, seen that even if we take empiricism as the proposal of a general meaning-criterion in terms of translatability into a chosen artificial language, we run into trouble. We have thus come to restrict the empiricist's job to providing merely an adequate sufficient condition of significance on an observational basis, in the form of an observational system capable of housing our scientific beliefs.

Even this restricted task has, however, turned out to have quite difficult obstacles before it. While the inclusion of needed disposition terms seemed to us not as formidable a problem as hitherto thought, we found theoretical terms to be generally resistant to straightforward empiricist interpretation. Considering this difficulty in the light of a number of recent approaches to philosophy of science, we found that the pragmatic rejection of our restricted empiricism does not constitute a refutation, while intrumentalism's easy solution fulfills such empiricism in only the most trivial sense. Taking empiricism's task as the provision of an appropriate modification of scientific discourse itself rather than simply of our notions of belief, we found the possibility of syntactic reinterpretation promising, though less intuitively satisfying then a direct reinterpretation of the object-language of science proper. Our final examination of Craig's method for eliminating theoretical terms wholly from such language while preserving its observational segment intact led to the conclusion that this method is, in itself, incapable of achieving the goal of our restricted empiricism.

It appears, in sum, that even a modest empiricism is presently a hope for clarification and a challenge to constructive investigation rather than a well-grounded doctrine, unless we construe it in a quite trivial way. Empiricists are perhaps best thought of as those who share the hope and accept the challenge—who refuse to take difficulty as a valid reason either for satisfaction with the obscure or for abandonment of effort.

NOTES

1. The labels I introduce here and in following sections are related to, but not intended as names of, specific philosophies associated with familiar historical movements or individual thinkers. They refer rather to characteristic trends, somewhat oversimplified and idealized, perhaps in comparison to actually held philosophies. Nevertheless, they are influential and salient trends and will be recognizable, I hope, as elements of much of the recent literature in philosophy of science and theory of knowledge. As recent illustrations (in a loose sense) of pragmatism, see R. Carnap, "Empiricism, Semantics, and Ontology," *Revue Internationale de Philosophie*, XI (1950), reprinted in L. Linsky, *Semantics and the Philosophy of Language* (Urbana, Il., 1952); and W. V. Quine, "Two Dogmas of Empiricism," *Philosophical Review* (1951), included in W. V. Quine, *From a Logical Point of View* (Cambridge, Mass., 1953); of instrumentalistic fictionalism, see S. E. Toulmin, *The Philosophy of Science* (London, 1953). Regarding vacillation between the latter two trends, see Nagel's discussion of Dewey in *Sovereign Reason* (Glencoe, Ill., 1954), pp. 110–15.

2. N. Goodman and W. V. Quine, "Steps Toward a Constructive Nominalism," *Journal of Symbolic Logic*, XII (1947), 105–22.

3. W. Craig, "On Axiomatizability Within a System," *Journal of Symbolic Logic*, XVIII (1953), 31, text and n. 9. See also W. Craig, "Replacement of Auxiliary Expressions," *Philosophical Review*, LXV (1956), 38–55.

4. C. G. Hempel, "Implications of Carnap's Work for the Philosophy of Science," to appear in the forthcoming Carnap volume of the *Library of Living Philosophers*.

5. This point was made by Goodman in correspondence with the present writer.

6. "A Purely Syntactical Definition of Confirmation," *Journal of Symbolic Logic*, VIII (1943), 122–43. See also C. G. Hempel, "Studies in the Logic of Confirmation," *Mind*, N. S. LIV (1945), 1–26 and 87–121, especially 107 ff.

7. It would also be desirable to clarify the notion of induction sufficiently to enable critical evaluation of the assumption that where the Craigian equivalent fails to reflect the confirmation relationships of the original theoretical system, it is itself confirmed to a lesser degree by

the available evidence on which that original rests. Perhaps even more interesting would be the examination of the analogous assumption for the strengthened functional equivalent discussed immediately below.

8. This formula is due to Professor Hempel, who suggested it to me as an improvement over my original version.

The Observational-Theoretical Distinction

Grover Maxwell

THE ONTOLOGICAL STATUS OF THEORETICAL ENTITIES

That anyone today should seriously contend that the entities referred to by scientific theories are only convenient fictions, or that talk about such entities is translatable without remainder into talk about sense contents or everyday physical objects, or that such talk should be regarded as belonging to a mere calculating device and, thus, without cognitive content—such contentions strike me as so incongruous with the scientific and rational attitude and practice that I feel this paper *should* turn out to be a demolition of straw men. But the instrumentalist views of outstanding physicists such as Bohr and Heisenberg are too well known to be cited, and in a recent book of great competence, Professor Ernest Nagel concludes that "the opposition between [the realist and the instrumentalist] views [of theories] is a conflict over preferred modes of speech" and "the question as to which of them is the 'correct position' has only terminological interest."[1] The phoenix, it seems, will not be laid to rest.

The literature on the subject is, of course, voluminous, and a comprehensive treatment of the problem is far beyond the scope of one essay. I shall limit myself to a small number of constructive arguments (for a radically realistic interpretation of theories) and to a critical examination of some of the more crucial assumptions (sometimes tacit, sometimes explicit) that seem to have generated most of the problems in this area.[2]

THE PROBLEM

Although this essay is not comprehensive, it aspires to be fairly self-contained. Let me, therefore, give a pseudohistorical introduction to the problem with a piece of science fiction (or fictional science).

In the days before the advent of microscopes, there lived a Pasteur-like scientist whom, following the usual custom, I shall call Jones. Reflecting on the fact that certain diseases seemed to be transmitted from one person to another by means of bodily contact or by contact with articles handled previously by an afflicted person, Jones began to speculate about the mechanism of the transmission. As a "heuristic crutch," he recalled that there is an obvious *observable* mechanism for transmission of certain afflictions (such as body lice), and he postulated that all, or most, infections diseases were spread in a similar manner but that in most cases the corresponding "bugs" were too small to be seen and, possibly, that some of them lived inside the bodies of their hosts. Jones proceeded to develop his theory and to examine its testable consequences. Some of these seemed to be of great importance for preventing the spread of disease.

After years of struggle with incredulous recalcitrance, Jones managed to get some of his preventative measures adopted. Contact with or proximity to diseased persons was avoided when possible, and articles which they handled were "disinfected" (a word coined by Jones) either by means of high temperatures or by treating them with certain toxic preparations which Jones termed "disinfectants." The results were spectacular: Within ten years the death rate had declined 40 percent. Jones and his theory received their well-deserved recognition.

However, the "crobes" (the theoretical term coined by Jones to refer to the disease-producing organisms) aroused considerable anxiety among many of the philosophers and philosophically inclined scientists of the day. The expression of this anxiety usually began something like this: "In order to account for the facts, Jones must assume that his crobes are too small to be seen. Thus the very postulates of his theory preclude their being observed; they are *unobservable in principle*." (Recall that no one had envisaged such a thing as a microscope.) This common prefatory remark was then followed by a number of different "analyses" and "interpretations" of Jones's theory. According to one of these, the tiny organisms were merely convenient fictions—*façons de parler*—extremely useful as heuristic devices for facilitating (in the "context of discovery") the thinking of scientists but not to be taken seriously in the sphere of cognitive knowledge (in the "context of justification"). A closely related view was that Jones's theory was merely an instrument, useful for organizing observation statements and (thus) for producing desired results, and that, therefore, it made no more sense to ask what was the nature of the entities to which it referred than it did to ask what was the nature of the entities to which a hammer or any other tool referred.[3] "Yes," a philosopher might have said, "Jones's theoretical expressions are just meaningless sounds or marks on paper which, when correlated with observation sentences by appropriate syntactical rules, enable us to predict successfully and otherwise organize data in a convenient fashion." These philosophers called themselves "instrumentalists."

According to another view (which, however, soon became unfashionable), although expressions containing Jones's theoretical terms were genuine sentences, they were translatable without remainder into a set (perhaps infinite) of observation sentences. For example, "There are crobes of disease X on this article" was said to translate into something like this: "If a person handles this article without taking certain precautions, he will (probably) contract disease X; and if this article is first raised to a high temperature, then if a person handles it at any time afterward, before it comes into contact with another person with disease X, he will (probably) not contract disease X; and. . . ."

Now virtually all who held any of the views so far noted granted, even insisted, that theories played a useful and legitimate role in the scientific enterprise. Their concern was the elimination of "pseudo problems" which might arise, say, when one began wondering about the "reality of supraempirical entities," etc. However, there was also a school of thought, founded by a psychologist named Pelter, which differed in an interesting manner from such positions as these. Its members held that while Jones's crobes might very well exist and enjoy "full-blown reality," they should not be the concern of medical research at all. They insisted that if Jones had employed the correct methodology, he would have discovered, even sooner and with much less effort, all of the observation laws relating to disease contraction, transmission, etc. without introducting superfluous links (the crobes) into the causal chain.

Now, lest any reader find himself waxing impatient, let me hasten to emphasize that this crude parody is not intended to convince anyone, or even to cast serious doubt upon sophisticated varieties of any of the reductionistic positions caricatured (some of them not too severely, I would contend) above. I am well aware that there are theoretical entities and theoretical entities, some of whose conceptual and theoretical statuses differ in important respects from Jones's crobes. (I shall discuss some of these later.) Allow me, then, to bring the Jonesean prelude to our examination of observability to a hasty conclusion.

Now Jones had the good fortune to live to see the invention of the compound microscope. His crobes were "observed" in great detail, and it became possible to identify the specific kind of *microbe* (for so they began to be called) which was responsible for each different disease. Some philosophers freely admitted error and were converted to realist positions concerning theories. Others resorted to subjective idealism or to a thoroughgoing phenomenalism, of which there were two principal varieties. According to one, the one "legitimate" observation language had for its descriptive terms only those which referred to sense data. The other maintained the stronger thesis that *all* "factual" statements were *translatable* without remainder into the sense-datum language. In either case, any two non-sense data (e.g., a theoretical entity and what would ordinarily be called an "observable physical object") had virtually the same status. Others contrived means of modifying their views much less drastically. One group maintained that Jones's crobes actually never had been unobservable in principle, for, they said, the theory did not imply the impossibility of finding a means (e.g., the microscope) of observing them. A more radical contention was that the crobes were not observed at all; it was argued that what was seen by means of the microscope was just a shadow or an image rather than a corporeal organism.

THE OBSERVATIONAL-THEORETICAL DICHOTOMY

Let us turn from these fictional philosophical positions and consider some of the actual ones to which they roughly correspond. Taking the last one first, it is interesting to note the following passage from Bergmann: "But it is only fair to point out that if this . . . methodological and terminological analysis [for the thesis that there are no atoms] . . . is strictly adhered to, even stars and microscopic objects are not physical things in a literal sense, but merely by courtesy of language and pictorial imagination. This might seem awkward. But when I look through a microscope, all I see is a patch of color which creeps through the field like a shadow over a wall. And a shadow, though real, is certainly not a physical thing."[4]

I should like to point out that it is also the case that if this analysis is strictly adhered to, we cannot observe physical things through opera glasses, or even through ordinary spectacles, and one begins to wonder about the status of what we see through an ordinary windowpane. And what about distortions due to temperature gradients—however small and, thus, always present—in the ambient air? It really *does* "seem awkward" to say that when people who wear glasses describe what they see they are talking about shadows, while those who employ unaided vision talk about physical things—or that when we look through a windowpane, we can only *infer* that it is raining, while if we raise the window, we may "observe directly" that it is. The point I am making is that there is, in principle, a continuous series beginning with looking through a vacuum and containing these as members: looking through a windowpane, looking through glasses, looking through binoculars, looking through a low-power microscope, looking through a high-power microscope, etc., in the order given. The important consequence is that, so far, we are left without criteria which would enable us to draw a nonarbitary line between "observation" and "theory." Certainly, we will often find it convenient to draw such a to-some-extent-arbitrary line; but its position will vary widely from context to context. (For example, if we are determining the resolving characteristics of a certain microscope, we would certainly draw the line beyond ordinary spectacles, probably beyond simple magnifying glasses, and possibly beyond another microscope with a lower power of resolution.) But what ontological ice does a mere methodologically convenient observational-theoretical dichotomy cut? Does an entiry attain physical thinghood and/or "real existence" in one context only to lose it in another? Or, we may ask, recalling the continuity from observable to unobservable, is what is seen through spectacles a "little bit less real" or does it "exist to a slightly less extent" than what is observed by unaided vision?[5]

However, it might be argued that things seen through spectacles and binoculars look like ordinary physical objects, while those seen through microscopes and telescopes look like shadows and patches of light. I can only reply that this does not seem to me to be the case, particularly when looking at the moon, or even Saturn, through a telescope or when looking at a small, though "directly observable," physical object through a low-power microscope. Thus, again, a continuity appears.

"But," it might be objected, "theory tells us that what we see by means of a microscope is a real image, which is certainly distinct from the object on the stage." Now first of all, it should be remarked that it seems odd that one who is espousing an austere empiricism which requires a sharp observational-language/theoretical-language distinction (and one in which the former language has a privileged status) should need a theory in order to tell him what is observable. But, letting this pass, what is to prevent us from saying that we still observe the object on the stage, even though a "real image" may be involved? Otherwise, we shall be strongly tempted by phenomenalistic demons, and at this point we are considering a physical-object observation language rather than a sense-datum one. (Compare the traditional puzzles: Do I see one physical object or two when I punch my eyeball? Does one object split into two? Or do I see one object and one image? Etc.)

Another argument for the continuous transition from the observable to the unobservable (theoretical) may be adduced from theoretical considerations themselves. For example, contemporary valency theory tells us that there is a virtually continuous transition from very small molecules (such as those of hydrogen) through "medium-sized" ones (such as those of the fatty acids, polypeptides, proteins, and viruses) to extremely large ones (such as crystals of the salts, diamonds, and lumps of polymeric plastic). The molecules in the last-mentioned group are macro, "directly observable" physical objects but are, nevertheless, genuine, single molecules; on the other hand, those in the first mentioned group have the same perplexing properties as subatomic particles (de Broglie waves, Heisenberg indeterminacy, etc.). Are we to say that a large protein molecule (e.g., a virus) which can be "seen" only with an electron microscope is a little less real or exists to somewhat less an extent than does a molecule of a polymer which can be seen with an optical microscope? And does a hydrogen molecule partake of only an infinitesimal portion of existence or reality? Although there certainly is a continuous transition from observability to unobservability, any talk of such a continuity from full-blown existence to nonexistence is, clearly, nonsense.

Let us now consider the next to last modified position which was adopted by our fictional philosophers. According to them, it is only those entities which are *in principle* impossible to observe that present special problems. What kind of impossibility is meant here? Without going into a detailed discussion of the various types of impossibility, about which there is abundant literature with which the reader is no doubt familiar, I shall assume what usually seems to be granted by most philosophers who talk of entities which are unobservable in principle, i.e., that the theory(s) itself (coupled with a physiological theory of perception, I would add) entails that such entities are unobservable.

We should immediately note that if this analysis of the notion of unobservability (and, hence, of observability) is accepted, then its use as a means of delimiting the observation language seems to be precluded for those philosophers who regard theoretical expressions as elements of a calculating device—as meaningless strings of symbols. For suppose they wished to determine whether or not "electron" was a theoretical term. First, they must see whether the theory entails the sentence "Electrons are unobservable." So far, so good, for their calculating devices are said to be able to select genuine sentences, provided they contain no theoretical terms. But what about the selected "sentence" itself? Suppose that "electron" is an observation term. It follows that the expression is a genuine sentence and asserts that electrons are unobservable. But this entails that "electron" is not an observation term. Thus if "electron" is an observation term, then it is *not* an observation term. Therefore it is not an observation term. But then it follows that "Electrons are unobservable" is not a genuine sentence and does not assert that electrons are unobservable, since it is a meaningless string of marks and does not assert anything whatever. Of course, it could be stipulated that when a theory "selects" a meaningless expression of the form "*X*'s are unobservable," then *X* is to be taken as a theoretical term. But this seems rather arbitrary.

But, assuming that well-formed theoretical expressions are genuine sentences, what shall we say about unobservability in principle? I shall begin by putting my head on the block and argue that the present-day status of, say, electrons is in many ways similar to that of Jones's crobes before microscopes were invented. I am well aware of the numerous theoretical arguments for the impossibility of observing electrons. But suppose new entities are discovered which interact with electrons in such a mild manner that if an electron is, say, in an eigenstate of position, then, in certain circumstances, the interaction does not disturb it. Suppose also that a drug is discovered which vastly alters the human perceptual apparatus—perhaps even activates latent capacities so that a new sense modality emerges. Finally, suppose that in our altered state we are able to perceive (not necessarily visually) by means of these new entities in a manner roughly analogous to that by which we now see by means of photons. To make this a little more plausible, suppose that the energy eigenstates of the electrons in some of the compounds present in the relevant perceptual organ are such that even the weak interaction with the new entities alters them and also that the cross sections, relative to the new entities, of the electrons and other particles of the gases of the air are so small that the chance of any interaction here is negligible. Then we might be able to "observe directly" the position and possibly the approximate diameter and other properties of some electrons. It would follow, of course, that quantum theory would have to be altered in some respects, since the new entities do not conform to all its principles. But however improbable this may be, it does not, I maintain, involve any logical or conceptual absurdity. Furthermore, the modification necessary for the inclusion of the new entities would not necessarily change the meaning of the term *electron.*[6]

Consider a somewhat less fantastic example, and one which does not involve any change in physical theory. Suppose a human mutant is born who is able to "observe" ultraviolet radiation, or even X rays, in the same way we "observe" visible light.

Now I think that it is extremely improbable that we will ever observe electrons directly (i.e., that it will ever be reasonable to assert that we have so observed them). But this is neither here nor there; it is not the purpose of this essay to predict the future development of scientific theories, and, hence, it is not its business to decide what actually is observable or what will become observable (in the more or less intuitive sense of "observable" with which we are now working). After all, we are operating, here, under the assumption that it is theory, and thus science itself, which tells us what is or is not, in this sense, observable (the "in principle" seems to have become superflous) And this is the heart of the matter; for it follows that, at least for this sense of "observable," there are no a priori or philosophical criteria for separating the observable from the unobservable. By trying to show that we can talk about the *possibility* of observing electrons without committing logical or conceptual blunders, I have been trying to support the thesis that any (nonlogical) term is a *possible* candidate for an observation term.

There is another line which may be taken in regard to delimitation of the observation language. According to it, the proper term with which to work is not "observ*able*" but, rather, "observ*ed*." There immediately comes to mind the tradition beginning with Locke and Hume (No idea without a preceding impression!), running through Logical Atomism and the Principle of Acquaintance, and ending (Perhaps) in contemporary positivism. Since the numerous facets of this tradition have been extensively examined and criticized in the literature, I shall limit myself here to a few summary remarks.

Again, let us consider at this point only observation languages which contain ordinary physical-object terms (along with observation predicates, etc., of course). Now, according to this view, all descriptive terms of the observation language must refer to that which has been observed. How is this to be interpreted? Not too narrowly, presumably, otherwise each language user would have a different observation language. The name of my Aunt Mamie, of California, whom I have never seen, would not be in my observation language, nor would "snow" be an observation term for many Floridians. One could, of course, set off the observation language by means of this awkward restriction, but then, obviously, not being the referent of an observation term would have no bearing on the ontological status of Aunt Mamie or that of snow.

Perhaps it is intended that the referents of observation terms must be members of a *kind*, some of whose members have been observed, or instances of a *property*, some of whose instances have been observed. But there are familiar difficulties here. For example, given any entity, we can always find a kind whose only member is the entity in question; and surely expressions such as "men over 14 feet tall" should be counted as observational even though no instances of the "property" of being a man over 14 feet tall have been observed. It would seem that this approach must soon fall back upon some notion of simples or determinables vs. determinates. But is it thereby saved? If it is held that only those terms which refer to observed simples or observed determinates are observation terms, we need only remind ourselves of such instances as Hume's notorious missing shade of blue. And if it is contended that in order to be an observation term an expression must at least refer to an observed determinable, then we can always find such a determinable which is broad enough in scope to embrace any entity whatever. But even if these difficulties can be circumvented, we see (as we knew all along) that this approach leads inevitably into phenomenalism, which is a view with which we have not been concerning ourselves.

Now it is not the purpose of this essay to give a detailed critique of phenomenalism. For the most part, I simply assume that it is untenable, at least in any of its translatability varieties.[7] However, if there are any unreconstructed phenomenalists among the readers, my purpose, insofar as they are concerned, will have been largely achieved if they will grant what I suppose most of them would stoutly maintain anyway, i.e., that theoretical entities are no worse off than so-called observable physical objects.

Nevertheless, a few considerations concerning phenomenalism and related matters may cast some light upon the observational-theoretical dichotomy and, perhaps, upon the nature of the "observation language." As a preface, allow me some overdue remarks on the latter. Although I have contended that the line between the observable and the unobservable is diffuse, that it shifts from one scientific problem to another and that it is constantly being pushed toward the "unobservable" end of the spectrum as we develop better means of observation—better instruments—it would, nevertheless, be fatuous to minimize the importance of the observation base, for it is absolutely necessary as a confirmation base for statements which do refer to entities which are unobservable at a given time. But we should take as its basis and its unit not the "observational term" but, rather, the quickly decidable sentence. (I am indebted to Feyerabend, *loc. cit.*, for this terminology.) A quickly decidable sentence (in the technical sense employed here) may be defined as a singular, nonanalytic sentence such that a reliable, reasonably sophisticated language user can very quickly decide[8] whether to assert it or deny it when he is reporting on an occurrent situation. *Observation term* may now be defined as a "descriptive (nonlogical) term which may occur in a quickly decidable sentence," and *observation sentence* as a "sentence whose only descriptive terms are observation terms."

Returning to phenomenalism, let me emphasize that I am not among those philosophers who hold that there are no such things as sense contents (even sense data), nor do I believe that they play no important role in our perception of "reality." But the fact remains that the referents of most (not all) of the statements of the linguistic framework used in everyday life and in science are not sense contents but, rather, physical objects and other publicly observable entities. Except for pains, odors, "inner states," etc., *we do not usually observe sense contents;* and although there is good reason to believe that they play an indispensable role in observation, *we are usually not aware of them when we* (visually or tactilely) *observe physical objects.* For example, when I observe a distorted, obliquely reflected image in a mirror, I may seem to be seeing a baby elephant standing on its head; later I discover it is an image of Uncle Charles taking a nap with his mouth open and his hand in a peculiar position. Or, passing my neighbor's home at a high rate of speed, I observe that he is washing a car. If asked to report these observations I could quickly and easily report a baby elephant and a washing of a car; I probably would not, without subsequent observations, be able to report what colors, shapes, etc. (i.e., what sense data), were involved.

Two questions naturally arise at this point: How is it that we can (sometimes) quickly decide the truth or falsity of a pertinent observation sentence? and, What role do sense contents play in the appropriate tokening of such sentences? The heart of the matter is that these are primarily scientific-theoretical questions rather than "purely logical," "purely conceptual," or "purely epistemological." If theoretical physics, psychology, neurophysiology, etc., were sufficiently advanced, we could give satisfactory answers to these questions, using, in all likelihood, the physical-thing language as our observation language and *treating sensations, sense contents, sense data, and "inner states" as theoretical* (yes, theoretical!) *entities.*[9]

It is interesting and important to note that, even before we give completely satisfactory answers to the two questions considered above, we can, with due effort and reflection, train ourselves to "observe directly" what were once theoretical entities—the sense contents (color sensations, etc.)—involved in our perception of physical things. As has been pointed out before, we can also come to observe other kinds of entities which were once theoretical. Those which most readily come to mind involve the use of instruments as aids to observation. Indeed, using our painfully acquired theoretical knowledge of the world, we come to see that we "directly observe" many kinds of so-called theoretical things. After listening to a dull speech while sitting on a hard bench, we begin to become poignantly aware of the presence of a considerably strong gravitational field, and as Professor Feyerabend is fond of pointing out, if we were carrying a heavy suitcase in a changing gravitational field, we could observe the changes of the $G_{\mu\nu}$ of the metric tensor.

I conclude that our drawing of the observational-theoretical line at any given point is an accident and a function of our physiological make-up, our current state of knowledge, and the instruments we happen to have available and, therefore, that it has no ontological significance whatever.

. . .

NOTES

1. E. Nagel, *The Structure of Science* (New York: Harcourt, Brace, and World, 1961), Chap. vi.

2. For the genesis and part of the content of some of the ideas expressed herein, I am indebted to a number of sources; some of the more influential are H. Feigl, "Existential Hypotheses," *Philosophy of Science*, XVII (1950), 35–62; P. K. Feyerabend, "An Attempt at a Realistic Interpretation of Experience," Proceeding of the Aristotelian Society, LVIII (1958), 144–70; N. R. Hanson, *Patterns of Discovery* (Cambridge: Cambridge University Press, 1958); E. Nagel, *loc. cit.*; Karl Popper, *The Logic of Scientific Discovery* (London: Hutchinson, 1959); M. Scriven, "Definitions, Explanations, and Theories," in *Minnesota Studies in the Philosophy of Science*, eds. H. Feigl, M. Scriven, and G. Maxwell, Vol. II (Minneapolis: University of Minnesota Press, 1958); Wilfrid Sellars, "Empiricism and the Philosophy of Mind," in *Minnesota Studies in the Philosohy of Science*, eds. H. Feigl and M. Scriven, Vol. I (Minneapolis: University of Minnesota Press, 1956), and "The Language of Theories," in *Current Issues in the Philosophy of Science*, eds. H. Feigl and G. Maxwell (New York: Holt, Rinehart, and Winston, 1961).

3. I have borrowed the hammer analogy from E. Nagel, "Science and [Feigl's] Semantic Realism," *Philosophy of Science*, XVII (1950), 174–81 but it should be pointed out that Professor Nagel makes it clear that he does not necessarily subscribe to the view which he is explaining.

4. G. Bergmann, "Outline of an Empiricist Philosophy of Physics," *American Journal of Physics*, II (1943), 248–58, 335–42, reprinted in *Readings in the Philosophy of Science*, eds. H. Feigl and M. Brodbeck (New York: Appleton-Century-Crofts, 1953), pp. 262–87.

5. I am not attributing to Professor Bergmann the absurd views suggested by these questions. He seems to take a sense-datum language as his observation language (the base of what he called "the empirical hierarchy"), and, in some ways, such a position is more difficult to refute than one which purports to take an "observable-physical-object" view. However, I believe that demolishing the straw men with which I am now dealing amounts to desirable preliminary "therapy." Some nonrealist interpretations of theories which embody the presupposition that the observable-theoretical distinction is sharp and ontologically crucial seem to me to entail positions which correspond to such straw men rather closely.

6. For arguments that it is possible to alter a theory without altering the meaning of its terms, see my "Meaning Postualtes in Scientific Theories," in *Current Issues in the Philosophy of Science*, eds. Feigl and Maxwell.

7. The reader is no doubt familiar with the abundant literature concerned with this issue. See, for example, Sellars's "Empiricism and the Philosophy of Mind," which also contains references to other pertinent works.

8. We may say "noninferentially" decide, provided this is interpreted liberally enough to avoid starting the entire controversy about observability all over again.

9. Cf. Sellars, "Empiricism and the Philosophy of Mind." As Professor Sellars points out, this is the crux of the "other-minds" problem. Sensations and inner states (relative to an intersubjective observation language, I would add) are theoretical entities (and they "really exist") and *not* merely actual and/or possible behavior. Surely it is the unwillingness to countenance theoretical entities—the hope that every sentence is translatable not only into some observation language but into the physical-thing language—which is responsible for the "logical behaviorism" of the neo-Wittgensteinians.

*Peter Achinstein**

THE PROBLEM OF THEORETICAL TERMS

Philosophers with quite different viewpoints have considered it important to distinguish two sorts of terms employed by scientists. While various labels have been suggested I shall use the expressions *theoretical* and *nontheoretical* to represent the intended distinction. Those who propose it provide examples of terms which fall into these respective categories, and, although there is by no means general agreement on all classifications, there is substantial accord on many examples. What follows is a list of some of the illustrations cited:

theoretical terms	
electric field	mass
electron	electrical resistance
atom	temperature
molecule	gene
wave function	virus
charge	ego

non-theoretical terms	
red	floats
warm	wood
left of	water
longer than	iron
hard	weight
volume	cell nucleus

Some philosophers base this distinction on a concept of "observability." Others appeal to a notion of "conceptual organization" or "theory dependence." My purpose here is to examine this distinction and suggest reasons for doubting the claim that there is some unique criterion or set of criteria which underlies

*The author is indebted to the National Science Foundation for support of research.

Reprinted from the *American Philosophical Quarterly*, II, No. 3 (July, 1965) by permission of the editor and author.

it. Rather, I shall argue, the notions of "observability," "conceptual organization," and "theory dependence" introduced by these authors generate many distinctions which result in a number of different ways of classifying terms on the above lists. Since the alleged distinction between theoretical and non-theoretical terms has played an important role in the philosophy of science, as well as epistemology, an examination of its basis may show the need for reformulating some rather persistent issues and indicate the sort of steps which might profitably be taken.

I

The proposal first to be considered is that the above classification rests upon a distinction between entities or properties which are observable and those which are not. Thus, according to Carnap, we have on the one hand "terms designating observable properties and relations," and on the other, "terms which may refer to unobservable events, unobservable aspects or features of events."[1] Carnap does not go on to explain what he means by "observable" and "unobservable" and presumably believes that his readers will understand in at least a general way the distinction intended. Unfortunately the situation is more complex than he seems willing to admit, since the terms "observable" and "unobservable" can be employed for the purpose of making a substantial number of different points.

Consider the case of visual observation. Just what it is that I can appropriately claim to have observed depends very importantly on the particular context in which the claim is made.[2] Suppose that while sitting by the roadside at night I am asked what I observe on the road ahead. I might, in one and the same situation, reply in a number of different ways; for example: a car, the front of a car, a pair of automobile headlights, two yellowish lights, etc. Or, when driving on a dirt road in the daytime I might, in one and the same situation, claim to be observing a car, a trail produced by a car, or just a cloud of dust. Two points deserve emphasis here. First, in each case what I will actually say that I have observed depends upon a number of factors, such as the extent of my knowledge and training, how much I am prepared to maintain about the object under the circumstances, and the type of answer I suppose my questioner to be interested in. Second, in both examples I might claim to have observed a car ahead, though in the first the only visible parts of the car are its headlights, and in the second I see none of its parts at all, not even a speck in the distance which *is* the car. Nor must such claims necessarily be deemed imprecise, inaccurate, ambiguous, or in any way untoward. In the particular circumstances it may be perfectly clear just what I am claiming, though someone ignorant of the context might misconstrue my claim and expect me to know more than I do, such as the color, the shape, or even the make of the car.

Both of these points can be illustrated by reference to scientific contexts.

Suppose that an experimental physicist, acquainted with the sorts of tracks left by various subatomic particles in cloud chambers, is asked what he is now observing in the chamber. He might reply in a number of ways, e.g., electrons passing through the chamber, tracks produced by electrons, strings of tiny water droplets which have condensed on gas ions, or just long thin lines. Similarly, to the question "What does one observe in a cathode-ray experiment?" the physicist might answer: electrons striking the fluorescent zinc sulfide screen, light produced when molecules of zinc sulfide are bombarded, a bright spot, etc. In each case what the physicist actually claims to have observed will depend upon how much he knows and is prepared to maintain, the knowledge and training of the questioner, and the sort of answer he thinks appropriate under the circumstances. Furthermore, just as in the previous automobile examples, there are situations in which the physicist, concerned mainly with indicating the occurrence of certain events and with proper identification, will report observing various particles pass through the chamber—electrons if the tracks are long and thin, alpha particles if they are shorter and heavier. Whether he chooses to describe the situation in this way will depend upon the sort of factors noted above.

Analogous considerations hold for other terms on the "theoretical" list. Thus the physicist may report having observed the electric field in the vicinity of a certain charge, or he may describe what he did as having observed the separation of leaves in an electroscope; he may report to be observing the rise in temperature of a given substance, or simply the increase in length of the column of mercury in the thermometer, etc. Accordingly, those who seek to compile a list of "observational" terms must not do so on the basis of an assumption to the effect that there exists some unique way of describing what is observed in a given situation. Nor can a classification of electrons, fields, temperature, etc., as *un*observable be founded simply on a claim that one cannot ever report observing such items. Or, at least, if such a claim is made it will need to be expanded and defended in a manner not attempted by Carnap.

Suppose then, Carnap were to acknowledge that scientists often describe what they have observed in different ways, and that physicists do speak of observing such things as subatomic particles in cloud chambers and electric fields in the vicinity of charges, when the main concern is to report the occurrence of a certain event or the presence of a certain type of entity. Still, he might urge, contextual considerations of the sort mentioned will be irrelevent when we consider, strictly speaking, what is really observed in such cases. For what the physicist (really) observes is not the electron itself (but only its track, or a flash of light), just as in the automobile examples, what is (really) observed is not the car itself (but only its headlights, or a dust trail). And when the physicist detects the presence of an electric field, what he observes is not the field itself, but (say), only the separation of leaves in the electroscope. In general, it might be said, the distinction desired can be drawn on the basis of the claim that items on the "theoretical" list are not themselves (really) observable.[3]

I do not want to deny that electrons, fields, temperature, etc., can be described in this way. Indeed, such a description may be invoked when the physicist begins to explain just what claims he is making when he speaks of observation in each of these cases. However, as will presently be indicated, when an expression such as "not itself (really) observable" is employed it makes sense only with reference to a specific context of observation and some particular contrast. And I want to show that this fact precludes the general sort of distinction desired.

Suppose that in the second automobile example I report that I cannot (really) observe the car itself. What claim am I making? I might be saying that all I can observe is a dust trail and no speck in the distance which I can identify as the car; or that I can observe only a speck, but not the body of the car; or again, that I can see the car in the mirror but not with the naked eye. And obviously other contrasts could be cited which would make the point of my assertion clear.

Consider now the case of a virus examined by means of an electron microscope. Suppose I say that the microbiologist does not (really) observe the virus itself. What am I claiming? I might be saying that since (let us suppose) he employs a staining technique, he sees not the virus but only the staining material known to be present in certain parts of the specimen. Or, I might simply be saying that what he observes is the image of the object as presented by the microscope, and not the object itself. On the other hand, comparing electron microscopy with X-ray diffraction, I might claim that in the former case he is able to observe the virus itself, whereas in the latter case he observes only the effect of X rays on the virus. On different occasions any one of these contrasts, and others, could underlie the claim that the virus itself cannot (or can) really be observed.

Similar considerations are relevant in understanding a corresponding claim regarding electrons. In the context of cloud chamber experiment, if I assert that electrons themselves cannot (really) be observed I might be saying that though a track is visible there is no speck which can be identified as the electron, in the way that if a jet airplane is close enough, one can see not only its trail but identify a certain speck in the distance as the airplane itself. On the other hand, I might wish to contrast the case of the electron with that of the neutron and claim that whereas electrons themselves *can* be observed in a cloud chamber, neutrons cannot. The point of this contrast is that neutrons, being neutral in charge, cannot cause ionization in their passage through the chamber, and hence will not produce a track, the way electrons will.[4]

In general, then, it is not sufficient simply to refer to the previous list of so-called theoretical terms and claim that the distinguishing feature of the items designated by these terms is that they are not themselves (really) observable. What must be done is to indicate for each item the point of such a classification; and this is most readily accomplished by contrasting the sense in which it is said to be (really) unobservable itself with the sense in which something else is claimed to be observable. Now I do not deny that one could supply a context for each item on the list in which it would be appropriate to speak of that item in the manner proposed. The important point is simply that these contexts, and the sorts of contrasts they may involve, will in general be quite different and will yield different classifications. Here are a few contrasts some of which have already been noted, namely those between:

(a) Objects such as electrons and alpha particles which are detected by means of their tracks, and objects such as cars and airplanes which can be seen together with or apart from their tracks.
(b) Objects such as neutrons and neutrinos which do not leave tracks in a cloud chamber, and those such as electrons and alpha particles which do.
(c) Objects such as smaller molecules which it is necessary to stain in order to observe with the electron microscope, and larger objects for which staining is unnecessary.
(d) Objects such as individual atoms which are too small to scatter electrons appreciably and hence cannot be seen with the electron microscope, and larger objects such as certain molecules which have significant scattering power and hence can be observed.
(e) Objects requiring illumination by electron beams which then must be transformed into light via impact with a suitable screen, and objects visible with ordinary light.
(f) Objects which can only be observed by the production of images in microscopes, and those which can be seen with the naked eye.
(g) Properties such as electrical resistance whose magnitudes must be (or are generally) calculated on the basis of measuring a number of other quantities, and those properties for which this is usually not necessary.
(h) Properties such as temperature for which some instrumentation is usually required (for determining differences), and those such as color for which it is not.
(j) Objects such as electric fields which are not the sorts of things to which mass and volume are ascribed, and objects such as solids to which they are.

Each of these contrasts, as well as others which could readily be cited, *might* be used to generate some sort of observational distinction. Yet the same item would be classified differently depending upon the particular distinction invoked. Electrons would be unobservable under (a) and (d) but observable under (b); heavy molecules would be unobservable under (e) and (f) but observable under (c) and (d); temperature would be unobservable under (h) but observable under (g). Also, under certain contrasts some items cannot be classified at all —electrons under (c), (e), and (f); heavy molecules under (a) and (b); temperature under (a)–(f). Accordingly, if contrasts of the type cited above must be invoked to provide significance for the claim that a certain entity or property is itself (really) unobservable, then the sort of distinction required by Carnap seems difficult if not impossible to draw.

II

I want now to consider two qualifications which some authors place on the notion of observation. The first is that one should not speak simply of observability, but of *direct* observability. Hempel, e.g., writes:

> In regard to an observational term it is possible, under suitable circumstances, to decide by means of direct observation whether the term does or does not apply to a given situation. . . . Theoretical terms, on the other hand, usually purport to refer to not directly observable entities and their characteristics.[5]

In offering this criterion, Hempel, like Carnap, mentions no special or technical sense which he attaches to the phrase "direct observation." Nor does he elaborate upon its meaning, except to cite a few examples.[6] He admits that his characterization does not offer a precise criterion and that there will be borderline cases. Yet the problem involved is more complex than Hempel seems willing to allow and, contrary to his suggestion, does not just turn on the question of drawing a more precise "dividing line." For the expression "(not) directly observable," like "(un) observable itself," is one whose use must be tied to a particular context and to some intended contrast. Thus, if the physicist claims that electrons cannot be observed directly, he may simply mean that instruments such as cloud chambers, cathode-ray tubes, or scintillation counters are necessary. Here direct observability has to do with observation by the unaided senses. Or, he might mean that when one observes an electron in a cloud chamber one sees only its track but not, e.g., a speck which one would identify as the electron itself. Again the nuclear physicist might claim that particles such as neutrons and neutrinos are not directly observable, meaning that such particles cannot themselves produce tracks in a cloud chamber, unlike electrons and alpha particles, which, under this contrast, would be deemed directly observable. Another type of situation in which the expression "direct observation" might be invoked involves a contrast between properties, such as electrical resistance, whose magnitudes must be calculated by first measuring other quantities, and those, such as length, or temperature, for which this is often not necessary.[7]

In short, many contrasts can be invoked by the notion of direct observation,[8] and a given item will be classified in different ways depending upon the particular one intended. An appeal to direct observation by itself does little to advance the cause of generating a unique distinction, and when such an appeal is spelled out in individual cases various distinctions emerge.

The second qualification sometimes placed on observability concerns the *number* of observations necessary correctly to apply a term or expression. Thus, in "Testability and Meaning," Carnap writes:

> A predicate '*P*' of a language *L* is called *observable* for an organism (e.g., a person) *N*, if, for suitable argument, e.g., '*b*', *N* is able under suitable circumstances to come to a decision with the help of a few observations about the full sentence, say '*P*(*b*)', i.e., to a confirmation of either '*P*(*b*)' or '-*P*(*b*)' of such high degree that he will either accept or reject '*P*(*b*)'.[9]

Carnap does not, however, explain his qualification in sufficient detail and leaves some important questions unanswered. Is the number of observations to mean the number of times the object must be observed (or, if an experiment is in question, the number of times the experiment needs to be repeated) before a property can definitely be ascribed to the object? Or does Carnap perhaps mean the number of different characteristics of the object which need to be observed? Again, he may be thinking of the amount of preliminary investigation necessary before a final observation can be made.[10] Or perhaps all of these considerations are relevant.

Yet whether an observation or experiment will need to be repeated, or many different characteristics of the item in question examined, or considerable preliminary investigation undertaken, depends not only on the nature of the object or property under examination, but also on the particular circumstances of the investigator and his investigation. One relevant factor will be the type of instrument employed and how easily the scientist has learned to manipulate it. The physicist familiar with electroscopes need make few, if any, repetitions of an experiment with this instrument to determine the presence of an electric charge and hence of an electric field. Nor need he observe many characteristics of the field (e.g., its intensity and direction at a certain point) in order to determine its presence. And he will not always need to make extensive preliminary observations on the instrument but only a few. Yet, charges and electric fields are alleged to be unobservable. Another factor determining the facility with which an observer will identify an object or property is the extent of his knowledge regarding the particular circumstances of the observation. If the physicist knows that a certain radioactive substance has been placed in a cloud chamber he may readily be able to identify the particles whose tracks are visible in the chamber. Whereas, if he knows nothing about the circumstances of the experiment, and he is simply shown a photograph of its results, successful identification may be a more complicated task. Rapid classification, then, depends in considerable measure upon particular features of the context of observation and the knowledge of the investigator. Under certain "suitable circumstances" (to use Carnap's phrase), quite a number of terms classified by him as non-observational can be correctly applied "with the help of [just] a few observations."

Furthermore, if nonobservability in Carnap's sense is held to be sufficient for a *theoretical* classification, additional difficulties emerge. For many fairly ordinary expressions are such that in numerous circumstances more than a "few observations" might well be required before correct application is possible; for example, "is chopped sirloin," "is a brige which will collapse," "was composed by Corelli." Yet these do not really seem to be the sorts of expressions the authors in question wish to call theoretical. On the other hand, if, following Scheffler,[11] Carnap's criterion for "nonobservability" is to be construed simply as a necessary but not a sufficient condition for being classified as theoretical, then unless further criteria are proposed (which they are not either by Carnap or Scheffler) we will have no general basis for separating theoretical from non-theoretical terms.[12] And if the above criterion concerning the number of observations is to be construed as a sufficient one for a *non*-theoretical classification (as Carnap and Scheffler suggest), then, as we have seen, many terms classified by these authors as theoretical will, in numerous situations, require reclassification.

III

I have considered attempts to base a theoretical–non–theoretical distinction upon some notion of observation. Generally speaking, the thesis that a list of "observational" terms can be compiled is defended by those envisaging the possibility of an "empiricist language." One of the underlying assumptions of this program appears to be that there exists a unique (or at least a most suitable) way of describing what is, or can be (really, directly) observed—a special "physical object" or "sense datum" vocabulary eminently fit for this task. Yet, as emphasized earlier, there are numerous ways to describe what one (really, directly) observes in a given situation, some more infused with concepts employed in various theories than others. This does not, of course, preclude the possibility of classifying certain reports as observational in a given case. The point is simply that there is no special class of terms which must be used in describing what is observed. Words from the previous "theoretical" list, such as "electron," "field," and "temperature," are frequently employed for this purpose.

Still, it might be urged, even though terms on both lists can be used in descriptions of what is (really, directly) observed, those on the first list are more "theory-dependent" than those on the second. And while it may not be possible to draw the intended large-scale distinction on the basis of observation, it is nevertheless feasible and important to separate terms on the basis of their "theoretical" character. It is to the latter position, which has been defended by Hanson and Ryle, that I now wish to turn.

According to Hanson a distinction should be drawn between terms which "carry a conceptual pattern with them," and terms which "are less rich in theory, and hence less able to serve in explanations of causes";[13] or, in Ryle's words, between expressions which are "more or less heavy with the burden of [a particular] theory . . . [and those which] carry none [of the luggage] from that theory."[14] As an example of a term which carries with it a conceptual pattern Hanson cites the word "crater":

> Galileo often studied the Moon. It is pitted with holes and discontinuities; but to say of these that they are craters—to say that the lunar surface is craterous—is to infuse theoretical astronomy into one's observations. . . . To speak of a concavity as a crater is to commit oneself as to its origin, to say that its creation was quick, violent, explosive. . . . .[15]

"Crater," then, carries with it a conceptual pattern not borne by (non-theoretical) terms such as "hole," "discontinuity," or "concavity."

Two notions underlie these suggestions, one stressed more by Hanson, the other by Ryle. The first is that a theoretical term is one whose application in a given situation can organize diffuse and seemingly unrelated aspects of that situation into a coherent, intelligible pattern; terms which carry no such organizing pattern Hanson sometimes calls "phenomenal." The second notion is that theoretical terms are such that "knowing their meanings requires some grasp of the theory" in which they occur.[16] "The special terms of a science," Ryle asserts, "are more or less heavy with the burden of the theory of that science. The technical terms of genetics are theory-laden, laden, that is, not just with theoretical luggage of some sort of other but with the luggage of genetic theory."[17]

Despite the fine examples Hanson cites, and the use he makes of the notion of "organizing patterns" in supplying trenchant criticisms of various philosophical positions, surely the first proposal fails to provide a sufficient characterization of those terms Hanson calls "theory-laden" (or "theoryloaded"). For almost any term can be employed in certain situations to produce the type of pattern envisaged. Indeed, Hanson himself offers many examples of this; quite early in his book, for instance, he presents a drawing whose meaning is incomprehensible until it is explained that it represents a bear climbing a tree. In this context, the expression "bear climbing a tree" is one which organizes the lines into an intelligible pattern. Moreover, one could describe contexts in which Hanson's "phenomenal" terms such as "hole," "concavity," "solaroid disc," etc., might be employed to organize certain initially puzzling data. Conversely, there are situations in which terms such as "crater," "wound," "volume," "charge," and "wave-length," which Hanson calls "theory-loaded," are used to describe data which are initially puzzling and require "conceptual organization." Hanson, at one point, grants that terms can have this dual function:

> It is not that certain words are absolutely theory-loaded, whilst others are absolutely sense-datum words. Which are the data words and which are the theory-words is a contextual question. Galileo's scar may at some times be a datum requiring explanation, but at other times it may be part of the explanation of his retirement.[18]

Yet this admission is a large one. For it means that the rendering of "conceptual organization" is not a special feature of terms on the "theoretical" list which sets them apart from those on the "nontheoretical" list. Whether a term provides an organizing pattern for the data depends on the particular situation in which it is employed. In some contexts, the term "electron" will serve to organize the data (e.g., tracks in a cloud chamber); in others the term "electron" will be employed to describe certain data requiring organization (e.g., the discontinuous radiation produced by electrons in the atom). But even a presumably non-theoretical expression such as "(X is) writing a letter" can be used in certain contexts to organize data thereby explaining a piece of behavior, and in others to describe something which itself demands explanation.

Hanson does make reference to the "width" of terms, claiming that some expressions are "wider" theoretically than others and hence presumably "carry a [greater] conceptual pattern with them."[19] So despite the fact that most terms are capable of serving explanatory functions, distinctions might still be drawn on the basis of "width." Yet it is not altogether clear how this metaphor should be unpacked. Sometimes the suggestion appears to be that one term will be "wider" than another if it can be used to explain situations whose descriptions contain the latter term. Thus, referring to the Galileo example quoted above, the term "scar" would be wider than the term "retirement" because it can be used to explain something designated by the latter. Yet this is not altogether satisfactory since we might imagine a case in which a man's retirement constituted part of an explanation of a scar he incurred. Again, the term "electron" can be employed in explanations of magnetic fields; yet the presence of a magnetic field can explain motions of electrons.

Perhaps, however, the reference to "width" should be understood in connection with the thesis, proposed by Ryle (and shared by Hanson), that certain terms depend for their meaning upon a particular theory, whereas others do not, or at least are much less dependent. By way of explanation Ryle cites as an analogy the situation in games of cards. To understand the expression "straight flush" one must know at least the rudiments of poker, whereas this is not so with the expression "Queen of Hearts," which is common to all card games and carries with it none of the special "luggage" of any of them. In a similar manner certain terms used by scientists are such that to understand them one must have at least some knowledge of the particular theory in which they appear. These are the "theory-laden" terms.[20] Other expressions utilized by scientists can be understood wothout recourse to any specific theoretical system.

Before examining this proposal some preliminary points should be noted. First, Ryle often seems to be suggesting that a theory-laden term is one which "carries the luggage" of one particular theory. Yet quite a few of the terms which he and Hanson classify as "theory-laden"—terms such as "temperature," "wave-length," "electron"—appear in, and might be thought to be infused with the concepts of, *many* scientific theories. Such terms are not restricted to just one theory, as with respect to games, "straight flush" is to poker. Second, it is certainly not a feature characteristic only of terms Ryle calls "theory-laden" —or only of such terms and those from card games—that they must be understood by reference to some scheme, system of beliefs, or set of facts. Following Ryle's lead one might draw up many different sorts of classifications, such as "university-laden" terms ("hour examination," "credit," and "tutorial") which cannot be understood without some knowledge of universities and their procedures; or, referring to scientific contexts, "instrument-laden" terms, such as "dial," "on," "off," which presumably would not appear on the "theoretical" list, yet require at least some knowledge of instruments and their uses. Thus Ryle's proposal must not be construed simply as a criterion for distinguishing terms which must be understood in the context of a set of beliefs from those which can be understood independently of any such set, though Hanson's "theory-laden" vs. "sense-datum" labels might misleadingly suggest this. Third, one must always specify the theory with respect to which a given term is or is not "theory-laden." And, it would appear, a term might receive this classification with reference to one theory but not another, though it occurs in both. For in one theory its meaning might not be understood unless the principles of the theory are known, whereas this would not necessarily be so in the case of the other theory (or at least there could be significant differences of degree). For example, "mass" might be considered "theory-laden" with respect to Newtonian mechanics but not with respect to the Bohr theory of the atom in which it also appears, since it can be understood independently of the latter. Thus, presumably, not every term occurring in a given theory will be "theory-laden," just as not every term found in standard formulations of the rules and principles of poker—such as "sequence," and "card"—will be "poker-laden," to use Ryle's expression. But if a theory must always be specified with respect to which a given term is deemed "theory-laden," and if a term can be classified in this way with reference to one theory and not to another in which it appears, then lists of the sort compiled at the beginning of this paper cannot be legitimately constructed. The most that could be done would be to cite particular theories and for each one compose such lists indicating which terms are to be considered theoretical and which not *for that theory.*[21]

Yet even this task may be deemed incapable of fulfillment once we examine Ryle's claim that expressions are theory-laden if they are such that "knowing their meanings requires some grasp of the theory." For there are many different ways in which terms might be said to be dependent upon principles of a given theory, and it is not altogether clear whether any or all of these should be classified as cases of "meaning-dependence." Since, we have already seen, the issue of whether a term is "theory-laden" must be considered always with reference to a specific theory, let us suppose that we are given such a theory. Within it we could expect to encounter the following sorts of terms or expressions which might be considered theory-dependent in some sense:

(1) A term or expression whose *definition* cannot be stated without formulating some law or principle of the theory in question. For example, "Newton's gravitational constant" could only be defined by invoking the law of universal gravitation. The expression "Bohr atom," frequently employed in atomic physics, is defined as one satisfying the postulates of the Bohr theory. "Electrical resistance" is defined by reference to Ohm's law, etc.

(2) A term or expression whose definition can be stated without formulating laws of the theory, but whose use must be *justified* by invoking some of these laws. In electrostatics an "electrostatic unit charge" (esu) is defined as one which when placed in a vacuum one centimeter away from a like equal charge will repel it with a force of one dyne. Such a definition proves useful provided that the force with which like charges repel each other in a vacuum depends upon their distance, which it does according to Coulomb's law.

(3) A term whose definition can be stated without formulating laws of the theory, yet which denotes some more or less complex expression which appears in a formula whose *derivation* in the theory will not be understood unless certain laws of that theory are known. Quite often various expressions utilized in the theory will not be considered thoroughly understood unless one knows "where they come from," i.e., how certain formulas containing these expressions are derived from more fundamental principles of the theory. This

is true, for example, of the term "enthalpy" in thermodynamics, which is defined as $U + pV$, where U is the internal energy of a system, p its pressure, and V its volume. One standard method for introducing this term is by considering a process of constant pressure and applying the first law of thermodynamics, arriving at an expression containing $U + pV$.

(4) A term or expression referring to something *x* which the theory is designed to describe and explain in certain ways, and for which the question "What is (an) *x?*" could be answered, at least in part, by considering principles of that theory. Very often this question, rather than "What does the term *x* mean?" will be asked, and an answer given not by reference to some formal definition of *x* (if indeed one exists) but to principles of the theory which characterize features of *x*. For example, suppose one were to ask, "What are electrons?" Many sorts of replies could of course be given depending upon the knowledge and interests of the inquirer. Part of the answer might involve references to the Bohr or quantum theories which describe the various energy states of electrons within the band theory of solids which uses quantum mechanical results in describing properties which electrons manifest in conductors; to the theory of the chemical bond which describes the sharing of electrons by atoms, etc. By characterizing various properties of electrons, theories such as these provide answers to the question "What are electrons?" and in this sense the term "electron" might be considered theory-dependent with respect to each.

(5) A term or expression whose *role* in the theory can only (or best) be appreciated by considering laws or principles in which it appears.[22] In most theories the roles of constituent terms can be examined from several points of view. One might simply consider whether and how a given term is needed for the purpose of *formulating* some of the principles of the theory (for example, how *h* (Planck's constant) is used in the formulation of two fundamental postulates of the Bohr theory). Once a theory has been stated one might ask whether and how a certain term affords a *simplification* or *concise expression* of other principles (as how the term '*s*' (entropy) facilitates the formulation of an equation combining the first and second laws of thermodynamics); or how it is used in *proofs* of important theorems. From a wider viewpoint, the role of a term might also be studied by considering how principles in which it functions serve to *explain* various phenomena (as how "resonance potential" in the Bohr theory is used in the explanation of electron transitions to different energy levels). Conversely, one might consider the manner in which principles of the theory are used to explain various phenomena which the term itself designates.[23] Thus, the role played by the expression "discrete spectral lines" in the Bohr theory might be specified by showing how the postulates of that theory serve to explain the sort of phenomenon referred to by this expression.

The five sorts of theory-dependence mentioned reflect various factors which may be relevant in understanding a given term and represent at least some of the ways in which expressions employed by a theory might be deemed "theory-laden" in Ryle's sense, i.e., dependent (at least in part) on the theory for their meaning. No doubt others could be listed and within those already mentioned further distinctions drawn. Yet each type of dependence cited, if employed to generate a classification of terms, might well yield different results. On the basis of the first sort of dependence noted we could distinguish (a) terms whose definitions are usually given by reference to laws of the theory in question, from (b) terms usually defined independently of such laws. (With respect to thermodynamics, "entropy" might be considered an expression of the former sort, whereas "enthalpy" would be placed in the latter category.) On the basis of the second dependence we could distinguish (a) terms (such as "electrostatic unit charge") whose definitions require justification by appeal to some of the principles of the theory, from (b) terms (such as "electrostatic force") which are not specifically defined in that theory, or whose definitions require no special defense by reference to principles of the theory. Referring to the third type of dependence we might distinguish between (a) expressions usually introduced by reference to formulas in the theory whose derivations will not be understood unless principles of that theory are known, and (b) expressions appropriated with unchanged definitions from other theories.[24] Consideration of the fourth type of dependence might prompt a distinction between (a) terms referring to entities and properties which the theory describes in such a way that the question "What is (an) x?" could be answered by considering principles of that theory, and (b) terms occurring in that theory for which this question would not usually be answered by reference to the theory itself but perhaps to others. With respect to the Bohr theory, terms such as "electron" and "atomic nucleus" would fall into the first category, terms such as "velocity," "acceleration," and "mass" would not. In the case of the fifth dependence several classifications seem possible depending upon the particular type of role considered. Thus, if we group together those terms whose (principal) role is construed as that of simply enabling certain postulates to be formulated, and call these "theory-laden," in general we might expect to get a different classification from that obtained by grouping together terms introduced mainly for the purpose of simplifying certain formulations. In the Bohr theory Planck's constant h might be classified in the first manner, h (h divided by 2 *pi*) in the second. And if we consider all of the different roles cited earlier and attempt to draw a distinction between (a) terms at least some of whose roles cannot be fully understood unless principles of the theory are known, and (b) terms for which this is not necessary, we will find hardly any distinction left to be drawn. For if the roles of a term or expression employed within a given theory are explained by showing how it enables certain principles to be formulated, or how it simplifies formulations, or how it functions in proofs of various theorems, or how it is employed in certain postulates for purposes of explanation, or how it is used to refer to something explained by the theoretical postulates, then obviously principles of the theory will need to be cited. Yet surely almost any term employed in a given theory can be shown to play at least one of these roles, and thus when all such terms are grouped together as "theory-laden" the class of non-theoretical terms all but vanishes.

In short, the various types of dependence noted earlier generate several distinctions between expressions in a given theory. In the case of some of these the class of expressions to be considered "theory-laden" will be quite small (e.g., those given explicit definitions by reference to laws of the theory); in other cases it might be larger (e.g., expressions designating entities or properties which are such that the question "What is (an) x?" could be answered by reference to principles of the theory); and in still other cases it would include practically every term employed by the theory (expressions serving at least one role which cannot be fully understood without a knowledge of some of the principles of the theory). Moreover, as should be evident from some of the examples cited, a term classified as "theory-laden" in one of these senses would not necessarily be so classified in another. For these reasons a criterion of theory-dependence of the sort proposed by Ryle and Hanson not only precludes the construction of theory-*in*dependent lists of the type given at the beginning of this paper, but, even with respect to a specific theory, can give rise to various distinctions under which terms may be classified differently. On the other hand if all the various senses in which a term might be dependent upon a given theory are lumped together and a term classified as "theory-laden" if it conforms to any of these, then such a label would become useless for distinguishing between terms in a given theory since it would be applicable to almost all of them.

Our conclusions here are relevent also for those seeking to draw the broader distinction between terms laden with the concepts and principles of some theory or other and terms dependent upon no theory whatever. Since there are many sorts of theory-dependence, various distinctions become possible. And should any one of the criteria outlined earlier be considered sufficient to render a term "theoretical," few if any terms will escape this classification.[25]

IV

We have been considering the doctrine that expressions employed by scientists can be divided into two sets. On one view the principle of division rests upon observation; on the other, upon conceptual organization or theory-dependence. What has been shown is not that divisions are impossible to make, but rather that the proposed criteria are capable of generating distinctions of many different sorts, each tied, in most cases rather specifically, to a particular context of observation or to a particular theory. Questions such as "Does the term refer to something which can be observed?" and "Does its meaning depend upon some theory?" are too nebulous to provide illuminating classifications.

This means that certain problems raised by philosophers need serious rethinking. For example, authors of logical empiricist persuasion introduce the following issue: If theoretical terms in science do not refer to what is observable, how can they be said to have meaning? The type of answer given consists in treating such terms as uninterpreted symbols which gain meaning in the context of a theory by being related via "correspondence rules" to observational terms. The problem is then to make this idea precise by providing a "criterion of meaningfulness" for theoretical terms.[26] Yet in the absence of some definite basis for drawing the intended large-scale distinction between theoretical and observational terms such problems cannot legitimately be raised, at least not in this form. Again Ryle, beginning with the question, "How is the World of Physics related to the Everyday World?" suggests the need to restate it in a clearer way as, in effect, "How are the 'theory-laden' concepts of physics related to others?" Yet unless his notion of theory-dependence is made precise and is shown to generate some rather definite distinction one will be in doubt about which terms he is referring to and what special characteristic he is attributing to them. This by no means precludes questions concerning the manner in which terms employed by scientists are tied to observation and to theories. Such questions, however, should not be raised about observation *in general,* or theories *in general,* but about particular sorts of observations and about terms in specific theories.

The concept *electron,* for instance, is tied to observation in a manner different from that of *temperature, field,* or *molecule.* And while there are some similarities there are also important differences, so that epithets such as "not directly observable" are bound to prove unhelpful. The philosopher of science genuinely concerned with the relation between theory and observation must begin by examining individual cases. Taking a clue from earlier discussion, one illuminating procedure consists in invoking a series of contrasts. How, for example, is the observation of electrons similar to and also unlike the observation of high flying jets, large molecules, neutrons, chairs, and tables? Each such contrast can be invoked to bring out some quite definite point concerning the relation between electrons and observation. This constitutes one important way of dealing with certain questions raised by logical empiricists, though obviously in a much more specific form than they envisage. Employing a series of contrasts we can also proceed to discover how given concepts are theory-dependent, thus turning our attention, though again in a specific way, to issues raised by Ryle and Hanson.

It is important to consider the manner in which entities studied by the physicist are tied to observation and to theory. Yet the philosopher of science must not at the outset assume that items on the "theoretical" list constructed earlier are related to observation (or theory) in the same way, or even in the same *general* way. Some items can be grouped together as similar in certain observational respects (for instance, electrons and neutrons as being too small to observe with electron microscopes), though in other such respects there will be important differences (e.g., ionizing effects). If categories are invoked to mark the similarities which do exist they will need to be a good deal more specific than the "theoretical" and "nontheoretical" classifications too often presupposed by epistemologists and philosophers of science.

NOTES

1. Rudolf Carnap, "The Methodological Character of Theoretical Concepts," in *Minnesota Studies in the Philosophy of Science*, eds. Herbert Feigl and Michael Scriven, Vol. I (Minneapolis, 1956), 38–76.

2. This is emphasized by J. L. Austin in *Sense and Sensibilia* (Oxford, 1962), pp. 97ff., in a discussion in which he is concerned with criticizing the doctrine that verbs of perception have different senses.

3. Cf. R. B. Braithwaite, "Models in the Empirical Sciences," in *Logic, Methodology, and Philosophy of Science*, eds. E. Nagel, P. Suppes, and A. Tarski (Stanford, 1962), p. 227: ". . . in all interesting cases the initial hypotheses of the theory will contain concepts which are . . . not themselves observable (call these theoretical concepts); examples are electrons, Schrödinger wave functions, genes, ego-ideals."

4. Another contrast sometimes invoked in atomic physics has to do with objects such as electrons for which (in accordance with the Heisenberg uncertainty relationships) the product of the uncertainties in position and simultaneous velocity is much greater than that for objects of considerably larger mass such as atoms and molecules. A particle of the former sort may be classified as (itself) unobservable where it is this particular contrast which is intended.

5. Carl G. Hempel, "The Theoretician's Dilemma," in *Minnesota Studies in the Philosophy of Science*, eds. H. Feigl, M. Scriven, and G. Maxwell, Vol. II (Minneapolis, 1958), 37–98.

6. Observations of "readings of measuring instruments, changes in color or odor accompanying a chemical reaction, utterances made. . . ."

7. Indeed, it is for this very reason that in thermodynamics pressure, volume, and temperature are frequently called directly observable properties of a thermodynamic system, whereas other thermodynamic properties such as internal energy and entropy are not.

8. Somewhat simpler ones may of course be presupposed when the expression "directly observable" is employed in more everyday situations. For example, a contrast between an object such as the bank robber's coat which is readily accessible to view and his revolver which is hidden under it during the robbery; or between an item such as this side of the moon's surface which can readily be observed and the far side which from our vantage point is always hidden from view.

9. Reprinted in *Readings in the Philosophy of Science*, eds. Herbert Feigl and May Brodbeck (New York, 1953), pp. 47–92. Quotation from p. 63. Similar qualifications on observability have been expressed more recently by Grover Maxwell, "The Ontological Status of Theoretical Entities," in *Minnesota Studies in the Philosophy of Science*, eds. H. Feigl and G. Maxwell, Vol. III (Minneapolis, 1962), and by Israel Scheffler, *The Anatomy of Inquiry* (New York, 1963), p. 164.

10. According to Carnap when instruments are used we have "to make a great many preliminary observations in order to find out whether the things before us are instruments of the kind required." ("Testability and Meaning," p. 64.)

11. *Op. cit.*, pp. 164ff.

12. Scheffler concludes (*op. cit.*, p. 164) that the only thing left to do is simply specify an exhaustive list of primitive terms to be called "observational" (and presumably a corresponding list of those to be called "theoretical"). But this leaves the question of the basis for this separation unanswered.

13. N. R. Hanson, *Patterns of Discovery* (Cambridge, England, 1958), p. 60.

14. Gilbert Ryle, *Dilemmas* (Cambridge, England, 1956), pp. 90–91.

15. *Op. cit.*, p. 56.

16. Ryle, *op. cit.*, p. 90.

17. *Ibid.*

18. Hanson, *op. cit.*, pp. 59–60.

19. See *op. cit.*, p. 61.

20. Cf. Hanson, *op. cit.*, pp. 61–62: "'Revoke,' 'trump,' 'finesse' belong to the parlance system of bridge. The entire conceptual pattern of the game is implicit in each term. . . . Likewise with 'pressure,' 'temperature,' 'volume,' 'conductor,' 'charge' . . . in physics . . . To understand one of these ideas thoroughly is to understand the concept pattern of the discipline in which it figures."

21. Ryle at many points does seem to be pressing for a distinction between terms appearing within the context (broadly speaking) of a given theory (or system) which are "theory-laden" and those which may also appear in the same general context but are not. This is evidenced by the sorts of examples he chooses ("light wave" vs. "blue"; "straight flush" vs. "Queen of Hearts"), and also by the questions he raises ("How are the special terms of Bridge or Poker [e.g., "trump"] logically related to the terms in which the observant child describes the cards that are shown to him [e.g., "hearts"]?"). So, relativizing the distinction to terms employed in the context of a particular theory is not completely foreign to Ryle's thought, though on some occasions he does suggest that he intends a broader distinction, as between terms "laden" with some theory or other and those dependent on no theory whatever.

22. Cf. Ryle, "The Theory of Meaning," in *The Importance of Language*, ed. Max Black (Englewood Cliffs, N.J., 1962), p. 161.

23. Hanson, it might be noted, suggests at one point that terms referring to something explained by a given theory are dependent in meaning upon that theory. (Thus, he claims, Tycho and Kepler, because they had different theories about the movement of the sun, attached different meanings to "sun." *Op. cit.*, p. 7.) This thesis is also defended by P. K. Feyerabend, "Explanation, Reduction, and Empiricism," in *Minnesota Studies in the Philosophy of Sciene*, eds. H. Feigl and G. Maxwell, Vol. III (Minneapolis, 1962).

24. In thermodynamics "enthropy" would fall into the first category; "pressure" into the second.

25. Especially if, following Hanson, a term which refers to something explained by a theory is to be considered "theory-laden." See n. 23.

26. See, e.g., Carnap, "The Methodological Character of Theoretical Concepts"; Hempel, *op. cit.*; William W. Rozeboom, "The Factual Content of Theoretical Concepts," in *Minnesota Studies in the Philosophy of Science*, Vol. III, *op. cit.*

The Role of Models

N. R. Campbell

WHAT IS A THEORY?

. . .

WHAT I DO MEAN BY A THEORY

I have now stated what I do not mean by a theory and a hypothesis; it remains to state what I do mean.

A theory is a connected set of propositions which are divided into two groups. One group consists of statements about some collection of ideas which are characteristic of the theory; the other group consists of statements of the relation between these ideas and some other ideas of a different nature. The first group will be termed collectively the "hypothesis" of the theory; the second group the "dictionary." The hypothesis is so called, in accordance with the sense that has just been stated, because the propositions composing it are incapable of proof or of disproof by themselves; they must be significant, but, taken apart from the dictionary, they appear arbitrary assumptions. They may be considered accordingly as providing a "definition by postulate" of the ideas which are characteristic of the hypothesis. The ideas which are related by means of the dictionary to the ideas of the hypothesis are, on the other hand, such that something is known about them apart from the theory. It must be possible to determine, apart from all knowledge of the theory, whether certain propositions involving these ideas are true or false. The dictionary relates some of these propositions of which the truth or falsity is known to certain propositions involving the hypothetical ideas by stating that if the first set of propositions is true then the second set is true and vice versa; this relation may be expressed by the statement that the first set implies the second.

In scientific theories (for it seems that there may be sets of propositions

---

From N. R. Campbell, *Physics: The Elements* (Cambridge University Press). Reprinted by permission of the publisher.

having exactly the same features in departments of knowledge other than science) the ideas connected by means of the dictionary to the hypothetical ideas are always concepts in the sense of chapter II, that is collections of fundamental judgements related in laws by uniform association; and the propositions involving these ideas, of which the truth or falsity is known, are always laws. Accordingly those ideas involved in a theory which are not hypothetical ideas will be termed concepts; it must be remembered that this term is used in a very special sense; concepts depend for their validity on laws, and any proposition in which concepts are related to concepts is again a law. Whether there is any necessary limitation on the nature of the ideas which can be admitted as hypothetical ideas is a question which requires much consideration; but one limitation is obviously imposed at the outset by the proviso that propositions concerning them are arbitrary, namely that they must not be concepts. As a matter of fact the hypothetical ideas of most of the important theories of physics, but not of other sciences, are mathematical constants and variables. (Except when the distinction is important, the term "variable" will be used in this chapter to include constants.)

The theory is said to be true if propositions concerning the hypothetical ideas, deduced from the hypothesis, are found, according to the dictionary, to imply propositions concerning the concepts which are true, that is to imply laws; for all true propositions concerning concepts are laws. And the theory is said to explain certain laws if it is these laws which are implied by the propositions concerning the hypothetical ideas.

An illustration will make the matter clearer. To spare the feelings of the scientific reader and to save myself from his indignation, I will explain at the outset that the example is wholly fantastic, and that a theory of this nature would not be of the slightest importance in science. But when it has been considered we shall be in a better position to understand why it is so utterly unimportant, and in what respects it differs from valuable scientific theories.

The hypothesis consists of the following mathematical propositions:

(1) $u, v, w, \ldots$ are independent variables.
(2) a is a constant for all values of these variables.
(3) b is a constant for all values of these variables.
(4) $c = d$, where c and d are dependent variables.

The dictionary consists of the following propositions:

(1) The assertion that $(c^2 + d^2)a = R$, where R is a positive and rational number, implies the assertion that the resistance of some definite piece of pure metal is R.
(2) The assertion that $cd/b = T$ implies the assertion that the temperature of the same piece of pure metal is T.

From the hypothesis we deduce:

$$(c^2 + d^2)a \Big/ \frac{cd}{b} = 2ab = \text{constant}.$$

Interpreting this proposition by means of the dictionary we arrive at the following law:

> The ratio of the resistance of a piece of pure metal to its absolute temperature is constant.

This proposition is a true law (or for our purpose may be taken as such). The theory is therefore true and explains the law.

This example, absurd though it may seem, will serve to illustrate some of the features which are of importance in actual theories. In the first place, we may observe the nature of the propositions, involving respectively the hypothetical ideas and the concepts, which are stated by the dictionary to imply each other. When the hypothetical ideas are mathematical variables, the concepts are measurable concepts (an idea of which much will be said hereafter), and the propositions related by mutual implication connect the variables, or some function of them, to the same number as these measurable concepts. When such a relation is stated by the dictionary it will be said for brevity that the function of the hypothetical ideas "is" the measurable concept; thus, we shall say that $(c^2 + d^2)\, a$ and cd/b "are" respectively the resistance and temperature. But it must be insisted that this nomenclature is adopted only for brevity; it is not meant that in any other sense of that extremely versatile word "is" $(c^2 + d^2)\, a$ is the resistance; for there are some senses of that word in which a function of variables can no more "be" a measurable concept than a railway engine can "be" the year represented by the same number.

If an hypothetical idea is directly stated by the dictionary to be some measurable concept, that idea is completely determined and every proposition about its value can be tested by experiment. But in the example which has been taken this condition is not fulfilled. It is only functions of the hypothetical ideas which are measurable concepts. Moreover since only two functions, which involve four mathematical variables and between which one relation is stated by the hypothesis, are stated to be measurable concepts, it is impossible by a determination of those concepts to assign definitely numerical values to them. If some third function of them had been stated to be some third measurable concept, then it would have been possible to assign to all of them numerical values in a unique manner. If further some fourth function has been similarly involved in the dictionary, the question would have arisen whether the values determined from one set of three functions is consistent with those determined from another set of three.

These distinctions are important. There is obviously a great difference between a theory in which some proposition based on experiment can be asserted about each of the hypothetical ideas, and one in which nothing can be said about these ideas separately, but only about combinations of them. There is also a difference between those in which several statements about those ideas can be definitely shown to be consistent and those in which such statements are merely known not to be inconsistent. In these respects actual theories differ in almost all possible degrees; it very often happens that some of the hypothetical ideas can be directly determined by experiment while others cannot; and in such cases there is an important difference between the two classes of ideas. Those which can be directly determined are often confused with the concepts to which they are directly related, while those which cannot are recognized as distinctly theoretical. But it must be noticed that a distinction of this nature has no foundation. The ideas of the hypothesis are never actually concepts; they are related to concepts only by means of the dictionary. Whatever the nature of the dictionary, all theories have this in common that no proposition based on experimental evidence can be asserted concerning the hypothetical ideas except on the assumption that the propositions of the theory are true. This is a most important matter which must be carefully borne in mind in all our discussions.

It will be observed that in our example there are no propositions in the dictionary relating any of the independent variables of the hypothesis to measurable concepts. This feature is characteristic of such theories. The nature of the connection between the independent variables and the concepts is clear from the use made, in the deduction of the laws, of the fact that a and b are constants, not varying with the independent variables. The conclusion that the electrical resistance is proportional to the absolute temperature would not follow unless $(c^2 + d^2)\, a$ were the resistance in the same state of the system the same as that in which cd/b is the temperature; and on the other hand it would not follow if a and b were not the same constants in all the propositions of the dictionary. Accordingly the assertion that a or b is a constant must imply that it is the same so long as the state of the system to which the concepts refer is the same; the independent variables on the contrary may change without a corresponding change in the state of the system. If therefore there is to be in the dictionary a proposition introducing the independent variables, it must state that a change in the independent variables does *not* imply a change in the state of the system; the omission of these variables from the dictionary must be taken to mean a definite negative statement. On the other hand, the independent variables may bear some relation to measurable concepts, so long as these concepts are not properties of the system. Thus, in almost all theories of this type, one of the independent variables is called the "time," and the use of this name indicates that it is related in some manner to the physically measurable "time" since some agreed datum. What exactly is this relation we shall have to inquire in the third part of this volume, but it is to be noted that a relation between one of the independent variables and physically measured time is not inconsistent with the statement that a change in this variable does not imply any change in the state of the system; for it is one of the essential properties of a system that its state should be, in a certain degree and within certain limits, independent of the time.

In some theories again, there are dependent variables which are not mentioned in the dictionary. But in such cases the absence of mention is not to be taken as involving the definite assertion that there is no relation between these variables and the concepts. It must always be regarded as possible that a further development of the theory may lead to their introduction into the dictionary.

AN EXAMPLE OF PHYSICAL THEORIES

The fantastic example on which this discussion has been based was introduced in order that, in defining a theory and examining some features of its formal constitution, we might be free from associated ideas which would be sure to arise if the example were taken from any actual theory. It is easier thus to realize the difference between the hypothesis of the theory and the dictionary, and between the nature of the ideas which are characteristic of those two parts of the theory, or to recognize that numerical values can be attributed by experiment to the hypothetical ideas only in virtue of the propositions of the theory. But now we have to consider whether there are any actual scientific propositions which have this formal constitution, and, if there are, whether the application of the term theory to them accords with the usual practice; further we have to decide, if we answer these questions in the affirmative, what it is that gives them a value so very much greater than that of the absurd example which has been used so far. For this purpose an actual scientific proposition will be taken which is generally considered to have considerable value and is always called a theory; and it will be shown that it has the formal constitution which has just been explained. It will thus appear that in one instance at least our definition accords with ordinary usage.

The theory which will be selected is the dynamical theory of gases. We shall start with it in its very simplest form, in which it explains only the laws of Boyle and Gay-Lussac. For such explanation no account need be taken of collisions between the molecules, which may therefore be supposed to be of infinitely small size. Though the theory in this form is known now not to be true, it will be admitted that it is as much a theory in this form as in its more complex modern form. By starting with the simplest form we shall abbreviate our original discussion and at the same time permit the interesting process of the development of a theory to be traced. And when the development of the theory is mentioned, it should be explained that the development traced will not be that which has actually occurred but that which might have occurred; no attention is paid to merely historical considerations. One further word of warning should be given at the outset. Objections have at times been raised to this theory, and to all of similar type, by those who would admit theories of a somewhat different nature. By taking the dynamical theory of gases as an example I am not overlooking these objections or assuming in any way that all scientific theories are essentially the same in nature as the example; we shall discuss these matters later.

Let us then attempt to express the theory in the form which has been explained. The hypothesis of the theory may be stated as follows:

(1) There is a single independent variable t.

(2) There are three constants, m, v, and l, independent of t.

(3) There are $3n$ dependent variables (x_s, y_s, z_s) $(s = 1 \text{ to } n)$ which are continuous functions of t. They form a continuous three-dimensional series and are such that $(x_s^2 + y_s^2 + z_s^2)$ is invariant for all linear transformations of

the type $x' = ax + by + cz$. (This last sentence is merely a way of saying that (x, y, z) are related like rectangular coordinates; but since any definitely spatial notions might give the idea that the properties of the (x, y, z) were somehow determined by experiment, they have been avoided.)

(4) $(d/dt)(x_s, y_s, z_s)$ is constant, except when (x_s, y_s, z_s) is 0 or l; when it attains either of these values it changes sign.

(5) $\frac{1}{n}\sum_1^n \left(\frac{dx_s}{dt}\right)^2 = v^2$, and similar propositions for y_s and z_s.

The dictionary contains the following propositions:

(1) l is the length of the side of a cubical vessel in which a "perfect" gas is contained.

(2) nm is the mass of the gas, M.

(3) $(1/\alpha)mv^2$ is T, the absolute temperature of the gas, where α is some number which will vary with the arbitrary choice of the degree of temperature.[1]

(4) Let $\Delta m(dx_s/dt)$ be the change in $m(dx_s/dt)$ which occurs when x_s attains the value l; let $\sum_\gamma \Delta m(dx_s/dt)$ be the sum of all values of $\Delta m(dx_s/dt)$ for which t lies between t and $t + \gamma$; let

$$p_a, p_b, p_c = \mathrm{Lt}_{n\to\infty \; \gamma\to\infty} \sum_{s=n}^{s=n} \frac{1}{\gamma} \sum_\gamma \Delta m \frac{d(x_s, y_s, z_s)}{dt},$$

then p_a, p_b, p_c are the pressures P_a, P_b, P_c on three mutually perpendicular walls of the cubical containing vessel.

From the propositions of the hypothesis it is possible to prove that:

$$p_a = p_b = p_c = \frac{1}{3l^3} nmv^2.$$

But l^3 is V, the volume of the gas. If we interpret this proposition according to the dictionary we find:

$$P_a = P_b = P_c = \frac{T}{V} \cdot \frac{an}{3},$$

which is the expression of Boyle's and Gay-Lussac's Laws, since $an/3$ is constant.

The theory is here expressed in a form exactly similar to that of our original example, and it will now be seen that this form is not wholly artificial, but has a real significance. In explaining the laws by the theory, we do actually deduce propositions from the hypothesis and interpret them in experimental terms by means of the dictionary. Moreover the distinction between the various kinds of variable in respect of their connection with measurable concepts is apparent. l is directly connected by the dictionary to a measurable concept, and the attribution to it of a numerical value requires nothing but a knowledge of the dictionary; the hypothesis is not involved. At the other extreme, the variables or constants n, m, x_s, y_s, z_s cannot be given numerical values by experiment even with help of the hypothesis; only functions of these variables and not the variables separately can be determined. Between these two extremes lies the constant v. We have deduced from the hypothesis that $v^2 = 3l^3p_a/nm$. The right-hand side of this equation can be given by experiment a numerical value, namely $3VP_a/M$, by means of the dictionary, so that v can also be evaluated. But this evaluation depends wholly on the acceptance of the propositions of the hypothesis; apart from those propositions a statement that v has a certain numerical value does not assert anything which can be proved by experiment.

Having thus shown that the dynamical theory of gases is a theory in our sense, we must now ask what is the difference between this valuable theory and the trivial example with which we began? It lies, of course, in the fact that the propositions of the hypothesis of the dynamical theory of gases display an analogy which the corresponding propositions of the other theory do not display. The propositions of the hypothesis are very similar in form to the laws which would describe the motion of a large number of infinitely small and highly elastic bodies contained in a cubical box. If we had such a number of particles, each of mass m, occupying points in a box of side l represented by the coordinates (x_s, y_s, z_s), and initially in motion, then their momentum would change sign at each impact on the walls of the box. l^2 (p_a, p_b, p_c) would be the rate of change of momentum at the walls of the box and would, accordingly, be the average force exerted upon those walls. And so on; it is unnecessary to state the analogy down to its smallest details. All these symbols, $m, l, t, x, y, z, \ldots$ would denote the numerical values of actually measurable physical concepts, and it would be a law that they were related in the way described; if they were actually measured and the resulting numerical values inserted in the equations stated, those equations would be satisfied.

Further the propositions of the dictionary are suggested by the analogy displayed by the propositions of the hypothesis. p is called the "pressure," and the pressure of the gas P is specially related to the variable p, because p, in the law to which the hypothesis is analogous, would be the average pressure on the walls of the box actually observed. Similar considerations suggested the establishment of the relation between nm and the total mass of the gas, and between l^3 and its volume. The basis of the relation established between T and mv^2 is rather more complex, and its full consideration must be left till we deal in detail with the theory as a part of actual physics; but again it lies in an analogy. Speaking roughly, we may say that the relation is made because, in the law of the elastic particles, mv^2 would be a magnitude which would be found to remain constant so long as the box containing the particles was isolated from all exterior interference, while in the case of the gas the temperature is found so to remain constant during complete isolation.

THE IMPORTANCE OF THE ANALOGY

We see then that the class of physical theories of which the theory of gases is a type has two characteristics. First they are of the form which has been described, consisting of a hypothesis and a dictionary; if they are to be true, they must be such that laws which are actually found to be true by observation can be deduced from the hypothesis by means of logical reasoning combined with translation through the dictionary. But in order that a theory may be valuable it must have a second characteristic; it must display an analogy. The propositions of the hypothesis must be analogous to some known laws.

This manner of expressing the formal constitution of a theory is probably not familiar to most readers, but there is nothing new in the suggestion that analogy with laws plays an important part in the development of theories. No systematic writer on the principles of science is in the least inclined to overlook the intimate connection between analogy and theories or hypotheses. Nevertheless it seems to me that most of them have seriously misunderstood the position. They speak of analogies as "aids" to the formations of hypotheses (by which they usually mean what I have termed theories) and to the general progress of science. But in the view which is urged here analogies are not "aids" to the establishment of theories; they are an utterly essential part of theories, without which theories would be completely valueless and unworthy of the name. It is often suggested that the analogy leads to the formulation of the theory, but that once the theory is formulated the analogy has served its purpose and may be removed and forgotten. Such a suggestion is absolutely false and perniciously misleading. If physical science were a purely logical science, if its object were to establish a set of propositions all true and all logically connected but characterized by no other feature, then possibly this view might be correct. Once the theory was established and shown to lead by purely logical deduction to the laws to be explained, then certainly the analogy might be abandoned as having no further significance. But, if this were true, there would never have been any need for the analogy to be introduced. Any fool can invent a logically satisfactory theory to explain any law. There is as a matter of fact no satisfactory physical theory which explains the variation of the resistance of a metal with the temperature. It took me about a quarter of an hour to elaborate the theory given on page 253; and yet it is, I maintain, formally as satisfactory as any theory in physics. If nothing but this were required we should never lack theories to explain our laws; a schoolboy in a day's work could solve the problems at which generations have laboured in vain by the most trivial process of trial and error. What is wrong with the theory of page 253, what makes it absurd and unworthy of a single moment's consideration, is that it does not display any analogy; it is just because an analogy has not been used in its development that it is so completely valueless.

Analogy, so far from being a help to the establishment of theories, is the greatest hindrance. It is never difficult to find a theory which will explain the laws logically; what is difficult is to find one which will explain them logically and at the same time display the requisite analogy. Nor is it true that, once the theory is developed, the analogy becomes unimportant. If it were found that the analogy was false it would at once lose its value; if it were presented to someone unable to appreciate it, for him the theory would have little value. To regard analogy as an aid to the invention of theories is as absurd as to regard melody as an aid to the composition of sonatas. If the satisfaction of the laws of harmony and the formal principles of development were all that were required of music, we could all be great composers; it is the absence of the melodic sense which prevents us all from attaining musical eminence by the simple process of purchasing a textbook.

The reason why the perverse view that analogies are merely an incidental help to the discovery of theories has ever gained credence lies, I believe, in a false opinion as to the nature of theories. I said just now that it was a commonplace that analogies were important in the framing of hypotheses, and that the name "hypotheses" was usually given in this connection to the propositions (or sets of propositions) which are here termed theories. This statement is perfectly true, but it is not generally recognized by such writers that the "hypotheses" of which they speak are a distinct class of propositions, and especially that they are wholly different from the class of laws; there is a tendency to regard a "hypothesis" merely as a law of which full proof is not yet forthcoming.

If this view were correct, it might be true that the analogy was a mere auxiliary to the discovery of laws and of little further use when the law was discovered. For once the law had been proposed the method of ascertaining whether or no it were true would depend in no way on the analogy; if the "hypothesis" were a law, its truth would be tested like that of any other law by examining whether the observations asserted to be connected by the relation of uniformity were or were not so connected. According as the test succeeded or failed, the law would be judged true or false; the analogy would have nothing to do with the matter. If the test succeeded, the law would remain true, even if it subsequently appeared that the analogy which suggested it was false; and if the test failed, it would remain untrue, however complete and satisfactory the analogy appeared to be.

A THEORY IS NOT A LAW

But a theory is not a law; it cannot be proved, as a law can, by direct experiment; and the method by which it was suggested is not unimportant. For a theory may often be accepted without the performance of any additional experiments at all; so far as it is based on experiments, those experiments are often made and known before the theory is suggested. Boyle's Law and Gay-Lussac's Law were known before the dynamical theory of gases was framed; and the theory was accepted, or partially accepted, before any other experimental laws which can be deduced from it were known. The theory was an addition to scientific knowledge which followed on no increase of experimental knowledge and on the establishment of no new laws; it cannot therefore have required for its proof new experimental knowledge. The reasons why it was accepted as providing something valuable which was not contained in Boyle's and Gay-Lussac's

Laws were not experimental, The reason for which it was accepted was based directly on the analogy by which it was suggested; with a failure of the analogy, all reason for accepting it would have disappeared.

The conclusion that a theory is not a law is most obvious when it is such that there are hypothetical ideas contained in it which are not completely determined by experiment, such ideas for example as the m, n, x, y, z in the dynamical theory of gases in its simple form. For in this case the theory states something, namely propositions about these ideas separately, which cannot be either proved or disproved by experiment; it states something, that is, which cannot possibly be a law, for all laws, though they may not always be capable of being proved by experiment, are always capable of being disproved by it. It may be suggested that it is only because the theory which has been taken as an example is of this type that it has been possible to maintain that it is not a law. In the other extreme when all the hypothetical ideas are directly stated by the dictionary to "be" measurable concepts, the conclusion is much less obvious; for then a statement can be made about each of the hypothetical ideas which, if it is not actually a law, can be proved and disproved by experiment. This condition is attained only in theories of a special, though a very important type, which will receive attention presently.

The case which demands further consideration immediately is that in which the dictionary relates functions of some, but not all, of the hypothetical ideas to measurable concepts, and yet these functions are sufficiently numerous to determine all the hypothetical ideas. In this case it is true that propositions can be stated about each of the hypothetical ideas which can be proved or disproved by experiment. Thus, in our example, if one litre of gas has a volume mass of 0.09 gm. when the pressure is a million dynes per $cm.^2$ then, in virtue of this experimental knowledge, it can be stated that v is 1.8×10^5 cm./sec. A definite statement can be made about the hypothetical idea v on purely experimental grounds. If the dictionary mentioned sufficient functions of the other ideas, similar definite experimental statements might be made about them. If the theory can thus be reduced to a series of definite statements on experimental grounds, ought it not to be regarded as a law, or at least as a proposition as definitely experimental as a law?

I maintain not. A proposition or set of propositions is not the same thing as another set to which they are logically equivalent and which are implied by them. They may differ in meaning. By the meaning of a proposition I mean (the repetition of the word is useful) the ideas which are called to mind when it is asserted. A theory may be logically equivalent to a set of experimental statements, but it means something perfectly different; and it is its meaning which is important rather than its logical equivalence. If logical equivalence were all that mattered, the absurd theory of page 253 would be as important as any other; it is absurd because it means nothing, evokes no ideas, apart from the laws which it explains. A theory is valuable, and is a theory in any sense important for science, only if it evokes ideas which are not contained in the laws which it explains. The evocation of these ideas is even more valuable than the logical equivalence to the laws. Theories are often accepted and valued greatly, by part of the scientific world at least, even if it is known that they are not quite true and are not strictly equivalent to any experimental laws, simply because the ideas which they bring to mind are intrinsically valuable. It is because men differ about intrinsic values that it has been necessary to insert the proviso, "By part of the scientific world at least"; for ideas which may be intrinsically valuable to some people may not be so to others. It is here that theories differ fundamentally from laws. Laws mean nothing but what they assert. They assert that certain judgments of the external world are related by uniformity, and they mean nothing more; if it is shown that there may be a case in which these judgments are not so related, then what the law asserts is false, and, since nothing remains of the law but this false assertion, the law has no further value. We can get agreement concerning this relation and we can therefore get agreement as to the value of laws.

THE DEVELOPMENT OF THEORIES

The distinction between what a theory means and what it asserts is of the utmost importance for the comprehension of all physical science. And it is in order to insist on this distinction that the case has been considered when all the hypothetical ideas can be determined by experiment, although not all of them are stated by the dictionary to "be" concepts. As a matter of fact I do not think this case ever occurs, though we cannot be certain of that conclusion until all physics has been examined in detail. There is always, or almost always, some hypothetical idea, propositions concerning which cannot be proved or disproved by experiment; and a theory always asserts, as well as means, something which cannot be interpreted in terms of experiment. Nevertheless it is true that a theory is the more satisfactory the more completely the hypothetical ideas in it can be experimentally determined; those ideas may be valuable even if nothing can be stated definitely about them, but they are still more valuable if something can be stated definitely. Thus, in our example, the theory is valuable even though we cannot determine m or n; but it will be more valuable if they can be determined. Accordingly when a theory containing such undetermined ideas is presented and appears to be true, efforts are always directed to determine as many as possible of the undetermined ideas still remaining in it.

The determination of the hypothetical ideas is effected, as we have noticed before, by the addition of new propositions to the hypothesis or to the dictionary, stating new relations of the hypothetical ideas to each other or to the concepts. The process demands some attention because it is intimately connected with a very important property of theories, namely their power to predict laws in much the same way as laws predict events. In passing it may be noted that a failure to distinguish a law from an event and a consequent confusion of two perfectly distinct kinds of prediction has also tended to obscure the difference between a theory and a law.

There is an important difference between the addition of new propositions to the hypothesis and to the dictionary. The hypothesis gives the real meaning of the theory and involves the analogy which confers on it its value; the dictionary uses the analogy, and the propositions contained in it are usually suggested by the analogy, but it adds nothing to it. Accordingly a change in the hypothesis involves to some extent a change in the essence of a theory and makes it in some degree a new theory; an addition to the dictionary does not involve such a change. If, then, a new law can be deduced by the theory by a simple addition to the dictionary, that law has been in the fullest and most complete sense predicted by the theory; for it is a result obtained by no alteration of the essence of the theory whatever. On the other hand if, in order to explain some new law or in order to predict a new one, a change in the hypothesis is necessary, it is shown that the original theory was not quite complete and satisfactory. The explanation of a new law and the determination of one more hypothetical idea by addition to the dictionary is thus a very powerful and convincing confirmation of the theory; a similar result by an addition to the hypothesis is, in general, rather evidence against its original form.

But the degree in which the necessity for an alteration in the hypothesis militates against the acceptance of the theory depends largely on the nature of the alteration. If it arises directly and immediately out of the analogy on which the hypothesis is based, it scarcely is an alteration. Thus, in the theory of gases in the form in which it has been stated so far, the only dynamical proposition (or more accurately the only proposition analogous to a dynamical law) which has been introduced is that the momentum is reversed in sign at an impact with the wall, while its magnitude is unchanged. But in dynamical systems this condition is fulfilled only if the systems are conservative; it is natural therefore to extend the hypothesis and to include in it any other propositions concerning the hypothetical ideas which are analogous to other laws[2] of a conservative system. Such an extension involves no essential alteration of the theory, but it permits the explanation of additional laws and thus provides arguments for rather than against the theory. For example, if the extension is made (the new propositions are so complex if they are stated in a full analytical form that space need not be wasted in stating them) the effect on the behavior of the gas of the motion of the walls of the vessel can be deduced and the laws of adiabatic expansion predicted. Here no addition is made to the dictionary; only the hypothesis is altered. But if the dictionary is altered, the establishment of a complete analogy between the hypothesis and the laws of a conservative system leads immediately to the view that $\frac{1}{2}nmv^2$ is the energy of the gas and to the explanation of all the laws, involving specific heats, which follow directly from Boyle's and Gay-Lussac's laws combined with the doctrine of energy and with the proposition that quantity of heat is energy. All these ideas are essentially contained in the original theory which can hardly be said to be altered by explicit statement of them. As a matter of fact they seem to have been stated in the earliest forms of the dynamical theory, although they were not necessary to explain the laws of Boyle and Gay-Lussac.

However further inquiry shows that a more important alteration is necessary. So long as we are considering only perfect gases (and it must be insisted that some gases over certain ranges are experimentally perfect) and no measurable concepts other than pressure, volume, temperature, and quantity of heat, the theory in its original form, with all its natural implications, explains all the experimental laws. The only objection to it is that the constants m, n, x, y, z remain undetermined. But if we attempt to explain the laws of viscosity or conduction of heat, we meet with new objections. The dynamical analogy leads immediately to an entry in the dictionary relating viscosity to the hypothetical ideas; for viscosity consists experimentally in the transfer of momentum from one to the other of two parallel planes in relative motion. But in the system of elastic particles there would also be such a transfer of momentum if the sides of the box in which they were contained were moving relatively to each other, and the known laws of such a system show that there is a relation between this transfer of momentum and the masses and velocities of the particles together with the distance and relative velocity of the sides; the transfer of momentum is a function of these magnitudes. Accordingly it is suggested that a similar function of the corresponding variables of the hypothesis should be related by means of the dictionary to the viscosity of the gas.

We can now deduce the relation which should exist according to the theory between the pressure, density, and temperature of the gas and its viscosity. The relation predicted does not accord with that determined experimentally; in particular it is found that the theory predicts that the coefficient of viscosity will be determined by the size and shape of the containing vessel, whereas experiment shows that it depends, in a given gas, only on the density and temperature.[3] Here the addition of an entry to the dictionary has led to a new law, but a law which is false. The theory is not true; it must be altered; and it can only be altered by changing the hypothesis. The change which is made is, of course, the introduction of a new hypothetical idea, a mathematical constant σ, and a consequent modification of the equations relating the variables and constants. The hypothesis thus modified is analogous to the laws of a system of elastic particles which are spheres of finite size, and the part which σ plays in those equations is the same as that of the diameter of the spheres in those laws. With this modification, and with such change in the dictionary as necessarily accompanies it, the relation between the coefficient of viscosity and the density and temperature predicted by the theory becomes in accordance with that experimentally determined.[4] The theory is once more satisfactory; and though in its earlier form the theory must be rejected, we do not regard the whole theory as false, because the new ideas introduced into the hypothesis are such an extremely natural extension of the old. If the analogy is based on the behavior of elastic bodies, it is extremely natural to attribute to them finite dimensions.

Thermal conductivity is related to transfer of energy as the coefficient of viscosity is to transfer of momentum. The addition of an entry in the dictionary to introduce thermal conductivity is therefore suggested in just the same manner as the entry to introduce viscosity, for in the system of elastic particles which provides the analogy of the hypothesis energy as well as momentum would be transferred between the walls. No addition to the hypothesis is necessary to deduce the relation between thermal conductivity, density and temperature; the relation predicted turns out to accord with experiment. Here an addition to the dictionary alone has predicted a true law, and the theory is correspondingly strengthened.

But in spite of these atterations the objection still remains that n, m, x, y, z are undetermined; indeed an additional undetermined idea, σ, has been introduced. This addition does not, however, make matters worse, for in the original theory n and m were so connected that the determination of one would involve the determination of the other; whereas now n, m, σ are found to be so connected that a determination of one would determine all three. The determination is effected by the application of the theory to gases which are not perfect. The introduction of σ alters somewhat the laws predicted by the theory for the relation between pressure, volume and temperature; they are no longer exactly those of Boyle and Gay-Lussac. It is found experimentally that these laws are not actually experimentally true; by comparing the deviations found experimentally with those, involving σ, predicted by the theory a new relation between n, m, σ and experimentally determined magnitudes is established. These relations, in addition to those arising from viscosity or thermal conductivity, enable each of these three hypothetical ideas to be determined. Here we have (or should have, if the statements made were correct) the most powerful confirmation of the theory; not only is a new law predicted without addition to the hypothesis and confirmed by experiment, but undetermined ideas in the hypothesis are determined. If no further discrepancies between theory and experiment were found, when yet other propositions were introduced into the dictionary, for the completion and final establishment of the theory, only the determination of x, y, z would be necessary.

But in this case, and in some others similar to it, special considerations make the determination of these variables less important than usual. In order to determine them completely it is necessary to know their values and the values of their first differential coefficients with respect to t for some value of t; their values for all other t's can then be deduced. But it can be shown that this knowledge is not required if it is required to determine only the limit to which some function of these variables tends as n tends to infinity. Whatever values of (x, y, z) and $(d/dt)(x, y, z)$ are associated with the value $t = t_0$, consistent with the relation between $(d/dt)(x, y, z)$ and v which is asserted by the hypothesis, then the value for $t = T + t_0$ of

$$\mathrm{Lt}_{n\to\infty} \sum_1^n f\left(x_s, y_s, z_s, \frac{dx_s}{dt}, \frac{dy_s}{dt}, \frac{dz_s}{dt}\right),$$

where f is any function, will tend to the same limit as T tends to infinity.[5] Or, expressing the matter in terms of the analogy, the properties of any infinite collection of the particles will be the same whatever were the positions of the particles and their velocities at a previous period infinitely distant. Now all the propositions of the dictionary which involve (x, y, z) at all, involve them in the form of the limit of some function as n tends to infinity. So long as we can imagine that the values which should be attributed to them, when experiments on the gas are made, correspond to a value t_0 such that the state of the system is unchanged for all values of t between t_0 and $t_0 - T$, where T may be greater than any assigned quantity, then the laws predicted by the theory will be the same whatever values are assumed to correspond to $t_0 - T$. For various reasons into which we need not inquire for the moment, we are prepared to make the assumption contained in the last sentence. Accordingly though we cannot determine the variables we can, by assigning any "initial" value to them that we please (i.e. values at $t_0 - T$), find values for them in the conditions of experiment which are indistinguishable by any experiment that we can perform from the values resulting from the assumption of any other initial values. That is to say, even if we could determine the variables, the deductions which could be made from the theory would be precisely the same as the deductions made from the theory with the assumed initial values; for if the values could be determined, these values would be associated with some initial values, and these initial values would lead to the same result as those which have been assumed.

We are therefore reconciled to the impossibility of determining these variables because we know that, if we could determine them, it would not make the slightest difference to the theory. Nevertheless, it causes us, I think, some slight mental discomfort; we feel that the theory would be even more satisfactory if they could be determined. Now the determination is impossible only so long as all the propositions in the dictionary introduce only the limits of functions when n is infinite; it would not be impossible if an entry in the dictionary could be made which introduced a function for a finite value of n. In recent years such an entry has been made in connection with the phenomenon known as Brownian motion; and the entry, without additions to the hypothesis, leads to the explanation of new laws and enables the determination of the variables to be made for certain systems. It was felt that the importance of the theory was thereby increased; and M. Perrin, on whose work the advance largely depended, wrote a book describing it entitled *Brownian Motion and Molecular Reality*. He felt that his researches had made molecules real in a way that they had not been real before.

. . .

NOTES

1. The occurrence of a needs some remark. Is it a hypothetical idea or a measurable concept? It is neither. We shall consider its nature when we deal with temperature, but it may be stated here briefly why a number of this kind occurs in this entry in the dictionary and not in the others. The reason is this. Experiment shows that pv is proportional to T. For various reasons, which we shall discuss, we desire that the factor of proportionality shall

not change if the unit of mass or the unit of pressure is changed: but we do not object to its changing when the degree of temperature changes. If we gave the factor a definite value once and for all, the degree of temperature would have to change when the units of mass and pressure changed; we wish to avoid this necessity and do so by changing the value of a when we change the degree. The value of a is therefore as purely arbitrary as the choice of a unit in any system of measurement.

2. The "laws" of a conservative system are not really laws, but for the present they may pass as such.

3. The theory which neglects the size of the molecules leads to the familiar result

$$\eta = \tfrac{1}{3}\rho v\lambda,$$

but λ, the mean free path, will be the distance that the molecules travel between the walls of the vessel, and will depend on its size and shape instead of simply on the properties of the gas, as it will if the free path is between successive collisions with molecules.

4. Of course this statement is not true, so far as the temperature is concerned. Agreement between theory and experiment in respect of the variation of viscosity with temperature can be obtained, if at all, only by giving to the molecules some form more complex than spheres, and by introducing forces between the molecules when not in contact. It is not my object here to expound the dynamical theory, but only to use it as an example; that use is not affected by supposing things to turn out more simply than they actually do. The further statements which will be made presently and are equally untrue will not be specifically noticed. The instructed reader will not require the notice; the uninstructed will be merely confused by it.

5. Certain very exceptional values of the variables and certain exceptional functions should strictly be excepted from this statement.

R. B. Braithwaite

MODELS IN THE EMPIRICAL SCIENCES

Any science which has passed beyond the interlinked stages of the classification of observable properties and relations, and of the establishment of empirically testable generalizations about these observable concepts, attempts to explain these generalizations by showing that they are logical consequences of more general hypotheses. A *scientific theory* is a deductive system consisting of certain *initial hypotheses* at the summit and empirically testable generalizations at the base. The deductive structure of a theory is shown explicitly by expressing the theory by means of a *formal axiomatic system* or *calculus* consisting of a sequence of sentences (or formulae) in which the initial hypotheses are represented by sentences called *axioms* and what are deduced from these hypotheses by sentences which are the *theorems* of the calculus. The calculus may also have axioms representing propositions, logical or mathematical, of the *basic logic* of the theory; such axioms will not concern us here. The rules of derivation of the calculus correspond to the deductive principles of the theory's basic logic. A theory expressed by a calculus will be called a *formalized theory*. The theory will be called a *semi-formalized theory* if (as is usually the case) it is expressed not by a whole calculus but by the parts of a calculus which provide the steps in the deductions which are not thought to be so obvious that they need not be explicitly stated.

In psychology and the social sciences the word *model* is frequently used merely as a synonym for a formalized or semi-formalized theory. For example, Richard Stone's *Three Models of Economic Growth*, which he has presented in section VIII of this Congress, are in fact three alternative theories for explaining the phenomena of economic growth expressed in mathematical form so that

---

Reprinted from *Logic, Methodology, and Philosophy of Science*, edited by Ernest Nagel, Patrick Suppes, and Alfred Tarski with the permission of the publishers, Stanford University Press.

mathematical techniques can be used in the deductions. As another example, R. R. Bush and F. Mosteller's book expounding their statistical learning theory is entitled *Stochastic Models for Learning*. There are, I think, three reasons why social scientists tend to use the word *model* to describe what is in fact a formalized or semi-formalized theory:

(1) The theory may seem such a small one, comprising so few deductive steps or covering such a limited topic in the field, that the word *theory* may seem too grand to apply to it.

(2) Theories that are even semi-formalized are so rare in the social sciences (except in economics) that a special term may seem to be appropriate to emphasize that the deductive system of the theory is being, at least in part, explicitly presented.

(3) The word *model* may be used instead of *theory* to indicate that the theory is only expected to hold as an approximation, or that employing it depends upon various simplifying assumptions. In particular, a system in which the hypotheses and the conclusions deduced from them are all thought only to hold *ceteris paribus* may be called a model to show ignorance of the conditions which would make this qualification unnecessary.

None of these seems to me a good reason for refusing to call a scientific deductive system a theory, even if it holds only for an "isolated" field and is only a little theory. It would be better to call it a *theoruncula* or (affectionately) a *theorita*, using a Latin or a Spanish diminutive, than to call it a model. In any case, to use the word *model* instead of *theory* presents no problems specific to the notion of model.

These problems arise only when *model* is used in a sense which distinguishes it from the theory for which it is a model. I take this sense to be that in which a *model for a theory* T is another theory M which corresponds to the theory T in respect of deductive structure. By *correspondence in deductive structure* between M and T is meant that there is a one-one correlation between the concepts of T and those of M which gives rise to a one-one correlation between the propositions of T and those of M which is such that if a proposition in T logically follows from a set of propositions in T, the correlate in M of the first proposition in T logically follows from the set of correlates in M of the propositions of the set in T.

Since the deductive structure of T is reflected in M, a calculus which expresses T can also be interpreted as expressing M: A theory and a model for it can both be expressed by the same calculus. Thus an alternative and equivalent explication of *model for a theory* can be given by saying that a model is another interpretation of the theory's calculus. Note that the interpretation need not consist of true propositions: It need not be what Alonzo Church calls a *sound* interpretation. The initial propositions of the model, which are correlated with the initial hypotheses of the theory, need not be true, or thought to be true; all that is required is that the rest of the model propositions must be logical consequences of the set of them. Scientists frequently use quite imaginary models: The nineteenth-century mechanical models for optical theory treated of fluids never found in heaven or earth. Mathematical logicians who use the word *model* for an interpretation of a calculus restrict it to a sound interpretation, and frequently to a sound interpretation which employs logical concepts only. This is perfectly reasonable for the purposes for which they want the notion, but it is too restrictive for philosophy of science. And consequently the requirements of categoricalness, etc., which play such an important role in mathematical logicians' concern with models have little relevance to models in the scientists' sense.

Sometimes the working of an automatic computing machine is thought of as modelling a scientific theory. When the mode of operation of the computer is irrelevant to such a modelling, the computer is merely being used *qua* "black box" to provide outputs interpreted as logical consequences of the inputs fed into it. Here, if the deductive apparatus of the theory is already incorporated in the working of the machine, the machine will in fact be operating the calculus representing the theory, the input being the axioms and the outputs the various theorems of the formal axiomatic system. So the working of the machine should be regarded as equivalent more properly to the calculus itself than to an interpretation of the calculus as a theory. Bush and Mosteller's "stat-rat" is the device of using a computer capable of applying Monte Carlo methods and programming it (among other things) with the probability-parameters which their learning theory attributes to an imagined rat. The output is interpreted as propositions about the long-term or statistical behaviour of rats; but the stat-rat itself corresponds to the Bush-Mosteller formal axiomatic system rather than to the theory expressed by this system by which they would explain a rat's behaviour.

But when the mode of operation of the computer is part of its modelling function, the physical processes in the computer will be regarded not as providing the deductive steps in the scientific theory but as echoing, in temporal sequence, the succession of processes in the field which is the subject-matter of the theory. For example, when a (binary) digital computer is proposed for modelling the working of a brain, the switching operations of the computer are taken as corresponding to the "firings" of synapses in the brain. Strictly speaking, the computer as a piece of hardware will be a model of a brain in the vulgar sense of, e.g., a clockwork model locomotive: What in my sense is a model for a theory of cerebral functioning will be the theory of the internal functioning of the computer, which will include propositions relating what happens in it at one time to what happens at a later time, corresponding to hypotheses about the temporal relationships of events in the brain.

Let us pass now to the most interesting question about models for scientific theories: Why use models at all in scientific thinking?

If the nonlogical concepts occurring in the initial hypotheses of the theory are all observable properties or relations, there is no reason whatever for thinking up a model for the theory except either to establish the theory's logical consistency, which finding a sound interpretation for its calculus will do, or for purely didactic purposes. Someone may find it easier to appreciate the deductive structure of the theory in an illustrative model, just as many of us find it helpful to think of spatial diagrams when considering the logical relationships of classes. But usually, and in all interesting cases, the initial hypotheses of the theory will contain concepts which are not purely logical but which are not themselves observable (call these *theoretical concepts*); examples are electrons, Schrödinger wave-functions, genes, ego-ideals. A fundamental problem for philosophy of science is how these theoretical concepts should be understood. One school of thought, represented most prominently by N. R. Campbell, holds that the only way to understand these concepts, and hence to understand the explanatory hypotheses of the scientific theory, is to represent the theory by a model whose deductive arrangement corresponds to that of the theory but in which all the concepts concerned are familiar concepts. (Campbell would also require that the propositions of the model should be, in fact, true propositions.)

Familiar concepts, but not necessarily observable properties or relations. What the modellists (as I shall call them) think necessary is that the correlates, in the model, of the theoretical concepts of the theory should be understood, and they may be understood as being theoretical concepts of a simpler theory which is already understood. Attempts to understand electromagnetic theory by constructing mechanical models for it did not depend upon supposing that all the mechanical concepts involved, e.g., kinetic energy, potential energy, action, were observable, but only that the theory of mechanics using these as theoretical concepts had been previously understood so that all its concepts were familiar. The modellist doctrine of scientific explanation is a stage-by-stage doctrine. Theory T_1 has to be understood by constructing a model for it, all of whose concepts are observable; theory T_2 can then be understood by constructing a model for it with the concepts of theory T_1, and so on.

But the modellist may say more than that understanding the model yields an understanding of the theory and of its theoretical concepts which is in some way fuller than that provided by an understanding based solely on the bare theory. He may say that a model is predictive in a way in which the bare theory is not, in that it yields new generalizations about observable properties which the theory itself does not provide. These new generalizations will be empirically testable, that is, they can be used for making new predictions; so the modelled theory will be a stronger theory than the theory in its bare form. I will discuss this thesis before passing to the more general question of understanding.

Let the observable properties concerned in the generalizations which the old theory was put forward to explain (the *old* generalizations) be $A_1, A_2 \ldots A_n$, and let $B_1, B_2 \ldots B_n$ be the familiar properties which are their correlates in the model. Let $L_1, L_2 \ldots L_m$ be the familiar properties which are correlates in the model of the theoretical concepts of the theory, i.e., the L's are the model-interpretations of the theoretical terms which occur in the axioms of the calculus representing the theory.

There are four ways in which consideration of the model may serve to yield new generalizations about observables in the theory:

(1) Considering the initial propositions in the model may show that is is possible to deduce from them propositions concerned with, e.g., B_1, B_2, B_3 which do not correspond to any of the old generalizations occurring in the theory. Then passing from the model to the theory will enable one to assert that the corresponding generalizations about A_1, A_2, A_3 are consequences of the initial propositions of the theory. Since these generalizations are not among those upon which the theory was founded, they will be new generalizations and can be used to make predictions that would otherwise not have been made by the theory. These new generalizations will be concerned with properties which were concerned in the old generalizations. Call this *first-type predictive novelty.*

(2) Considering the familiar properties L_1, L_2, ... L_m occurring in the initial propositions of the model may show that there is a new familiar property B_{n+1} which is such that generalizations in which it occurs, together with one or more of the old properties B_1, B_2, ... B_n, follow from the initial propositions of the model, together with a new initial proposition relating B_{n+1} to L_1, L_2, ... L_m. Passing then from the model to the theory will suggest looking for an observable property A_{n+1}, which is the theory-correlate of the B_{n+1} in the model. If such a property can be found, the addition to the theory's initial hypotheses of one relating A_{n+1} to the theoretical concepts will enable new generalizations to be deduced relating A_{n+1} to some of A_1, A_2 ... A_n. This *second-type predictive novelty* will yield new generalizations about a new observable property.

(3) Considering the familiar properties L_1, L_2, ... L_m occurring in the initial propositions of the model may suggest propositions about some of these familiar properties which would, if added to the initial propositions in the model, enable new generalizations about, e.g., B_1, B_2, B_3 to be deduced in the model. Passing then from the model to the theory will suggest that if corresponding new initial hypotheses are added to the theory, the extended theory will yield new testable generalizations about A_1, A_2, A_m. As with the first type this *third-type predictive novelty* will yield new generalizations about properties which were concerned in the old generalizations.

(4) So far there has been no extension to L_1, L_2, ... L_m or to the corresponding theoretical concepts of the theory. But the model may suggest a new familiar property L_{m+1} which might enter in combination with some of L_1, L_2, ... L_m into new initial propositions of the model from which new generalizations about the B's can be deduced. There will usually be no object in adding L_{m+1} unless there can also be added new familiar properties B_{n+1}, B_{n+2}, etc., which enter into the new generalizations. In this case, of course, more initial propositions will have to be added relating the new B's to some or all of the L's. Passing from the model to the theory will then suggest looking for observable properties A_{n+1}, A_{n+2}, etc., which are such that new generalizations relating them to some of A_1, A_2, ... A_n can be deduced from the initial hypotheses of the theory with the addition of new hypotheses corresponding to the new initial propositions of the model which contain L_{n+1}. This *fourth-type predictive novelty* will yield new generalizations about new observable properties by extending the theory to incorporate hypotheses about new theoretical concepts.

The relations between these four types of predictive novelty can be summarized by saying that, in the first type there is no change in the initial hypotheses of the theory; in the second type the change is the addition of an extra hypothesis relating a new observable property to the theoretical concepts; in the third type the change is the addition of extra hypotheses relating together theoretical concepts of the theory; and in the fourth type there is the addition of extra hyphotheses containing new theoretical concepts.

I have described these four types of predictive novelty as what might come out of consideration of a model. But how much, in fact, does the employment of a model assist in providing or suggesting any of these types of predictive novelty? For the first type not at all, since there is no addition to the axioms of the calculus which expresses both the model and the theory, and any proposition that can be deduced within the model can be deduced with equal ease within the theory itself without calling upon the model for assistance. And it is almost the same in the case of the second type. For here the only addition to the axioms of the calculus provided by the new initial proposition of the model is one to be interpreted in the theory as relating a new observable property to the theoretical concepts. The difficulty here is almost always that of discovering such an observable property and not in determining the initial hypothesis that would relate it, when found, to the theoretical concepts.

It is with regard to the third and fourth types of predictive novelty that a model may plausibly be claimed to be of genuine assistance. For here there are additions to the axioms of the calculus concerned with relationships between the theoretical terms, and in the fourth type also an addition to the number of the theoretical terms themselves. Since the model interprets the theoretical terms of the calculus as familiar concepts, there may well be propositions (true or false) relating these familiar concepts together (or relating them to new familiar concepts) which are not included in the model's initial propositions, but which thinking of the model immediately brings to mind. The model may then be said to *point* towards its extension in a way which thinking of the calculus in isolation would not do. This pointing will be most striking when the propositions of the model are known to be true (so that the model is known to be a sound interpretation of the calculus): For then there may well be other known relations between the familiar concepts of the model (or between other familiar concepts and those of the model) which point to corresponding relations between the theoretical concepts of the theory (or of an extended theory). In this case the pointing is frequently held to provide an argument—an argument by analogy—for inferring from the known features of the model to unknown features of the theory.[1] Indeed it is sometimes held that analogy provides a good reason for believing that *any* feature of the extended model will be reproduced in the correspondingly extended theory, until an empirically testable generalization, which is a consequence of supposing that the extended theory has this feature, is refuted by experience.

This claim cannot be allowed. Analogy can provide no more than suggestions of how the theory might be extended; and the history of science tells us that while some analogical suggestions have led to valuable extensions of the theory, others have led to dead ends.

The pointing of a model towards extensions of it can only provide suggestions for extensions of the theory (except in the case of first-type novelty, but here the extensions are as latent in the theory as in the model). Whether the theory will bear extensions in its initial hypotheses will have to be decided by testing against experience the testable generalizations deduced in the extended theory, and in this testing a model is of no use whatever. The thesis that a modelled theory has, *ipso facto*, greater predictive power than the bare theory cannot be sustained.

The modellist, however, will retort with a series of questions, which modulate into the crucial question of what it is to *understand* a theory. How can the theory as a whole be empirically tested unless its testable generalizations are known to follow from the initial hypotheses? How can this be known unless the initial hypotheses are understood? How can initial hypotheses concerned with theoretical concepts be understood unless we understand what these theoretical concepts are? And how can theoretical concepts be understood except by taking the theory to correspond with a model all of whose concepts are already familiar?

To answer these questions requires giving an adequate account of the functioning of theoretical concepts which makes no use whatever of models. Such an account (which I will call the *contextualist* account) has been given by many recent writers: Quine in his recent book[2] refers to works by Carnap, Einstein, Frank, Hempel and myself, and this list could be expanded. Briefly, the contextualists hold that the way in which theoretical concepts function in a scientific theory is given by an interpretation of the calculus expressing the theory which works from the bottom upwards. The final theorems of the calculus are interpreted as expressing empirically testable generalizations, the axioms of the calculus are interpreted as propositions from which these generalizations logically follow, and the theoretical terms occurring in the calculus are given a meaning implicitly by their context, i.e., by their place within the calculus. So an understanding of a theoretical concept in a scientific theory is an understanding of the role which the theoretical term representing it plays in the calculus expressing the theory; and the empirical nature of the theoretical concept is based upon the empirical interpretation of the final theorems of the calculus.

If such a contextualist account of the meaning of theoretical terms is adequate, thinking of a model for a theory is quite unnecessary for a full understanding of the theory. But we contextualists can easily explain why it is that most people find models so helpful. For the contextualist account makes explicit reference to the calculus expressing the theory; so a full understanding of a theoretical concept requires what Quine calls a "semantic ascent" from thinking of things to thinking of the symbolism or language which represents the things. In the case of explanation by a scientific theory, this requires considering the uninterpreted calculus, which is to express the theory *before* interpreting it to express the theory. Many people find the notion of an uninterpreted calculus difficult to digest. For them it is easier to think of a scientific deductive system by first thinking of a model for it, all of whose concepts are familiar, and then by formalizing (or semi-formalizing) the model by expressing it in a calculus (or part of a calculus). The calculus is thus always thought of as interpreted, the model being its original interpretation. However, to pass from the calculus expressing the model to the same calculus expressing the theory, it is necessary first to "disinterpret" it (to use Quine's felicitous neologism) by ignoring the interpretations of its non-logical terms as standing for familiar concepts, retaining only the interpretations of its logical features. The calculus thus disinterpreted emprirically will then be reinterpreted (from the base upwards) to serve to express the scientific theory. The psychological advantage of proceeding through the two stages of disinterpretation and subsequent reinterpretation is that at no point need a completely uninterpreted calculus be considered.

Nevertheless the essential step, philosophically speaking, is the same for the modellist as for the contextualist—to interpret the calculus, from the base upwards, so that it expresses the theory. The modellist does this by reinterpreting an originally interpreted calculus which he has previously disinterpreted; the contextualist does it by interpreting an originally uninterpreted calculus. The modellist does not escape the semantic ascent: He dodges having to speak explicitly of a calculus by talking of model and theory having the same deductive structure, but to understand this presupposes a semantic ascent. So the modellist cannot justly claim that his account provides a deeper understanding of scientific explanation than that provided by the contextualist account, when the use of his model (apart from its psychological function in illustrating the theory) is to approach indirectly a difficulty which the contextualist attacks directly and clearsightedly by talking of his calculus. The philosopher of science, concerned with the problem of how it is that we understand a theory containing theoretical concepts, cannot avoid a semantic ascent.

NOTES

1. See M. B. Hesse, *Forces and Fields* (1961), p. 27, on the "surplus content" of models.
2. W. V. Quine, *Word and Object* (1960), p. 16n.

Marshall Spector

MODELS AND THEORIES

I. INTRODUCTION

In this paper I will attempt to show that an important currently held view as to the nature of a model for a physical theory is in error. Indirectly, I shall be concerned to show that a certain related thesis about the nature of physical theories themselves is infected with serious difficulties. This will be done in the contest of a modification of the former view which I shall offer.

It will be essential to have before us a careful statement of the latter thesis before we begin, for it seems to me that much of its apparent plausibility rests on the fact that presentations of it are not always very clear. (In fact, I believe that several influential attempts to *refute* this analysis of physical theories also rest on unclear and misleading presentations of it.)

According to this thesis, a physical theory is to be analysed as an empirically interpreted hypothetico-deductive system or formal calculus—in Rudolf Carnap's terms, a "semantical system."[1] Its most basic assumption is that a general distinction can be drawn between two types of terms occurring in physical theories—*observation* terms and *theoretical* terms.[2] The former, terms like "green," "desk," "longer than," refer to observable objects, properties, relations, and events, and can be understood independently of any physical theory. The latter, expressions like "electron," "magnetic field," "spin angular momentum," refer to unobservable (= theoretical) objects, properties, etc., and can be understood only in the context of the theories in which they occur. The attractiveness of the view I shall be examining lies in its claim to present a general, well-articulated schema for showing just how we are to understand theories which seem to talk about objects which we have never observed, and perhaps never will directly observe. In outline, the analysis is as follows:

It is maintained that we can distinguish two components in any theory.

Reprinted from the *British Journal for the Philosophy of Science* (1965) by permission of the author and the publishers, Cambridge University Press.

The first is its *calculus*, which is the logical skeleton of the theory, considered as devoid of any empirical meaning. This will consist of a set of primitive formulas, i.e., sentences which are taken as postulates on the calculus; and other formulas which are obtained by derivation from the postulates in acordance with specified rules of transformation. Two types of terms appearing in the calculus may be distinguished: the primitive terms, i.e., terms which are not defined on the basis of other terms within the calculus; and non-primitive terms, i.e., those which are introduced on the basis of the primitives. (This distinction is not identical with that between observation terms and theoretical terms.) When this calculus, or syntactical system, is given an empirical interpretation, or meaning, it becomes a system of empirical statements having the structure of a hypothetic deductive system. The primitive formulas become empirical hypotheses, and the derived formulas become empirical statements which will be true if the hypotheses are true.

To illustrate some of these ideas, consider the kinetic theory of gases. The postulates of this theory will contain such expressions as "molecule," "mass of a molecule," and "position of a molecule." These might be considered as primitives. Other expressions, such as "momentum of a molecule" and "mean kinetic energy of a group of molecules," will be introduced on the basis of the primitives. One of the postulates might be "All gases are composed of molecules." A typical derived formula might be, "If the pressure of a gas is increased while its temperature remains constant, its volume will decrease."

The second component of a theory, the empirical interpretation, is given to the calculus by *semantical rules* for terms of the calculus, i.e., rules which are formulated in a suitable metalanguage (usually "ordinary" English, or German, etc.) and provides the meaning of the terms by stating what properties, relations, or individuals the terms designate.[3] An example is, "The term *P* of the calculus designates the pressure of a sample of gas."

Now the authors we shall be considering maintain that:

(a) not all terms of the calculus of a theory *need* be given semantical rules, and
(b) not all terms of the calculus *can* be given semantical rules.

(It has not been clearly recognised that these are *distinct* claims. The importance of noting this distinction will become clear as we proceed.) Only the observation terms of the unanalysed theory are thus "directly interpreted." That is, if we look upon the calculus of a semantical system as the uninterpreted logical skeleton of a theory, for which the semantical system provides a *reconstruction*, only those terms in the calculus which represent the observation terms of the unreconstructed theory will be given semantical rules in the completed reconstruction. Theoretical terms of the unanalysed theory will not be given semantical rules in the semantical system reconstruction. It is claimed that such terms *cannot* be "understood in themselves," but must be understood—given their meaning—in an "indirect" manner, through the role they play in the theory. Such terms obtain an empirical meaning if and only if they appear in sentences in the calculus which also contain terms which *are* given semantical rules—the observation terms. Such sentences are known as *correspondence rules*.

An example of a correspondence rule is the postulate stated earlier: "All gases are composed of molecules." Symbolically, this would read $(x)(Gx \supset Qx)$. The term G is given the semantical rule "G designates the property of being a sample of gas," and the term Q (which is a symbolic translation of the theoretical expression "is composed of molecules") is not given a semantical rule, but is said to obtain a partial meaning indirectly by virtue of its occurrence in a sentence which contains a term (G) whose meaning is given directly and completely by a semantical rule. (It is important to notice that *all* the terms of a theory are found in the calculus, including the observation terms. If this is not kept in mind, there arises a tendency to confuse semantical rules with correspondence rules.)

This, in outline, is the analysis of physical theories which underlies the analysis of the concept of a model for a theory which I shall be examining. For reasons which should be apparent, I shall refer to it as the *partial interpretation thesis*.[4]

The clearest and most precise explication of the concept of a model for a physical theory offered by a proponent of the partial interpretation thesis is that given by Braithwaite,[5] so the burden of my analysis will be directed towards his position. A model for a theory, according to Braithwaite, is to be understood as *another interpretation* of the theory's calculus, in which the *theoretical* terms are directly interpreted (by semantical rules).[6] It is sometimes also stated that these direct interpretations must be in terms of familiar concepts. But this need not be stated as a separate condition if we remember Braithwaite's (and Carnap's) injunction to the effect that this is the only way that direct interpretations can usefully be given.[7] Now if we use the term *model* to designate not a domain of non-linguistic entities, but rather the statements about such entities, we can say that a theory and a model for the theory are two sets of statements which share the same calculus, but with the epistemological order of the two reversed. As Braithwaite puts it:

> A theory and a model for it . . . have the same formal structure, since theory and model are both represented by the same calculus. . . . But the theory and the model have different epistemological structures; in the model the logically prior premisses determine the meaning of the terms occurring in the representation in the calculus of the conclusions; in the theory the logically posterior consequences determine the meaning of the theoretical terms occurring in the representation in the calculus of the premisses.[8]

Braithwaite offers this explication of the concept of a model not as an arbitrary definition, but rather as "an attempt to make more precise the notion of a model for a scientific theory widely current in discussions of the philosophy of science." [*ibid.*]

Braithwaite also points out, quite emphatically, what models allegedly are *not*. Braithwaite claims that one of the chief "dangers" in the use of models is tendency that: "The theory will be identified with a model for it, so that the objects with which the model is concerned . . . will be supposed actually to be the same as the theoretical concepts of the theory." [*op. cit.* p. 93.]

Nagel also warns against confusing the model with the theory itself. After commenting on the possibility that a model may be "an obstacle to the fruitful development of a theory," he writes: "The only point that can be affirmed with confidence is that a model for a theory is not the theory itself."[9] Braithwaite goes on to say that:

> Thinking of scientific theories by means of models is always *as-if* thinking; hydrogen atoms behave (in certain respects) as if they were solar systems each with an electronic planet revolving round a protonic sun. But hydrogen atoms are not solar systems; it is only useful to think of them as if they were such systems if one remembers all the time that they are not.[10]

According to the explication offered by Braithwaite, then, the objects of the model cannot be identified with the theoretical objects of the theory. Such an identification would be a logical error; the possibility of this identification (i.e. the question of the "reality" of the objects of the model) cannot even arise.[11]

However, as I shall argue, these questions *do* legitimately arise for some systems which physicists would recognise as models; and since Braithwaite's explication of the concept of a model cannot allow for this, it is inadequate. To show this, I shall begin by distinguishing four types of domains. They will be quite different in important respects; yet Braithwaite's explication will not be able to distinguish them. Then I will offer a modification of Braithwaite's explication which will be able to distinguish these types of domains. I shall conclude with an analysis of the effects of this modification on the partial interpretation thesis itself.

2. FOUR "BRAITHWAITEAN" MODELS

(1) Suppose we have a geometrical interpretation of the calculus of some physical theory. (Braithwaite frequently uses geometrical interpretations of his "factor-theories" as examples of models.) Such an interpretation would be a model for the theory in Braithwaite's sense, and yet the possible identification of the "objects" of this system with the theoretical objects of the theory would not even arise for the physicist. There would be no question of the lines, triangles, circles, etc., being identified with, or being similar to, the theoretical objects of the theory. Only the formal structure would be relevant. If there are, in fact, triangles, etc., which satisfy the postulates of a given physical theory (its calculus), then the objects of this system are "real"; but this does not mean that the theoretical objects of the theory are triangles, etc. This sort of identification would indeed be a mistake, but a strange sort of mistake that only a modern-day Pythagorean might make. (It is certainly quite different from the mistake of identifying the theoretical objects of the kinetic theory of gases with, say, billiard

balls.) In this sense, a model has nothing whatsoever to do with the *domain* of the theory. Here, the system is another interpretation of the theory's calculus in a very strong sense of "another." The objects of the "model" are of a different logical type from the objects of the domain of the theory. A physicist would probably not call this a model at all.

(2) It has been recognised that there is a rather thorough-going analogy between some of the laws of acoustical theory and those of electric circuit theory. Put another way, there is an important correspondence between acoustical systems and electric circuits. Let us consider a specific instance of this. If we have a series circuit consisting of a resistance R, a capacitance C, and an inductance L, with a periodic electromotance ϵ equal to $E \cos(wt)$, the charge on the capacitor, q, will saitsfy the equation

$$L\frac{d^2q}{dt^2} + R\frac{dq}{dt} + \frac{1}{C}q = E \cos(wt).$$

Now consider a Helmholtz resonator, which is a "flask" of volume V with a neck of length d, radius a, and cross-sectional area $S = \pi a^2$. A sound wave of amplitude P impinges on the resonator opening, so that the driving force at the neck is given by $SP \cos(wt)$. The air displacement z at the neck will then satisfy the equation

$$(pd'S)\frac{d^2z}{dt^2} + \left(\frac{pck^2S^2}{2\pi}\right)\frac{dz}{dt} + \left(\frac{pc^2S^2}{V}\right)z = SP \cos(wt),$$

where $d' = d + 16a/2\pi$, $p =$ mean air density, $c =$ sound wave velocity, and $k = w/c$. These two equations are of the same form:

$$a\frac{d^2\xi}{dt^2} + \beta\frac{d\xi}{dt} + \gamma\xi = \delta \cos(wt).$$

For other acoustical systems, it is also possible to find electric circuit "analogues." In fact, this an be done in a general way, and certain combinations of acoustical parameters are given names taken from electric circuit theory. In the system just described, for example, the quantity $pck^2/2\pi$ is called the "acoustical resistance," R; V/pc^2 is called the "compliance," C; and pd'/S is called the "inertance," M. Moreover, these names are not given merely on the basis of the expressions appearing in the same place in the acoustical equation as the corresponding expressions in the equation for the series circuit; the analogy is more than formal. Thus, the inductance of an electric circuit is a measure of the tendency of the circuit to resist changes in current ($= dq/dt$), while the inertance of the acoustical system is a measure of the tendency of the system to resist changes in air velocity ($= dz/dt$). Similarly, the capacitance of a circuit is a measure of the ability of charge to "pile up," so to speak, in one place in the circuit, while the compliance of an acoustical system, e.g., the resonator, is a measure of just how far back the air in the neck will "allow itself" to be pushed. (If it is claimed that the analogy is nevertheless a formal one only, this would strengthen the point I shall presently make.)

Finally, the analogy is actually used to solve practical problems. It is not merely a heuristic aid—allowing one theory to be taught in terms of the other. To quote from a textbook in acoustics:

> Consideration of the equivalent electric circuit offers many advantages in solving practical engineering problems of applied acoustics. For example, many acoustical systems are so complicated that their mathematical analysis is very difficult, if not impossible, and their design by a cut-and-try [sic] experimental method is extremely tedious, as each change involves constructing a new part. On the other hand, if it is possible to set up an equivalent electric circuit, the electrical constants of this network may readily be varied [actually—not just in the mathematical analysis] to obtain the desired experimental characteristics, and the constants of the mechanical system may then be calculated from their electrical equivalents. This technique has been used in the design of loudspeaker systems and other acoustical devices.[12]

Now, given all of this, is electric circuit theory a model for acoustical theory? (Or, are electric circuits models for acoustical theory?) Usually, the term *analogue* is used here, though one would be understood if the above relation were referred to as a modelling relation. The important point is, again, that there is no question of the identification of the subject matter of the two theories. Acoustics deals with small vibrations in air (generally, elastic fluids), whereas electric circuit theory deals with the movement of electric charge in certain types of systems. The analogy is perhaps not completely formal but the substantive similarities are rather weak.

(3) Here I have in mind systems which physicists would certainly recognise as models—systems for which the question of "reality" *does* arise. But here, the physicist knows that the objects of the model cannot be identified with the theoretical objects of the theory for *definite physical reasons*, although there is some *substantive similarity*. As an example of this, consider the following passage from the arch-modellist, William Thompson (Lord Kelvin). (I have italicised the parts to which I wish attention drawn.)

> To think of ponderable matter, imagine for a moment that we make a *rude* mechanical *model*. Let this be . . . [Here Kelvin describes a rather elaborate contraption, as Duhem might have called it] . . . you will have a *crude model*, as it were, of what Helmholtz makes the subject of his paper on anomalous dispersion. . . . If we had only dispersion to deal with there would be no difficulty in getting a full explanation by putting this *not* in a *rude* mechanical model form, but in a form which would commend itself to our judgment as presenting the *actual* mode of action of the particles, of gross matter, whatever they may be upon the luminiferous ether. . . . It seems that there must be something in this molecular hypothesis, and that as a mechanical symbol, it is certainly *not a mere* hypothesis, but a *reality*. But alas for the difficulties of the undulatory theory of light. . . .[13]

Kelvin has offered a model, which is only a "rude" model, *because* it will not work in certain important, fully specified, situations. If it were not for these other circumstances, which contradict the phenomena that could be expected on the basis of the model, this would not be a rude model, but a "reality." Consider also the following passage:

> The luminiferous ether we must imagine to be a substance which *so far as* luminiferous vibrations are concerned moves *as if* it were an elastic solid. That it moves *as if* it were an elastic solid *in respect to* the luminiferous vibrations is the fundamental assumption of the wave theory of light. [*loc. cit.* p. 9 (Italics mine).]

Kelvin does not identify his models with the domain of the relevant theory for *empirical* reasons—not because of a general philosophical point about the relation between models and theories. For Kelvin goes on to show that there are *other specific* phenomena with respect to which the ether does *not* behave like an elastic solid (for example, the fact that material bodies move through it, showing that it sometimes behaves as if it were a fluid).[14] I think it is quite clear from the quoted passages that if it were not for these other, specific, intransigent phenomena, Kelvin would have dropped the as-if, rude-model, terminology.[15]

Notice that there is no question of identifying the ether with the actual gadgetry in a constructed model, just as for a nineteenth-century physicist there is no question of identifying the molecules of the kinetic theory of gases with actual billiard balls (which are made of ivory, have numbers on them, etc.). The proposed identity would be between the molecules and elastic spheres, which are exemplified by billiard balls; or between the ether and an array of springs, etc., which is exemplified by the gadget on the laboratory bench.

Failure to make this distinction is what lends plausibility to Braithwaite's statement about hydrogen atoms and solar systems quoted near the end of the introduction to this paper. Of course hydrogen atoms are not solar systems; the nucleus of the hydrogen atom is not a star and the electron is not a planet—just as gas molecules are not billiard balls. But is it as obvious that the hydrogen atom is not a system in which something is going around something else? (Unfortunately there is no standard general term for such systems, analogous to the term "elastic sphere" to describe billiard balls and other such objects.) Bohr believed this to be the case; we no longer do for definite physical reasons, although the similarity is still there—hence the "as-if."

(4) Finally, there are systems which are no longer *merely* models. Here, the question of the reality of the model (i.e., the identification of the objects of the model with the theoretical objects of the theory) is not only significant, but is answered in the affirmative. A system may originally be introduced as a model in sense (3), but may eventually be modified to such an extent that we will finally speak of the identity of the objects of this system with the theoretical objects of the theory. When this happens, we may or may not continue to speak of the system as being a *model*. Actual usage may depend on factors such as the historical development of the theory. For example, we may still speak of the discrete particle model for the kinetic theory of gases, even though we have identified the concept of a molecule from the kinetic theory with the concept of a discrete particle (or "object"). Here, the model is not *another* interpretation

of the theory's calculus, but a *filling out of the original interpretation.* (This point will be made clearer in the following sections.)

Now, the first serious problem with Braithwaite's explication of the concept of a model as simply another interpretation of the theory's calculus is that it cannot distinguish between the four types of systems just discussed—and these *are* distinguished by physicists. For Braithwaite, all four would be models. However, we found that the first would not be recognised as such by physicists. The second would usually be described as an analogue. The fourth would probably not be described as a model either, but for a quite different reason—it is too good, as it were. "Questions of reality" arise for some of these systems, but not for others; whereas for Braithwaite, it is claimed that they should not arise for any of them. In other words, Braithwaite is correct in saying that identity of logical structure alone is not enough to be able to claim identity between the objects of the model and the theoretical objects of the theory. But I take this as showing that Braithwaite's explication of the concept of a model is inadequate, and not as an insight into the nature of models for physical theories.

It could be objected at this point that reliance on what physicists say in such situations is not sufficient for establishing a point about the logic of the situation (or for refuting such a point, as I am attempting to do). Perhaps not; I shall have more to say about this below. But suppose that we could emend Braithwaite's explication of the concept of a model in such a way as to take account of what physicists say—which *would* leave a place for the distinctions drawn above. And suppose that this emendation would be such that it could be accepted by the partial interpretation theorists as being within the letter and spirit of their analysis of the structure of physical theories. Clearly this would be a gain in understanding. I shall, in the next section, offer such a modification of Braithwaite's explication. I shall also attempt to show that this modification will be able to overcome certain further difficulties inherent in Braithwaite's explication.

3. AN EMENDATION OF BRAITHWAITE'S EXPLICATION WITH AN EXAMPLE

As long as only identity of formal structure is required between a theory and a model for it, there can be no relevant connection between the domain or subject matter of the theory and the domain of the model. A geometrical model for the calculus of a physical theory provides a striking example of this, and the electric circuit analogue for acoustical theory provides an example where the domains are both of physical objects, yet quite different. In each case, we do not have a model in the physicist's sense, and there is no question of the identification of the objects of the model with the theoretical objects of the theory.

If, however, the observable properties of the domain of the theory—the designata of the observation terms—are similar to the properties of the model represented by these same terms when the calculus is interpreted in the domain of the model, then the possibility arises of comparing the properties of the model represented by the theoretical terms of the calculus with the theoretical objects of the theory. That is, *we can argue by analogy to the nature of the theoretical properties.* A good example of this is provided by the elementary kinetic theory of gases and its usual model. It will be useful to compare in some detail the interpretations of the calculus of the kinetic theory of gases in the theory itself and in the elastic sphere model for the theory. (Notice that this distinction itself sounds somewhat strange, for this theory is usually presented by means of the model. The reasons for this will become clear as we proceed.)

In the theory, certain *defined* terms of the calculus are interpreted as designating the observable properties of a sample of gas in a container at equilibrium—volume, pressure, and temperature.[16] In the model, on the other hand, certain *primitive* terms of the calculus are interpreted as designating a group of elastic spheres in a container and some of their properties. Some of the primitive formulas express propositions describing these elastic spheres (their masses, for example), and others are interpreted as expressing the laws of classical dynamics (the laws governing their motion).

Now we notice that the designata of some of the defined terms of the calculus, when used to represent the theory, are similar (in this case, identical) to the corresponding designata of these terms, when the calculus is used to represent statements about the model. Thus, for example, there is an expression in the calculus, which by the interpretation given the calculus in the model, represents the total rate of momentum transfer per unit area to the walls of the container in which the elastic spheres are moving. This expression, when the calculus is used to represent the theory, appears in a correspondence rule of biconditional form (a definition, in Carnap's sense) with the observation term P, which designates the pressure of the gas. But according to classical dynamics, rate of momentum transfer per unit area is equal to force per unit area, which is (by definition) the pressure on the wall of the container. Thus we not only have a formal identity—a shared calculus, but also a *substantive* identity of two properties, one from the domain of the theory and one from the domain of the model.[17]

The same is trivially true in the case of volume, thus giving us two sets of *identical properties*—out of a possible three.[18] We are left with temperature in the theory and mean kinetic energy of the elastic spheres in the model as the designata of a defined term (T) in the calculus when used in the theory and in the model, respectively. Are these last two properties also similar? Reasoning by analogy on the basis of the first two similarities (identities, in this case), we would expect so, which would amount to suspecting that gases are, in fact, *composed of* elastic spheres (suspecting that molecules are elastic spheres—that the model "corresponds with reality"). At first sight, however, it would seem that the temperature of a gas and the mean kinetic energy of a swarm of elastic spheres are quite dissimilar. But we are not concerned here with temperature as a felt quality of bodies or as a sensation (hot-cold). Rather, we are concerned with the level of a column of mercury (for example—assuming that we are using a mercury thermometer) in a capillary tube. Now we notice that if liquids were also composed of elastic spheres,[19] they would expand if placed in contact with a gas, so composed, if (and only if) the mean kinetic energy of the particles of the gas were higher than in the liquid immediately before contact. In fact, an equilibrium state would soon be reached in which the mean kinetic energies of the two would be the same, at which time the liquid would cease expanding. All of this follows from classical dynamics. But this is just what is observed. When a mercury thermometer is inserted in a gas, the column will rise (or fall) for a time, and reach a stationary level.

On the basis of this, and the two identities mentioned earlier, we can conclude at least that gases behave just as we would expect them to behave if they were in fact composed of small elastic spheres in incessant motion. That is, we can tentatively identify the objects of the model with the theoretical objects of the theory. Gases behave as if they were composed of elastic spheres. The "as-if" does not here indicate that we have contrary information (cf. Kelvin) but rather that we may still feel that there are other tests to which we would like to put this hypothesis before committing ourselves. If we had not the slightest idea of what other tests would be relevant, or if we could satisfy ourselves that other tests were impossible, there would be no point in the "as-if." In this case, there are such further tests, and they follow directly from a consideration of the elastic sphere model. (The theory itself—considered as the partially interpreted calculus with semantical rules for the observation terms only—affords no reason whatsoever to expect these other phenomena.) I have in mind the so-called "molecular beam" experiments.

We open a slit in one wall of the container of gas, and by a suitable experimental arrangement[20] we can see whether the results are those which would be expected (quantitatively as well as qualitatively) if a stream of small (unobserved) particles having a certain mean kinetic energy were to issue from the slit in accordance with the dynamics of such particles. As is well known, such experiments do in fact confirm the hypothesis that gases are composed as described—that we can identify the theoretical objects of the theory with the objects of the model. The measured velocities of the particles are what they would have to be in order to identify the temperature of the gas with the mean kinetic energies of these particles, thus completing the analogy.

Imagine a physicist who now says: "Well, it has seemed as if gases might be composed of these small elastic spheres, considering the similarities you mentioned; now I'm convinced that they really are. After all, the results of this experiment too were exactly what could be expected if they were." Has he committed an error? Has he succumbed to one of the "dangers" involved in model thinking? If he had relied only on the *formal* characteristics of the model, he would indeed have made a mistake, although in this case a fortunate one. But the argument was based on the *substantive* characteristics of the model, and as such is perfectly good analogical reasoning. And it would still have been good reasoning even if the molecular beam experiment had given wholly different results, thus refuting the identification of the theoretical objects with the objects of the model. (Although we might then doubt the reliability of the apparatus rather than the model!) His mistake would then merely be that he had chosen a false hypothesis—but not a meaningless one, or an improbable one, or one which indicated a basic misunderstanding of the very use of models. The identification, on the basis of the first two sets of identical properties alone, is at least probable, and the results of the molecular beam experiment increase the probability.

Notice that it is not necessary at this point to produce a satisfactory theory of confirmation. Any explication of the notion of the probability of a scientific hypothesis which did not reflect the above would have to be rejected, as the reasoning sketched is as good a paradigm as can be found of one type of sound scientific reasoning.[21]

Thus, for a model to qualify as a "candidate for reality" (or, for a system to qualify as a model in the physicist's sense) it must be more than just some other interpretation of the theory's calculus. There must be a substantive similarity between the designata of the defined terms of the calculus when used to represent the observable properties in the domain of the theory and when used to represent the (formally) corresponding properties in the domain of the model. If there is, moreover, an identity of these properties, the model is no longer "merely" a model; the theoretical objects and properties simply *are*. . . .

This, then is the emendation of Braithwaite's explication which will allow one to distinguish between the four types of domains sketched earlier, and which will accommodate an important type of reasoning based on models which physicists do in fact use. This emended form of Braithwaite's explication can also account for some of the uses to which models are put by physicists in modifying and extending a theory. I shall attempt to show this in the next section. But at the same time, it will become apparent that this emendation cannot be accepted by one who holds the partial interpretation thesis. We shall see that the modification which I have suggested is in contradiction with one of the most basic assumptions of this type of analysis of the structure of physical theories.

4. MODELS AND MODIFICATIONS OF A THEORY

The laws deduced from the simplified kinetic theory of gases sketched above are not as accurate as we would like them to be, and we would like to modify the theory to remedy this situation. In this case, the elastic sphere model "points to its own extension,"[22] providing leads as to how it can reasonably be modified. For example, we might argue as follows:

If gases are composed of elastic spheres, perhaps we should take their radii into account (i.e., they are not just point masses). Also, they may exert forces on each other when not in actual contact. If we make these assumptions, and modify the calculus of the theory accordingly, we will be able to derive a formula which, when interpreted term by term in the domain of the theory, is the van der Waals equation of state,

$$\left(P + \frac{a}{V^2}\right)(V - b) = cT,$$

where a is a constant for a particular type of gas depending on the force law, and b is a constant for each type of gas depending on the radii of the spheres. This is a more accurate description of the behaviour of gases than the original ideal gas law, $PV = kT$.

The new primitive formulas of the calculus (or, the primitive formulas of the new calculus) which allowed for the derivation of this law were essentially "read off" from the model. By this I mean that a domain of objects was described which obeys the laws of another familiar theory—the dynamics of rigid spheres attracting each other in accordance with a stated force law (involving a). The statements of the description, together with the statements of the laws of this other theory, upon disinterpretation, become the primitive formulas of the modified calculus. (In the original model, the other theory was the dynamics of mass-points interacting only by contact. This was an idealisation, in an obvious sense, hence the name "ideal gas law" for the equation of state derived from the associated theory. Note also how this is a reduction of part of thermodynamics to classical dynamics.)

In this modification of the model, the identity among derived properties in the model and in the theory, spoken of earlier, still obtains. Thus if we had started with the van der Waals equation of state, we could have used the same sort of reasoning by analogy described earlier to establish, or make probable, the identification of the primitive properties of this model with the theoretical properties of the theory. And once again, this sort of reasoning would not be valid if we had considered only the formal similarity (as Braithwaite correctly recognises when he argues against identifying the model with the theory). Moreover, if we had started with the van der Waals equation, it would have been impossible to *construct* the calculus of the kinetic theory without thinking of the model. (The very form in which the equation is stated betrays its origin in the model.) And if perchance someone hit upon the right modification of the original calculus without considering the substantive aspects of the model, he would have no justification for believing it probable.

This difference between what can be done with the "theory itself" and with the model is even more striking when we consider the partial interpretation thesis view that theoretical terms obtain their meaning by being connected with observation terms via correspondence rules. For now we have a much more complicated calculus just in order that the directly interpreted formula relating P, V, and T can be more accurate for a certain range of conditions. Now suppose (contrary to fact) that the ideal gas law were more accurate than the van der Waals law. Before dropping the modified theory in favour of the simpler one, physicists would, I think, look for experimental errors, because the modified model, associated with the modified theory, appears to be a *more plausible* representation of the theoretical objects than is the original model. But on the basis of the partial interpretation thesis, and its associated explication of the notion of a model, this would be an irrational procedure, stemming from a "misunderstanding" of what a model is.

Also, the derivation of the results of the molecular beam experiment could not be carried out with either the simplified or the full theory without a change in the observation language itself. We would need new terms in the calculus designating the distance between the container of gas and a sheet of film located inside a rotating cylinder, the angular velocity of the cylinder, the degree of blackening of the film, etc.[23] But if we introduce these new terms and the new correspondence rules needed for them, and at the same time accept the partial interpretation thesis as to how theoretical terms obtain their empirical meaning, we will have the very strange result that the meanings of the terms *molecule*, *mass*, *velocity*, etc., have changed. After all, the theoretical terms are supposed to obtain their meaning from the observation terms through correspondence rules, and we have added new observation terms and correspondence rules. In the model, on the other hand, there is no such change in meaning. "Elastic sphere" means the same when the model is first conceived as it does after we realise that this model points to a new *test* of the theory in the molecular beam experiment. But we found that we could identify the objects of the model with the theoretical objects of the theory, through analogical reasoning based on substantive similarities in the two domains. Therefore, on the basis of this identification we are forced to conclude (what is more plausible on its own merit) that the addition of the new test with its corresponding set of new observation terms and correspondence rules does not change the meaning of the theoretical terms of the theory, but rather gives new ways to test the truth of the theoretical statements (whose meanings are given in some other way—directly, as will soon become apparent).

The view that the meanings of the theoretical terms change with the addition of new observation terms and corresondence rules has a similar strange consequence when we consider the sort of unification of physical theories that is accomplished by reasoning based on the use of a common model. For example, Niels Bohr, in his 1913 paper, "On the Constitution of Atoms and Molecules,"[24] using reasoning based on a model of the atom (notice that it is not called a model for atomic theory) which can be considered as a detailing of the elastic sphere model of kinetic theory, explains why certain lines in the spectrum of hydrogen gas had been missing, and predicts under what conditions we can expect to observe them. According to the partial interpretation thesis, even though the Bohr theory contains the terms *atom*, *mass*, etc., these terms have a different meaning from that in the kinetic theory of gases. The observation language of the Bohr theory is different from that of the kinetic theory, as are (of course) the correspondence rules; and his reasoning, which depends on the identity of the meanings of these terms in the two theories, would be invalid.

In cases like this, it is the model which is the heart of the physicist's investigations. If we disinterpet the calculus that is read off from the model, and reinterpret it "from the bottom up," and call the result the theory itself, looking now upon the model as merely another possible interpretation of the theory's calculus, we must accept the conclusion that arguments such as Bohr's are colossal logical blunders—fortunate blunders—showing a lack of understanding of how theoretical terms get their meaning, and a lack of understanding of what a physical theory is.

Now it may well be the case that much of what physicists *say* about the methodology of their science is of dubious value—I do not wish to argue this point here. But the partial interpretation thesis, coupled with Braithwaite's explication of the concept of a model (which leaves out substantive similarities), is not only in contradiction with this; it is also in contradiction with actual theory construction and theory modification—what physicists *do*. Now Braithwaite, Carnap, and Nagel say that they are not interested in constructing a logic of discovery. Rather, they are interested in analysing the final product. But their position is not neutral with respect to the former; it instead declares certain types of reasoning which are paradigms of physical genius to be (fortunate) logical mistakes of a very basic sort. (The parenthetical adjective alone should make one suspicious.)

We could not even describe the kinetic theory of gases as a reduction of thermodynamics to (statistical) dynamics. We could only notice that there is a formal similarity between some of the primitive formulas of the kinetic theory and the laws of classical dynamics. But these laws would not have the same meaning in each case. In the kinetic theory, the meanings of the terms *force*, *mass*, *momentum*, etc., would have to be analysed in terms of the observation terms *pressure*, *temperature*, and *volume*.

What has happened here? I think the basic point is that the suggested emendation of Braithwaite's explication of the notion of a model, i.e., taking into account substantive features of the model, as plausible as it may seem, cannot be accepted by one who holds the partial interpretation thesis. That is, even though my suggested emendation eliminates the difficulties I have been pointing out, and apparently does so while remaining within the letter and spirit of the programme of the partial interpretation theorists, the emendation contradicts a basic assumption of the partial interpretation thesis. For analogical reasoning from *substantive* similarities in the designata of terms in the derived formulas in the model and in the theory to substantive similarities in the theoretical (primitive) properties amounts to a direct interpretation of the theoretical terms in the theory. Thus, in the case of an *identity* of derived properties, the completion of the analogy is tantamount to *giving semantical rules for the theoretical terms*. (If there is only a similarity, in which case we have one type of "as-if thinking," we are giving qualified semantical rules, so to speak—cf. the earlier quotations from Kelvin.) But according to the partial interpretation thesis, we *cannot* give a consistent direct interpretation (semantical rules) for the theoretical terms of the calculus when it is used to express the theory, for "we could not understand them" (Carnap); the theoretical terms are "not understood in themselves," but only as part of the whole system (Braithwaite). Theoretical terms gain what empirical meaning they have *only* "from below"; the "zipper" moves from the bottom up in giving meaning to the theoretical terms. Thus, if this thesis is held as to how theoretical terms become meaningful, one is forced to consider a model as some *other* interpretation of the calculus. To accept our results is to accept the possibility of giving meaning directly for theoretical terms.

5. CONCLUSIONS

Our suggested emendation of Braithwaite's explication of the concept of a model has implied that theoretical terms *can* be given direct interpretations—semantical rules, where the metalanguage used to state them *is* "understood." It is in fact the same metalanguage used to interpret the observation terms of the calculus. That is, if we wish to analyse physical theories in terms of interpreted formal calculi, we *can* give semantical rules for theoretical terms (thus, for *all* terms of the calculus). Notice that this still gives the meanings of the theoretical terms "on the basis of observations," but by statements outside the calculus rather than by correspondence rules.

I have also shown that theoretical terms *must* be given semantical rules, to allow for certain types of physical reasoning. Certain extensions and modifications of a theory were based on the use of a model *in the emended sense*, which we saw is tantamount to interpreting the theoretical terms of the theory. But this refutes what I referred to as assumptions (a) and (b) of the partial interpretation thesis in the introduction to this paper.[25]

All of this also implies that theoretical or unobservable *objects* may (in some cases) be described by observational *predicates*. Let us see why this is so. If a semantical rule is to be successful in giving meaning to a theoretical term of the calculus, it must be stated in a metalanguage which is "already understood." That is, it must supply a designatum which is understood independently of the theory being reconstructed. Otherwise it would be, as Carnap has put it, a useless transcription "from a symbol in a symbolic calculus to a corresponding word expression in a calculus of words."[26] This requirement has usually been put in the form of demanding that the properties, things, and relations which are to be the designata of the terms of the calculus must be observable, i.e., the metalanguage must be a theory-uninfected observation language. (This observation language is not to be confused with the observation language of the theory's calculus, which is part of the object language being analysed.) It would seem that familiarity should be sufficient; but let us grant the stronger requirement of observability for the present. Now, if we give semantical rules for the theoretical predicates (say) of the calculus, and do so in accordance with the just mentioned condition on the metalanguage, we will have the result that *unobservable objects may be characterised by observational predicates.* More accurately, we will have a sentence of the form *P*(*a*), with *a* designating an unobservable (= theoretical) object and *P* designating a property of which it is possible to observe instances applying to observable objects (e.g., "having weight" as applied to an electron or atom).

This is the same conclusion reached earlier from a consideration of the concept of a model for a theory. We saw that the identification of the objects (and properties) of the model with the designata of the theoretical terms of "the theory itself" was tantamount to giving semantical rules for theoretical terms and thus applying observational predicates to unobservable objects. In the example of the kinetic theory of gases, this amounted to maintaining that the unobservable atoms could have observational properties—or at least familiar properties (mass, velocity) from another theory (classical dynamics).

Moreover, we saw that there is nothing unintelligible about unobservable objects being characterised by observational predicates. The reasoning that leads a physicist to impute observational properties to unobservable objects was seen to be perfectly acceptable analogical reasoning.

These conclusions, however, conflict with the most basic presupposition of the proponents of the semantical system approach. This is the assumption that we can distinguish in a general manner two types of terms or concepts (and statements) in physical theories: The observation terms, which designate observable properties, relations, events, and objects *only*; and the theoretical terms, which purport to designate unobservable properties, relations, events, and objects *only*. This "dual language model" of the vocabulary of science has been expressed by Carnap as follows:

> [We accept] the customary and useful [division of] the language of science into two parts, the observation language and the theoretical language [, where] the observation language uses terms designating observable properties for the description of observable things and events [, and] the theoretical language [uses] terms which may refer to unobservable aspects or features of events, e.g., to microparticles . . .[27]

We have seen, therefore, that what are perhaps the three most basic assumptions of the partial interpretation thesis are in error.

Now these assumptions involve two crucial notions—that of an observation term, and that of a theoretical term. In this paper, I have treated these notions as if they were clear and unproblematic. They are not, however, and a full evaluation of the partial interpretation thesis—and the semantical system approach itself—requires a careful analysis of them. This I hope to do in future papers.

NOTES

1. The oldest and most precise formulation of this position is found in Rudolf Carnap, "Foundations of Logic and Mathematics," I, No. 3, of the *International Encyclopedia of Unified Science* (Chicago, 1939). Later statements of this view are found in other writings of

Carnap; in various papers by Carl Hempel; in Ernest Nagel's *The Structure of Science* (New York, 1961); in R. B. Braithwaite's *Scientific Explanation* (Cambridge, 1953); Arthur Pap's *Introduction to the Philosophy of Science* (Glencoe, Ill., 1962); Ernest Hutten's *The Language of Modern Physics* (London, 1956); Peter Caws's *The Philosophy of Science* (Princeton, 1965). Further references may be found in the works just cited.

2. See Carnap, *loc. cit.* p. 203; Carnap, "The Methodological Character of Theoretical Concepts," in *Minnesota Studies in the Philosophy of Science*, I, 38; Nagel, *op. cit.* pp. 81 ff.; Braithwaite, *op. cit.* p. 51. The same assumption is made by the other authors mentioned in n. 1. Most of them hasten to say that the distinction may not be a sharp one.

3. See Carnap, "Foundations of Logic and Mathematics," *loc. cit.* p. 153; Carnap, *Meaning and Necessity* (Chicago, 1946), pp. 4 f. See also the introductory sections of Alonzo Church, *Introduction to Mathematical Logic* (Princeton, 1956).

4. I have borrowed this expression from Professor Peter Achinstein of the Johns Hopkins University.

5. See his book, *Scientific Explanation*, as well as his more recent paper, "Models in the Empirical Sciences," in *Logic, Methodology, and Philosophy of Science*, eds. Nagel, Suppes, and Tarski (Stanford, 1960).

6. It will not usually matter whether we use the term *model* to refer to the domain of objects which the interpreted calculus makes statements about, or to the interpretative statements themselves.

7. Cf. "Foundations of Logic and Mathematics," p. 204. The metalanguage must be already understood. The familiarity of the concepts need not entail their observational character. They may be from another, "better understood" theory; but eventually, the chain must end with observational concepts. (See Braithwaite, "Models in the Empirical Sciences," *loc. cit.* p. 227.)

8. *Scientific Explanation*, p. 90.

9. *The Structure of Science*, p. 116.

10. *Scientific Explanation*, p. 31.

11. Compare Nagel, *op. cit.* p. 116.

12. L. E. Kinsler and A. R. Frey, *Fundamentals of Acoustics* (New York, 1950), p. 33.

13. *Baltimore Lectures on Molecular Dynamics and the Wave Theory of Light* (London, 1904), pp. 12–14.

14. Thus it was much like the product once on the market in the United States known as "Silly Putty." This was a substance which could be moulded like clay, but which would also bounce if dropped several feet. It behaved like a fluid under some circumstances (slowly applied force), but like an elastic solid under others (rapidly applied force). Ice is another example; glaciers flow, although ice is "usually" brittle.

15. See also Edmund Whittaker, "Models of the Aether," *A History of Theories of Aether and Electricity* (New York, 1960), I, Chap. ix. Whittaker describes a series of (mechanical) models of the ether, and it is apparent that the "question of reality" *did* arise for these models and was taken quite seriously. In each of these cases, there was some *substantive* similarity between the domain of the model and the domain of the theory (electrodynamics). And in each case, the model was considered only as an "as-if" because it didn't work, i.e., certain phenomena (though not all) expected on the basis of the model were not observed.

It is also interesting to note that in many of the models cited by Whittaker, part of the reason for failure was a lack of complete *formal* identity. Thus, according to Braithwaite's explication, certain paradigm cases of models employed by physicists would have to be described as *not* being models at all! But I shall let this pass.

16. If pressure and temperature are not considered as sufficiently elementary or observable, the analysis could be carried out in terms of the observed heights of mercury columns; but this would be at the expense of clarity without changing the results.

17. Here, the identity is established on the basis of classical dynamics, rather than being an observed identity. But, as Braithwaite has noted, the familiarity need not necessarily involve observability. (See n. 7, p. 292 supra.)

18. We are at present interested in only three "observable" properties of the theory—those which enter into the ideal gas law, $PV = kT$.

19. But where the spheres also exert an attractive force on one another. This is admittedly crude, but putting in all of the details would only complicate matters without helping (or hindering) the point I wish to make.

20. See F. W. Sears, *Thermodynamics* (New York, 1955), p. 241—or almost any other elementary book on thermodynamics.

21. This type of reasoning is fruitfully compared with C. S. Peirce's "abductive reasoning"; "The surprising fact, *C*, is observed; but if *A* were true, *C* would be a matter of course. Hence, there is reason to suspect *A* is true." [*Collected Papers of C. S. Peirce* (Cambridge, Mass., 1935), V, para. 189.]

Einstein's explanation of the Brownian motion on the basis of the kinetic theory of gases is a more striking example of this. It was this which convinced many doubters that the atomic theory was not just a "convenient fiction."

22. This is Braithwaite's phrase in "Models in the Empirical Sciences," *loc. cit.* p. 229.

23. See Sears, *op. cit.*

24. *Philosophical Magazine*, XXVI, 1913, pp. 9–10

25. I should point out here that in one sense the conclusions I have so far drawn, and those I shall presently draw, are not entirely new. [See, for example, R. Harré, *An Introduction to the Logic of the Sciences* (London, 1960) Chap. iv; M. B. Hesse, *Models and Analogies in Science* (London, 1963); N. R. Campbell, *Physics, The Elements* (Cambridge, 1920) Chap. vi (recently reprinted in a paper-back edition by Dover entitled *Foundations of Science*); H. Putnam, "What Theories Are Not," in *Proceedings of the Congress of Logic, Methodology, and Philosophy of Science* (Stanford, 1962) pp. 240–51.]

However, I believe that I have argued for these conclusions in an importantly novel way. I have been concerned to show how these conclusions can be *generated out of* the "partial interpretation" analysis of the structure of physical theories and models. I do not believe that this has been done by the authors cited above. (Hesse comes the closest to this sort of undertaking.) My *general* point of view, which may be called, roughly, "realistic," is of course not new at all.

26. "Foundations of Logic and Mathematics," p. 210.

27. "The Methodological Character of Theoretical Concepts," in *Minnesota Studies in the Philosophy of Science*, I (Minneapolis, 1956), 38.

Theories and Observational Generalizations

Ernest Nagel

MECHANISTIC EXPLANATION AND ORGANISMIC BIOLOGY

Vitalism of the substantival type sponsored by Driesch and other biologists during the preceding and early part of the present century is now a dead issue in the philosophy of biology—an issue that has become quiescent less, perhaps, because of the methodological and philosophical criticism that has been levelled against the doctrine than because of the infertility of vitalism as a guide in biological research and because of the superior heuristic value of alternative approaches for the investigation of biological phenomena. Nevertheless, the historically influential Cartesian conception of biology as simply a chapter of physics continues to meet resistance; and outstanding biologists who find no merit in vitalism believe there are conclusive reasons for maintaining the irreducibility of biology to physics and for asserting the intrinsic autonomy of biological method. The standpoint from which this thesis is currently advanced commonly carries the label of "organismic biology"; and though the label covers a variety of special biological doctrines that are not all mutually comparible, those who fall under it are united by the common conviction that biological phenomena cannot be understood adequately in terms of theories and explanations which are of the so-called "mechanistic type." It is the aim of the present paper to examine this claim.

It is, however, not always clear what thesis organismic biologists are rejecting when they declare that "mechanistic" explanations are not fully satisfactory in biology. In one familiar sense of "mechanistic," a theory is mechanistic if it employs only such concepts which are distinctive of the science of mechanics. It is doubtful, however, whether any professed mechanist in biology would

From *Philosophy and Phenomenological Research*, I, No. 3, March, 1951. Reprinted with permission of the author and the editor.

today explicate his position in this manner. Physicists themselves have long since abandoned the seventeenth-century hope that a universal science of nature would be developed within the framework of the fundamental conceptions of mechanics. And no one today, it is safe to say, subscribes literally to the Cartesian program of reducing all the sciences to the science of mechanics and specifically to the mechanics of contact-action. On the other hand, it is not easy to state precisely what is the identifying mark of a mechanistic explanation if it is not to coincide with an explanation that falls within the science of mechanics. In a preliminary way, and for lack of anything better and clearer, I shall adopt in the present paper the criterion proposed long ago by Jacques Loeb, according to whom a mechanist in biology is one who believes that all living phenomena "can be unequivocally explained in physico-chemical terms," that is, in terms of theories that have been originally developed for domains of inquiry in which the distinction between the living and nonliving plays no role, and that by common consent are classified as belonging to physics and chemistry.

As will presently appear, this brief characterization of the mechanistic thesis in biology does not suffice to distinguish in certain important respects mechanists in biology from those who adopt the organismic standpoint; but the above indication will do for the moment. It does suffice to give point to one further preliminary remark which needs to be made before I turn to the central issue between mechanists and organismic biologists. It is an obvious commonplace, but one that must not be ignored if that issue is to be justly appraised, that there are large sectors of biological study in which physico-chemical explanations play no role at present and that a number of outstanding biological theories have been successfully exploited which are not physico-chemical in character. For example, a vast array of important information has been obtained concerning embryological processes, though no explanation of such regularities in physico-chemical terms is available; and neither the theory of evolution even in its current form, nor the gene theory of heredity is based on any definite physico-chemical assumptions concerning living processes. Accordingly, organismic biologists possess at least some grounds for their skepticism concerning the inevitability of the mechanistic standpoint; and just as a physicist may be warranted in holding that some given branch of physics (e.g., electro-magnetic theory) is not reducible to some other branch (e.g., mechanics), so an organismic biologist may be warranted in holding an analogous view with respect to the relation of biology and physico-chemistry. If there is a genuine issue between mechanists and organismic biologists, it is not prima facie a pseudo-question.

However, organismic biologists are not content with making the obviously justified observation that only a relatively small sector of biological phenomena has thus far been explained in physico-chemical terms; they also maintain that *in principle* the mode of analysis associated with mechanistic explanations is inapplicable to some of the major problems of biology, and that therefore mechanistic biology cannot be adopted as the ultimate ideal in biological research. What are the grounds for this contention and how solid is the support which organismic biologists claim for their thesis?

The central theme of organismic biology is that living creatures are not assemblages of tissues and organs functioning independently of one another, but are integrated structures of parts. Accordingly, living organisms must be studied as "wholes," and not as the mere "sums" of parts. Each part, it is maintained, has physico-chemical properties; but the interrelation of the parts involves a distinctive organization, so that the study of the physico-chemical properties of the parts taken in isolation of their membership in the organized whole which is the living body fails to provide an adequate undetstanding of the facts of biology. In consequence, the continuous adaptation of an organism to its environment and of its parts to one another so as to maintain its characteristic structure and activities cannot be described in terms of physical and chemical principles. Biology must employ categories and a vocabulary which are foreign to the sciences of the inorganic, and it must recognize modes and laws of behavior which are inexplicable in physico-chemical terms.

There is time to cite but one brief quotation from the writings of organismic biologists. I offer the following from E. S. Russell as a typical statement of this point of view:

> Any action of the whole organism would appear to be susceptible of analysis to an indefinite degree—and this is in general the aim of the physiologist, to analyze, to decompose into their elementary processes the broad activities and functions of the organism.
>
> But . . . by such a procedure something is lost, for the action of the whole has a certain unifiedness and completeness which is left out of account in the process of analysis. . . . In our conception of the organism we must . . . take into account the unifiedness and wholeness of its activities [especially since] the activities of the organism all have reference to one or other of three great ends [development, maintenance, and reproduction], and both the past and the future enter into their determination. . . .
>
> . . . It follows that the activities of the organism as a whole are to be regarded as of a different order from physico-chemical relations, both in themselves and for the purposes of our understanding. . . .
>
> . . . Bio-chemistry studies essentially the *conditions* of action of cells and organisms, while organismal biology attempts to study the actual modes of action of whole organisms, regarded as conditioned by, but irreducible to, the modes of action of lower unities. . . . (*Interpretation of Development and Heredity*, pp. 171–72, 187–88.)

Accordingly, while organismic biology rejects every form of substantival vitalism, it also rejects the possibility of physico-chemical explanation of vital phenomena. But does it, in point of fact, present a clear alternative to physico-chemical theories of living processes, and, if so, what types of explanatory theories does it recommend as worth exploring in biology?

(1) At first blush, the sole issue that seems to be raised by organismic biology is that commonly discussed under the heading of "emergence" in other branches of science, including the physical sciences; and, although other questions are involved in the organismic standpoint, I shall begin with this aspect of the question.

The crux of the doctrine of emergence, as I see it, is the determination of the conditions under which one science can be reduced to some other one, i.e., the formulation of the logical and empirical conditions which must be satisfied if the laws and other statements of one discipline can be subsumed under, or explained by, the theories and principles of a second discipline. Omitting details and refinements, the two conditions which seem to be necessary and sufficient for such a reduction are briefly as follows. Let S_1 be some science or group of sciences such as physics and chemistry, hereafter to be called the "primary discipline," to which a second science, S_2, for example biology, is to be reduced. Then (i) every term which occurs in the statements of S_2 (e.g., terms like *cell, mytosis, heredity*, etc.) must be either explicitly definable with the help of the vocabulary specific to the primary discipline (e.g., with the help of expressions like *length, electric charge, osmosis*); or well-established empirical laws must be available with the help of which it is possible to state the sufficient conditions for the applications of all expressions in S_2, exclusively in terms of expressions occurring in the explanatory principles of S_1. For example, it must be possible to state the truth-conditions of a statement of the form *x is a cell* by means of sentences constructed exclusively out of the vocabulary belonging to the physico-chemical sciences. Though the label is not entirely appropriate, this first condition will be referred to as the condition of definability. (ii) Every statement in the secondary discipline, S_2, and especially those statements which formulate the laws established in S_2, must be derivable logically from some appropriate class of statements that can be established in the primary science, S_1—such classes of statements will include the fundamental theoretical assumptions of S_1. This second condition will be referred to as the condition of derivability.

It is evident that the second condition cannot be fulfilled unless the first one is, although the realization of the first condition does not entail the satisfaction of the second one. It is also quite beyond dispute that in the sense of reduction specified by these conditions biology has thus far not been reduced to physics and chemistry, since not even the first step in the process of reduction has been completed—for example, we are not yet in the position to specify exhaustively in physico-chemical terms the conditions for the occurrence of cellular division.

Accordingly, organismic biologists are on firm ground if what they maintain is that all biological phenomena are not explicable thus far physico-chemically, and that no physico-chemical theory can possibly explain such phenomena until the descriptive and theoretical terms of biology meet the condition of definability. On the other hand, nothing in the facts surveyed up to this point warrants the conclusion that biology is *in principle* irreducible to physico-chemistry. Whether biology is reducible to physicochemistry is a question that only further experimental and logical research can settle; for the supposition that each of the two conditions for the reduction of biology to physico-chemistry may some day be satisfied involves no patent contradiction.

(2) There are, however, other though related considerations underlying the organismic claim that biology is intrinsically autonomous. A frequent argument

used to support this claim is based on the fact that living organisms are hierarchically organized and that, in consequence, modes of behaviour characterizing the so-called "higher levels" of organization cannot be explained in terms of the structures and modes of behavior which parts of the organism exhibit on lower levels of the hierarchy.

There can, of course, be no serious dispute over the fact that organisms do exhibit structures of parts that have an obvious hierarchical organization. Living cells are structures of cellular parts (e.g., of the nucleus, cytoplasm, central bodies, etc.), each of which in turn appears to be composed of complex molecules; and, except in the case of unicellular organisms, cells are further organized into tissues, which in turn are elements of various organs that make up the individual organism. Nor is there any question but that parts of an organism which occupy a place at one level of its complex hierarchical organization stand in relations and exhibit activities which parts occupying positions at other levels of organization do not manifest: A cat can stalk and catch mice, but though its heart is involved in these activities, that organ cannot perform these feats; again, the heart can pump blood by contracting and expanding its muscular tissues, but no tissue is able to do this; and no tissue is able to divide by fission, though its constituent cells may have this power; and so on down the line. If such facts are taken in an obvious sense, they undoubtedly support the conclusion that behavior on higher levels of organization is not explained by merely citing the various behaviors of parts on lower levels of the hierarchy. Organismic biologists do not, of course, deny that the higher level behaviors occur only when the component parts of an organism are appropriately organized on the various levels of the hierarchy; but they appear to have reason on their side in maintaining that a knowledge of the behavior of these parts, when these latter are not component elements in their structured living organism, does not suffice as a premise for deducing anything about the behavior of the whole organism in which the parts do stand in certain specific and complex relations to one another.

But do these admitted facts establish the organismic thesis that mechanistic explanations are not adequate in biology? This does not appear to be the case, and for several reasons. It should be noted, in the first place, that various forms of hierarchical organization are exhibited by the materials of physics and chemistry, and not only by those of biology. On the basis of current theories of matter, we are compelled to regard atoms as structures of electric charges, molecules as organizations of atoms, solids and liquids as complex systems of molecules; and we must also recognize that the elements occupying positions at different levels of the indicated hierarchy generally exhibit traits and modes of activity that their component parts do not possess. Nonetheless, this fact has not stood in the way of establishing comprehensive theories for the more elementary physical particles, in terms of which it has been possible to explain some, if not all, of the physico-chemical properties exhibited by things having a more complex organization. We do not, to be sure, possess at the present time a comprehensive and unified theory which is competent to explain the whole range of physico-chemical phenomena at all levels of complexity. Whether such a theory will ever be achieved is certainly an open question. But even if such an inclusive theory were never achieved, the mere fact that we can now explain some features of relatively highly organized bodies on the basis of theories formulated in terms of relations between relatively more simply structured elements—for example, the specific heats of solids in terms of quantum theory or the changes in phase of compounds in terms of the thermodynamics of mixtures—should give us pause in accepting the conclusion that the mere fact of the hierarchical organization of biological materials precludes the possibility of a mechanistic explanation.

This observation leads to a second point. Organismic biologists do not deny that biological organisms are complex structures of physico-chemical processes, although like everyone else they do not claim to know in minute detail just what these processes are or just how the various physico-chemical elements (assumed as the ultimate parts of living creatures) are related to one another in a living organism. They do maintain, however, (or appear to maintain) that even if our knowledge in this respect were ideally complete, it would still be impossible to account for the characteristic behavior of biological organisms—their ability to maintain themselves, to develop, and to reproduce—in mechanistic terms. Thus, it has been claimed that even if we were able to describe in full detail in physico-chemical terms what is taking place when a fertilized egg segments, we would, nevertheless, be unable to explain mechanistically the fact of segmentation—in the language of E. S. Russell, we would then be able to state the physico-chemical *conditions* for the occurrence of segmentation, but we would still be unable to "explain the *course* which development takes." Now this claim seems to me to rest on a misunderstanding, if not on a confusion. It is entirely correct to maintain that a knowledge of the physico-chemical composition of a biological organism does not suffice to explain mechanistically its mode of action—anymore than an enumeration of the parts of a clock and a knowledge of their distribution and arrangement suffices to explain and predict the mode of behavior of the time piece. To do the latter one must *also* assume some theory or set of laws (e.g., the theory of mechanics) which formulates the way in which certain elementary objects behave when they occur in certain initial distributions and arrangements, and with the help of which we can calculate and predict the course of subsequent development of the mechanism. Now it may indeed be the case that our information at a given time may suffice to describe physico-chemically the constitution of a biological organism; nevertheless, the established physico-chemical theories may not be adequate, even when combined with a physico-chemical description of the initial state of the organism, for deducing just what the course of the latter's development will be. To put the point in terms of the distinction previously introduced, the condition of definability may be realized without the condition of derivability being fulfilled. But this fact must not be interpreted to mean that it is possible under any circumstances to give explanations without the use of some theoretical assumptions, or that because one body of physico-chemical theory is not competent to explain certain biological phenomena it is *in principle impossible* to construct and establish mechanistic theories which might do so.

(3) I must now examine the consideration which appears to constitute the main reason for the negative attitude of organismic biologists toward mechanistic explanations. Organismic biologists have placed great stress on what they call the "unifiedness," the "unity," the "completeness," or the "wholeness" of organic behavior; and, since they believe that biological organisms are complex systems of mutually determining and interdependent processes to which subordinate organs contribute in various ways, they have maintained that organic behavior cannot be analyzed into a set of independently determinable component behaviors of the parts of an organism, whose "sum" may be equated to the total behavior of the organism. On the other hand, they also maintain that mechanistic philosophies of organic action are "machine theories" of the organism, which assume the "additive point of view" with respect to biological phenomena. What distinguishes mechanistic theories from organismic ones, from this perspective, is that the former do while the latter do not regard an organism as a "machine," whose "parts" are separable and can be studied in isolation from their actual functioning in the whole living oragnism, so that the latter may then be understood and explained as an aggregate of such independent parts. Accordingly, the fundamental reasons for the dissatisfaction which organismic biologists feel toward mechanistic theories is the "additive point of view" that allegedly characterizes the latter. However, whether this argument has any merit can be decided only if the highly ambiguous and metaphorical notion of "sum" receives at least partial clarification; and it is to this phase of the question that I first briefly turn.

(i) As is well known, the word "sum" has a large variety of different uses, a number of which bear to each other certain formal analogies while others are so vague that nothing definite is conveyed by the word. There are well-defined senses of the term in various domains of pure mathematics, e.g., arithmetical sum, algebraic sum, vector sum, and the like; there are also definite uses established for the word in the natural sciences, e.g., sum of weights, sum of forces, sum of velocities, etc. But with notable exceptions; those who have employed it to distinguish wholes which are sums of their parts from wholes which supposedly are not, have not taken the trouble to indicate just what would be the sum of parts of a whole which allegedly is not equal to that whole.

I therefore wish to suggest a sense for the word "sum" which seems to me relevant to the claim of organismic biologists that the total behavior of an organism is not the sum of the behavior of its parts. That is, I wish to indicate more explicitly than organismic biologists have done—though I hasten to add that the proposed indication is only moderately more precise than is customary—what it is they are asserting when they maintain, for example, that the behavior of the kidneys in an animal body is more than the "sum" of the behaviors of the tissues, blood stream, blood vessels, and the rest of the parts of the body involved in the functioning of the kidneys.

Let me first state the suggestion in schematic, abstract form. Let T be a definite body of theory which is capable of explaining a certain indefinitely large class of statements concerning the simultaneous or successive occurrence of some set of properties $P_1, P_2, \ldots P_k$. Suppose further that it is possible with the help of the Theory *T* to explain the behavior of a set of individuals *i* with respect to their manifesting these properties *P* when these individuals form a closed system s_1 under circumstances C_1; and that it is also possible with the help of *T* to explain the behavior of another set of individuals *j* with respect to their manifesting these properties *P* when the individuals *j* form a closed system s_2 under circumstances C_2. Now assume that the two sets of individuals *i* and *j* form an enlarged closed system s_3 under circumstances C_3, in which they exhibit certain modes of behavior which are formulated in a set of laws *L*. Two cases may now be distinguished: (a) It may be possible to deduce the laws *L* from *T* conjoined with the relevant initial conditions which obtain in C_3; in this case, the behavior of the system s_3 may be said to be the sum of the behaviors of its parts s_1 and s_2; or (b) the laws *L* cannot be so deduced, in which case the behavior of the system s_3 may be said *not* to be the sum of the behaviors of its parts.

Two examples may help to make clearer what is here intended. The laws of mechanics enable us to explain the mechanical behaviors of a set of cogwheels when they occur in certain arrangements; those laws also enable us to explain the behavior of freely-falling bodies moving against some resisting forces, and also the behavior of compound pendula. But the laws of mechanics also explain the behavior of the system obtained by arranging cogs, weights, and pendulum in certain ways so as to form a clock; and, accordingly, the behavior of a clock can be regarded as the sum of the behavior of its parts. On the other hand, the kinetic theory of matter as developed during the nineteenth century was able to explain certain thermal properties of gases at various temperatures, including the relations between the specific heats of gases; but it was unable to explain the relations between the specific heats of solids—that is, it was unable to account for these relations theoretically when the state of aggregation of molecules is that of a solid rather than a gas. Accordingly, the thermal behavior of solids is not the sum of the behavior of its parts.

Whether the above proposal to interpret the distinction between wholes which are and those which are not the sums of their parts would be acceptable to organismic biologists, I do not know. But, while I am aware that the suggestion requires much elaboration and refinement to be an adequate tool of analysis, in broad outline it represents what seems to me to be the sole intellectual content of what organismic biologists have had to say in this connection. However, if the proposed interpretation of the distinction is accepted as reasonable, then one important consequence needs to be noted. For, on the above proposal, the distinction between wholes which are and those which are not sums of parts is clearly *relative to some assumed body of theory T*; and, accordingly, though a given whole may not be the sum of its parts relative to one theory, it may indeed be such a sum relative to another. Thus, though the thermal behavior of solids is not the sum of the behavior of its parts relative to the classical kinetic theory of matter, it is such a sum relative to modern quantum mechanics. To say, therefore, that the behavior of an organism is not the sum of the behavior of its parts, and that its total behavior cannot be understood adequately in physico-chemical terms even though the behavior of each of its parts is explicable mechanistically, can only mean that no body of general theory is now available from which statements about the total behavior of the organism are derivable. The assertion, even if true, does *not* mean that it is *in principle* impossible to explain such total behavior mechanistically, and it supplies no competent evidence for such a claim.

(ii) There is a second point related to the organismic emphasis on the "wholeness" of organic action upon which I wish to comment briefly. It is frequently overlooked, even by those who really know better, that no theory, whether in the physical sciences or elsewhere, can explain the operations of any concrete system, unless various restrictive or boundary conditions are placed on the generality of the theory and unless, also, specific initial conditions, relevantly formulated, are supplied for the application of the theory. For example, electrostatic theory is unable to specify the distribution of electric charges on the surface of a given body unless certain special information, not deducible from the fundamental equation of the theory (Poisson's equation) is supplied. This information must include statements concerning the shape and size of the body, whether it is a conductor or not, the distribution of other charges (if any) in the environment of the body, and the value of the dialectric constant of the medium in which the body is immersed.

But though this point is elementary, organismic biologists seem to me to neglect it quite often. They sometimes argue that though mechanistic explanations can be given for the behaviors of certain parts of organisms when these parts are studies in abstraction or isolation from the rest of the organism, such explanations are not possible if those parts are functioning conjointly and in mutual dependence as actual constituents of a living organism. This argument seems to me to have no force whatever. What it overlooks is that the initial and boundary conditions which must be supplied in explaining physico-chemically the behavior of an organic part acting in isolation are, in general, *not sufficient* for explaining mechanistically the conjoint functioning of such parts. For when these parts are assumed to be acting in mutual dependence, the environment of each part no longer continues to be what it was supposed to be when it was acting in isolation. Accordingly, a necessary requirement for the mechanistic explanation of the unified behavior of organisms is that boundary and initial conditions bearing on the actual relations of parts as parts of living organisms be stated in *physico-chemical* terms. Unless, therefore, appropriate data concerning the physico-chemical constitution and arrangement of the various parts of organisms are specified, it is not surprising that mechanistic explanations of the total behavior of organisms cannot be given. In point of fact, this requirement has not yet ben fulfilled even in the case of the simplest forms of living organisms, for our ignorance concerning the detailed physico-chemical constitution of organic parts is profound. Moreover, even if we were to succeed in completing our knowledge in this repsect—this would be equivalent to satisfying the condition of definability stated earlier—biological phenomena might still not be all explicable mechanistically: For this further step could be taken only if a comprehensive and independently warranted physico-chemical theory were available from which, together with the necessary boundary and initial conditions, the laws and other statements of biology are derivable. We have certainly failed thus far in finding mechanistic explanations for the total range of biological phenomena, and we may never succeed in doing so. But, though we continue to fail, then if this paper is not completely in error, the reasons for such failure are not the a priori arguments advanced by organismic biology.

(4) One final critical comment must be added. It is important to distinguish the question whether mechanistic explanations of biological phenomena are possible, from the quite different though related problem whether living organisms can be effectively synthesized in a laboratory out of nonliving materials. Many biologists apparently deny the first possibility because of their skepticism concerning the second, even when their skepticism does not extend to the possibility of an artificial synthesis of every chemical compound that is normally produced by biological organisms. But the two questions are not related in a manner so intimate; and though it may never be possible to create living organisms by artificial means, it does not follow from this assumption that biological phenomena are incapable of being explained mechanistically. We do not possess the power to manufacture nebulae or solar systems, though we do have available physico-chemical theories in terms of which the behaviors of nebulae and solar systems are tolerably well understood; and, while modern physics and chemistry are beginning to supply explanations for the various properties of metals in terms of the electronic structure of their atoms, there is no compelling reason to suppose that we shall one day be able to manufacture gold by putting together artificially its subatomic constituents. And yet the general tenor, if not the explicit assertions, of some of the literature of organismic biology is that the possibility of mechanistic explanations in biology entails the possibility of taking apart and putting together in overt fashion the various parts of living organisms to reconstitute them as unified creatures. But in point of fact, the condition for achieving mechanistic explanations is quite different from that necessary for the artificial manufacture of living organisms. The former involves the construction of factually warranted *theories* of physico-chemical processes; the latter depends on the availability of certain physico-chemical substances and on the invention of effective techniques of control. It is no doubt unlikely that living organisms will every be synthesized in the laboratory except with the help of mechanistic theories of organic processes—in the absence of such theories, the artificial creation of living things would at best be only a fortunate accident. But, however this may be, these conditions are logically independent of each other, and either might be realized without the other being satisfied.

(5) The central thesis of this paper is that none of the arguments advanced by organismic biologists establish the inherent impossibility of physico-chemical explanations of vital processes. Nevertheless, the stress which organismic biologists have placed on the facts of the hierarchical organization of living things and on the mutual dependence of their parts is not without value. For though organismic biology has not demonstrated what it proposes to prove, it has

succeeded in making the heuristically valuable point that the explanation of biological processes in physico-chemical terms is not a necessary condition for the fruitful study of such processes. There is, in fact, no more good reason for dissatisfaction with a biological theory (e.g., modern genetics) because it is not explicable mechanistically than there is for dissatisfaction with a physical theory (e.g., electro-magnetism) because it is not reducible to some other branch of that discipline (e.g., to mechanics). And a wise strategy of research may, in fact, require that a given discipline be cultivated as an autonomous branch of science, at least during a certin period of its development, rather than as a mere appendage to some other and more inclusive discipline. The protest of organismic biology against the dogmatism frequently associated with mechanistic approaches to biology is salutary.

On the other hand, organismic biologists sometimes write as if any analysis of living processes into the behaviors of distinguishable parts of organisms entails a radical distortion of our understanding of such processes. Thus Wildon Carr, one proponent of the organismic standpoint, proclaimed that "Life is individual; it exists only in living beings, and each living being is indivisible, a whole not constituted of parts." Such pronouncements exhibit a tendency that seems far more dangerous than is the dogmatism of intransigent mechanists. For it is beyond serious question that advances in biology occur only through the use of an abstractive method, which proceeds to study various aspects of organic behavior in relative isolation of other aspects. Organismic biologists proceed in this way, for they have no alternative. For example, in spite of his insistence on the indivisible unity of the organism, J. S. Haldane's work on respiration and the chemistry of the blood did not proceed by considering the body as a whole, but by studying the relations between the behavior of one part of the body (e.g., the quantity of carbon dioxide taken in by the lungs) and the behavior of another part (the chemical action of the red blood cells). Organismic biologists, like everyone else who contributes to the advance of science, must be selective in their procedure and must study the behavior of living organisms under specialized and isolating conditions—on pain of making the free but unenlightening use of expressions like "wholeness" and "unifiedness" substitutes for genuine knowledge.

John G. Kemeny and
Paul Oppenheim

ON REDUCTION

THE CONCEPT OF REDUCTION

The label "reduction" has been applied to a certain type of progress in science. As this process has been the subject of much philosophical controversy, it is the task of the philosopher of science to give a rational reconstruction of the essential features of reduction. We will discuss the basic features of this process informally, we will review two previous attempts to make the concept precise, and we will offer certain improvements which we hope will bring the philosophical characterization of reduction closer to what actually happens in science.

Scientific progress may broadly be divided into two types: (1) an increase in factual knowledge, by the addition to the total amount of scientific observations; (2) an improvement in the body of theories, which is designed to explain the known facts and to predict the outcome of future observations. An especially important case of the second type is the replacement of an accepted theory (or body of theories) by a new theory (or body of theories) which is in some sense superior to it. Reduction is an improvement in this sense.

What are the special features of reduction? Since it is to be progress in science, we must certainly require that the new theory should fulfill the role of the old one, i.e., that it can explain (or predict) all those facts that the old theory could handle. Secondly, we do not recognize the replacement of one theory by another as progress unless the new theory compares favorably with the old one in a feature that we can *very roughly* describe as its simplicity. (We will try to make this more precise later on.) And the special feature of reduction is that it accomplishes all this and at the same time allows us to effect an economy in the theoretical vocabulary of science.

These points can be made clearer by describing one of the goals of the ra-

---

AUTHORS' NOTE. The authors wish to thank C. G. Hempel for his constructive criticism and for many helpful suggestions.

Reprinted from *Philosophical Studies* (1967), by permission of the editor.

tional reconstruction of the process of reduction. The dispute between the "mechanist" and "vitalist" can be stated as the question of whether biology can be reduced to physics (or physico-chemistry). In terms of what we have said above, we can reformulate this as a dispute as to the possibility of forming a theory in physics which can take the place of the totality of biological theories. Such a theory would have to explain all biological phenomena (or at least all phenomena explaïnable by means of biological theories), without introducing undue complexity into the theoretical structure of science. If we succeeded in doing this, then all the theoretical terms of biology could be eliminated from the vocabulary of science, thus effecting a considerable economy. This would be an example of reduction, the reduction of biology to physics.[1] We will not take any stand on this dispute; our interest in the present paper is only to make the process of reduction clear.

There are many examples of reductions that have been achieved. For example, a great part of classical chemistry has been reduced to atomic physics; and the classical theory of heat has been reduced to statistical mechanics. These examples have been discussed at some length elsewhere.[2] The difficulty lies in finding the essential features that such historical examples have in common.

It is hopeless to attempt a rigorous logical analysis of science in the form in which we normally find it. It is not the role of the working scientist to make his fundamental assumptions clear, and it is not reasonable to expect that he will proceed according to rigorous logical rules. Hence, it is customary for the philosopher of science to consider science in an idealized form.

We will suppose that science, or at least those branches under discussion, have been formulated within a formalized language. The statements of the scientist are of two types: (1) those reporting the results of his observations, or of the observations of other scientists that he is willing to accept, (2) generalizations, hypotheses, general laws, etc., that he believes are correct. The former are his *observational statements*, the latter his *theoretical statements*. Each will be expressed by a number of sentences in the formalized language. And a set of sentences can always be replaced by the single sentence that is the conjunction of the members of the set, hence we do not distinguish between an observation (a theory) and a body of observations (theories).

There is a part of the total extra-logical vocabulary of science that is used in recording observations, while the remaining terms are used only in the theories. These latter terms form the *theoretical vocabulary*. We will also speak of the theoretical vocabulary of a theory, which consists of that part of the entire theoretical vocabulary of science that is used in the formulation of this particular theory. There has been a good deal of controversy as to just what the observational and theoretical terms are. Fortunately for this discussion we need to suppose only that a separation has been made; it is irrelevant for us how the two are distinguished. Given such a criterion, we can recognize as (potential) observational statements all statements of the language containing only observational terms.

It is further important to note that from a logical point of view there is no difference between explanation and prediction. The distinction is a pragmatic one, depending on whether the fact deduced is already known or not yet observed. Hence we will use "explain" to cover both processes.

A certain oversimplification is involved in this schematic representation. We intentionally overlook the fact that most observations involve a margin of error. We will also assume that the existing theories are consistent with all observations, which is unfortunately not always the case. This point will occupy us further, later on.

In a reduction we are presented with two theories, T_1 and T_2, and with the observational knowledge of today represented by the complex sentence ${}^{t}O$. (The superscript is added to emphasize that this is the body of observations available at a given time t.) The theoretical vocabulary of T_2, Voc(T_2), contains terms which are not in Voc(T_1). (And, of course, we must require that these terms be not definable in terms of Voc(T_1). We add this additional requirement once and for all—it will not be explicitly mentioned from here on.) But it turns out that T_1 can explain all that T_2 can, and it is no more complex. Hence we drop T_2 from our body of theories, and strike out all terms in Voc(T_2) which are not in Voc(T_1). Then we say that T_2 has been reduced to T_1.[3]

As an important special case Voc(T_1) may be a subset of Voc(T_2). We will call this process *internal reduction*. We must further point out that some writers have not applied the term "reduction" to all such processes, but only to the case where T_1 and T_2 are the bodies of theories of two branches of science, B_1 and B_2, respectively. In this important special case we will speak of the *reduction of branch* B_2 *to branch* B_1. And we can again distinguish the *internal* reduction of a branch, where it is reduced to a subbranch. (It should be pointed out that implicit in the last two definitions is the identification of a branch by its theoretical vocabulary.) Thus we distinguish four related processes, all of which are called reduction.

This discussion has been, of course, highly informal. We gave only the circumstances under which reduction may occur. We must now ask what exactly constitutes reduction.

PREVIOUS DEFINITIONS

Let us start by examining the two definitions that have so far been worked out in detail. Woodger offers us a definition of the internal reduction of a theory, and Nagel defines the reduction of a branch to another. We will sum up the essential points of these two definitions, in our teminology.[4] We will speak of W-reduction and N-reduction, to identify these definitions.

Definition 1. T_2 is W-reduced to T_1 if:

(A) Voc(T_1) is a proper subset of Voc(T_2).

(B) For every term P in Voc(T_2) but not in Voc(T_1) there is a biconditional $(x)[Px \equiv Mx]$ such that:

(1) M is in terms of Voc(T_1).
(2) The biconditional is well established.

(C) The "translation" of T_2 by means of the biconditionals, i.e., the result of replacing each P in T_2 by the corresponding M—follows from T_1.

In the Nagel definition, defining the reduction of a branch, reference must be made to the theories of the branch, i.e., the theories at the given time. Hence the parameter t enters.

Definition 2. B_2 is N-reduced to B_1 at time t if:

(A) The theoretical vocabulary of B_2 at time t contains terms not in the theoretical vocabulary of B_1 at time t.

(B) For every such term P there is a biconditional $(x)[Px \equiv Mx]$ such that:
(1) M is in terms of the theoretical vocabulary of B_1.
(2) The biconditional is well established.

(C) The translation of the theories of B_2 at t by means of these biconditionals follows from the theories of B_1 at t.

The connection between the two definitions is the following: B_2 is N-reduced to B_1 at time t if and only if the conjunction of the theories at t of B_1 W-reduces the conjunction of the theories at t of the combined branches.

We can simplify this by introducing the following notation: Let Wred (T_1, T_2) and Nred(B_1, B_2, t) stand for the terms defined in definitions 1 and 2, respectively. And let ${}^tT_B$ be the conjunction of all theories of the branch B at time t. Then Nred(B_1, B_2, t) if and only if Wred(${}^tT_{B1}$, ${}^tT_{B1} \cdot {}^tT_{B2}$).

Because of this close connection between the two definitions, the following analysis will apply to both concepts.

Our first question is whether (C) in definition 2 is independent of the rest. Nagel calls (B) the condition of "definability" and (C) the condition of "derivability." He points out that derivability is impossible without definability.[5] He then goes on to say that, although definability is a necessary condition, it is not in itself sufficient to assure derivability. But the opposite seems to be the case.

Let T be the translation of ${}^tT_{B2}$ by means of the biconditionals. Then T follows from accepted theories (the theories of B_2) by means of further well-established theories (the biconditionals), and hence T must be recognized as a well-established theory. Furthermore, T makes use of the theoretical vocabulary of B_1 only, hence it is one of the well-established theories of branch B_1. Thus it must either be one of the conjunctive terms of ${}^tT_{B1}$ or at least it must follow from this. In either case T, which is the translation of ${}^tT_{B2}$ is a consequence of ${}^tT_{B1}$. Hence derivability is automatically assured.

There appear to be two ways of salvaging Nagel's claim that derivability is an independent condition. The first one is a pragmatic one, namely an appeal to the fact that *T* need not appear in any scientific publication as a theory of B_1.

According to this method a theory is recognized as a theory of the branch today only if it has actually been written down, and was so labeled. Thus it is entirely possible that there is a consequence of the accepted theories of science today which belongs to that branch but is not called a theory of this branch. Nagel seems to have had this in mind; but we feel that such pragmatic considerations are irrelevant to a logical analysis.

The other alternative is to impose an additional condition on the reducing theory, which is not automatically satisfied by T. An example will help to clarify the situation.

Classical chemistry may be thought of as dealing with various elements, so that its theoretical vocabulary will have such terms as "hydrogen gas" and "oxygen gas" in it. As typical of its theories we may take the very simple example, "Two volumes of hydrogen gas and one volume of oxygen gas combined to form a very stable liquid." Let us now go to a stage of atomic physics in which we have a simple model of the atom (such as Bohr's), but not elaborate theory like that of quantum mechanics. We may take as a primitive of this stage of atomic physics the concept of "atomic number," defined by the positive charge of the nucleus of the given atom. Our biconditionals identify hydrogen gas with gas whose atoms have number 1, and oxygen gas with gas whose atoms have number 8. The translation of our sample theory is, "Two volumes of gas whose atoms have number 1 and one volume of gas whose atoms have number 8 combine to form a very stable liquid." (For the sake of the argument let us take "stable" and "liquid" as observational terms.) This is certainly a theoretical statement from atomic physics, and it would be most unnatural to deny it the status of an accepted theory of this field. Hence our chemical theory is replaced by a physical one. And yet we hesitate to call this process reduction.

If we are asked why we hesitate to accept the process, we will given an answer somewhat as follows: "The reducing theory is no better than the one supposedly reduced. What gain is there for science if a simpler theory is replaced by a more complex one?" This sort of consideration was briefly mentioned in our introduction, and we must now try to make it more precise. As a first approximation we might say that the reducing theory should be simpler than the theory reduced. But this is not the complete answer. If the reducing theory is much stronger, it would seem reasonable to allow it some additional complexity. What our intuition tells us is that we must be satisfied that any loss in simplicity is compensated for by a sufficient gain in the strength of the body of theories. We need some measure that combines strength and simplicity, in which additional complexity is balanced by additional strength. Let us express this combined concept by talking about how well a theory is *systematized*. (We realize that this concept is in need of precise definition, but at least we can make it intuitively clear.[6]) We will then require that the reducing theory be at least as well systematized as the theory reduced.

Without the requirement of systematization we cannot even understand the need for theories. After all ${}^{t}O$, the totality of our observations to date, can (trivially) explain all that we have observed. Any theory can be replaced, for the purposes of explanation and prediction, by the set of all observational statements that follow from it. This set is normally infinite, but it is recursively enumerable. What can a theory add to this? It is important to realize that it cannot *add* anything. The role of a theory is not to give us more facts but to organize facts into a practically manageable form. In place of an infinite set of observation statements we are given a reasonably simple theory. Such a theory has the same explanatory ability as the long (or infinite) list of statements, but no one will deny that it is vastly simpler and hence preferable to such a list.

Only thus do we see the need for introducing theoretical terms. Anything that we want to say about actual observations can be said without theoretical terms, but their introduction allows a much more highly systematized treatment of our total knowledge. Nevertheless, since theoretical terms are in a sense a luxury, we want to know if we can get along without some of them. It is, then, of considerable interest to know that a set of theoretical terms is superfluous since we can replace the theories using these by others in which they do not occur, without sacrificing the degree of systematization achieved by science to this day.

Hence we suggest that in definition 2, C be replaced by:

(N-C′) There is a theory T_1 of B_1 which is at least as well systematized as $^tT_{B2}$, and from which the translation of $^tT_{B2}$ by means of the biconditionals follows.

Because of the connection between W-reduction and N-reduction, it is desirable to modify the former also, replacing C by:

(W-C′) T_1 is at least as well systematized as T_2, and the translation of the latter by means of the biconditionals follows from the former.

We might also consider strengthening both of these conditions by inserting "is better systematized" in place of "is at least as well systematized." In either case we note that C′ is now independent of the other conditions, since there is no reason why the translation of a theory should be as well systematized as the theory itself; on the contrary, in all cases we can think of, the translation is more complex than the original theory.[7]

The second point to be made is a very minor one. There is some suggestion on the parts of both Nagel and Woodger that the biconditionals be "tested." It is very difficult to see how such biconditionals could be tested in general without using other theories. They connect highly abstract terms from the theoretical vocabularly, whose connections to experience may be very complex. For example, P alone may not be connected to experience, but only in some complex combination with other theoretical terms Q, R, . . . Then what would a test of $(x)[Px \equiv Mx]$ be? On the other hand, according to the definition we have two well-established theories (e.g., T_1 and T_2 in definition 1), which are independent without the biconditionals, but of which one is deducible from the other by means of the biconditionals. Since these theories have their own evidence, this ability to connect the evidence of T_2 to T_1 should be sufficient justification for the biconditionals. It may even be the only possible type of justification. This, if correct, means that requirement B2 is unnecessary in both definitions. This would be very satisfying also, since B2 introduces a pragmatic element into the definitions.

Our last point is the most important one. There are two oversimplifications in both the Woodger and Nagel definitions: (1) They assert that the translation of T_2, and precisely that, follows from T_1; (2) they believe that the connection between the two theories is by means of simple biconditionals.

The first of these points ignores the fact that the old theory usually holds only within certain limits, and even then only approximately. For example, in the reduction of Kepler's laws to Newton's we must restrict ourselves to the case of a large central mass with sufficiently small masses, sufficiently far apart, around it. And even then the laws hold only approximately—as far as we can neglect the interaction of the planets. While these points are of fundamental importance, there is no way of taking them into account as long as we tacitly assume that our theories are correct. If we abandon this (contrary-to-fact) assumption, then the problem of reduction becomes hopelessly complex. Hence we will go along with the previous authors in their first oversimplification.[8]

However, there seems to be no justification for the second oversimplification. Any actual example has to be stretched considerably if it is to exemplify connections by means of biconditionals, and most examples will under no circumstances fall under this pattern. It seems to us that an entirely unnecessary feature was introduced here.

NEW DEFINITIONS

As we see it, the essence of reduction cannot be understood by comparing only the two theories; we must bring in the observations. It is not the case that the vocabulary of T_2 is in any simple way connected with the vocabulary of T_1, but only that T_1 can fulfill the role T_2 played, i.e., that it can explain all that T_2 can and normally more. This suggests the following modification of the given definitions. We take as fundamental the concept of the reduction of a theory T_2 by means of T_1 *relative* to ovservational data O. We abbreviate this as Red(T_1, T_2, O).

Definition 3. Red(T_1, T_2, O) if:

(1) Voc(T_2) contains terms not in Voc(T_1).[9]

(2) Any part of O explainable by means of T_2 is explanable by T_1.[10]

(3) T_1 is at least as well systematized as T_2.

We must ask, however, what we can do about the undesired parameter O. One possibility is to put ^{t}O for O, that is, relativize reduction with respect to the present moment. Thus we get the concept that T_1 *today* reduces T_2.

Definition 4. Red^t(T_1, T_2) if and only if Red(T_1, T_2, O^t).

And the corresponding internal reduction concept:

Definition 5. Intred ${}^t(T_1, T_2)$ if and only if $Red^t(T_1, T_2)$ and $Voc(T_1)$ is a proper subset of $Voc(T_2)$.

The above is a very restricted notion of reduction, but no doubt some authors have this in mind. If we do not want to put a particular value of O into definition 3, we must eliminate the undesired variable by quantification. And this seems to lead to a very fruitful approach.

Definition 6. $Red(T_1, T_2)$ if for every O consistent with T_2, $Red(T_1, T_2, O)$.[11]

Definition 7. $Intred(T_1, T_2)$ if $Red(T_1, T_2)$ and $Voc(T_1)$ is a proper subset of $Voc(T_2)$.

There is a natural way of basing the reduction of branches on these concepts, namely by taking the theoretical body of the branch (at a given time) as the T_2 in question.[12]

Definition 8. $Red(B_1, B_2, t)$ if and only if there is a theory T_1 in B_1 at t such that $Red(T_1, {}^tT_{B2})$.

Definition 9. $Intred(B_1, B_2, t)$ if and only if $Red(B_1, B_2, t)$ and B_1 is a branch of B_2.

We would like, of course, to make these definitions independent of t. There is the obvious procedure of choosing t as "the present moment," which appears to be Nagel's procedure. But there is no analogue of the quantification of O. It is reasonable to require that the reducing theory should serve in lieu of the reduced one no matter what the facts turn out to be, but it is not reasonable to require that it should reduce every theory—true or false. Hence in this case we must relativize to the present moment.

Definitions 6–9 represent our proposed definition of reduction.

Let us study definitions 6 and 7 (definitions 8 and 9 are just special cases). Condition 1 in the former requires that $Voc(T_2)$ have certain terms not in Voc (T_1), and in the latter definition this is strengthened by requiring that in addition $Voc(T_1)$ be a subset of $Voc(T_2)$. The other two conditions are the same for both definitions. Condition 3 requires that T_1 be at least as well systematized as T_2. We could, if desired, strength this to "better systematized." Condition 2 is the only one to which the universal quantifier applies. Thus we have: For any observational statement O, consistent with T_2, if T_2 can explain the same part.

This formulation allows a variety of interpretations of "explain." But let us use the commonest one.[13] T can explain a part of O if there are two "non-overlapping" parts, O_1 and O_2, such that $T \cdot O_1$ implies O_2. Thus the above condition becomes: If, for two given "non-overlapping" observational statements (such that T_2 is consistent with both) $T_2 \cdot O_1$ implies O_2, then so does

$T_1 \cdot O_1$. In particular, O_1 may be empty (analytically true). Thus a necessary condition is that if T_2 implies an observational statement, then so does T_1. But this is also sufficient; because if $T_2 \cdot O_1$ implies O_2, then T_2 implies $O_1 \supset O_2$, hence T_1 implies $O_1 \supset O_2$, hence $T_1 \cdot O_1$ implies O_2. This establishes the following theorem:

Theorem 1. Red(T_1, T_2) if and only if (1) Voc(T_2) contains terms not in Voc(T_1), (2) every observational statement implied by T_2 is also implied by T_1, and (3) T_1 is at least as well systematized as T_2. We have Intred(T_1, T_2) if and only if the same three conditions hold, and in addition Voc(T_1) is a subset of Voc(T_2).

In many special cases (in particular when the universe of discourse is finite) there will be a strongest observational consequence of T_2, say T^*_2. In these cases we may replace the second condition by the requirement that T_1 implies T^*_2.

Next we want to show that the Woodger and Nagel definitions are special cases of ours. At least this is the case if the definitions are modified by changing C to C′. and since there is some doubt about the desirability of requirement B2, we will prove our results irrespective of whether B2 is included or not.

Theorem 2. If in definition 1 condition C is replaced by W-C′, then it defines a special case of definition 7. And the same holds even if B2 is omitted.

Proof: The only condition of Theorem 1 that is not obviously fulfilled is the second condition. Let T'_2 be the translation of T_2 by means of the biconditionals. Take any interpretation of Voc(T_1) that makes T_1 true. Since T'_2 follows from T_1, it too must be true under this interpretation. Now extend the interpretation to Voc(T_2) by making the P in each biconditional synonymous with its M. This makes all the biconditionals true. And since according to these T_2 and T'_2 are equivalent, T_2 must also be true under the interpretation. Thus any observational statement implied by T_2 must be true. This shows that under any interpretation making T_1 true, an observational statement following from T_2 must also be true. Which proves that any observational statement following from T_2 must also follow from T_1. (And B2 was not used.) Q.E.D.

Quite analogously we can prove:

Theorem 3. If in definition 2 condition C is replaced by N-C′, then it defines a special case of definition 8. And the same holds even if B2 is omitted.

The question arises of why a method of translation should be essential for Woodger and Nagel but not for us. The answer lies in the fact that they attempt to establish a direct connection between two theories, while our connection is indirect. If one theory is to follow from another, it must be translatable into the vocabulary of the latter.[14] But it is entirely possible that a theory should be able to explain all facts that another can, without there being any method of translation. Of course, each set of theoretical terms must be connected to observational terms, and hence to each other, but this connection is normally much weaker than a full translation.

Naturally, we do not exclude the possibility that reduction may be accomplished by means of a translation. That is why the previous definitions (at least in the modified form) are special cases of our corresponding definitions. But we maintain that they cover what is an extremely special case.

We have offered four definitions for types of reductions, but it must not be supposed that there is any sharp distinction between them. Generally a case of reduction is classified as of one of the four types, but that does not exclude that it could have been classified otherwise. Given a case of $\text{Red}(T_1, T_2)$, we could just as well have described it as the internal reduction $\text{Intred}(T_1, T_1 \cdot T_2)$. Or we could have taken the branches B_1 and B_2 which have T_1 and T_2 as their theories (i.e., if we are willing to call them branches), and speak of $\text{Red}(B_1, B_2, t)$. Or we could take B_3 to be the union of B_1 and B_2 and speak of $\text{Intred}(B_1, B_3, t)$. It all depends on what we wish to recognize as a single theory (or body of theories) and what we call a branch. It is rather that there is one process of reduction that is describable from four different points of view.

Take as an example the reduction of pre-Newtonian mechanics (including celestial mechanics) to Newton's theories. We may think of each body of theories as reduced to the newly created body, and then take this as an example of definition 6. Or we may think of the totality of previous theories (which presumably used all of Newton's theoretical terms) as reduced, in which case it is an example of definition 7. And in this case there is pretty nearly a method of translation, so this is as near as we ever come to an example of definition 1. But when we pass to the reduction of Newtonian mechanics to relativity theory, there is nothing that even remotely resembles the translation by means of biconditionals.

Similarly for the reduction of branches. The more or less complete reduction of chemistry to physics has been used as a standard example. It is optional whether we take this as an example of definition 8 or whether we speak of the reduction of physico-chemistry to physics, in which case it is an example of definition 9. And while some attempts have been made to show that something like biconditional translation does exist, this certainly is not the case if quantum mechanics is taken as the reducing science.

In conclusion, we hope to have shown that the two previous definitions were too narrow in that they excluded most actual cases of reduction. We have presented reasons for the belief that narrowness is an unavoidable fault of any definition trying to establish direct connections between theories. Of course, we are open to correction on this point. But if this is right, then our much wider alternative is the natural choice.

Independently of this we argued for the inclusion of an additional condition, namely that the reducing theory be as well (or better) systematized as the theory reduced. It is on these two points that our definition must be judged.

It must be pointed out that our definition shares with the previous ones a serious oversimplification. We are ignoring the fact that the theory to be reduced may be only approximately true, and only with certain restrictions at that. We certainly would not want to require that an incorrect prediction of T_2 should be a consequence of T_1. We might suggest that it is some modification T'_2 of T_2 that is actually reduced to T_1.[15] But such a T'_2 is not usually formed, and it may be very difficult to formulate it. In addition, we would still be ignoring the fact that T'_2 holds only approximately.

FURTHER RESEARCH

There are several related concepts which deserve consideration. The authors hope to discuss them in a future paper. They are here listed in the hope that others may be interested in some of these problems.

While definitions 6 and 7 seem to answer all questions concerning the reduction of one theory to another, there are some questions about the reduction of branches not answered by definitions 8 and 9. We often hear speculation about whether a branch B_2, which is not today reduced by B_1, is in some sense "reducible" to it. Presumably this involves some hypothesis about the future development of science. We should also consider cases in which although B_2 is not reduced to B_1 some part of it is. This would lead to a concept of "partial reduction" and possibly even a numerical measure of the "reductive power" of a branch of science.

There is an especially important special case of reduction that deserves independent treatment. This is reduction by means of a micro-theory.

There is also a clear connection to Feigl's "levels of explanation,"[16] and it would appear to be fruitful to carry his ideas further by means of the definitions here offered.

NOTES

1. In this example the economy is achieved by eliminating some biological terms. This reduces the number of theoretical terms in science as a whole. But it is not necessarily the case that the reducing branch has fewer terms than the branch reduced. It *may* be the case that physics has fewer terms than biology, but this question is never considered in the reduction-literature.

2. There is an excellent discussion of the reduction of theories of heat to statistical mechanics by E. Nagel in his "The Meaning of Reduction in the Natural Sciences," in *Science and Civilization*, ed. R. C. Stauffer (Madison: University of Wisconsin Press, 1949).

3. While the terms of $Voc(T_2)$ now become logically superfluous, there may be good extra-logical grounds for keeping them. The replaceable terms may be less abstract than the terms of $Voc(T_1)$, and hence more convenient to work with. It may also be the case that T_1 is less well confirmed than T_2, and hence we hesitate to make a final replacement—at least till further evidence is available.

4. Cf. J. H. Woodger, *Biology and Language* (Cambridge: Cambridge University Press,

1952), and E. Nagel, "Mechanistic Explanation and Organismic Biology," *Philosophy and Phenomenological Research*, XI, No. 3, March, 1951. We hope that we have done no injustice in translating these definitions into the terminology of the present paper.

5. Nagel makes use of the same type of argument as was used by Poincaré to show that science cannot give us value-judgments, since a value-statement cannot follow from a factual premise: The argument is based on the supposed fact that a term missing from the premises cannot occur in the conclusion of an argument, in any essential form. But from the premise P(a) we can conlude $P(a) \vee Q(a)$, showing that the argument is not strictly correct. (This counter-example was shown to the authors by C. G. Hempel, who is apparently the first one to have noticed the loophole.) Nevertheless, while the argument is not correct, the conclusion is essentially all right. While predicates not in the premises (and independent of terms in the premises) may occur in the conclusion, they play an inessential role. For example, if such a predicate is replaced everywhere by its negation, the conclusion still follows.

6. While the concept of the simplicity of a theory is still in need of considerable study, some partial results toward its explication are available. In some cases there will be complete agreement among scientists as to which of two theories is simpler; for example, if one can be stated in a single line, while the other—using the same vocabulary—requires several pages to state, then there would be little room for argument. The first two steps toward an explication were taken by Karl Popper in *Logik der Forschung* (Vienna: Julius Springer, 1935), Secs. 39–464, and by John G. Kemeny in "The Use of Simplicity in Induction," *Philosophical Review*, DCXXV, No. 3, July, 1953. This concept of simplicity is a purely syntactic one applicable to theories as a whole. It must be distinguished from certain nonsyntactic concepts, as well as from Nelson Goodman's measure of the simplicity of sets of predicate.

7. Since in a translation atomic sentences are, in general, replaced by molecular ones, there is little doubt as to the general validity of this claim.

8. The authors are indebted to C. G. Hempel for clarifying their thinking on this point.

9. That this condition occurs in all definitions of reduction is due to the historic usage of "reduction."

10. The usage of "explain" will be discussed below. There is a definite connection between this condition and the concept of "systematic power." The present condition implies that $sp(T_1, O) >$ or $= sp(T_2, O)$, but it is stronger than the latter. Cf. "Systematic Power," by the present authors, in *Philosophy of Science*, XXII, No. 1, January, 1955.

11. "Every O" means, of course, every conceivable set of observational data, which in turn means that O can be any observational statement. We require that O be consistent with T_2, since otherwise we get queer explanations in which the "initial conditions" are inconsistent with the theory, and hence anything follows from the two together. There is no harm, however, in allowing O to be inconsistent with T_1. Similarly, it does no harm to apply the definitions to a self-contradictory T_1. The resulting "reductions" are merely uninteresting.

12. "Branch" is here taken in a syntactic, rather than pragmatic, sense. B_1 is a branch of B_2 if its theoretical vocabulary is a proper subset of the theoretical vocabulary of B_2.

13. Cf. C. G. Hempel and Paul Oppenheim, "The Logic of Explanation," *Philosophy of Science*, XV (1948). While this definition is adequate for many purposes, it is oversimplified in some ways. Theorem 1 depends on the form of this definition, but our general approach is consistent with any explicatum of "explain."

14. Cf. n. 6.

15. For example, T_2' may state that T_2 holds under specified conditions.

16. Cf. H. Feigl's contribution to the symposium on operationism in the *Psychological Review*, LII, No. 5, September, 1945.

P. K. Feyerabend

HOW TO BE A GOOD EMPIRICIST—A PLEA FOR TOLERANCE IN MATTERS EPISTEMOLOGICAL*

> *"Facts?" he repeated. "Take a drop more grog, Mr. Franklin, and you'll get over the weakness of believing in facts! Foul play, Sir!"*
>
> Wilkie Collins
> Moonstone

1. CONTEMPORARY EMPIRICISM LIABLE TO LEAD TO ESTABLISHMENT OF A DOGMATIC METAPHYSICS

Today empiricism is the professed philosophy of a good many intellectual enterprises. It is the core of the sciences, or so at least we are taught, for it is responsible both for the existence and for the growth of scientific knowledge. It has been adopted by influential schools in aesthetics, ethics, and theology. And within philosophy proper the empirical point of view has been elaborated in great detail and with even greater precision. This predilection for empiricism is due to the assumption that only a thoroughly observational procedure can exclude fanciful speculation and empty metaphysics as well as to the hope that an empiristic attitude is most liable to prevent stagnation and to further the progress of knowledge. It is the purpose of the present paper to show that empiricism in the form in which it is practiced today cannot fulfill this hope.

---

*Revised copy of paper originally appearing in *Inquiry*.

From *Philosophy of Science: The Delaware Seminar, II*. Reprinted by permission of the University of Delaware Press.

Putting it very briefly, it seems to me that the contemporary doctrine of empiricism has encountered difficulties, and has created contradictions which are very similar to the difficulties and contradictions inherent in some versions of the doctrine of democracy. The latter are a well-known phenomenon. That is, it is well known that essentially totalitarian measures are often advertised as being a necessary consequence of democratic principles. Even worse—it not so rarely happens that the totalitarian character of the defended measures is not explicitly stated but covered up by calling them "democratic," the word *democratic* now being used in a new, and somewhat misleading, manner. This method of (conscious or unconscious) verbal camouflage works so well that it has deceived some of the staunchest supporters of true democracy. What is not so well known is that modern empiricism is in precisely the same predicament. That is, some of the methods of modern empiricism which are introduced in the spirit of anti-dogmatism and progress are bound to lead to the establishment of a dogmatic metaphysics and to the construction of defense mechanisms which make this metaphysics safe from refutation by experimental inquiry. It is true that in the process of establishing such a metaphysics the words *empirical* or *experience* will frequently occur; but their sense will be as distorted as was the sense of "democratic" when used by some concealed defenders of a new tyranny.[1] This, then, is my charge: Far from eliminating dogma and metaphysics and thereby encouraging progress, modern empiricism has found a new way of making dogma and metaphysics respectable, viz., the way of calling them "well-confirmed theories," and of developing a method of confirmation in which experimental inquiry plays a large though well controlled role. In this respect, modern empiricism is very different indeed from the empiricism of Galileo, Faraday, and Einstein, though it will of course try to represent these scientists as following its own paradigm of research, thereby further confusing the issue.[2]

From what has been said above it follows that the fight for tolerance in scientific matters and the fight for scientific progress must still be carried on. What has changed is the denomination of the enemies. They were priests, or "school-philosophers," a few decades ago. Today they all themselves "philosophers of science," or "logical empiricists."[3] There are also a good many scientists who work in the same direction. I maintain that all these groups work against scientific progress. But whereas the former did so openly and could be easily discerned, the latter proceed under the flag of progressivism and empiricism and thereby deceive a good many of their followers. Hence, although their presence is noticeable enough they may almost be compared to a fifth column, the aim of which must be exposed in order that its detrimental effect be fully appreciated. It is the purpose of this paper to contribute to such an exposure.

I shall also try to give a positive methodology for the empirical sciences which no longer encourage dogmatic petrification in the name of experience. Put in a nutshell, the answer which this method gives to the question in the title is: You can be a good empiricist only if you are prepared to work with many alternative theories rather than with a single point of view and "experience." This plurality of theories must not be regarded as a preliminary stage of knowledge which will at some time in the future be replaced by the One True Theory. Theoretical pluralism is assumed to be an *essential feature* of all knowledge that claims to be objective. Nor can one rest content with a plurality which is merely abstract and which is created by denying now this and now that component of the dominant point of view. Alternatives must rather be developed in such detail that problems already "solved" by the accepted theory can again be treated in a new and perhaps also more detailed manner. Such development will of course take time, and it will not be possible, for example, at once to construct alternatives to the present quantum theory which are comparable to its richness and sophistication. Still, it would be very unwise to bring the process to a standstill in the very beginning by the remark that some suggested new ideas are undeveloped, general, metaphysical. *It takes time to build a good theory* [a triviality that seems to have been forgotten by some defenders of the Copenhagen point of view of the quantum theory]; and it also takes time to develop an alternative to a good theory. The *function* of such concrete alternatives is, however, this: They provide means of criticizing the accepted theory in a manner which goes *beyond* the criticism provided by a comparison of that theory "with the facts": However closely a theory seems to reflect the facts, however universal its use, and however necessary its existence seems to be to those speaking the corresponding idiom, its factual adequacy can be asserted only *after* it has been confronted with alternatives *whose invention and detailed development must therefore precede any final assertion of practical success and factual adequacy*. This, then, is the methodological justification of a plurality of *theories:* Such a plurality allows for a much sharper criticism of accepted ideas than does the comparison with a domain of "facts" which are supposed to sit there independently of theoretical considerations. The function of unusual *metaphysical* ideas which are built up in a nondogmatic fashion and which are then developed in sufficient detail to give an (alternative) account even of the most common experimental and observational situations is defined accordingly: They play a decisive role in the criticism and in the development of what is generally believed and "highly confirmed"; and they have therefore to be present at *any* stage of the development of our knowledge.[4] A science that is free from *metaphysics* is on the best way to become a *dogmatic* metaphysical system. So far the summary of the method I shall explain, and defend, in the present paper.

It is clear that this method still retains an essential element of *empiricism*: The decision between alternative theories is based upon *crucial experiments*. At the same time it must *restrict* the range of such experiments. Crucial experiments work well with theories of a low degree of generality whose principles do not touch the principles on which the ontology of the chosen observation language is based. They work well if such theories are compared with respect to a much more general background theory which provides a stable meaning for the observation sentences. However, this background theory, like any other theory, is itself in need of criticism. Criticism must use alternatives. Alternatives will be the more efficient the more radically they differ from the point of view to be investigated. It is bound to happen, then, that the alternatives do not share a single statement with the theories they criticize. Clearly, a crucial experiment is now impossible. It is impossible, not because the experimental device is too complex, or because the calculations leading to the experimental prediction are too difficult; it is impossible because there is no statement capable of expressing what emerges from the observation. This consequence, which severely restricts the domain of empirical discussion, cannot be circumvented by any of the methods which are currently in use and which all try to work with relatively stable observation languages. It indicates that the attempt to make empiricism a universal basis of all our factual knowledge cannot be carried out. The discussion of this situation is beyond the scope of the present paper.

On the whole, the paper is a concise summary of results which I have explained in a more detailed fashion in the following essays: "Explanation, Reduction, and Empiricism"; "Problems of Microphysics"; "Problems of Empiricism"; "Linguistic Philosophy and the Mind-Body Problem."[5] All the relevant acknowledgements can be found there. Let me only repeat here that my general outlook derives from the work of K. R. Popper (London) and David Bohm (London) and from my discussions with both. It was severely tested in discussion with my colleague, T. S. Kuhn (Berkeley). It was the latter's skillful defense of a scientific conversatism which triggered two papers, including the present one. Criticism by A. Naess (Oslo), D. Rynin (Berkeley), Roy Edgley (Bristol), and J. W. N. Watkins (London) have been responsible for certain changes I made in the final version.

2. TWO CONDITIONS OF CONTEMPORARY EMPIRICISM

In this section I intend to give an outline of some assumptions of contemporary empiricism which have been widely accepted. It will be shown in the sections to follow that these apparently harmless assumptions which have been explicitly formulated by some logical empiricists, but which also seem to guide the work of a good many physicists, are bound to lead to exactly the results I have outlined above: Dogmatic petrification and the establishment, on so-called "empirical grounds," of a rigid metaphysics.

One of the cornerstones of contemporary empiricism is its *theory of explanation*. This theory is an elaboration of some simple and very plausible ideas first proposed by Popper[6] and it may be introduced as follows: Let T and T' be two different scientific theories, T' the theory to be explained, or the explanandum, T the explaining theory, or the explanans. Explanation (of T') consists in the *derivation* of T' from T and initial conditions which specify the domain D' in which T' is applicable. Prima facie, this demand of derivability seems to be a very natural one to make for "otherwise the explanans would not constitute adequate grounds for the explanation" (Hempel[7]). It implies two things: First, that the consequences of a satisfactory explanans, T, inside D' must be compatible with the explanadum, T'; and secondly, that the main descriptive terms of these consequences must either coincide, with respect to their meanings, with the main descriptive terms of T', or at least they must be related to them via an empirical hypothesis. The latter result can also be formulated by saying that the meaning of T' must be unaffected by the explanation. "It is of the utmost importance," writes Professor Nagel,[8] emphasizing this point, "that the expressions peculiar to a science will possess meanings that are fixed by its *own* procedures, and are therefore intelligible in terms of its own rules of usage, whether or not the science has been, or will be [explained in terms of] the other discipline."

Now if we take it for granted that more general theories are always introduced with the purpose of explaining the existent successful theories, then every new theory will have to satisfy the two conditions just mentioned. Or, to state it in a more explicit manner:

(1) Only such theories are then admissible in a given domain which either *contain* the theories already used in this domain, or which are at least *consistent* with them inside the domain;[9] and
(2) meanings will have to be invariant with respect to scientific progress; that is, all future theories will have to be phrased in such a manner that their use in explanations does not affect what is said by the theories, or factual reports to be explained.

The two conditions I shall call the *consistency condition* and the *condition of meaning invariance*, respectively.

Both conditions are *restrictive* conditions and therefore bound profoundly to influence the growth of knowledge. I shall soon show that the development of actual science very often violates them and that it violates them in eactly those places where one would be inclined to perceive a tremendous progress of knowledge. I shall also show that neither condition can be justified from the point of view of a tolerant empiricism. However, before doing so I would like to mention that both conditions have occasionally entered the domain of the sciences and have been used here in attacks against new developments and even in the process of theory construction itself. Especially today, they play a very important role in the construction as well as in the defense of certain points of view in microphysics.

Taking first an earlier example, we find that in his *Wärmelehre*, Ernst Mach[10] makes the following remark:

> Considering that there is, in a purely mechanical system of absolutely elastic atoms no real analogue for the *increase of entropy*, one can hardly suppress the idea that a violation of the second law . . . should be possible if such a mechanical system were the *real* basis of thermodynamic processes.

And referring to the fact that the second law is a highly confirmed physical law, he insinuates (in his *Zwei Aufsaetze*[11]) that for this reason the mechanical hypothesis must not be taken too seriously. There were many similar objections against the kinetic theory of heat.[12] More recently, Max Born has based his arguments against the possibility of a return to determinism upon the consistency condition and the assumption which we shall here take for granted, that wave mechanics is incompatible with determinism:

> If any future theory should be deterministic it cannot be a modification of the present one, but must be entirely different. How this should be possible without sacrificing a whole treasure of well established results [i.e., without contradicting highly confirmed physical laws and thereby violating the consistency condition] I leave the determinist to worry about.[13]

Most members of the so-called Copenhagen school of quantum theory would argue in a similar manner. For them the idea of complementarity and the formalism of quantization expressing this idea do not contain any hypothetical element as they are "uniquely determined by the facts."[14] Any theory which contradicts this idea is factually inadequate and must be removed. Conversely, an explanation of the idea of complementarity is acceptable only if it either contains this idea, or is at least consistent with it. This is how the consistency condition is used in arguments against theories such as those of Bohm, de Broglie, and Vigier.[15]

The use of the consistency condition is not restricted to such general remarks, however. A decisive part of the existing quantum theory *itself*, viz., the projection postulate,[16] is the result of the attempt to give an account of the definiteness of macro objects and macro events that is in accordance with the consistency condition. The influence of the condition of meaning invariance goes even further.

> The Copenhagen-interpretation of the quantum theory [writes Heisenberg[17]] starts from a paradox. Any experiment in physics, whether it refers to the phenomena of daily life or to atomic events is to be described in the terms of classical physics . . . *We cannot and should not replace these concepts by any others* [my italics]. Still the application of these concepts is limited by the relation of uncertainty. We must keep in mind this limited range of applicability of the classical concepts while using them, but we cannot, and should not try to improve them.

This means that the meaning of the classical terms must remain invariant with respect to any future explanation of microphenomena. Microtheories have to be formulated in such a manner that this invariance is guaranteed. The principle of correspondence and the formalism of quantization connected with it were explicitly devised for satisfying this demand. Altogether, the quantum theory seems to be the first theory after the downfall of the Aristotelian physics that has been quite explicitly constructed with an eye both on the consistency condition and the condition of (empirical) meaning invariance. In this respect it is very different indeed from, say, relativity which violates both consistency and meaning invariance with respect to earlier theories. Most of the arguments used for the defense of its customary interpretation also depend on the validity of these two conditions and they will collapse with their removal. An examination of these conditions is therefore very topical and bound deeply to affect present controversies in microphysics. I shall start this investigation by showing that some of the most interesting developments of physical theory in the past have violated both conditions.

3. THESE CONDITIONS NOT INVARIABLY ACCEPTED BY ACTUAL SCIENCE

The case of the consistency condition can be dealt with in a few words: It is well known (and has also been shown in great detail by Duhem[18]) that Newton's theory is inconsistent with Galileo's law of the free fall and with Kepler's laws; that statistical thermodynamics is inconsistent with the second law of the phenomenological theory; that wave optics is inconsistent with geometrical optics; and so on. Note that what is being asserted here is *logical* inconsistency; it may well be that the differences of prediction are too small to be detectable by experiment. Note also that what is being asserted is not the inconsistency of, say, Newton's theory and Galileo's law, but rather the inconsistency of *some consequences* of Newton's theory in the domain of validity of Galileo's law, and Galileo's law. In this last case the situation is especially clear. Galileo's law asserts that the acceleration of the free fall is a constant, whereas application of Newton's theory to the surface of the earth gives an acceleration that it not a constant but *decreases* (although imperceptibly) with the distance from the center of the earth. Conclusion: If actual scientific procedure is to be the measure of method, then the consistency condition is inadequate.

The case of meaning invariance requires a little more argument, not because it is intrinsically more difficult, but because it seems to be much more closely connected with deep-rooted prejudices. Assume that an explanation is required, in terms of the special theory of relativity, of the classical conservation of mass in all reactions in a closed system S. If $m', m'', m''', \ldots, m^i, \ldots$ are the masses of the parts $P', P'', P''', \ldots, P^i, \ldots$ of S, then what we want is an explanation of

$$\sum m^i = \text{const.} \tag{1}$$

for all reactions inside S. We see at once that the consistency condition cannot be fulfilled: According to special relativity $\sum m^i$ will vary with the velocities of the parts relative to the coordinate system in which the observations are carried out, and the total mass of S will also depend on the relative potential energies of the parts. However, if the velocities and the mutual forces are not too large, then the variation of $\sum m^i$ predicted by relativity will be so small as to be undetectable by experiment. Now let us turn to the *meanings* of the terms in the relativistic law and in the corresponding classical law. The first indication of a possible change of meaning may be seen in the fact that in the classical case the mass of an aggregate of parts equals the sum of the masses of the parts:

$$M(\sum P^i) = \sum M(P^i).$$

This is not valid in the case of relativity where the relative velocities and the relative potential energies contribute to the mass balance. That the relativistic concept and the classical concept of mass are very different indeed becomes clear if we also consider that the former is a *relation*, involving relative velocities, between an object and a coordinate system, whereas the latter is a *property* of the object itself and independent of its behavior incoordinate systems. True, there have been attempts to give a relational analysis even of the classical concept (Mach). None of these attempts, however, leads to the relativisic idea with its velocity dependence on the coordinate system, which idea must therefore be added even to a *relational* account of classical mass. The attempt to identify the classical mass with the relativistic rest mass is of no avail either. For although both may have the same numerical value, the one is still dependent on the coordinate system chosen (in which it is at rest and has that specific value), whereas the other is not so dependent. We have to conclude, then, that $(m)_c$ and $(m)_r$ mean very different things and that $(\sum m^i)_c =$ const. and $(\sum m^i)_r =$ const. are very different assertions. This being the case, the derivation from relativity of either equation (1) or of a law that makes slightly different quantitative predictions with $\sum m^i$ used in the classical manner, will be possible only if a further premise is added which establishes a relation between the $(m)_c$ and the $(m)_r$. Such a "bridge law"—and this is a major point in Nagel's theory of reduction—is a hypothesis

> according to which the occurrence of the properties designated by some expression in the premises of the [explanans] is a sufficient, or a necessary and sufficient condition for the occurrence of the properties designated by the expressions of the [explanandum].[19]

Applied to the present case this would mean the following: Under certain conditions the occurrence of relativistic mass of a given magnitude is accompanied by the occurrence of classical mass of a corresponding magnitude; this assertion is inconsistent with another part of the explanans, viz., the theory of relativity. After all, this theory asserts that there are no invariants which are directly connected with mass measurements and it thereby asserts that $(m)_c$ does not express real features of physical systems. Thus we inevitably arrive at the conclusion that mass conservation cannot be explained in terms of relativity (or "reduced" to relativity) without a violation of meaning invariance. And if one retorts, as has been done by some critics of the ideas expressed in the present paper,[20] that meaning invariance is an essential part of both reduction and explanation, then the answer will simply be that equation (1) can neither be explained by, nor reduced to relativity. Whatever the *words* used for describing the situation, the *fact* remains that actual science does not observe the requirement of meaning invariance.

This argument is quite general and is independent of whether the terms whose meaning is under investigation are observable or not. It is therefore stronger than may seem at first sight. There are some empiricists who would admit that the meaning of theoretical terms may be changed in the course of scientific progress. However, not many people are prepared to extend meaning *variance* to observational terms also. The idea motivating this attitude is, roughly, that the meaning of observational terms is uniquely determined by the procedures of observation such as looking, listening, and the like. These procedures remain unaffected by theoretical advance.[21] Hence, observational meanings, too, remain unaffected by theoretical advance. What is overlooked, here, is that the "logic" of the observational terms is not exhausted by the procedures which are connected with their application "on the basis of observation." As will turn out later, it also depends on the more general ideas that determine the "ontology" (in Quine's sense) of our discourse. These general ideas may change without any change of observational procedures being implied. For example, we may change our ideas about the nature, or the ontological status (property, relation, object, process, etc.) of the color of a self-luminescent object without changing the methods of ascertaining that color (looking, for example). Clearly, such a change is bound profoundly to influence the meanings of our observational terms.

All this has a decisive bearing upon some contemporary ideas concerning the interpretation of scientific theories. According to these ideas, theoretical terms receive their meanings via correspondence rules which connect them with an observational language *that has been fixed in advance* and independently of the structure of the theory to be interpreted. Now, our above analysis would seem to show that *if we interpret scientific theories in the manner accepted by the scientific community*, then most of these correspondence rules will be either false, or nonsensical. They will be *false* if they *assert* the existence of entities denied by the theory; they will be *nonsensical* if they *presuppose* this existence. Turning the argument around, we can also say that the attempt to interpret the calculus of some theory that has been voided of the meaning assigned to it by the scientific community with the help of the double language system will lead to a very different theory. Let us again take the theory of relativity as an example: It can be safely assumed that the physical thing language of Carnap, and any similar language that has been suggested as an observation language, is not Lorentz-invariant. The attempt to interpret the *calculus* of relativity on *its* basis therefore cannot lead to the *theory* of relativity as it was understood by Einstein. What we shall obtain will be at the very most *Lorentz's interpretation* with its inherent asymmetries. This undesirable result cannot be evaded by the *demand* to use a different and more adequate observation language. The double language system assumes that theories which are not connected with some observation language do not posess an interpretation. The demand assumes that they do, and asks to choose the observation language most suited to it. It reverses the relation between theory and experience that is characteristic for the double language method of interpretation, which means, it gives up this method. Contemporary empiricism, therefore, has not led to any satisfactory account of the meanings of scientific theories.[22]

What we have shown so far is that the two conditions of section 2 are frequently violated in the course of scientific practice and especially at periods of scientific revolution. This is not yet a very strong argument. True: There are empirically inclined philosophers who have derived some satisfaction from the assumption that they only make explicit what is implicitly contained in scientific practice. It is therefore quite important to show that scientific practice is not what it is supposed to be by them. Also, strict adherence to meaning invariance and consistency would have made impossible some very decisive advances in physical theory such as the advance from the physics of Aristotle to the physics of Galileo and Newton. However, how do we know (independently of the fact that they do exist, have a certain structure, and are very influential—a circumstance that will have great weight with opportunists only[23]) that the sciences are a desirable phenomenon, that they contribute to the advancement of knowledge, and that their analysis will therefore lead to reasonable methodological demands? And did it not emerge in the last section that meaning invariance and the consistency condition *are* adopted by some scientists? Actual scientific practice, therefore, cannot be our last authority. We have to find out whether consistency and meaning invariance are *desirable* conditions and this quite independently of who accepts and praises them and how many Nobel prizes have been won with their help.[24] Such an investigation will be carried out in the next sections.

4. INHERENT UNREASONABLENESS OF CONSISTENCY CONDITION

Prima facie, the case of the consistency condition can be dealt with in very few words. Consider for that purpose a theory T' that successfully describes the situation in the domain D'. From this we can infer (a) that T' agrees with a *finite* number of observations (let their class be F); and (b) that it agrees with these observations inside a margin M of error only.[25] Any alternative that contradicts T' outside F and inside M is supported by exactly the same observations and therefore acceptable if T' was acceptable (we shall assume that F are the only observations available). The consistency condition is much less tolerant. It eliminates a theory not because it is in disagreement with the *facts*; it eliminates it because it is in disagreement with *another theory*, with a theory, moreover, whose confirming instances it shares. *It thereby makes the as yet untested part of that theory a measure of validity*. The only difference between such a measure and a more recent theory is age and familiarity. Had the younger theory been there first, then the consistency condition would have worked in its favor. In this respect the effect of the consistency condition is rather similar to the effect of the more traditional methods of transcendental deduction, analysis of essences, phenomenological analysis, linguistic analysis. It contributes to the preservation of the old and familiar not because of any inherent advantage in it—for example, not because it has a better foundation in observation than has the newly suggested alternative, or because it is more elegant—but just because it is old and familiar. This is not the only instance where on closer inspection a rather surprising similarity emerges between modern empiricism and some of the school philosophies it attacks.

Now it seems to me that these brief considerations, although leading to an interesting *tactical* criticism of the consistency condition, do not yet go to the heart of the matter. They show that an alternative of the accepted point of view which shares its confirming instances cannot be *eliminated* by factual reasoning. They do not show that such an alternative is *acceptable*; and even less do they show that it *should be used*. It is bad enough, so a defender of the consistency condition might point out, that the accepted point of view does not possess full empirical support. Adding new theories *of an equally unsatisfactory character* will not improve the situation; nor is there much sense in trying to *replace* the accepted theories by some of their possible alternatives. Such replacement will be no easy matter. A new formalism may have to be learned and familiar problems may have to be calculated in a new way. Textbooks must be rewritten, university curricula readjusted, experimental results reinterpreted. And what will be the result of all the effort? Another theory which, from an empirical point of view, has no advantage whatever over and above the theory it replaces. The only real improvement, so the defender of the consistency condition will continue, derives from the *addition of new facts*. Such new facts will either support the current theories, or they will force us to modify them by indicating precisely where they go wrong. In both cases they will precipitate real progress and not only arbitrary change. The proper procedure must therefore consist in the confrontation of the accepted point of view with as many relevant facts as possible. The exclusion of alternatives is then required for reasons of expediency: Their invention not only does not help, but it even hinders progress by absorbing time and manpower that could be devoted to better things. And the function of the consistency condition lies precisely in this. It eliminates such fruitless discussion and it forces the scientist to concentrate on the facts which, after all, are the only acceptable judges of a theory. This is how the practicing scientist will defend his concentration on a single theory to the exclusion of all empirically possible alternatives.[26]

It is worthwhile repeating the reasonable core of this argument: Theories should not be changed unless there are pressing reasons for doing so. The only pressing reason for changing a theory is disagreement with facts. Discussion of incompatible facts will therefore lead to progress. Discussion of incompatible alternatives will not. Hence, it is sound procedure to increase the number of relevant facts. It is not sound procedure to increase the number of factually adequate, but incompatible alternatives. One might wish to add that formal improvements such as increase of elegance, simplicity, generality, and coherence should not be excluded. But once these improvements have been carried out, the collection of facts for the purpose of test seems indeed to be the only thing left to the scientist.

5. RELATIVE AUTONOMY OF FACTS

And this it is—provided these facts *exist, and are available independently of whether or not one considers alternatives to the theory to be tested*. This assumption on which the validity of the argument in the last section depends in a most decisive manner I shall call the assumption of the relative autonomy of facts, or the autonomy principle. It is not asserted by this principle that the discovery and description of facts is independent of *all* theorizing. But it *is* asserted that the facts which belong to the empirical content of some theory are available whether or not one considers alternatives to *this* theory. I am not aware that this very important assumption has ever been explicitly formulated as a separate postulate of the empirical method. However, it is clearly implied in almost all investigations which deal with questions of confirmation and test. All these investigations use a model in which a *single* theory is compared with a class of facts (or observation statements) which are assumed to be "given" somehow. I submit that this is much too simple a picture of the actual situation. Facts and theories are much more intimately connected than is admitted by the autonomy principle. Not only is the description of every single fact dependent on *some* theory (which may, of course, be very different from the theory to be tested). There exist also facts which cannot be unearthed except with the help of alternatives to the theory to be tested, and which become unavailable as soon as such alternatives are excluded. This suggests that the methodological unit to which we must refer when discussing questions of test and empirical content is constituted by a *whole set of partly overlapping, factually adequate, but mutually inconsistent theories*. In the present paper only the barest outlines will be given of such a test model. However, before doing this I want to discuss an example which shows very clearly the function of alternatives in the discovery of facts.

As is well known, the Brownian particle is a perpetual motion machine of the second kind and its existence refutes the phenomenological second law. It therefore belongs to the domain of relevant facts for this law. Now, could this relation between the law and the Brownian particle have been discovered in a *direct* manner, i.e., could it have been discovered by an investigation of the observational consequences of the phenomenological theory that did not make use of an alternative account of heat? This question is readily divided into two: (1) Could the *relevance* of the Brownian particle have been discovered in this manner? (2) Could it have been demonstrated that it actually *refutes* the second law? The answer to the first question is that we do not know. It is impossible to say what would have happened had the kinetic theory not been considered by some physicists. It is my guess, however, that in this case the Brownian particle would have been regarded as an oddity much in the same way in which some of the late Professor Ehrenhaft's astounding effects[27] are regarded as an oddity, and that it would not have been given the decisive position it assumes in contemporary theory. The answer to the second question is simply—No. Consider what the discovery of the inconsistency between the Brownian particle and the second law would have required! It would have required (a) measurement of the exact *motion* of the particle in order to ascertain the changes of its kinetic energy plus the energy spent on overcoming the resistance of the fluid; and (b) it would have required precise measurement of temperature and heat transfer in the surrounding medium in order to ascertain that any loss occurring here was indeed compensated by the increase of the energy of the moving particle and the work done against the fluid. Such measurements are beyond experimental possibilities.[28] Neither is it possible to make precise measurements of the heat transfer; nor can the path of the particle be investigated with the desired precision. Hence a "direct" refutation of the second law that considers only the phenomenological theory and the "facts" of Brownian motion is impossible. And, as is well known, the actual refutation was brought about in a very different manner. It was brought about via the kinetic theory and Einstein's utilization of it in the calculation of the statistical properties of the Brownian motion.[29] In the course of this procedure the phenomenological theory (T') was incorporated into the wider context of statistical physics (T) *in such a manner that the consistency condition was violated*; and *then* a crucial experiment was staged (investigations of Svedberg and Perrin).

It seems to me that this example is typical for the relation between fairly general theories, or points of view, and "the facts." Both the relevance and the refuting character of many very decisive facts can be established only with the help of other theories which, although factually adequate, are yet not in agreement with the view to be tested. This being the case, the production of such refuting facts may have to be preceded by the invention and articulation of alternatives to that view. Empiricism demands that the empirical content of whatever knowledge we possess be increased as much as possible. Hence *the invention of alternatives in addition to the view that stands in the center of discussion constitutes an essential part of the empirical method.* Conversely, the fact that the consistency condition eliminates alternatives now shows it to be in disagreement with empiricism and not only with scientific practice. By excluding valuable tests it decreases the empirical content of the theories which are permitted to remain (and which, as we have indicated above, will usually be the theories which have been there first); and it especially decreases the number of those facts which could show their limitations. This last result of a determined application of the consistency condition is of very topical interest. It may well be that the refutation of the quantum-mechanical uncertainties presupposes just such an incorporation of the present theory into a wider context which is no longer in accordance with the idea of complementarity and which therefore suggests new and decisive experiments. And it may also be that the insistence, on the part of the majority of contemporary physicists, on the consistency condition will, if successful, forever protect these uncertainties from refutation. This is how modern empiricism may finally lead to a situation where a certain point of view petrifies into dogma by being, in the name of experience, completely removed from any conceivable criticism.

6. THE SELF-DECEPTION INVOLVED IN ALL UNIFORMITY

It is worthwhile to examine this apparently empirical defense of a dogmatic point of view in somewhat greater detail. Assume that physicists have adopted, either consciously or unconsciously, the idea of the uniqueness of complementarity and that they therefore elaborate the orthodox point of view and refuse to consider alternatives. In the beginning such a procedure may be quite harmless. After all, a man can do only so many things at a time and it is better when he pursues a theory in which he is interested rather than a theory he finds boring. Now assume that the pursuit of the theory he chose has led to successes and that the theory has explained in a satisfactory manner circumstances that had been unintelligible for quite some time. This gives empirical support to an idea which to start with seemed to possess only this advantage: It was interesting and intriguing. The concentration upon the theory will now be reinforced, the attitude towards alternatives will become less tolerant. Now if it is true, as has been argued in the last section, that many facts become available only with the help of such alternatives, then the refusal to consider them *will result in the elimination of potentially refuting facts*. More especially, it will eliminate facts whose discovery would show the complete and irreparable inadequacy of the theory.[30] Such facts having been made inaccessible, the theory will appear to be free from blemish and it will seem that "all evidence points with merciless definiteness in the . . . direction . . . [that] all the processes involving . . . unknown interactions conform to the fundamental quantum law" (n. 14, p. 44). This will further reinforce the belief in the uniqueness of the current theory and in the complete futility of any account that proceeds in a different manner. Being now very firmly convinced that there is only one good microphysics, the physicists will try to explain even adverse facts in its terms, and they will not mind when such explanations are sometimes a little clumsy. By now the success of the theory has become public news. Popular science books (and this includes a good many books on the philosophy of science) will spread the basic postulates of the theory; applications will be made in distant fields. More than ever the theory will appear to possess tremendous empirical support. The chances for the consideration of alternatives are now very slight indeed. The final success of the fundamental assumptions of the quantum theory and of the idea of complementarity will seem to be assured.

At the same time it is evident, on the basis of the considerations in the last section, that this appearance of success *cannot in the least be regarded as a sign of truth and correspondence with nature*. Quite the contrary, the suspicion arises that the absence of major difficulties is a result of the decrease of empirical content brought about by the elimination of alternatives, and of facts that can be discovered with the help of these alternatives only. In other words, *the suspicion arises that this alleged success is due to the fact that in the process of application to new domains the theory has been turned into a metaphysical system*. Such a system will of course be very "successful" not, however, because it agrees so well with the facts, but because no facts have been specified that would constitute a test and because some such facts have even been removed. Its "success" *is entirely manmade*. It was decided to stick to some ideas and the result was, quite naturally, the survival of these ideas. If now the initial decision is forgotten, or made only implicitly, then the survival will seem to constitute independent support, it will reinforce the decision, or turn it into an explicit one, and in this way close the circle. This is how empirical "evidence" may be *created* by a procedure which quotes as its justification the very same evidence it has produced in the first place.

At this point an "empirical" theory of the kind described (and let us always remember that the basic principles of the present quantum theory and especially the idea of complementarity are uncomfortably close to forming such a theory) becomes almost indistinguishable from a myth. In order to realize this, we need only consider that on account of its all-pervasive character a myth such as the myth of witchcraft and of demonic possession will possess a high degree of confirmation on the basis of observation. Such a myth has been taught for a long time; its content is enforced by fear, prejudice, and ignorance as well as by a jealous and cruel priesthood. It penetrates the most common idiom, infects all modes of thinking and many decisions which mean a great deal in human life. It provides models for the explanation of any conceivable event, conceivable, that is, for those who have accepted it.[31] This being the case, its key terms will be fixed in an unambiguous manner and the idea (which may have led to such a procedure in the first place) that they are copies of unchanging entities and that change of meaning, if it should happen, is due to human mistake—this idea will now be very plausible. Such plausibility reinforces all the maneuvres which are used for the preservation of the myth (elimination of opponents included). The conceptual apparatus of the theory and the emotions connected with its application having penetrated all means of communication, all actions, and indeed the whole life of the community, such methods as transcendental deduction, analysis of usage, phenomenological analysis which are means for further solidifying the myth will be extremely successful (which shows, by the way, that all these methods which have been the trademark of various philosophical schools old and new, have one thing in common: They tend to *preserve* the *status quo* of the intellectual life).[32] Observational results too, will speak in favor of the theory as they are formulated in its terms. It will seem that at last the truth has been arrived at. At the same time it is evident that all contact with the world has been lost and that the stability achieved, the semblance of absolute truth, *is nothing but the result of an absolute conformism.*[33] For how can we possibly test, or improve upon, the truth of a theory if it is built in such a manner that any conceivable event can be described, and explained, in terms of its principles? The *only* way of investigating such all-embracing principles is to compare them with a different set of *equally all-embracing* principles—but this way has been excluded from the very beginning. The myth is therefore of no objective relevance, it continues to exist solely as the result of the effort of the community of believers and of their leaders, be these now priests or Nobel prize winners. *Its "success" is entirely manmade.* This, I think, is the most decisive argument against any method that encourages uniformity, be it now empirical or not. Any such method is in the last resort a method of deception. It enforces an unenlightened conformism, and speaks of truth; it leads to a deterioration of intellectual capabilities, of the power of imagination, and speaks of deep insight; it destroys the most precious gift of the young, their tremendous power of imagination, and speaks of education.

To sum up: *Unanimity of opinion may be fitting for a church, for the frightened victims of some (ancient, or modern) myth, or for the weak and willing followers of some tyrant; variety of opinion is a feature necessary for objective knowledge; and a method that encourages variety is also the only method that is compatible with a humanitarian outlook.* To the extent to which the consistency condition (and, as will emerge, the condition of meaning invariance) delimits variety, it contains a theological element (which lies, of course, in the worship of "facts" so characteristic for nearly all empiricism).

7. INHERENT UNREASONABLENESS OF MEANING INVARIANCE

What we have achieved so far has immediate application to the question whether the meaning of certain key terms should be kept unchanged in the course of the development and improvement of our knowledge. After all, the meaning of every term we use depends upon the theoretical context in which it occurs. Hence, if we consider two contexts with basic principles which either contradict each other, or which lead to inconsistent consequences in certain domains, it is to be expected that some terms of the first context will not occur in the second context with exactly the same meaning. Moreover, if our methodology demands the use of mutually inconsistent, partly overlapping, and empirically adequate theories, then it thereby also demands the use of conceptual systems which are mutually *irreducible* (their primitives cannot be connected by bridge laws which are meaningful *and* factually correct) and it demands that meanings of terms be left elastic and that no binding commitment be made to a certain set of concepts.

It is very important to realize that such a tolerant attitude towards meanings, or such a change of meaning in cases where one of the competing conceptual systems has to be abandoned need not be the result of directly accessible observational difficulties. The law of inertia of the so-called *impetus theory* of the later Middle Ages[34] and Newton's own law of inertia are in perfect quantitative agreement: Both assert that an object that is not under the influence of any outer force will proceed along a straight line with constant speed. Yet despite this fact, the adoption of Newton's theory entails a conceptual revision that forces us to abandon the inertial law of the impetus theory, not because it is quantitatively incorrect but *because it achieves the correct predictions with the help of inadequate concepts*. The law asserts that the *impetus* of an object that is beyond the reach of outer forces remains constant.[35] The impetus is interpreted as an inner *force* which pushes the object along. Within the impetus theory such a force is quite conceivable as it is assumed here that forces determine *velocities* rather than accelerations. The concept of impetus is therefore formed in accordance with a law (forces determine velocities), and this law is inconsistent with the laws of Newton's theory and must be abandoned as soon as the latter is adopted. This is how the progress of our knowledge may lead to conceptual revisions for which no direct observational reasons are available. The occurrence of such changes quite obviously refutes the contention of some philosophers that the invariance of *usage* in the trivial and uninteresting contexts of the private lives of not too intelligent and inquisitive people indicates invariance of *meaning* and the superficiality of all scientific changes. It is also a very decisive objection against any crudely operationalistic account of both observable terms and theoretical terms.

What we have said applies even to singular statements of observation. Statements which are empirically adequate, and which are the result of observation (such as "here is a table") may have to be reinterpreted, not because it has been found that they do not adequately express what is seen, heard, felt, but because of some changes in sometimes very remote parts of the conceptual scheme to which they belong. Witchcraft is again a very good example. Numerous eyewitnesses claim that they have actually *seen* the devil, or *experienced* demonic influence. There is no reason to suspect that they were lying. Nor is there any reason to assume that they were sloppy observers, for the phenomena leading to the belief in demonic influence are so obvious that a mistake is hardly possible (possession; split personality; loss of personality; hearing voices; etc.). These phenomena are well known today.[36] In the conceptual scheme that was the one generally accepted in the fifteenth and sixteenth centuries, the only way of describing them, or at least the way that seemed to express them most adequately, was by reference to demonic influences. Large parts of this conceptual scheme were changed for philosophical reasons and also under the influence of the evidence accumulated by the sciences. Descartes's materialism played a very decisive role in discrediting the belief in spatially localizable spirits. The language of demonic influences was no part of the new conceptual scheme that was created in this manner. It was for this reason that a reformulation was needed, and a reinterpretation of even the most common "observational" statements. Combining this example with the remarks at the beginning of the present section, we now realize that according to the method of classes of alternative theories a lenient attitude must be taken with respect to the meanings of all the terms we use. We must not attach to great an importance to "what we mean" by a phrase, and we must be prepared to change whatever little we have said concerning this meaning as soon as the need arises. Too great concern with meanings can only lead to dogmatism and sterility. Flexibility, and even sloppiness in semantical matters is a prerequisite of scientific progress.[37]

8. SOME CONSEQUENCES

Three consequences of the results so far obtained deserve a more detailed discussion. The first consequence is an evaluation of *metaphysics* which differs significantly from the standard empirical attitude. As is well known, there are empiricists who demand that science start from observable facts and proceed by generalization, and who refuse the admittance of metaphysical ideas at any point of this procedure. For them, only a system of thought that has been built up in a purely inductive fashion can claim to be genuine knowledge. Theories which are partly metaphysical, or "hypothetical," are suspect, and are best not used at all. This attitude has been formulated most clearly by Newton[38] in his reply to Pardies's second letter concerning the theory of colors:

> If the possibility of hypotheses is to be the test of truth and reality of things, I see not how certainty can be obtained in any science; since numerous hypotheses may be devised, which shall seem to overcome new difficulties.

This radical position, which clearly depends on the demand for a theoretical monism, is no longer as popular as it used to be. It is now granted that metaphysical considerations may be of importance when the task is to *invent* a new physical theory; such invention, so it is admitted, is a more or less irrational act containing the most diverse components. Some of these components are, and perhaps must be, metaphysical ideas. However, it is also pointed out that as soon as the theory has been developed in a formally satisfactory fashion and has received sufficient confirmation to be regarded as empirically successful, it is pointed out that in the very same moment it can *and must* forget its metaphysical past; metaphysical speculation must *now* be replaced by empirical argument.

> On the one side I would like to emphasize [writes Ernst Mach on this point[39]] that *every and any* idea is admissible as a means for research, provided it is helpful; still, it must be pointed out, on the other side, that it is very necessary from time to time to free the presentation of the *results* of research from all inessential additions.

This means that empirical considerations are still given the upper hand over metaphysical reasoning. Especially in the case of an inconsistency between metaphysics and some highly confirmed empirical theory it will be decided, *as a matter of course*, that the theory or the result of observation must stay, and that the metaphysical system must go. A very simple example is the way in which materialism is being judged by some of its opponents. For a materialist the world consists of material particles moving in space, of collections of such particles. Sensations, as introspected by human beings, do not look like collections of particles, and their observed existence is therefore assumed to refute and thereby to remove the metaphysical doctrine of materialism. Another example which I have analyzed in "Problems of Microphysics" is the attempt to eliminate certain very general ideas concerning the nature of microentities on the basis of the remark that they are inconsistent "with an immense body of experience" and that "to object to a lesson of experience by appealing to metaphysical preconceptions is unscientific."

The methodology developed in the present paper leads to a very different evaluation of metaphysics. Metaphysical systems are scientific theories in their most primitive stage. If they *contradict* a well-confirmed point of view, then this indicates their usefulness as an alternative to this point of view. Alternatives are needed for the purpose of criticism. Hence, metaphysical systems which contradict observational results of well-confirmed theories *are most welcome* starting points of such criticism. Far from being misfired attempts at anticipating, or circumventing, empirical research which were deservedly exposed by a reference to experience, they are the only means at our disposal for examining those parts of our knowledge which have already become observational and which are therefore inaccessible to a criticism "on the basis of observation."

A second consequence is that a new attitude has to be adopted with respect to the *problem of induction*. This problem consists in the question of what justification there is for asserting the truth of a statement S given the truth of another statement, S', whose content is smaller than the content of S. It may be taken for granted that those who want to justify the truth of S also assume that after the justification the truth of S will be *known*. Knowledge to the effect that S implies the *stability* of S (we must not change, remove, criticize, what we know to be true). The method we are discussing at the present moment cannot allow such stability. It follows that the problem of induction at least in some of its formulations, is a problem whose solution leads to undesirable results. It may therefore be properly termed a pseudo problem.

The third consequence, which is more specific, is that *arguments from synonymy* (or from coextensionality), far from being that measure of adequacy as which they are usually introduced, are liable severely to impede the progress of knowledge. Arguments from synonymy judge a theory or a point of view not by its capability to mimic the world but rather by its capability to mimic the descriptive terms of another point of view which for some reason is received favorably. Thus for example, the attempt to give a materialistic, or else a purely physiological, account of human beings is criticized on the grounds that materialism, or physiology, cannot provide synonyms for "mind," "pain," "seeing red," "thinking of Vienna," in the sense in which these terms are used either in ordinary English (provided there is a well-established usage concerning these terms, a matter which I doubt) or in some more esoteric mentalistic idiom. Clearly, such criticism silently assumes the principle of meaning invariance, that is, it assumes that the meanings of at least some fundamental terms must remain unchanged in the course of the progress of our knowledge. It cannot therefore be accepted as valid.[40]

However, we can, and must, go still further. The ideas which we have developed above are strong enough not only to *reject* the demand for synonymy, wherever it is raised, but also to *support* the demand for irreducibility (in the sense in which this notion was used at the beginning of section 7). The reason is that irreducibility is a presupposition of high critical ability on the part of the point of view shown to be irreducible. An outer indication of such irreducibility which is quite striking in the case of an attack upon commonly accepted ideas is the feeling of *absurdity*: We deem absurd what goes counter to well-established linguistic habits. The absence, from a newly introduced set of ideas, of synonymy relations connecting it with part of the accepted point of view; the feeling of absurdity therefore indicate that the new ideas are fit for the purpose of criticism, i.e., that they are fit for either leading to a strong *confirmation* of the earlier theories, or else to a very revolutionary *discovery*: Absence of synonymy, clash of meanings, absurdity are desirable. Presence of synonymy, intuitive appeal, agreement with customary modes of speech, far from being *the* philosophical virtue, indicates that not much progress has been made and that the business of investigating what is commonly accepted *has not even started.*

9. HOW TO BE A GOOD EMPIRICIST

The final reply to the question put in the title is therefore as follows. A good empiricist will not rest content with the theory that is in the center of attention and with those tests of the theory which can be carried out in a direct manner. Knowing that the most fundamental and the most general criticism is the criticism produced with the help of alternatives, he will try to invent such alternatives.[41] It is, of course, impossible at once to produce a theory that is formally comparable to the main point of view and that leads to equally many predictions. His first step will therefore be the formulation of fairly general assumptions which are not yet directly connected with observations; this means that his first step will be the invention of a new *metaphysics*. This metaphysics must then be elaborated in sufficient detail in order to be able to compete with the theory to be investigated as regards generality, details of prediction, precision of formulation.[42] We may sum up both activities by saying that a good empiricist must be a critical metaphysician. Elimination of all metaphysics, far from increasing the empirical content of the remaining theories, is liable to turn these theories into dogmas. The consideration of alternatives together with the attempt to criticize each of them in the light of experience also leads to an attitude where meanings do not play a very important role and where arguments are based upon assumptions of fact rather than analysis of (archiac, although perhaps very precise) meanings. The effect of such an attitude upon the development of human capabilities should not be underestimated either. Where speculation and invention of alternatives is encouraged, bright ideas are liable to occur in great number and such ideas may then lead to a change of even the most "fundamental" parts of our knowledge, i.e., they may lead to a change of assumptions which either are so close to observation that their truth seems to be dictated by "the facts," or which are so close to common prejudice that they seem to be "obvious," and their negation "absurd." In such a situation it will be realized that neither "facts" nor abstract ideas can ever be used for defending certain principles, come what may. Wherever facts play a role in such a dogmatic defense, we shall have to suspect foul play (see the opening quotation)—the foul play of those who try to turn good science into bad, because unchangeable, metaphysics. In the last resort, therefore, being a good empiricist means being critical, and basing one's criticism not just on an abstract principle of skepticism but upon *concrete suggestions* which indicate in every single case how the accepted point of view might be further tested and further investigated and which thereby prepare the next step in the development of our knowledge.

For support of research the author is indebted to the National Science Foundation and the Minnesota Center for the Philosophy of Science.

NOTES

1. K. R. Popper, *The Open Society and Its Enemies* (Princeton, N. J.: Princeton University Press, 1953).

2. It is very interesting to see how many so-called empiricists, when turning to the past, completely fail to pay attention to some very obvious facts which are incompatible with their empiristic epistemology. Thus Galileo has been represented as a thinker who turned away from the empty speculations of the Aristotelians and who based his own laws upon facts which he had carefully collected beforehand. Nothing could be further from the truth. *The Aristotelians could quote numerous observational results in their favor*. The Copernican idea of the motion of the earth, on the other hand, did not possess independent observational support, at least not in the first 150 years of its existence. Moreover, it was inconsistent with facts and highly confirmed physical theories. And *this* is how modern physics started: not as an observational enterprise, *but as an unsupported speculation that was inconsistent with highly confirmed laws*. For details and further references see my "Realism and Instrumentalism," to appear in *The Critical Approach: Essays in Honor of Karl Popper*. (subnote 2*.)

* P. K. Feyerabend, "Realism and Instrumentalism," in *The Critical Approach: Essays in Honor of Karl Popper*, ed. M. Bunge (Glencoe, Illinois: The Free Press, to be published).

3. One might be inclined to add those who base their pronouncements upon an analysis of what they call "ordinary language." I do not think they deserve to be honored by a criticism. Paraphrasing Galileo, one might say that they "deserve not even that name, for they do not talk plainly and simply but are content to adore the shadows, philosophizing not with due circumspection but merely from having memorized a few ill-understood principles."

4. It is nowadays frequently assumed that "if one considers the history of a special branch of science, one gets the impression that non-scientific elements . . . relatively frequently occur in the earlier stages of development, but that they gradually retrogress in later stages and even tend to disappear in such advanced stages which become ripe for more or less thorough formalization." (H. J. Groenewold, *Synthese* (1957), p. 305). Our considerations in the text would seem to show that such a development is very undesirable and can only result in a well-formalized, precisely expressed, and completely petrified metaphysics.

5. These essays were published in Vol. III of the *Minnesota Studies in the Philosophy of Science*; in Vols. I and II of the *Pittsburgh Studies in the Philosophy of Science*; and in *Problems of Philosophy, Essays in Honor of Herbert Feigl*, respectively.

6. See subnote 6*. The decisive feature of Popper's theory, a feature which was not at all made clear by earlier writers on the subject of explanation, is the emphasis he puts on the initial conditions and the implied possibility of two kinds of laws, viz., (1) laws concerning the temporal sequence of events; and (2) laws concerning the space of initial conditions. In the case of the quantum theory, the laws of the second kind provide very important information about the nature of the elementary particles and it is to *them* and *not* to the laws of motion that reference is made in the discussions concerning the interpretation of the uncertainty relations. In general relativity, the laws formulating the initial conditions concern the structure of the universe at large and only by overlooking them could it be believed that a purely relational account of space would be possible. For the last point, cf. Hill, subnote 6†.

* K. R. Popper, *Logic of Scientific Discovery* (New York, 1959), Sec. 12. This is a translation of his *Logik der Forschung* published in 1935.

† E. L. Hill, "Quantum Physics and the Relativity Theory," in *Current Issues in the*

Philosophy of Science, eds. H. Feigl and G. Maxwell (New York: Holt, Rinehart and Winston, 1961).

7. C. G. Hempel, "Studies in the Logic of Explanation," reprinted in *Readings in the Philosophy of Science*, eds. H. Feigl and M. Brodbeck (New York, 1953), p. 321.

8. E. Nagel, "The Meaning of Reduction in the National Sciences," reprinted in *Philosophy of Science*, eds. A. C. Danto and S. Morgenbesser (New York, 1960), p. 301.

9. It has been objected to this formulation that theories which are consistent with a given explanadum may still contradict each other. This is quite correct, but it does not invalidate my argument. For as soon as a single theory is regarded as sufficient for explaining all that is known (and represented by the other theories in question), it will have to be consistent with all these other theories.

10. E. Mach, *Wärmelehre* (Leipzig, 1897), p. 364.

11. E. Mach, *Zwei Aufsaetze* (Leipzig, 1912).

12. For a discussion of these objections, cf. ter Haar's review article in *Reviews of Modern Physics* (1957).

13. M. Born, *Natural Philosophy of Cause and Chance* (New York: Oxford University Press, 1948), p. 109.

14. L. Rosenfeld, "Misunderstandings about the Foundations of the Quantum Theory," in *Observation and Interpretation* (London, 1957), p. 42.

15. Cf. the discussions in *Observation and Interpretation* (See n. 14).

16. For details and further literature, cf. Sec. 11 of my paper "Problems of Microphysics."

17. W. Heisenberg, *Physics and Philosophy* (New York, 1958), p. 44.

18. P. Duhem, *La Théorie Physique: Son Objet, Sa Structure* (Paris, 1914), Chaps. ix and x. See also K. R. Popper, "The Aim of Science," *Ratio*, Vol. I (1957).

19. E. Nagel, n. 8, p. 302.

20. Cf. Sec. 4.7 of M. Scriven's paper "Explanations, Predictions, and Laws," in Vol. III of the *Minnesota Studies in the Philosophy of Science*. Similar objections have been raised by Kraft (Vienna) and Rynin (Berkeley).

21. For an exposition and criticism of this idea, cf. my "Attempt at a Realistic Interpretation of Experience, "*Proceedings of the Aristotelian Society*, New Series, LVIII (1958), 143–70.

22. It must be admitted, however, that Einstein's original interpretation of the special theory of relativity is hardly ever used by contemporary physicists. For them the theory of relativity consists of two elements: (1) the Lorentz transformations; and (2) mass-energy equivalence. The Lorentz transformations are interpreted purely formally and are used to make a selection among possible equations. This interpretation does not allow to distinguish between Lorentz's original point of view and the entirely different point of view of Einstein. According to it Einstein achieved a very minor *formal* advance [this is the basis of Whittaker's attempt to "debunk" Einstein]. It is also very similar to what application of the double language model would yield. Still, an undesirable philosophical procedure is not improved by the support it gets from an undesirable procedure in physics. [The above comment on the contemporary attitude towards relativity was made by E. L. Hill in discussions at the Minnesota Center for the Philosophy of Science.]

23. In about 1925 philosophers of science were bold enough to stick to their theses even in those cases where they were inconsistent with actual science. They meant to be *reformers* of science, and not *imitators*. (This point was explicitly made by Mach in his controversy with Planck. Cf. again his *Zwei Aufsaetze*, n. 11.) In the meantime they have become rather tame (or beat) and are much more prepared to change their ideas in accordance with the latest

discoveries of the historians, or the latest fashion of the contemporary scientific enterprise. This is very regrettable, indeed, for it considerably decreases the number of the rational critics of the scientific enterprise. And it also seems to give unwanted support to the Hegelian thesis (which is now implicitly held by many historians and philosophers of science) that what exists has a "logic" of its own and is for that very reason reasonable.

24. Even the most dogmatic enterprise allows for discoveries (cf. the "discovery" of so-called "white Jews" among German physicists during the Nazi period). Hence, before hailing a so-called discovery, we must make sure that the system of thought which forms its background is not of a dogmatic kind.

25. The indefinite character of all observations has been made very clear by Duhem, n. 18, Chap. ix. For an alternative way of dealing with this indefiniteness, cf. S. Körner, *Conceptual Thinking* (New York, 1960).

26. More detailed evidence for the existence of this attitude and for the way in which it influences the development of the sciences may be found in Kuhn's book *Structure of Scientific Revolutions*, subnote 26*. The attitude is extremely common in the contemporary quantum theory. "Let us enjoy the successful theories we possess and let us not waste our time with contemplating what *would* happen if *other* theories were used"—this seems to be the motto of almost all contemporary physicists (cf. Heisenberg, n. 17, pp. 56, 144) and philosophers (cf. Hanson, subnote 26†). It may be traced back to Newton's papers and letters (to Hooke, and Pardies) on the theory of color. See also n. 23.

* T. Kuhn, *Structure of Scientific Revolutions* (Chicago: University of Chicago Press, 1962).

† N. R. Hanson, "Five Cautions for the Copenhagen Critics," *Philosophy of Science*, XXVI (1959), 325–37.

27. Having witnessed these effects under a great variety of conditions, I am much more reluctant to regard them as mere curiosities than is the scientific community of today. Cf. also my edition of Ehrenhaft's lectures, *Einzelne Magnetische Nord- und Südpole und deren Auswirkung in den Naturwissenschaften* (Vienna, 1947).

28. R. Fürth, *Zeitschrift für Physik*, 81 (1933), 143–62.

29. For these investigations, cf. A. Einstein, *Investigations on the Theory of the Brownian Motion*, subnote 29*, which contains all the relevant papers by Einstein and an exhaustive bibliography by R. Fürth. For the experimental work, cf. J. Perrin, *Die Atome*, subnote 29†. For the relation between the phenomenological theory and the kinetic theory, cf. also Smoluchowski, subnote 29** and Popper, subnote 29‡. Despite Einstein's epoch-making discoveries and von Smoluchowski's splendid presentation of their effect (for the latter cf. also subnote 29§), the present situation in thermodynamics is extremely unclear, especially in view of the continued presence of the ideas of reduction which we criticized in the text above. To be more specific, it is frequently attempted to determine the entropy balance of a complex *statistical* process by reference to the (refuted) *phenomenological* law after which procedure fluctuations are superimposed in a most artificial fashion. For details cf. Popper, *loc. cit.*

* A. Einstein, *Investigations on the Theory of the Brownian Motion* (New York, 1956).

† J. Perrin, *Die Atome*, (Leipzig, 1920).

** M. v. Smoluchowski, "Experimentell nachwiesbare, der üblichen Thermodynamik widersprechende Molekularphanomene," *Physikalische Zeitschrift*, XIII (1912), 1069.

‡ K. R. Popper, "Irreversibility, or, Entropy since 1905," *British Journal for the Philosophy of Science*, VIII (1957), 151.

§ *Oeuvres de Marie Smoluchowski*, II (Cracouvie, 1927), 226 ff., 316 ff., 462 ff., and 530 ff.

30. The quantum theory can be adapted to a great many difficulties. It is an open theory

in the sense that apparent inadequacies can be accounted for in an *ad hoc* manner, by *adding* suitable operators, or elements in the Hamiltonian, rather than by recasting the whole structure. A refutation of its basic formalism (i.e., of the formalism of quantization, and of non-commuting operators in a Hilbert space or a reasonable extension of it) would therefore demand proof to the effect that *there is no conceivable adjustment of the Hamiltonian, or of the operators used* which makes the theory conform to a given fact. It is clear that such a general statement can only be provided by an *alternative theory* which of course must be detailed enough to allow for independent and crucial tests.

31. For a very detailed description of a once very influential myth, cf. C. H. Lea, *Materials for a History of Witchcraft*, 3 Vols. (New York, 1957), as well as *Malleus Malleficarum*, translated by Montague Summers (who, by the way, counts it "among the most important, wisest [sic!], and weightiest books of the world") (London, 1928).

32. Quite clearly, analysis of usage, to take only one example, presupposes certain regularities concerning this usage. The more people differ in their fundamental ideas, the more difficult will it be to uncover such regularities. Hence, analysis of usage will work best in a closed society that is firmly held together by a powerful myth such as was the philosophy in the Oxford of about ten years ago.

33. Schizophrenics very often hold beliefs which are as rigid, all-pervasive, and unconnected with reality, as are the best dogmatic philosophies. Only such beliefs come to them naturally whereas a professor may sometimes spend his whole life in attempting to find arguments which create a similar state of mind.

34. For details and further references, cf. Sec. 6 of my "Explanation, Reduction, and Empiricism," *loc. cit.*

35. We assume here that a dynamical rather than a kinematic characterization of motion has been adopted. For a more detailed analysis, cf. again the paper referred to in the previous footnote.

36. For very vivid examples, cf. K. Jaspers, *Allgemeine Psychopathologie* (Berlin, 1959), pp. 75–123.

37. Mae West is by far preferable to the precisionists: "I ain't afraid of pushin' grammar around so long as it sounds good" (*Goodness Had Nothing to do With It*, New York, 1959, p. 19).

38. I. B. Cohen, ed., *Isaac Newton's Papers & Letters on Natural Philosophy* (Cambridge, Mass.: Harvard University Press, 1958), p. 106.

39. "Der Gegensatz zwischen der mechanischen und der phaenomenologischen Physik," *Wärmelehre* (Leipzig, 1896), pp. 362 ff.

40. For details concerning the mind-body problem, cf. my "Materialism and the Mind-Body Problem," *Review of Metaphysics* (Sept. 1963).

41. In my paper "Realism and Instrumentalism," subnote 2*, I have tried to show that this is precisely the method which has brought about such spectacular advances of knowledge as the Copernican Revolution, the transition to relativity and to quantum theory.

42. Cf. Sec. 13 of my "Realism and Instrumentalism."

Wilfrid Sellars

THE LANGUAGE OF THEORIES*

I. INTRODUCTION

(1) My purpose is to see what fresh light, if any, is thrown on old familiar puzzles about the empirical (or factual) meaningfulness of theoretical statements and the reality of theoretical entities by certain views on related topics which I have sketched in a number of recent papers.[1] These views concern (a) the interpretation of basic semantical categories; (b) the role of theories in scientific explanation.

(2) The term *theory*, it is generally recognized, covers a wide variety of explanatory frameworks resembling one another by that family resemblance which is easy to discern but most difficult to describe. Each type of theory presents its own problems to the philosopher of science, and although current literature shows an increasing tendency to reflect the realities of scientific practice rather than antecedent epistemological commitments, the type of theory with which I shall be concerned—namely, that which postulates unobserved entities to explain observable phenomena—is still suffering from the effects of a Procrustean treatment by positivistically oriented philosophies of science.

(3) I shall assume, at least to begin with, that *something* like the standard modern account of this type of theory is correct. And in view of the distinguished names associated with this account, it would be most surprising if it were not

---

*I wish to acknowledge the invaluable assistance I have received from friends and colleagues who gave me their comments on an earlier draft. I am particulary grateful to Professor Adolf Grünbaum for a page by page critique with respect to both exposition and substance, without which the paper would have fallen far shorter than it does of saying what I wanted it to say.

close to the truth. It is built upon a distinction between: (a) the vocabulary, postulates, and theorems of the theory as an uninterpreted calculus; (b) the vocabulary and inductively testable statements of the observation framework; (c) the "correspondence rules" which correlate, in a way which shows certain analogies to inference and certain analogies to translations, statements in the theoretical vocabulary with statements in the language of observation. Each of these categories calls for a brief initial comment.

(4) The theoretical language contains, in addition to that part of its vocabulary which ostensibly refers to unobserved entities and their properties, (a) logical and mathematical expressions which have their ordinary sense, and (b) the vocabulary of space and time. (Query: Can we say that the latter part of the theoretical vocabulary, too, has its ordinary sense? To use the material mode, are the space and time of kinetic theory the same as the space and time of the observable world, or do they merely "correspond" to them? In relativity physics it is surely the latter.)

(5) The nontheoretical language with which a given theory is connected by means of correspondence rules may itself be a theory with respect to some other framework, in which case it is nontheoretical only in a relative sense. This calls up a picture of levels of theory and suggests that there is a level which can be called nontheoretical in an absolute sense. Let us assume for the moment that there is such a level and that it is the level of the observable things and properties of the everyday world and of the constructs which can be explicitly defined in terms of them. If following Carnap we call the language appropriate to this level the *physical-thing language*, then the above assumption can be formulated as the thesis that the physical-thing language is a nontheoretical language in an absolute sense. The task of theory is then constructed to be that of explaining inductively testable generalizations formulated in the physical-thing language, which task is equated with *deriving* the latter from the theory by means of the correspondence rules.

(6) Correspondence rules typically connect defined expressions in the theoretical language with definable expressions in the language of observation. They are often said to give a "partial interpretation" of the theory in terms of observables, but this is at best a very misleading way of talking; for whatever may be true of "correspondence rules" in the case of physical geometry,[2] it is simply not true, in the case of theories which postulate unobserved micro-entities, that a correspondence rule stipulates that a theoretical expression is to *have the same sense* as the correlated expression in the observation language. The phrase "partial interpretation" suggests that the only sense in which the interpretation fails to be a translation is that it is partial; that is, that while *some* stipulations of identity of sense are laid down, they do not suffice to make possible a complete translation of the theoretical language into the language of observation. It is less misleading to say that while the correspondence rules *coordinate* theoretical and observational sentences, neither they nor the derivative rules which are their consequences place the primitives of the theory into one-one correspondence with observation language counterprarts. This way of

putting it does not suggest, as does talk of "partial interpretation," that if the partial correlation could be made complete, it would *be* a translation. (That a "complete correlation" could be *transformed* into a translation by reformulating correspondence rules as semantical stipulations is beside the point.)

(7) For the time being, then, we shall regard the correspondence rules of theories of the kind we are examining as a special kind of verbal bridge taking one from statements in the theoretical vocabularly to statements in the observation vocabulary and vice versa. The term "correspondence rule" has the advantage, as compared with "bridge law," "coordinating definition," or "interpretation," of being neutral as between various interpretations of the exact role played by these bridges in different kinds of theory.

(8) Puzzles about the meaning of theoretical terms and the reality of theoretical entities are so intimately bound up with the status of correspondence rules that to clarify the latter would almost automatically resolve the former. This fact is the key to my strategy in this paper. But before attempting to develop a suitable framework for this analytical task, a few remarks on contemporary treatments of correspondence rules are in order. Until recently it was customary, in schematic representations of theories, to keep the postulates and theorems of the theory, the empirical generalizations of the observation framework, and the correspondence rules linking theory with observation in three distinct compartments. This had the value of emphasizing the methodologically distinct roles of these three different types of statement. On the other hand this mode of representation carried with it the suggestion of an *ontological* (as contrasted with methodological) dualism of theoretical and observational universes of discourse which a more neutral presentation might obviate. Thus it has recently been the tendency to list the correspondence rules with the postulates of the theory, distinguishing them simply as those postulates which include observational as well as theoretical expressions.[3] This procedure can do no harm if the relevant methodological and semantical distinctions ultimately find adequate expression in some other way. And if, as I shall argue, it is a mistake to interpret the dualism of theoretical and observational frameworks as of more than methodological significance, this new mode of representation may well be the better way of picturing the situation.

. . .

II. MICROTHEORETICAL EXPLANATION

(33) It would seem, then, that if kinetic theory is a *good* theory, we are entitled to say that molecules exist. This confronts us with a classical puzzle. For, it would seem, we can also say that if our observation framework is a *good* one, we are entitled to say that horses, chairs, tables, etc., exist. Shall we then say that *both* tables and molecules exist? If we do, we are immediately faced with the problem as to how theoretical objects and observational objects "fit togeth-

er in one universe." To use Eddington's well-worn example, instead of the one table with which pretheoretical discourse was content, we seem forced to recognize two tables of radically different kinds. Do they both *really* exist? Are they, perhaps, *really* the same table? If only one of then *really* exists, which?

(34) It has frequently been suggested that a theory might be a *good* theory, and yet be *in principle* otiose. (Not otiose; *in principle otiose*.) By this is meant that the theory might be known on general grounds to be the sort of thing which, in the very process of being perfected, generates a *substitute* which, in the limiting case of perfection, would serve all the scientific purposes which the perfected theory could serve. The idea is, in brief, that the cash value at each moment of a developing theory is a set of propositions in the observation framework known as *the observational consequences of the theory*, and that once we separate out the heuristic or "pragmatic" role of a theory from its role in explanation, we see that the observational consequences of an ideally successful theory would serve all the scientific purposes of the theory itself.

(35) If we knew that theories of the kind we are considering were *in principle* otiose, we might well be inclined to say that there *really* are no such things as molecules, and even to abandon our habit of talking about theoretical expressions in semantical and quasi-semantical terms. We might refuse to say that theoretical terms express concepts, or name or denote objects; we might refuse to say that theoretical objects exist. And we might well put this by claiming that theoretical languages are *mere* calculational devices. This resolution of the initial puzzle has at least the merit of being neat and tidy. It seems to carve theoretical discourse at a joint, and to cut off a superfluous table with no loss of blood.

(36) I shall argue that this is an illusion. But what is the alternative? One possible line of thought is based on the idea that perhaps the observational level of physical things (which includes one of the tables) has been mistakenly taken to be an "absolute." It points out that if the framework of physical things were in principle subject to discard, the way would be left open for the view that perhaps there is only one table after all; this time, however, the table construed in theoretical terms.

(37) I think that this suggestion contains the germ of the solution to the puzzle; but only if it is developed in such a way as to free it from the misleading picture—the levels picture—which generates the puzzle. It is, I believe, a blind alley if, accepting this picture, it simply argues that the observational framework *is itself a theory*, and that the relation of the framework of microphysical theory to it is implicitly repeated in its relation to a more basic level—the level, say, of sense contents. For though this account might well enable one to dispense in principle with the physical objects which serve as mediating links between sense contents and molecules (the latter two being capable, in principle, of being directly connected), nevertheless we should still be left with *two* tables, a cloud of molecules on the one hand, and a pattern of actual and possible sense contents on the other.

(38) My line will be that the true solution of our puzzle is to be found by rejecting as above the unchallengeable status of the physical-thing framework,

without, however, construing this framework as a theory with respect to a more basic level. My strategy will be to bring out the misleading and falsifying nature of the *levels* picture of theories. Thus I shall not be concerned, save incidentally and by implication, with the widespread view that the relation of physical-object discourse to sense-impression discourse is analogous to that of microtheories in physics to the framework of physical things.

(39) There are two main sources of the temptation to talk of theories in terms of levels: (i) In the case of microtheories, there is the difference of size between macro- and micro-objects. With respect to this I shall only comment that the entities postulated by a theory need not be smaller than the objects of which the behavior is to be explained. Thus, it is logically possible that physical objects might be theoretically explained as singularities arising from the interference of waves of cosmic dimensions. (ii) The more important source of the plausibility of the *levels* picture is the fact that we not only explain *singular matters of empirical fact* in terms of *empirical generalizations*; we also, or so it seems, explain these generalizations themselves by means of *theories*. This way of putting it immediately suggests a hierarchy at the bottom of which are:

Explained Nonexplainers,

the intermediate levels being:

Explained Explainers,

and the top consisting of:

Unexplained Explainers.

Now there is clearly *something* to this picture. But it is radically misleading if (a) it finds too simple—too simple in a sense to be given presently—a connection between explaining an explanadum and finding a defensible general proposition under which it can either be subsumed,[4] or from which it can be derived with or without the use of correspondence rules; (b) it is supposed that whereas in the observation framework inductive generalizations serve as principles of explanation for particular matters of fact, microtheoretical principles are principles of explanation *not (directly) for particular matters of fact in the observation framework but for the inductive generalizations in this framework (the explaining being equated with deriving the latter from the former) which in their turn serve as principles of explanation for particular matters of fact.*[5]

(40) This latter point is the heart of the matter; for to conceive of the *explananda* of theories as empirical laws and of theoretical explanation as the derivation of empirical laws from theoretical postulates by means of logic, mathematics, and correspondence rules is to sever the vital tie between theoretical principles and particular matters of fact in the framework of observation. Indeed, the idea that the aim of theories is to explain *not* particular matters of fact *but rather* inductive generalizations is nothing more nor less than the idea that theories are in principle dispensable. For to suppose that particular observable matters of fact are the proper *explananda* of inductive generalizations in the observation framework is to suppose that even though theoretical considerations

may lead us to formulate new hypotheses in the observational framework for inductive testing and may lead us to modify, subject to inductive confirmation, such generalizations as have already received inductive support, the *conceptual framework* of the observation levels is autonomous and immune from theoretical criticism.

(41) The truth of the matter is that the idea that microtheories are designed to explain empirical laws rests on the above mentioned confusion between explanation and derivation. To avoid this confusion is to see that theories about observable things *do not explain empirical laws, they explain why observable things obey, to the extent that they do, these empirical laws*;[6] that is, they explain why individual objects of various kinds and in various circumstances in the observation framework behave in those ways in which it has been inductively established that they do behave. Roughly, it is because a gas is—in some sense of "is"—a cloud of molecules which are behaving in theoretically defined ways, and *hence*, in particular cases, places and times behaves in a certain way, that it obeys the Boyle-Charles law.

(42) Furthermore, theories not only explain why observable things obey certain laws, they also explain why in certain respects their behavior does not obey a law. This point can best be introduced by contriving an artificially simple example. It might, at a certain time, have been discovered that gold which has been put in *aqua regia* sometimes dissolves at one rate, sometimes at another, even though as far as can be observationally determined, the specimens and circumstances are identical. The microtheory of chemical reactions current at that time might admit of a simple modification to the effect that there are two structures of microentities each of which "corresponds" to gold as an observational construct, but such that pure samples of one dissolve, under given conditions of pressure, temperature, etc., at a different rate from samples of the other. Such a modification of the theory would explain the observationally unpredicted variation in the rate of dissolution of gold by saying that samples of observational gold are mixtures of these two theoretical structures in various proportions, and have a rate of dissolution which varies with the proportion. Of course, if the correspondence rules of the theory enables one to derive observational criteria for distinguishing between observational golds of differing theoretical compositions, one would be in a position to replace the statement that gold dissolves in *aqua regia* sometimes at one rate, sometimes at another, by laws setting fixed rates for two *varieties* of observational gold and their mixtures. But it is by no means clear that the correspondence rules (together with the theory) *must* enable one to do this in order for the theory to be a good theory. The theory must, of course, explain why observational chemical substances do obey *some* laws, and the theoretical account of the variation in the rate at which gold dissolves in *aqua regia* must cohere with its general explanation of chemical reactions, and not simply postulate that there is an unspecified dimension of variation in the microstructure of gold which corresponds to this observed variation. But this is a far cry from requiring that the theory lead to a confirmable set of empirical laws by which to replace the initial account of random variations.

(43) Thus, microtheories not only explain why observational constructs obey inductive generalizations, they explain what, as far as the observational framework is concerned, is a random component in their behavior, and, in the last analysis it is by doing the latter that microtheories establish their character as indispensable elements of scientific explanation and (as we shall see) as knowledge about what *really* exists. Here it is essential to note that in speaking of the departure from lawfulness of observational constructs I do not have in mind simply departure from all-or-none lawfulness ("strict universality"). Where microexplanation is called for, correct macroexplanation will turn out (its eyes sharpened by theoretical considerations) to be in terms of "satistical" rather than strictly universal generalizations. But this is only the beginning of the story, for the distinctive feature of those domains where microexplanation is appropriate is that in an important sense such regularities as are available are not statistical *laws*, because they are unstable, and this instability is explained by the microtheory.

The logical point I am making can best be brought out by imagining a domain of inductive generalizations about observables to be idealized by discounting errors of measurement and other forms of experimental error. For once these elements in the "statistics" have been discounted, our attention can turn to the logico-mathematical structure of these idealized statistical statements. And reflection makes clear that where microtheoretical explanation is to be appropriate, these statements must have (and this is a logical point) a mathematical structure which is not only compatible with, but calls for, an explanation in terms of "microvariables" (and hence *microinitial conditions*: the nonlawlike element adumbrated in the preceding paragraph) such as the microtheory provides. This point is but the converse of the familiar point that the irreducibly and lawfully statistical ensembles of quantum-mechanical theory are mathematically inconsistent with the assumption of hidden variables.

Thus, to say that "statistical lawfulness" can stand on its own feet may be to confuse the *empirical* ("pragmatic") point that we may have to settle for statistical knowledge with respect to a given observational domain with the *logical* point that there is a definable type of statistical lawfulness which *must* stand on its own feet if it is to stand at all.

To sum up the above results, microtheories explain why inductive generalizations pertaining to a given domain *and any refinement of them within the conceptual framework of the observaion language* are at best approximations to the truth. To this it is anticlimatic to add that theories explain why inductive generalizations hold only within certain boundary conditions, accounting for discontinuities which, as far as the observation framework is concerned, are brute facts.

(44) My contention, then, is that the widespread picture of theories as means of explaining observation-framework laws is a mistake, a mistake which cannot be corrected by extending the term *law* to include a spectrum of inductively established statistical uniformities ranging from 100 percent to 50–50. Postitively put, my contention is that theories explain not laws, but why the

objects of the domain in question obey the laws that they do to the extent that they do.

III. CORRESPONDENCE RULES AGAIN

(45) Suppose it to be granted that this contention is correct. What are its implications for the puzzles with which we began? The first point to be made is that if the basic schema of (micro-) theoretical explanation is:

> Molar objects of such and such kinds obey (approximately) such and such inductive generalizations because they *are* configurations of such and such theoretical entities

then our puzzles are focused, as it were, into the single puzzle of the force of the italicized word *are*. Prima facie it stands for identity, but how is this identity to be understood? Once again we are led to ask the methodological counterpart question, that is, what *is* a correspondence rule?

(46) One possible but paradoxical line of thought would be that an effective microtheory for a certain domain of objects for which inductive generalizations exist is from the standpoint of a philosopher interested in the ontology of science, a framework which aspires to *replace* the observation framework. The observation framework would be construed as a poorer explanatory framework with a better one available to replace it. But thus boldly conceived, this replacement would involve dropping both the empirical generalizations and the individual observational facts to be explained by the theory, and would seem to throw out the baby with the bath. The observation framework would be construed as a poorer explanatory framework with a better one available to replace it. Before we ask what could be meant by "replacing an observation framework by a theoretical framework," let us note one possible reaction to this suggestion. It might be granted that this is the sort of thing that is done when one theoretical framework is "reduced" to another, and that the notion of the replaceability of a microframework by a micro-microframework is a reasonable explanation of the force of such a statement as:

> Ions behave as they do because they *are* such and such configurations of subatomic particles.

Yet the parallel explanation of the force of "are" where the identity relates not theoretical entities with other theoretical entities, but theoretical entities with observational entities might be ruled out of court. Once again we would have run up against the thesis of the inviolability of observation concepts on which the rejection of the replacement idea is ultimately grounded. This thesis, however, is false.

(47) Nor is it satisfactory to interpret the proposal in paragraph 46 as follows: The framework of physical things is a candidate for replacement *on the ground that* it is actually a common sense theoretical framework, and *qua*

theoretical framework may be replaced by another. For unless the conception of the framework of physical things as a replaceable explanatory framework goes hand in hand with an abandonment of the levels picture of explanation, it leads directly to the idea that below the level of physical-thing discourse is a level of observation in a stricter sense of this term, and of confirmable inductive generalizations pertaining to the entities thus observed.

(48) The notion of such a level is a myth. The idea that sense contents exhibit a lawfulness which can be characterized without placing them in a content either of persons and physical things or of microneurological events is supported only by the conviction that it must be so if we are not to flaunt "established truths" about meaning and explanation. Since my quarrel on this occasion is with these "established truths," I shall not argue directly against the idea that there is an autonomous level of sense contents with respect to which the framework of physical things plays a role analogous to that of a theory in the levels of explanation sense.[7]

(49) My answer to the question of paragraph 45 requires that we distinguish between two interpretations of the idea that the framework of physical things is an explanatory framework capable in principle of being replaced by a better explanatory framework. One of these interpretations is the view on which I have just been commenting. The alternative, in general terms, will clearly be a view according to which the framework of physical things is a replaceable theory-like structure in a sense that does not involve a commitment to a deeper "level" of observation and explanation.

(50) The groundwork for such a view has already been laid with the above rejection of the idea that theories explain laws. But what is a correspondence rule if it is *not* a device for deriving laws from theoretical postulates? We have seen that a correspondence rule is not a partial definition of theoretical terms by observation terms. Nor, obviously, is it a definition of observation terms as currently used by means of theoretical terms. But might it not be construed as a *redefinition* of observation terms? Such a redefinition would, of course, be a dead letter unless it were actually carried out in linguistic practice. And it is clear that to be fully carried out in any interesting sense, it would not be enough that sign designs which play the role of observation terms be borrowed for use in the theoretical language as the defined equivalents of theoretical expressions. For this would simply amount to making these sign designs ambiguous. In their new use they would no longer be *observation* terms. The force of the "redefinition" must be such as to demand not only that the observation-sign design correlated with a given theoretical expression be syntactically interchangeable with the latter, *but that the latter be given the perceptual or observational role of the former so that the two expressions become synonymous by mutual readjustment*. And to this there is an obvious object: *The meaningful use of theories simply does not require this usurpation of the observational role by theoretical expressions*. Correspondence rules thus understood would remain dead letters.

(51) But if the above conception of correspondence rules as "redefinitions" will not do as it stands, it is nevertheless in the neighborhood of the truth; for

if correspondence rules cannot be regarded as implemented redefinitions, can they not be regarded as statements to the effect that certain redefinitions of observation terms would be in principle acceptable. This would be compatible with the fact that the redefinitions in question are implemented only in the syntactical dimension, no theoretical expressions actually acquiring the observational-perceptual roles they would have to have if they were to be synonyms of other expressions playing this role. This view has at least the merit of accounting for the peculiar character of correspondence rules as expressing more than a factual equivalence but less than an identity of sense. It would explain how theoretical complexes can be unobservable, yet "really" identical with observable things.

(52) On one classical interpretation, correspondence rules would appear in the material mode as statements to the effect that the same objects which have observational properties also have theoretical properties, thus identifying the denotation, but not the sense, of certain observational and theoretical expressions. On another classical interpretation, correspondence rules would appear in the material mode as asserting the coexistence of two sets of objects, one having observational properties, the other theoretical properties, thus identifying neither the denotation nor the sense of theoretical and observational expressions. According to the view I am proposing, correspondence rules would appear in the material mode as statements to the effect that the objects of the observational framework *do not really exist—there really are no such things*. They envisage the *abandonment* of a sense and its denotation.

(53) If we put this by saying that to offer the theory is to claim that the theoretical language could beat the observation language *at its own game* without loss of scientific meaning, our anxieties are aroused. Would not something be *left out if* we taught ourselves to use the language of physical theory as a framework in terms of which to make our perceptual responses to the world? I do not have in mind the role played by our observational concept in our practical life, our emotional and esthetic response. The repercussions here of radical conceptual changes such as we are envisaging would no doubt be tremendous. I have in mind the familiar question: Would not the abandonment of the framework of physical things mean the abandonment of the *qualitative* aspects of the world?

(54) To this *specific* question, of course, the answer is yes. But it would be a mistake to generalize and infer that *in general* the replacement of observation terms by theoretical constructs must "leave something out." Two points can be touched on briefly. (a) I have suggested elsewhere[8] that the sensible qualities of the common-sense world, omitted by the physical theory of material things, might reappear in a new guise in the microtheory of sentient organisms. This claim would appear in the material mode as the claim that the sensible qualities of things *really* are a dimension of neural activity. (b) There is an obvious sense in which scientific theory cannot leave out qualities, or, for that matter, relations. Only the most pythagoreanizing philosopher of science would attempt to dis-

pense with descriptive (that is, nonlogical) predicates in his formulation of the scientific picture of the world.

NOTES

1. "Empiricism and the Philosophy of Mind," in *Minnesota Studies in the Philosophy of Science*, eds. Herbert Feigl and Michael Scriven, Vol. I (Minneapolis: Univ. of Minn. Press, 1956); "Counterfactuals, Dispositions and the Causal Modalities," in *Minnesota Studies in the Philosophy of Science*, eds. H. Feigl, M. Scriven, and G. Maxwell, Vol. II (Minneopolis: Univ. of Minn. Press, 1958); "Grammar and Existence: A Preface to Ontology," *Mind*, LXIX (1960); "Some Reflections on Language Games," *Philos. of Science*, XXI (1954).

2. The case of geometry is not independent, for geometrical oncepts must be defined for micro-entities. Even if abstraction is made from this, there remains the problems of extending idealized congruences to situations in which the congruences are physically impossible—for example, the center of the sun.

3. This method of presentation is in certain respects analogous to that of drawing no formal distinction between definitions, on the one hand, and postulates and theorems of the form "$a = b$" on the other in the development of a calculus, leaving it to subsequent reflection to determine how the latter are to be parceled out into definitional and nondefinitional identities.

4. For a sustained critique of the subsumption picture of scientific explanation from a somewhat different point of view, see Michael Scriven's papers in *Minnesota Studies in the Philosophy of Science*, Vols. I and II, and his unpublished doctoral dissertation (Oxford) on explanation.

5. From a purely formal point of view, of course, one could derive ("explain") the observational consequence (C) of an observational antecedent (A) by using the theoretical theorem $A_T \longrightarrow C_T$ and the correspondence rules $A \longleftrightarrow A_T$ and $C \longleftrightarrow C_T$ without using the inductive generalization $A \longrightarrow C$. This, however, would only disguise the commitment to the autonomous or "absolute" (*not* unrevisible) status of inductive generalizations in the observation framework.

6. The same is true in principle—though in a way which is methodologically more complex —of micro-microtheories about microtheoretical objects.

7. For a reinterpretation of the status of "sense contents which frees them from the myth of the given," see my "Empiricism and the Philosophy of Mind," *op. cit.*, pp. 321ff.

8. "Empiricism and the Philosophy of Mind," pp. 60–63.

Postscript

Thomas S. Kuhn

THE FUNCTION OF DOGMA IN SCIENTIFIC RESEARCH*

At some point in his or her career every member of this Symposium has, I feel sure, been exposed to the image of the scientist as the uncommitted searcher after truth. He is the explorer of nature—the man who rejects prejudice at the threshold of his laboratory, who collects and examines the bare and objective facts, and whose allegiance is to such facts and to them alone. These are the characteristics which make the testimony of scientists so valuable when advertising proprietary products in the United States. Even for an international audience, they should require no further elaboration. To be scientific is, among other things, to be objective and open-minded.

Probably none of us believes that in practice the real-life scientist quite succeeds in fulfilling this ideal. Personal acquaintance, the novels of Sir Charles

---

*The ideas developed in this paper have been abstracted, in a drastically condensed form, from the first third of my forthcoming monograph, *The Structure of Scientific Revolutions*, which will be published during 1962 by the University of Chicago Press. Some of them were also partially developed in an earlier essay, "The essential tension: tradition and innovation in scientific research," which appeared in Calvin W. Taylor, ed., *The Third (1959) University of Utah Research Conference on the Identification of Creative Scientific Talent* (Salt Lake City, 1959).

On this whole subject see also I. B. Cohen, "Orthodoxy and scientific progress", *Proceedings of the American Philosophical Society*, XCVI (1952), 505–12, and Bernard Barber, "Resistance by scientists to scientific discovery," *Science*, CXXXIV (1961), 596–602. I am indebted to Mr. Barber for an advance copy of that helpful paper. Above all, those concerned with the importance of quasi-dogmatic commitments as a requisite for productive scientific research should see the works of Michael Polanyi, particularly his *Personal Knowledge* (Chicago, 1958) and *The Logic of Liberty* (London, 1951). The discussion which follows this paper will indicate that Mr. Polanyi and I differ somewhat about what scientists are committed to, but that should not disguise the very great extent of our agreement about the issues discussed explicitly below.

Snow, or a cursory reading of the history of science provides too much counter-evidence. Though the scientific enterprise may be open-minded, whatever this application of that phrase may mean, the individual scientist is very often not. Whether his work is predominantly theoretical or experimental, he usually seems to know, before his research project is even well under way, all but the most intimate details of the result which that project will achieve. If the result is quickly forthcoming, well and good. If not, he will struggle with his apparatus and with his equations until, if at all possible, they yield results which conform to the sort of pattern which he has foreseen from the start. Nor is it only through his own research that the scientist displays his firm convictions about the phenomena which nature can yield and about the ways in which these may be fitted to theory. Often the same convictions show even more clearly in his response to the work produced by others. From Galileo's reception of Kepler's research to Nägeli's reception of Mendel's, from Dalton's rejection of Gay Lussac's results to Kelvin's rejection of Maxwell's, unexpected novelties of fact and theory have characteristically been resisted and have often been rejected by many of the most creative members of the professional scientific community. The historian, at least, scarcely needs Planck to remind him that: "A new scientific truth is not usually presented in a way that convinces its opponents. . . ; rather they gradually die off, and a rising generation is familiarized with the truth from the start."[1]

Familiar facts like these—and they could easily be multiplied—do not seem to bespeak an enterprise whose practitioners are notably open-minded. Can they at all be reconciled with our usual image of productive scientific research? If such a reconciliation has not seemed to present fundamental problems in the past, that is probably because resistance and preconception have usually been viewed as extraneous to science. They are, we have often been told, no more than the product of inevitable *human* limitations; a proper scientific method has no place for them; and that method is powerful enough so that no mere human idiosyncrasy can impede its success for very long. On this view, examples of a scientific *parti pris* are reduced to the status of anecdotes, and it is that evaluation of their significance that this essay aims to challenge. Verisimilitude, alone, suggests that such a challenge is required. Preconception and resistance seem the rule rather than the exception in mature scientific development. Furthermore, under normal circumstances they characterize the very best and most creative research as well as the more routine. Nor can there be much question where they come from. Rather than being characteristics of the aberrant individual, they are community characteristics with deep roots in the procedures through which scientists are trained for work in their profession. Strongly held convictions that are prior to research often seem to be a precondition for success in the sciences.

Obviously I am already ahead of my story, but in getting there I have perhaps indicated its principal theme. Though preconception and resistance to innovation could very easily choke off scientific progress, their omnipresence is nonetheless symptomatic of characteristics upon which the continuing vitality of research depends. Those characteristics I shall collectively call the dogmatism of mature science, and in the pages to come I shall try to make the following points about them. Scientific education inculcates what the scientific community had previously with difficulty gained—a deep commitment to a particular way of viewing the world and of practising science in it. That commitment can be, and from time to time is, replaced by another, but it cannot be merely given up. And, while it continues to characterize the community of professional practitioners, it proves in two respects fundamental to productive research. By defining for the individual scientist both the problems available for pursuit and the nature of acceptable solutions to them, the commitment is actually constitutive of research. Normally the scientist is a puzzle-solver like the chess player, and the commitment induced by education is what provides him with the rules of the game being played in his time. In its absence he would not be a physicist, chemist, or whatever he has been trained to be.

In addition, commitment has a second and largely incompatible research role. Its very strength and the unanimity with which the professional group subscribes to it provides the individual scientist with an immensely sensitive detector of the trouble spots from which significant innovations of fact and theory are almost inevitably educed. In the sciences most discoveries of unexpected fact and all fundamental innovations of theory are responses to a prior breakdown in the rules of the previously established game. Therefore, though a quasidogmatic commitment is, on the one hand, a source of resistance and controversy, it is also instrumental in making the sciences the most consistently revolutionary of all human activities. One need make neither resistance nor dogma a virtue to recognize that no mature science could exist without them.

Before examining further the nature and effects of scientific dogma, consider the pattern of education through which it is transmitted from one generation of practitioners to the next. Scientists are not, of course, the only professional community that acquires from education a set of standards, tools, and techniques which they later deploy in their own creative work. Yet even a cursory inspection of scientific pedagogy suggests that it is far more likely to induce professional rigidity than education in other fields, excepting, perhaps, systematic theology. Admittedly the following epitome is biased toward the American pattern, which I know best. The contrasts at which it aims must, however, be visible, if muted, in European and British education as well.

Perhaps the most striking feature of scientific education is that, to an extent quite unknown in other creative fields, it is conducted through textbooks, works written especially for students. Until he is ready, or very nearly ready, to begin his own dissertation, the student of chemistry, physics, astronomy, geology, or biology is seldom either asked to attempt trial research projects or exposed to the immediate products of research done by others—to, that is, the professional communications that scientists write for their peers. Collections of "source readings" play a negligible role in *scientific* education. Nor is the science student encouraged to read the historical classics of his field—works in which he might encounter other ways of regarding the questions discussed in his text, but in which he would also meet problems, concepts, and standards of solution that his future profession had long-since discarded and replaced.[2] Whitehead somewhere caught this quite special feature of the sciences when he wrote, "A science that hesitates to forget its founders is lost."

An almost exlusive reliance on textbooks is not all that distinguishes scientific education. Students in other fields are, after all, also exposed to such books, though seldom beyond the second year of college and even in those early years not exclusively. But in the sciences different textbooks display different subject matters rather than, as in the humanities and many social sicences, exemplifying different approaches to a single problem field. Even books that compete for adoption in a single science course differ mainly in level and pedagogic detail, not in substance or conceptual struture. One can scarcely imagine a physicist's or chemist's saying that he had been forced to begin the education of his third-year class almost from first principles because its previous exposure to the field had been through books that consistently violated his conception of the discipline. Remarks of that sort are not by any means unprecedented in several of the social sciences. Apparently scientists agree about what it is that every student of the field must know. That is why, in the design of a preprofessional curriculum, they can use textbooks instead of eclectic samples of research.

Nor is the characteristic technique of textbook presentation altogether the same in the sciences as elsewhere. Except in the occasional introductions that students seldom read, science texts make little attempt to describe the *sorts* of problems that the professional may be asked to solve or to discuss the *variety* of techniques that experience has made available for their solution. Instead, these books exhibit, from the very start, concrete problem-solutions that the profession has come to accept as paradigms, and they then ask the student, either with a pencil and paper or in the laboratory, to solve for himself problems closely modelled in method and substance upon those through which the text has led him. Only in elementary language instruction or in training a musical instrumentalist is so large or essential a use made of "finger exercises." And those are just the fields in which the object of instruction is to produce with maximum rapidity strong "mental sets" or *Einstellungen*. In the sciences, I suggest, the effect of these techniques is much the same. Though scientific development is particularly productive of consequential novelties, scientific education remains a relatively dogmatic initiation into a pre-established problemsolving tradition that the student is neither invited nor equipped to evaluate.

The pattern of systematic textbook education just described existed in no place and in no science (except perhaps elementary mathematics) until the early nineteenth century. But before that date a number of the more developed sciences clearly displayed the special characteristics indicated above, and in a few cases had done so for a very long time. Where there were no textbooks there had often been universally received paradigms for the practice of individual sciences. These were scientific achievements reported in books that all the practitioners of a given field knew intimately and admired, achievements upon which they modelled their own research and which provided them with a measure of their own accomplishment. Aristotle's *Physica*, Ptolemy's *Almagest*, Newton's *Principia* and *Opticks*, Franklin's *Electricity*, Lavoisier's *Chemistry*, and Lyell's *Geology*—these works and many others all served for a time implicitly to define the legitimate problems and methods of a research field for succeeding generations of practitioners. In their day each of these books, together with others modelled closely upon them, did for its field much of what textbooks now do for these same fields and for others besides.

All of the works named above are, of course, classics of science. As such their role may be thought to resemble that of the main classics in other creative fields, for example the works of a Shakespeare, a Rembrandt, or an Adam Smith. But by calling these works, or the achievements which lie behind them, paradigms rather than classics, I mean to suggest that there is something else special about them, something which sets them apart both from some other classics of science and from all the classics of other creative fields.

Part of this "something else" is what I shall call the exclusiveness of paradigms. At any time the practitioners of a given specialty may recognize numerous classics, some of them—like the works of Ptolemy and Copernicus or Newton and Descartes—quite incompatible one with the other. But that same group, if it has a paradigm at all, can have only one. Unlike the community of artists—which can draw simultaneous inspiration from the works of, say, Rembrandt *and* Cézanne and which therefore studies both—the community of astronomers had no alternative to choosing *between* the competing models of scientific activity supplied by Copernicus and Ptolemy. Furthermore, having made their choice, astronomers could thereafter neglect the work which they had rejected. Since the sixteenth century there have been only two full editions of the *Almagest*, both produced in the nineteenth century and directed exclusively to scholars. In the mature sciences there is no apparent function for the equivalent of an art museum or a library of classics. Scientists know when books, and even journals, are out of date. Though they do not then destroy them, they do, as any historian of science can testify, transfer them from the active departmental library to desuetude in the general university depository. Up-to-date works have taken their place, and they are all that the further progress of science requires.

This characteristic of paradigms is closely related to another, and one that has a particular relevance to my selection of the term. In receiving a paradigm the scientific community commits itself, consciously or not, to the view that the fundamental problems there resolved have, in fact, been solved once and for all. That is what Lagrange meant when he said of Newton: "There is but one universe, and it can happen to but one man in the world's history to be the interpreter of its laws."[3] The example of either Aristotle or Einstein proves Lagrange wrong, but that does not make the fact of his commitment less consequential to scientific development. Believing that what Newton had done need not be done again, Lagrange was not tempted to fundamental reinterpretations of nature. Instead, he could take up where the men who shared his Newtonian paradigm had left off, striving both for neater formulations of that paradigm and for an articulation that would bring it into closer and closer agreement with observations of nature. That sort of work is undertaken only by those who feel that the model they have chosen is entirely secure. There is nothing quite like it in the arts, and the parallels in the social sciences are at best partial. Paradigms determine a developmental pattern for the mature sciences that is unlike the one familiar in other fields.

That difference could be illustrated by comparing the development of a paradigm-based science with that of, say, philosophy or literature. But the same effect can be achieved more economically by contrasting the early developmental pattern of almost any science with the pattern characteristic of the same field in its maturity. I cannot here avoid putting the point too starkly, but what I have in mind is this. Excepting in those fields which, like biochemistry, originated in the combination of existing specialties, paradigms are a relatively late acquisition in the course of scientific development. During its early years a science proceeds without them, or at least without any so unequivocal and so binding as those named illustratively above. Physical optics before Newton or the study of heat before Black and Lavoisier exemplifies the pre-paradigm developmental pattern that I shall immediately examine in the history of electricity. While it continues, until, that is, a first paradigm is reached, the development of a science resembles that of the arts and of most social sicences more closely than it resembles the pattern which astronomy, say, had already acquired in Antiquity and which all the natural sciences make familiar today.

To catch the difference between pre- and post-paradigm scientific development consider a single example. In the early eighteenth century, as in the seventeenth and earlier, there were almost as many views about the nature of electricity as there were important electrical experimenters, men like Hauksbee, Gray, Desaguliers, Du Fay, Nollet, Watson, and Franklin. All their numerous concepts of electricity had something in common—they were partially derived from experiment and observation and partially from one or another version of the mechanico-corpuscular philosophy that guided all scientific research of the day. Yet these common elements gave their work no more than a family resemblance. We are forced to recognize the existence of several competing schools and subschools, each deriving strength from its relation to a particular version (Cartesian or Newtonian) of the corpuscular metaphysics, and each emphasizing the particular cluster of electrical phenomena which its own theory could do most to explain. Other observations were dealt with by *ad hoc* elaborations or remained as outstanding problems for further research.[4]

One early group of electricians followed seventeenth-century practice, and thus took attraction and frictional generation as the fundamental electrical phenomena. They tended to treat repulsion as a secondary effect (in the seventeenth century it had been attributed to some sort of mechanical rebounding) and also to postpone for as long as possible both discussion and systematic research on Gray's newly discovered effect, electrical conduction. Another closely related group regarded repulsion as the fundamental effect, while still another took attraction and repulsion together to be equally elementary manifestations of electricity. Each of these groups modified its theory and research accordingly, but they then had as much difficulty as the first in accounting for any but the simplest conduction effects. Those effects provided the starting point for still a third group, one which tended to speak of electricity as a "fluid" that ran through conductors rather than as an "effluvium" that emanated from nonconductors. This group, in its turn, had difficulty reconciling its theory with a number of attractive and repulsive effects.[5]

At various times all these schools made significant contributions to the body of concepts, phenomena, and techniques from which Franklin drew the first paradigm for electrical science. Any definition of the scientist that excludes the members of these schools will exclude their modern successors as well. Yet anyone surveying the development of electricity before Franklin may well conclude that, though the field's practitioners were scientists, the immediate result of their activity was something less than science. Because the body of belief he could take for granted was very small, each electrical experimenter felt forced to begin by building his field anew from its foundations. In doing so his choice of supporting observation and experiment was relatively free, for the set of standard methods and phenomena that every electrician must employ and explain was extraordinarily small. As a result, throughout the first half of the century, electrical investigations tended to circle back over the same ground again and again. New effects were repeatedly discovered, but many of them were rapidly lost again. Among those lost were many effects due to what we should now describe as inductive charging and also Du Fay's famous discovery of the two sorts of electrification. Franklin and Kinnersley were suprised when, some fifteen years later, the latter discovered that a charged ball which was repelled by rubbed glass would be attracted by rubbed sealing-wax or amber.[6] In the absence of a well-articulated and widely received theory (a desideratum which no science possesses from its very beginning and which few if any of the social sciences have achieved today), the situation could hardly have been otherwise. During the first half of the eighteenth century there was no way for electricians to distinguish consistently between electrical and non-electrical effects, between laboratory accidents and essential novelties, or between striking demonstration and experiments which revealed the essential nature of electricity.

This is the state of affairs which Franklin changed.[7] His theory explained so many—though not all—of the electrical effects recognized by the various earlier schools that within a generation all electricians had been converted to some view very like it. Thought it did not resolve quite all disagreements, Franklin's theory was electricity's first paradigm, and its existence gives a new tone and flavour to the electrical researches of the last decades of the eighteenth century. The end of interschool debate ended the constant reiteration of fundamentals; confidence that they were on the right track encouraged electricians to undertake more precise, esoteric, and consuming sorts of work. Freed from concern with any and all electrical phenomena, the newly united group could pursue selected phenomena in far more detail, designing much special equipment for the task and employing it more stubbornly and systematically than electricians had ever done before. In the hands of a Cavendish, a Coulomb, or a Volta the collection of electrical facts and the articulation of electrical theory were, for the first time, highly directed activities. As a result the efficiency and effectiveness of electrical research increased immensely, providing evidence for a societal version of Francis Bacon's acute methodological dictum: "Truth emerges more readily from error than from confusion."

Obvisouly I exaggerate both the speed and the completeness with which the transition to a paradigm occurs. But that does not make the phenomenon itself less real. The maturation of electricity as a science is not coextensive with the entire development of the field. Writers on electricity during the first four decades of the eighteenth century possessed far more information about elec-electrical phenomena than had their sixteenth- and seventeenth-century predecessors. During the half-century after 1745 very few new sorts of electrical phenomena were added to their lists. Nevertheless, in important respects the electrical writings of the last two decades of the century seemed further removed from those of Gray, Du Fay, and even Franklin than are the writings of these early eighteenth-century electricians from those of their predecessors a hundred years before. Some time between 1740 and 1780 electricians, as a group, gained what astronomers had achieved in Antiquity, students of motion in the Middle Ages, of physical optics in the late seventeenth century, and of historical geology in the early nineteenth. They had, that is, achieved a paradigm, possession of which enabled them to take the foundation of their field for granted and to push on to more concrete and recondite problems.[8] Except with the advantage of hindsight, it is hard to find another criterion that so clearly proclaims a field of science.

These remarks should begin to clarify what I take a paradigm to be. It is, in the first place, a fundamental scientific achievement and one which includes both a theory and some exemplary applications to the results of experiment and observation. More important, it is an open-ended achievement, one which leaves all sorts of research still to be done. And, finally, it is an accepted achievement in the sense that it is received by a group whose members no longer try to rival it or to create alternates for it. Instead, they attempt to extend and exploit it in a variety of ways to which I shall shortly turn. That discussion of the work that pardigms leave to be done will make both their role and the reasons for their special efficacy clearer still. But first there is one rather different point to be made about them. Though the reception of a paradigm seems historically prerequisite to the most effective sorts of scientific research, the paradigms which enhance research effectiveness need not be and usually are not permanent. On the contrary, the developmental pattern of mature science is usually from paradigm to paradigm. It differs from the pattern characteristic of the early or pre-paradigm period not by the total elimination of debate over fundamentals, but by the drastic restriction of such debate to occasional periods of paradigm change.

Ptolemy's *Almagest* was not, for example, any less a paradigm because the research tradition that descended from it had ultimately to be replaced by an incompatible one derived from the work of Copernicus and Kepler. Nor was

Newton's *Opticks* less a paradigm for eighteenth-century students of light because it was later replaced by the ether-wave theory of Young and Fresnel, a paradigm which in its turn gave way to the electromagnetic displacement theory that descends from Maxwell. Undoubtedly the research work that any given paradigm permits results in lasting contributions to the body of scientific knowledge and technique, but paradigms themselves are very often swept aside and replaced by others that are quite incompatible with them. We can have no recourse to notions like the "truth" or "validity" of paradigms in our attempt to understand the special efficacy of the research which their reception permits.

On the contrary, the historian can often recognize that in declaring an older paradigm out of date or in rejecting the approach of some one of the pre-paradigm schools a scientific community has rejected the embryo of an important scientific perception to which it would later be forced to return. But it is very far from clear that the profession delayed scientific development by doing so. Would quantum mechanics have been born sooner if nineteenth-century scientists had been more willing to admit that Newton's corpuscular view of light might still have something significant to teach them about nature? I think not, although in the arts, the humanities, and many social sciences that less doctrinaire view is very often adopted toward classic achievements of the past. Or would astronomy and dynamics have advanced more rapidly if scientists had recognzed that Ptolemy and Copernicus had chosen equally legitimate means to describe the earth's position? That view was, in fact, suggested during the seventeenth century, and it has since been confirmed by relativity theory. But in the interim it was firmly rejected together with Ptolemaic astronomy, emerging again only in the very late nineteenth century when, for the first time, it had concrete relevance to unsolved problems generated by the continuing practice of nonrelativistic physics. One could argue, as indeed by implication I shall, that close eighteenth- and nineteenth-century attention either to the work of Ptolemy or to the relativistic views of Descartes, Huygens, and Leibniz would have delayed rather than accelerated the revolution in physics with which the twentieth century began. Advance from paradigm to paradigm rather than through the continuing competition between recognized classics may be a functional as well as a factual characteristic of mature scientific development.

Much that has been said so far is intended to indicate that—except during occasional extraordinary periods to be discussed in the last section of this paper—the practitioners of a mature scientific specialty are deeply committed to some one paradigm-based way of regarding and investigating nature. Their paradigm tells them about the sorts of entities with which the universe is populated and about the way the members of that population behave; in addition, it informs them of the questions that may legitimately be asked about nature and of the techniques that can properly be used in the search for answers to them. In fact, a paradigm tells scientists so much that the questions it leaves for research seldom have great intrinsic interest to those outside the profession. Though educated men as a group may be fascinated to hear about the spectrum of fundamental particles or about the processes of molecular replication, their interest is usually quickly exhausted by an account of the beliefs that already underlie research on these problems. The outcome of the individual research project is indifferent to them, and their interest is unlikely to awaken again until, as with parity nonconservation, research unexpectedly leads to paradigm-change and to a consequent alteration in the beliefs which guide research. That, no doubt, is why both historians and popularizers have devoted so much of their attention to the revolutionary episodes which result in change of paradigm and have so largely neglected the sort of work that even the greatest scientists necessarily do most of the time.

My point will become clearer if I now ask what it is that the existence of a paradigm leaves for the scientific community to do. The answer—as obvious as the related existence of resistance to innovation and as often brushed under the carpet—is that scientists, given a paradigm, strive with all their might and skill to bring it into closer and closer agreement with nature. Much of their effort, particularly in the early stages of a paradigm's development, is directed to articulating the paradigm, rendering it more precise in areas where the original formulation has inevitably been vague. For example, knowing that electricity was a fluid whose individual particles act upon one another at a distance, electricians after Franklin could attempt to determine the quantitative law of force between particles of electricity. Others could seek the mutual interdependence of spark length, electroscope deflection, quantity of electricity, and conductor-configuration. These were the sorts of problems upon which Coulomb, Cavendish, and Volta worked in the last decades of the eighteenth century, and they have many parallels in the development of every other mature science. Contemporary attempts to determine the quantum mechancial forces governing the interactions of nucleons fall precisely in this same category, paradigm-articulation.

That sort of problem is not the only challenge which a paradigm sets for the community that embraces it. There are always many areas in which a paradigm is assumed to work but to which it has not, in fact, yet been applied. Matching the paradigm to nature in these areas often engages much of the best scientific talent in any generation. The eighteenth-century attempts to develop a Newtonian theory of vibrating strings provide one significant example, and the current work on a quantum mechanical theory of solids provides another. In addition, there is always much fascinating work to be done in improving the match between a paradigm and nature in an area where at least limited agreement has already been demonstrated. Theoretical work on problems like these is illustrated by eighteenth-century research on the perturbations that cause planets to deviate from their Keplerian orbits as well as by the elaborate twentieth-century theory of the spectra of complex atoms and molecules. And accompanying all these problems and still others besides is a recurring series of instrumental hurdles. Special apparatus had to be invented and built to permit Coulomb's determination of the electrical force law. New sorts of telescopes were required for the observations that, when completed, demanded an improved

Newtonian perturbation theory. The design and construction of more flexible and more powerful accelerators is a continuing desideratum in the attempt to articulate more powerful theories of nuclear forces. These are the sorts of work on which almost all scientists spend almost all of their time.[9]

Probably this epitome of normal scientific research requires no elaboration in this place, but there are two points that must now be made about it. First, all of the problems mentioned above were paradigm-dependent, often in several ways. Some—for example the derivation of perturbation terms in Newtonian planetary theory—could not even have been stated in the absence of an appropriate paradigm. With the transition from Newtonian to relativity theory a few of them became different problems and not all of these have yet been solved. Other problems—for example the attempt to determine a law of electric forces—could be and were at least vaguely stated before the emergence of the paradigm with which they were ultimately solved. But in that older form they proved intractable. The men who described electrical attractions and repulsions in terms of effluvia attempted to measure the resulting forces by placing a charged disc at a measured distance beneath one pan of a balance. Under those circumstances no consistent or interpretable results were obtained. The prerequisite for success proved to be a paradigm that reduced electrical action to a gravity-like action between point particles at a distance. After Franklin electricians thought of electrical action in those terms; both Coulomb and Cavendish designed their apparatus accordingly. Finally, in both these cases and in all the others as well a commitment to the paradigm was needed simply to provide adequate motivation. Who would design and build elaborate special-purpose apparatus, or who would spend months trying to solve a particular differential equation, without a quite firm guarantee that his effort, if successful, would yield the anticipated fruit?

This reference to the anticipated outcome of a research project points to the second striking characteristic of what I am now calling normal, or paradigm-based, research. The scientist engaged in it does not at all fit the prevalent image of the scientist as explorer or as inventor of brand new theories which permit striking and unexpected predictions. On the contrary, in all the problems discussed above everything but the detail of the outcome was known in advance. No scientist who accepted Franklin's paradigm could doubt that there was a law of attraction between small particles of electricity, and they could reasonably suppose that it would take a simple algebraic form. Some of them had even guessed that it would prove to be an inverse square law. Nor did Newtonian astronomers and physicists doubt that Newton's law of motion and of gravitation could ultimately be made to yield the observed motions of the moon and planets even though, for over a century, the complexity of the requisite mathematics prevented good agreement's being uniformly obtained. In all these problems, as in most others that scientists undertake, the challenge is not to uncover the unknown but to obtain the known. Their fascination lies not in what success may be expected to disclose but in the difficulty of obtaining success at all. Rather than resembling exploration, normal research seems like the effort to assemble a Chinese cube whose finished outline is known from the start.

Those are the characteristics of normal research that I had in mind when, at the start of this essay, I described the man engaged in it as a puzzle-solver, like the chess player. The paradigm he has acquired through prior training provides him with the rules of the game, describes the pieces with which it must be played, and indicates the nature of the required outcome. His task is to manipulate those pieces within the rules in such a way that the required outcome is produced. If he fails, as most scientists do in at least their first attacks upon any given problem, that failure speaks only to his lack of skill. It cannot call into question the rules which his paradigm has supplied, for without those rules there would have been no puzzle with which to wrestle in the first place. No wonder, then, that the problems (or puzzles) which the practitioner of a mature science normally undertakes presuppose a deep commitment to a paradigm. And how fortunate it is that that commitment is not lightly given up. Experience shows that, in almost all cases, the reiterated efforts, either of the individual or of the professional group, do at last succeed in producing within the paradigm a solution to even the most stubborn problems. That is one of the ways in which science advances. Under those circumstances can we be surprised that scientists resist paradigm-change? What they are defending is, after all, neither more nor less than the basis of their professional way of life.

By now one principal advantage of what I began by calling scientific dogmatism should be apparent. As a glance at any Baconian natural history or a survey of the pre-paradigm development of any science will show, nature is vastly too complex to be explored even approximately at random. Something must tell the scientist where to look and what to look for, and that something, though it may not last beyond his generation, is the paradigm with which his education as a scientist has supplied him. Given that paradigm and the requisite confidence in it, the scientist largely ceases to be an explorer at all, or at least to be an explorer of the unknown. Instead, he struggles to articulate and concretize the known, designing much special-purpose apparatus and many special-purpose adaptations of theory for that task. From those puzzles of design and adaptation he gets his pleasure. Unless he is extraordinarily lucky, it is upon his success with them that his reputation will depend. Inevitably the enterprise which engages him is characterized, at any one time, by drastically restricted vision. But within the region upon which vision is focused the continuing attempt to match paradigms to nature results in a knowledge and understanding of esoteric detail that could not have been achieved in any other way. From Copernicus and the problem of precession to Einstein and the photoelectric effect, the progress of science has again and again depended upon just such esoterica. One great virtue of commitment to paradigms is that it frees sceintists to engage themselves with tiny puzzles.

Nevertheless, this image of scientific research as puzzle-solving or paradigm-matching must be, at the very least, thoroughly incomplete. Though the scientist may not be an explorer, scientists do again and again discover new and unexpected sorts of phenomena. Or again, though the scientist does not normally strive to invent new sorts of basic theories, such theories have repeatedly emerged from the continuing practice of research. But neither of these types of innovation would arise if the enterprise I have been calling normal science were always successful. In fact, the man engaged in puzzle-solving very often resists substantive novelty, and he does so for good reason. To him it is a change in the rules of the game and any change of rules is intrinsically subversive. That subversive element is, of course, most apparent in major theoretical innovations like those associated with the names of Copernicus, Lavoisier, or Einstein. But the discovery of an unanticipated phenomenon can have the same destructive effects although usually on a smaller group and for a far shorter time. Once he had performed his first follow-up experiments, Röntgen's glowing screen demonstrated that previously standard cathode ray equipment was behaving in ways for which no one had made allowance. There was an unanticipated variable to be controlled; earlier researches, already on their way to becoming paradigms, would require re-evaluation; old puzzles would have to be solved again under a somewhat different set of rules. Even so readily assimilable a discovery as that of Xrays can violate a paradigm that has previously guided research. It follows that, if the normal puzzle-solving activity were altogether successful, the development of science could lead to no fundamental innovations at all.

But of course normal science is not always successful, and in recognizing that fact we encounter what I take to be the second great advantage of paradigm-based research. Unlike many of the early electricians, the practitioner of a mature science knows with considerable precision what sort of result he should gain from his research. As a consequence he is in a particularly favourable position to recognize when a research problem has gone astray. Perhaps, like Galvani or Röntgen, he encounters an effect that he knows ought not to occur. Or perhaps, like Copernicus, Planck, or Einstein, he concludes that the reiterated failures of his predecessors in matching a paradigm to nature is presumptive evidence of the need to change the rules under which a match is to be sought. Or perhaps, like Franklin or Lavoisier, he decides after repeated attempts that no existing theory can be articulated to account for some newly discovered effect. In all of these ways and in others besides the practice of normal puzzle-solving science can and inevitably does lead to the isolation and recognition of anomaly. That recognition proves, I think, prerequisite for almost all discoveries of new sorts of phenomena and for all fundamental innovations in scientific theory. After a first paradigm has been achieved, a breakdown in the rules of the pre-established game is the usual prelude to significant scientific innovation.

Examine the case of discoveries first. Many of them, like Coulomb's law or a new element to fill an empty spot in the periodic table, present no problem. They were not "new sorts of phenomena," but discoveries anticipated through a paradigm and achieved by expert puzzle-solvers: That sort of discovery is a natural product of what I have been calling normal science. But not all discoveries are of that sort: Many could not have been anticipated by any extrapolation from the known; in a sense they had to be made "by accident." On the other hand the accident through which they emerged could not ordinarily have occurred to a man just looking around. In the mature sciences discovery demands much special equipment, both conceptual and instrumental, and that special equipment has invariably been developed and deployed for the pusuit of the puzzles of normal research. Discovery results when that equipment fails to function as it should. Furthermore, since some sort of at least temporary failure occurs during almost every research project, discovery results only when the failure is particularly stubborn or striking and only when it seems to raise questions about accepted beliefs and procedures. Established paradigms are thus often doubly prerequisite to discoveries. Without them the project that goes astray would not have been undertaken. And even when the project has gone astray, as most do for a while, the paradigm can help to determine whether the failure is worth pursuing. The usual and proper response to a failure in puzzle-solving is to blame one's talents or one's tools and to turn next to another problem. If he is not to waste time, the scientist must be able to discriminate essential anomaly from mere failure.

That pattern—discovery through an anomaly that calls established techniques and beliefs in doubt—has been repeated again and again in the course of scientific development. Newton discovered the composition of white light when he was unable to reconcile measured dispersion with that predicted by Snell's recently discovered law of refraction.[10] The electric battery was discovered when existing detectors of static charges failed to behave as Franklin's paradigm said they should.[11] The planet Neptune was dicovered through an effort to account for recognized anomalies in the orbit of Uranus.[12] The element chlorine and the compound carbon monoxide emerged during attempts to reconcile Lavoisier's new chemistry with laboratory observations.[13] The so-called noble gases were the products of a long series of investigations initiated by a small but persistent anomaly in the measured density of atmospheric nitrogen.[14] The electron was posited to explain some anomalous properties of electrical conduction through gases, and its spin was suggested to account for other sorts of anomalies observed in atomic spectra.[15] Both the neutron and the neutrino provide other examples, and the list could be extended almost indefinitely.[16] In the mature sciences unexpected novelties are discovered principally after something has gone wrong.

If, however, anomaly is significant in preparing the way for new discoveries, it plays a still larger role in the invention of new theories. Contrary to a prevalent, though by no means universal, belief, new theories are not invented to account for observations that have not previously been ordered by theory at all. Rather, at almost all times in the development of any advanced science, all the facts whose relevance is admitted seem either to fit existing theory well or to be in the process of conforming. Making them conform better provides many of the standard problems of normal science. And almost always committed scientists succeed in solving them. But they do not always succeed, and, when they fail repeatedly and in increasing numbers, then their sector of the scientific community encounters what I am elsewhere calling "crisis." Recognizing that something is fundamentally wrong with the theory upon which their work is based, scientists will attempt more fundamental articulations of theory than those which were admissible before. (Characteristically, at times of crisis, one encounters numerous different versions of the paradigm theory.)[17] Simultaneously they will often begin more nearly random experimentation within the area of difficulty hoping to discover some effect that will suggest a way to set the situation right. Only under circumstances like these, I suggest, is a fundamental innovation in scientific theory both invented and accepted.

The state of Ptolemaic astronomy was, for example, a recognized scandal before Copernicus proposed a basic change in astronomical theory, and the preface in which Copernicus described his reasons for innovation provides a classic description of the crisis state.[18] Galileo's contributions to the study of motion took their point of departure from recognized difficulties with medieval theory, and Newton reconciled Galileo's mechanics with Copernicanism.[19] Lavoisier's new chemistry was a product of the anomalies created jointly by the proliferation of new gases and the first systematic studies of weight relations.[20] The wave theory of light was developed amid growing concern about anomalies in the relation of diffraction and polarization effects to Newton's corpuscular theory.[21] Thermodynamics, which later came to seem a superstructure for existing sciences, was established only at the price of rejecting the previously paradigmatic caloric theory.[22] Quantum mechanics was born from a variety of difficulties surrounding black-body radiation, specific heat, and the photoelectric effect.[23] Again the list could be extended, but the point should already be clear. New theories arise from work conducted under old ones, and they do so only when something is observed to have gone wrong. Their prelude is widely recognized anomaly, and that recognition can come only to a group that knows very well what it would mean to have things go right.

Because limitations of space and time force me to stop at this point, my case for dogmatism must remain schematic. I shall not here even attempt to deal with the fine-structure that scientific development exhibits at all times. But there is another more positive qualification of my thesis, and it requires one closing comment. Though successful research demands a deep commitment to the status quo, innovation remains at the heart of the enterprise. Scientists are *trained* to operate as puzzle-solvers from established rules, but they are also *taught* to regard themselves as explorers and inventors who know no rules except those dictated by nature itself. The result is an acquired tension, partly within the individual and partly within the community, between professional skills on the one hand and professional ideology on the other. Almost certainly that tension and the ability to sustain it are important to science's success. In so far as I have dealt exclusively with the dependence of research upon tradition, my discussion is inevitably one-sided. On this whole subject there is a great deal more to be said.

But to be one-sided is not necessarily to be wrong, and it may be an essential preliminary to a more penetrating examination of the requisites for successful scientific life. Almost no one, perhaps no one at all, needs to be told that the vitality of science depends upon the continuation of occasional tradition-shattering innovations. But the apparently contrary dependence of research upon a deep commitment to established tools and beliefs receives the very minimum of attention. I urge that it be given more. Until that is done, some of the most striking characteristics of scientific education and development will remain extraordinarily difficult to understand.

NOTES

1. *Wissenschaftliche Selbstbiographie* (Leipzig, 1948), p. 22, my translation.

2. The individual sciences display some variation in these respects. Students in the newer and also in the less theoretical sciences, e.g., parts of biology, geology, and medical science, are more likely to encounter both contemporary and historical source materials than those in, say, astronomy, mathematics, or physics.

3. Quoted in this form by S. F. Mason, *Main Currents of Scientific Thought* (New York, 1956), p. 254. The original, which is identical in spirit but not in words, seems to derive from Delambre's contemporary éloge, *Memoires de . . . l'Institut . . . , année 1812*, Part II (Paris, 1816), p. xlvi.

4. Much documentation for this account of electrical development can be retrieved from Duane Roller and Duane H. D. Roller, *The Development of the Concept of Electric Charge: Electricity from the Greeks to Coulomb*, Harvard Case Histories in Experimental Science, VIII (Cambridge, Mass., 1954) and from I. B. Cohen, *Franklin and Newton: An Inquiry into Speculative Newtonian Experimental Science and Franklin's Work in Electricity as an Example Thereof* (Philadelphia, 1956). For analytic detail I am, however, very much indebted to a still unpublished paper by my student, John L. Heilbron, who has also assisted in the preparation of the three notes that follow.

5. This division into schools is still somewhat too simplistic. After 1720 the basic division is between the French school (Du Fay, Nollet, etc.) who base their theories on attraction-repulsion effects and the English school (Desaguliers, Watson, etc.) who concentrate on conduction effects. Each group had immense difficulty in explaining the phenomena that the other took to be basic. (See, for example, Needham's report of Lemonier's investigations, in *Philosophical Transactions*, XLIV (1746), 247. Within each of these groups, and particularly the English, one can trace further subdivision depending upon whether attraction or repulsion is considered the more fundamental electrical effect.

6. Du Fay's discovery that there are two sorts of electricity and that these are mutually attractive but self-repulsive is reported and documented in great experimental detail in the fourth of his famous memoirs on electricity: "De l'Attraction & Répulsion des Corps Electriques," *Memoires de . . . l'Academie . . . de l'annee 1733* (Paris, 1735), 457–76. These memoirs were well known and widely cited, but Desaguliers seems to be the only electrician who, for almost two decades, even mentions that some charged bodies will attract each other (*Philosophical Transactions . . .*, XLII, 1741–42, 140–43). For Franklin's and Kinnersley's "surprise" see I. B. Cohen, ed., *Benjamin Franklin's Experiments: A New Edition of Franklin's Experiments and Observations on Electricity* (Cambridge, Mass., 1941), 250–55. Note also that, though Kinnersley had *produced* the effect, neither he nor Franklin seems ever to have *recognized* that two resinously charged bodies would repel each other, a phenomenon directly contrary to Franklin's theory.

7. The change is not, of course, due to Franklin alone nor did it occur overnight. Other electricians, most notably William Watson, anticipated parts of Franklin's theory. More important, it was only after essential modifications, due principally to Aepinus, that Franklin's

theory gained the general currency requisite for a paradigm. And even then there continued to be two formulations of the theory: the Franklin-Aepinus one-fluid form and a two-fluid form due principally to Symmer. Electricians soon reached the conclusion that no electrical test could possibly discriminate between the two theories. Until the discovery of the battery, when the choice between a one-fluid theory began to make an occasional difference in the design and analysis of experiments, the two were equivalent.

8. Note that this first electrical paradigm was fully effective only until 1800, when the discovery of the battery and the multiplication of electro-chemical effects initiated a revolution in electrical theory. Until a new paradigm emerged from that revolution, the literature of electricity, particularly in England, reverted in many respects to the tone characteristic of the first half of the eighteenth century.

9. The discussion in this paragraph and the next is considerably elaborated in my paper, "The function of measurement in modern physical science," *Isis*, LII (1961), 161–93.

10. See my "Newton's optical papers" in *Isaac Newton's Papers & Letters on Natural Philosophy*, ed. I. B. Cohen (Cambridge, Mass., 1958), pp. 27–45.

11. Luigi Galvani, *Commentary on the Effects of Electricity on Muscular Motion*, trans. M. G. Foley with notes and an introduction by I. B. Cohen (Norwalk, Conn., 1954), pp. 27–29.

12. Angus Armitage, *A Century of Astronomy* (London, 1950), pp. 111–15.

13. For chlorine see Ernst von Meyer, *A History of Chemistry from the Earliest Times to the Present Day*, trans. G. M'Gowan (London, 1891), pp. 224–27. For carbon monoxide see Hermann Kopp, *Geschichte der Chemie* (Braunschweig, 1845), III, 294–96.

14. William Ramsay, *The Gases of the Atmosphere: the History of their Discovery*, London, 1896), Chaps. iv and v.

15. J. J. Thomson, *Recollections and Reflections* (New York, 1937), pp. 325–71; T. W. Chalmers, *Historic Researches: Chapters in the History of Physical and Chemical Discovery* (London, 1949), pp. 187–217; and F. K. Richtmeyer, E. H. Kennard and T. Lauritsen, *Introduction to Modern Physics*, 5th ed. (New York, 1955), p. 212.

16. *Ibid.*, pp. 466–70; and Rogers D. Rusk, *Introduction to Atomic and Nuclear Physics* (New York, 1958), pp. 328–30.

17. One classic example, for which see the reference cited below in the next note, is the proliferation of geocentric astronomical systems in the years before Copernicus's heliocentric reform. Another, for which see J. R. Partington and D. McKie, "Historical studies of the phlogiston theory," *Annals of Science*, II (1937), 361–404, III (1938), 1–58, 337–71, and IV (1939), 113–49, is the multiplicity of "phlogiston theories" produced in response to the general recognition that weight is always gained on combustion and to the experimental discovery of many new gases after 1760. The same proliferation of versions of accepted theories occurred in mechanics and electromagnetism in the two decades preceding Einstein's special relativity theory. (E. T. Whittaker, *History of the Theories of Aether and Electricity* (2nd ed.), 2 Vols. (London, 1951–53), I, Chap. xii, and II, Chap. ii. I concur in the widespread judgment that this is a very biased account of the genesis of relativity theory, but it contains just the detail necessary to make the point here at issue.)

18. T. S. Kuhn, *The Copernican Revolution: Planetary Astronomy in the Development of Western Thought* (Cambridge, Mass., 1957), pp. 133–40.

19. For Galileo see Alexandre Koyré, *Etudes Galiléennes*, 3 Vols. (Paris, 1939); for Newton see Kuhn, *op. cit.* pp. 228–60 and 289–91.

20. For the proliferation of gases see Partington, *A Short History of Chemistry* (2nd ed.) (London, 1948), Chap. vi; for the role of weight relations, see Henry Guerlac, "The origin of Lavoisier's work on combustion," *Archives internationales d'histoire des sciences*, XII (1959), 113–35.

21. Whittaker, *Aether and Electricity*, II, 94–109; William Whewell, *History of the Inductive Sciences* (revised ed.), 3 Vols. (London, 1847), II, 213–71; and Kuhn, "Function of measurement," p. 181 n.

22. For a general account of the beginnings of thermodynamics (including much relevant bibliography) see my "Energy conservation as an example of simultaneous discovery" in *Critical Problems in the History of Science*, ed. Marshall Clagett (Madison, Wisc., 1959), pp. 321–56. For the special problems presented to caloric theorists by energy conservation see the Carnot papers, there cited in n. 2, and also S. P. Thompson, *The Life of William Thomson, Baron Kelvin of Largs*, 2 Vols. (London, 1910), Chap. vi.

23. Richtmeyer et al., *Modern Physics*, pp. 89–94; 124–32, and 409–14; Gerald Holton, *Introduction to Concepts and Theories in Physical Science* (Cambridge, Mass., 1953), pp. 528–45.

part 3

THE CONFIRMATION OF SCIENTIFIC HYPOTHESES

INTRODUCTION

We have assumed so far that the scientist has at his disposal a collection of laws and theories that he can use to explain and/or predict what he observes. It is now necessary to analyze the process by which the scientist acquires this assemblage. Considering that at any given moment, there may be many alternative hypotheses that have been proposed, how does the scientist know which, if any, of these hypotheses are true? What justification can the scientist give for claiming that certain proposed laws and theories are correct? This is the problem of the confirmation of scientific hypotheses.

Many of the classical philosophers of science have proposed the following model (often called the *hypothetico-deductive model*) of the confirmation of scientific hypotheses: The scientist deduces from a hypothesis that he wants to test an observable consequence of that hypothesis. He then runs an experiment to see if this hypothesis holds. If it does not, he knows the hypothesis if false, but if it does, then the scientist has a confirming instance for his hypothesis. As the scientist finds more confirming instances for his hypothesis, the probability or degree of confirmation of that hypothesis rises, and when he has accumulated a sufficient number and variety of confirming instances, he is justified in accepting the hypothesis as being true.

This model for the confirmation of scientific hypotheses is clearly preferable to the simpler model of *induction by enumeration*. According to this latter model, hypotheses of the form "All *A*'s are *B*'s" are acceptable when one has examined a sufficient number of *A*'s and seen that they are all *B*'s. The trouble with this simpler model is that it is only applicable to hypotheses of a particular form which are such that one can determine by observation whether or not objects have properties *A* and *B*. The hypothetico-deductive model, on the other hand, seems applicable to all hypotheses. For this reason, most contemporary philosophers of science have adopted some version of the hypothetico-deductive model of confirmation, and recent work in the theory of confirmation has been devoted to an elaboration and formalization of this intuitive model.

Professor Hempel's "Studies in the Logic of Confirmation" offers an account of the nature of confirming instances. In particular, he defines a function that indicates whether some evidence confirms, disconfirms, or is irrelevant to a given hypothesis. Hempel offers several conditions that must be satisfied by any acceptable confirmation function, argues that various functions that have been proposed are incorrect precisely because they fail to satisfy these conditions, and then offers a new function, the *satisfaction function*, which is intuitively plausible and which does satisfy his conditions of adequacy.

In a recent paper, I argue that Hempel's set of conditions are not sufficient because there are legitimate and necessary evidential inferences that might not be allowed by a confirmation function that merely satisfied Hempel's conditions. In order to meet this problem, I suggest an additional condition that should be met by any satisfactory confirmation function and I suggest various ways for defining a function that satisfies this condition as well as Hempel's original conditions. The reader will have to decide for himself whether confirmation functions should satisfy this additional condition, and if so, how one might define a function that did this.

Hempel raises in his paper a related problem about the notion of a confirming instance which has been widely discussed in recent years. This is the "paradox of the ravens." Hempel shows that the adoption of a function which (a) satisfies his equivalence condition, and (b) satisfies our intuitive feeling that the observation of an *A* that is a *B* confirms the hypothesis that all *A*'s are *B*'s, leads to the seemingly paradoxical result that the observation of a white table confirms the hypothesis that all ravens are black. Hempel argues that this result should be accepted and that its paradoxical nature should be ascribed to a psychological illusion.

Many philosophers, however, have not been convinced by his arguments and are unwilling to accept this paradoxical result. They have therefore attempted to construct alternative accounts of confirming evidence that avoid this difficulty. One account, suggested by Nelson Goodman and developed in a somewhat different fashion by Richard Grandy, asserts that confirming evidence for a hypothesis must not merely be in accord with that hypothesis; it must also rule out alternatives to that hypothesis. Employing this general idea, Goodman and Grandy by different methods attempt to rule out at least some of the paradoxical confirmation relations that Hempel allows. A different account, offered by J.W.N. Watkins, is that a confirming instance is one in which the evidence, instead of merely being in accord with a hypothesis, arises out of a real test of that hypothesis. Watkins proceeds to argue that such an approach implies that the observation of a non-black non-raven can confirm the raven hypothesis, but only in those cases in which this confirmation is intuitively plausible.

It is clear by now that even the most basic notion of the hypothetico-deductive model, the notion of a confirming instance, is difficult to define precisely. But as mentioned above, the hypothetico-deductive model also employs the notion of the degree of confirmation of a hypothesis. It claims that the degree of confirmation of a hypothesis rises as we get additional confirming evidence for that hypothesis. Is it possible to offer a formal precise account of the notion of the degree of confirmation of a hypothesis, given a certain body of evidence? There have been many attempts to do this, but Professor Carnap's account has attracted the most interest. The two selections from his writtings reprinted in this section present both a formal and an informal account of his approach.

Carnap defines a function that states, for any given hypothesis and body of evidence, the degree of confirmation of that hypothesis on the basis of that body of evidence. He argues that such functions should be both regular and symmetrical (these two requirements are carefully explained in his second selection). There are, however, many functions that satisfy both of these requirements, and in the selections following, Carnap discusses two of them, c^* and c_w. He argues that c^* is preferable to c_w because the former, but not the latter, allows us to learn from experience, i.e., to modify our expectations in light of the observations we make. Carnap points out, however, that c^* is not the only regular and symmetrical function that enables us to learn from experience. It therefore remains an open question for the Carnapian school to find some justification for adopting one of these many possible functions as a definition of "degree of confirmation."

At the end of Carnap's second article, he shows how we can understand a variety of inductive inferences in light of his account of degrees of confirmation. One type of inductive inference presents a special problem for Carnap, however: This is the inference from an observed sample to a hypothesis of universal form, an inference that seems to be fundamental to the confirmation of scientific laws and theories. Unfortunately, the larger the universe in question, the smaller the value, in any of Carnap's systems, of the degree of confirmation of a universal hypothesis. In an infinite universe, all universal statements, no matter how much evidence we collect, have a zero degree of confirmation. It would seem, therefore, that if we adopt a Carnapian system of inductive logic, we must conclude that scientific laws and theories are never well-confirmed. Because this result has seemed so counter-intuitive to many philosophers, they have questioned, along with Professor Nagel, Carnap's whole approach.

Carnap, in attempting to resolve this difficulty, claims that the scientist is really concerned with the predictions he can derive from the law, but not with the law itself, and these predictions do not have the low (or zero) degree of confirmation that the law itself has. Nagel, in reply, presents several arguments to show that Carnap's solution does not work. As usual, the reader will have to decide who is right; but if he decides that Carnap is wrong, he must not then conclude that there is no way of constructing an inductive logic based on degrees of confirmation that allows for the confirmation of laws and theories. Professor Hintikka presents an alternative degree of confirmation function that allows universal statements, even in an infinite universe, to have a high degree of confirmation on the basis of the appropriate confirming evidence. The whole problem of the degree of confirmation of laws and theories is clearly open to much further analysis.

Many additional objections have been raised against Carnap's work, and the reader should consult Nagel's article for a clear account of them. But there is one fundamental problem that Goodman has raised which needs to be elaborated upon here. Consider the hypothesis "all emeralds are green." It seems natural to suppose that the observation of green emeralds confirms that hypothesis, and that, given the fact that people have observed emeralds under many diverse conditions and found them all green, the hypothesis in question has a high degree of confirmation. But now consider the hypothesis that "all emeralds are grue," where *grue* means "examined before *t* and green or not examined before *t* and blue" (where *t* is some time after now). Since all the emeralds that we have examined are grue as well as green, it looks as though we are forced to conclude against our intuitions that our observations of emeralds confirm the grue hypothesis as well as the green hypothesis, and that the grue hypothesis has the same high degree of confirmation as the green hypothesis. Goodman points out, however, that this highly counter-intuitive result leads to the disastrous consequence that the observations we make offer an equal degree of confirmation to any prediction we care to make about any object or event. Any theory of confirming evidence or of degrees of confirmation must, therefore, find a way to prevent hypotheses like the grue hypothesis from being confirmed in the same way that ordinary hypotheses are confirmed. Goodman calls this problem the *new riddle of induction.*

Most philosophers considering this problem have attempted to solve it by finding some significant distinction between the predicates "green" and "grue" which can serve as the basis for ruling out the confirmation of hypotheses containing predicates like "grue" in the distinguishing respect. Carnap, for example, has argued that hypotheses containing grue-like predicates involve a reference to specific temporal positons, and this enables us to distinguish them from legitimate hypotheses that can be confirmed. Goodman shows, however, that this distinction does not exist. Recently, Professors Peter Achinstein and Steven Barker have argued that there is still an important sense in which "grue," but not "green," is a temporal predicate. Their point is that in order to apply "grue," but not "green," to a particular object, one must know what the date is, and it is in this sense that "grue," but not "green," is a temporal predicate. Goodman, while rejecting this argument, agrees that the way to solve the problem is to distinguish between types of predicates. In his book, *Fact, Fiction, and Forecast*, he tries to draw the distinction in terms of the extent to which the predicates in question (or predicates co-extensive with them) have been previously used in other hypotheses.

Not all philosophers have agreed that the way to solve Goodman's problem is to distinguish between types of predicates. Perhaps the best critique of this approach is found in Professor Haskell Fain's recent article. He suggests, as an alternative approach, that the problem can be solved only if we allow for a difference in the initial probability of the two hypotheses, and this can only be justified in light of other information. Whether this alternative approach is superior to the standard approach is a question that the reader will have to decide for himself.

We have so far discussed two of the basic concepts employed in the hypothetico-deductive model, the concept of confirming evidence and the concept of degrees of confirmation. There is, however, a third basic concept employed in that model, viz., the concept of accepting a hypothesis as true. This concept has been discussed at great length in recent years, and we turn now to a consideration of that discussion. As described above, the model pictured the scientist accepting a hypothesis when he had accumulated sufficient confirming evidence and when the degree of confirmation of the hypothesis is high enough. But how much evidence is sufficient? When is the degree of confirmation high enough? These fundamental questions have been two point of departure for the recent discussion of the acceptance of scientific hypotheses.

Professor Richard Rudner claims that the answer to these questions depends upon the scientist's value judgments. In particular, the scientist must first decide how detrimental the consequences would be if he adopted the hypothesis in question and it turned out to be false. The worse the consequences, the higher should be the degree of confirmation required before the hypothesis is accepted as true. Rudner concludes, therefore, that the scientist, qua scientist, must make value judgments.

Replying to Rudner, Professor Richard Jeffreys argues that Rudner mistakenly views the scientist as accepting or rejecting hypotheses. Jeffreys argues that it would be irrational for the scientist to do this; he should merely accumulate evidence and assign, on the basis of this evidence, appropriate degrees of confirmation to alternative hypotheses. Then, in any given case, the practical decision maker can decide for himself whether or not there is a sufficient degree of confirmation to justify his supposing, in that particular case, that the hypothesis is true. It is the practical decision maker, and not the scientist, who must make value judgments.

Professor Isaac Levi, in his recent article, suggests a method of combining the stronger points of Rudner's and Jeffreys's postions. In agreement with Rudner and the whole tradition of the hypothetico-deductive model, Levi sees the scientist accepting or rejecting hypotheses. On the other hand, he agrees with Jeffrey in rejecting the claim that the scientist must make extra-scientific value judgments. Instead, Levi suggests that there may be standards of acceptance that are intrinsic to the scientific activity and which do not involve any reference to the consequences of actions based upon the assumption that the hypothesis in question is true. The reader is referred to the bibliographical essay for references to Levi's latest work in which he develops this compromise approach in greater detail.

It seems safe to conclude, on the basis of our discussion until now, that the hypothetico-deductive model, while clearly on the right course, needs a great deal of additional analysis and development before it can serve as a precise and formal model for the confirmation of scientific hypotheses. But even if it were fully developed, it could not be used as an answer to *all* the important philosophical questions about confirmation. In particular, it would leave unanswered the fundamental problem of justification. What reason is there to suppose that the use of the hypothetico-deductive model is more likely to lead to the adoption of correct scientific hypotheses than the use of some other method? This problem is, of course, nothing more than a contemporary version of the problem of justifying induction that has plagued philosophers since the time of David Hume.

Many philosophers feel that the problem of justifying our methods for confirming hypotheses cannot be fruitfully discussed until we have a much more precise account of the methods that are actually used in science. There are philosophers who feel otherwise, however, and the second part of this section considers their attempts to deal with the problem of induction. The selection from Bertrand Russell gives us a clear statement of the problem. Russell argues that our ordinary inductive arguments from data to hypotheses involve empirical presuppositions (he calls them the *principles of induction*) about the likelihood of unobserved cases being similar to the cases already observed. These presuppositions must themselves be justified by inductive inferences; thus any attempt to use them to justify inductive inferences begs the question.

Two attempts to solve this problem of justifying induction have attracted much interest in recent years. Professor Max Black claims that there is nothing wrong with using certain inductive inferences to justify the use of inductive inferences. He attempts to show this by contending that such an argument could meet a very demanding set of requirements for satisfactory inductive arguments. Professor Wesley Salmon purports to prove that our inductive methods are the only consistent methods for confirming hypotheses that lead to the right result if used long enough. Consequently, even though there can be no guarantee that any method will work in a given case, it is rational to adopt the only consistent method that must eventually work. The reader will have to decide for himself whether Black and/or Salmon have succeeded in justifying the methods for confirming hypotheses that they discuss.

One additional approach to the problem of justification should be considered. Several philosophers, and most notably Professor Peter Strawson, have argued that the whole problem of justifying induction is a pseudo-problem. Strawson claims that our normal canons for evaluating inductive inferences are ultimate principles of justification, and that there are no further principles of justification which we can use in the evaluation of our normal canons. But if there are no more ultimate principles of justification—if our normal canons actually help define rationality—then the question of their rationality cannot be meaningfully raised. Salmon, in his article, vehemently objects to this approach, arguing that there still is a perfectly legitimate question about the rationality of the canons that can be raised. As we have found so often in our discussion, problems in the philosophy of science often involve very fundamental philosophical issues—in this case, the very meaning of rationality—and the resolution of the former must often wait upon the resolution of the latter.

Finally, the postscript to this section contains a selection from the writings of Professor Norwood Hanson who is concerned with an issue that he feels practically all philosophers of science have unjustly neglected. The issue in question is the analysis of the process of scientific discovery. According to the standard account, this process can be analyzed psychologically, sociologically, and historically, but there is no possible logical account of this process. In other words, the standard account maintains that whereas we can offer a prescriptive account of the conditions under which a hypothesis once proposed is really confirmed, the most that we can offer is a descriptive account of the conditions under which scientists actually propose new hypotheses.

Opposing this point of view, Professor Hanson argues that the process of scientific discovery is amenable to logical analysis. He claims that there is an important difference between good reasons and bad reasons for suggesting a hypothesis, and the logic of scientific discovery is a prescriptive analysis of the conditions in which it would be reasonable to suggest a given hypothesis.

Hanson is quite aware of the following objection that might be raised: Any considerations that would make the suggestion of a given hypothesis plausible would also help confirm that hypothesis. Consequently, there is no special logic of scientific discovery; the only special analyses of that process are the descriptive historical, psychological, and sociological analyses.

Hanson's reply to this objection is that there are special considerations that make the suggestion of hypotheses plausible but which do not confirm those hypotheses. What are these special considerations? According to Hanson, they are analogical considerations which suggest that a certain type of hypothesis is likely to be correct. Although such considerations do not confirm any particular hypothesis, they do make plausible the suggestion of any hypothesis of that type. Hanson illustrates the type of analogical considerations he has in mind by an analysis of Kepler's work on planetary motion.

The reader will have to decide for himself whether Hanson has succeeded in showing that there is a special logic of discovery. If Hanson is right, then we will have to rethink the whole issue of the role of models and analogies in the development and understanding of a scientific theory. After all, Hanson seems to assign to analogies a far greater role in the process of theory construction than we have allowed for until now.

The Definition of a Confirming Instance

Carl G. Hempel

STUDIES IN THE LOGIC OF CONFIRMATION

To the memory of my wife,
Eva Ahrends Hempel

1. OBJECTIVE OF THE STUDY[1]

The defining characteristic of an empirical statement is its capability of being tested by a confrontation with experimental finding, i.e., with the results of suitable experiments or "focussed" observations. This feature distinguishes statements which have empirical content both from the statements of the formal sciences, logic and mathematics, which require no experimental test for their validation, and from the formulations of transempirical metaphysics, which do not admit of any.

The testability here referred to has to be understood in the comprehensive sense of "testability in principle"; there are many empirical statements which, for practical reasons, cannot be actually tested at present. To call a statement of this kind testable in principle means that it is possible to state just what experiential findings, if they were actually obtained, would constitute favourable evidence for it, and what findings or "data," as we shall say for brevity, would constitute unfavourable evidence; in other words, a statement is called testable in principle, if it is possible to describe the kind of data which would confirm or disconfirm it.

The concepts of confirmation and of disconfirmation as here understood are clearly more comprehensive than those of conclusive verification and falsification. Thus, e.g., no finite amount of experiential evidence can conclusively verify a hypothesis expressing a general law such as the law of gravitation, which covers an infinity of potential instances, many of which belong either to the as yet inaccessible future, or to the irretrievable past; but a finite set of relevant data may well be "in accord with" the hypothesis and thus constitute confirming

Reprinted from *Mind* (1945) by permission of the editor of *Mind*.

evidence for it. Similarly, an existential hypothesis, asserting, say, the existence of an as yet unknown chemical element with certain specified characteristics, cannot be conclusively proved false by a finite amount of evidence which fails to "bear out" the hypothesis; but such unfavourable data may, under certain conditions, be considered as weakening the hypothesis in question, or as constituting disconfirming evidence for it.[2]

While, in the practice of scientific research, judgments as to the confirming or disconfirming character of experiential data obtained in the test of a hypothesis are often made without hesitation and with a wide consensus of opinion, it can hardly be said that these judgments are based on an explicit theory providing general criteria of confirmation and of disconfirmation. In this respect, the situation is comparable to the manner in which deductive inferences are carried out in the practice of scientific research: This, too, is often done without reference to an explicitly stated system of rules of logical inference. But while criteria of valid deduction can be and have been supplied by formal logic, no satisfactory theory providing general criteria of confirmation and disconfirmation appears to be available so far.

In the present essay, an attempt will be made to provide the elements of a theory of this kind. After a brief survey of the significance and the present status of the problem, I propose to present a detailed critical analysis of some common conceptions of confirmation and disconfirmation and then to construct explicit definitions for these concepts and to formulate some basic principles of what might be called the logic of confirmation.

2. SIGNIFICANCE AND PRESENT STATUS OF THE PROBLEM

The establishment of a general theory of confirmation may well be regarded as one of the most urgent desiderata of the present methodology of empirical science.[3] Indeed, it seems that a precise analysis of the concept of confirmation is a necessary condition for an adequate solution of various fundamental problems concerning the logical structure of scientific procedure. Let us briefly survey the most outstanding of these problems.

(a) In the discussion of scientific method, the concept of relevant evidence plays an important part. And while certain "inductivist" accounts of scientific procedure seem to assume that relevant evidence, or relevant data, can be collected in the context of an inquiry prior to the formulation of any hypothesis, it should be clear upon brief reflection that relevance is a relative concept; experiential data can be said to be relevant or irrelevant only with respect to a given hypothesis; and it is the hypothesis which determines what kind of data or evidence are relevant for it. Indeed, an empirical finding is relevant for a hypothesis if and only if it constitutes either favourable or unfavourable evidence for it; in other words, if it either confirms or disconfirms the hypothesis.

Thus, a precise definition of relevance presupposes an analysis of confirmation and disconfirmation.

(b) A closely related concept is that of instance of a hypothesis. The so-called method of inductive inference is usually presented as proceeding from specific cases to a general hypothesis of which each of the special cases is an "instance" in the sense that it "conforms to" the general hypothesis in question, and thus constitutes confirming evidence for it.

Thus, any discussion of induction which refers to the establishment of general hypotheses on the strength of particular instances is fraught with all those logical difficulties—soon to be expounded—which beset the concept of confirmation. A precise analysis of this concept is, therefore, a necessary condition for a clear statement of the issues involved in the complex problem of induction and of the ideas suggested for their solution—no matter what their theoretical merits or demerits may be.

(c) Another issue customarily connected with the study of scientific method is the quest for "rules of induction." Generally speaking, such rules would enable us to "infer," from a given set of data, that hypothesis or generalization which accounts best for all the particular data in the given set. Recent logical analyses have made it increasingly clear that this way of conceiving the problem involves a misconception: While the process of invention by which scientific discoveries are made is as a rule *psychologically guided and stimulated* by antecedent knowledge of specific facts, its results are *not logically determined* by them; the way in which scientific hypotheses or theories are discovered cannot be mirrored in a set of general rules of inductive inference.[4] One of the crucial considerations which lead to this conclusion is the following: Take a scientic theory such as the atomic theory of matter. The evidence on which it rests may be described in terms referring to directly observable phenomena, namely to certain "macroscopic" aspects of the various experimental and observational data which are relevant to the theory. On the other hand, the theory itself contains a large number of highly abstract, nonobservational terms such as "atom," "electron," "nucleus," "dissociation," "valence," and others, none of which figures in the description of the observational data. An adequate rule of induction would therefore have to provide, for this and for every conceivable other case, mechanically applicable criteria determining unambiguously, and without any reliance on the inventiveness or additional scientific knowledge of its user, all those new abstract concepts which need to be created for the formulation of the theory that will account for the given evidence. Clearly, this requirement cannot be satisfied by any set of rules, however ingeniously devised; there can be no general rules of induction in the above sense; the demand for them rests on a confusion of logical and psychological issues. What determines the soundness of a hypothesis is not the way it is arrived at (it may even have been suggested by a dream or a hallucination), but the way it stands up when tested, i.e., when confronted with relevant observational data. Accordingly, the quest for rules of induction in the original sense of canons of scientific discovery has to be replaced, in the logic of science, by the quest for general objective criteria determining (a) whether, and—if possible—even (b) to what degree, a hypothesis *H* may be said to be corroborated by a given body of evidence *E*. This approach differs essentially from the inductivist conception of the problem in that it presupposes not only *E*, but also *H*, as given and then seeks to determine a certain logical relationship between them. The two parts of this latter problem can be restated in somewhat more precise terms as follows:

(A) To give precise definitions of the two non-quantitative relational concepts of confirmation and of disconfirmation, i.e., to define the meaning of the phrases "*E* confirms *H*" and "*E* disconfirms *H*." (When *E* neither confirms nor disconfirms *H*, we shall say that *E* is neutral, or irrelevant, with respect to *H*.)
(B) (1) To lay down criteria defining a metrical concept "degree of confirmation of *H* with respect to *E*," whose values are real numbers; or, failing this,
(2) To lay down criteria defining two relational concepts, "more highly confirmed than" and "equally well confirmed with," which make possible a nonmetrical comparison of hypotheses (each with a body of evidence assigned to it) with respect to the extent of their confirmation.

Interestingly, problem (B) has received much more attention in methodological research than problem (A); in particular, the various theories of the "probability of hypotheses" may be regarded as concerning this problem complex; we have here adopted[5] the more neutral term "degree of confirmation" instead of "probability" because the latter is used in science in a definite technical sense involving reference to the relative frequency of the occurrence of a given event in a sequence, and it is at least an open question whether the degree of confirmation of a hypothesis can generally be defined as a probability in this statistical sense.

The theories dealing with the probability of hypotheses fall into two main groups: The "logical" theories construe probability as a logical relation between sentences (or propositions; it is not always clear which is meant)[6]; the "statistical" theories interpret the probability of a hypothesis in substance as the limit of the relative frequency of its confirming instances among all relevant cases.[7] Now it is a remarkable fact that none of the theories of the first type which have been developed so far provides an explicit general definition of the probability (or degree of confirmation) of a hypothesis *H* with respect to a body of evidence *E*; they all limit themselves essentially to the construction of an uninterpreted postulational system of logical probability. For this reason, these theories fail to provide a complete solution of problem (B). The statistical approach, on the other hand, would, if successful, provide an explicit numerical definition of the degree of confirmation of a hypothesis; this definition would be formulated in terms of the numbers of confirming and disconfirming instances for *H* which constitute the body of evidence *E*. Thus, a necessary condition for an adequate interpretation of degrees of confirmation as statistical probabilities is the establishment of precise criteria of confirmation and disconfirmation, in other words, the solution of problem (A).

However, despite their great ingenuity and suggestiveness, the attempts which have been made so far to formulate a precise statistical definition of the degree of confirmation of a hypothesis seem open to certain objections,[8] and several authors[9] have expressed doubts as to the possibility of defining the degree of confirmation of a hypothesis as a metrical magnitude, though some of them consider it as possible, under certain conditions, to solve at least the less exacting problem (B) (2), i.e., to establish standards of nonmetrical comparison between hypotheses with respect to the extent of their confirmation. An adequate comparison of this kind might have to take into account a variety of different factors;[10] but again the numbers of the confirming and of the disconfirming instances which the given evidence includes will be among the most important of those factors.

Thus, of the two problems, (A) and (B), the former appears to be the more basic one, first, because it does not presuppose the possibility of defining numerical degrees of confirmation or of comparing different hypotheses as to the extent of their confirmation; and second because our considerations indicate that any attempt to solve problem (B)—unless it is to remain in the stage of an axiomatized system without interpretation—is likely to require a precise definition of the concepts of confirming and disconfirming instance of a hypothesis before it can proceed to define numerical degrees of confirmation or to lay down nonmetrical standards of comparison.

(d) It is now clear that an analysis of confirmation is of fundamental importance also for the study of the central problem of what is customarily called epistemology; this problem may be characterized as the elaboration of "standards of rational belief" or of criteria of warranted assertibility. In the methodology of empirical science this problem is usually phrased as concerning the rules governing the test and the subsequent acceptance or rejection of empirical hypotheses on the basis of experimental or observational findings, while in its "epistemological" version the issue is often formulated as concerning the validation of beliefs by reference to perceptions, sense data, or the like. But no matter how the final empirical evidence is construed and in what terms it is accordingly expressed, the theoretical problem remains the same: to characterize, in precise and general terms, the conditions under which a body of evidence can be said to confirm, or to disconfirm, a hypothesis of empirical character; and that is again our problem (A).

(e) The same problem arises when one attempts to give a precise statement of the empiricist and operationalist criteria for the empirical meaningfulness of a sentence; these criteria, as is well known, are formulated by reference to the theoretical testability of the sentence by means of experimental evidence;[11] and the concept of theoretical testability, as was pointed out earlier, is closely related to the concepts of confirmation and disconfirmation.[12]

Considering the great importance of the concept of confirmation, it is surprising that no systematic theory of the non-quantitative relation of confirmation seems to have been developed so far. Perhaps this fact reflects the tacit assumption that the concepts of confirmation and of disconfirmation have a sufficiently clear meaning to make explicit definitions unnecessary or at least comparatively trivial. And indeed, as will be shown below, there are certain features which are rather generally associated with the intuitive notion of confirming evidence, and which, at first, seem well-suited to serve as defining characteristics of confirmation. Closer examination will reveal the definitions thus obtainable to be seriously deficient and will make it clear that an adequate definition of confirmation involves considerable difficulties.

Now the very existence of such difficulties suggests the question whether the problem we are considering does not rest on a false assumption: Perhaps there are no objective criteria of confirmation; perhaps the decision as to whether a given hypothesis is acceptable in the light of a given body of evidence is no more subject to rational, objective rules than is the process of inventing a scientific hypothesis or theory; perhaps, in the last analysis, it is a "sense of evidence," or a feeling of plausibility in view of the relevant data, which ultimately decides whether a hypothesis is scientifically acceptable.[13] This view is comparable to the opinion that the validity of a mathematical proof or of a logical argument has to be judged ultimately by reference to a feeling of soundness or convincingness; and both theses have to be rejected on analogous grounds: They involve a confusion of logical and psychological considerations. Clearly, the occurrence or nonoccurrence of a feeling of conviction upon the presentation of grounds for an assertion is a subjective matter which varies from person to person, and with the same person in the course of time; it is often deceptive, and can certainly serve neither as a necessary nor as a sufficient condition for the soundness of the given assertion.[14] A rational reconstruction of the standards of scientific validation cannot, therefore, involve reference to a sense of evidence; it has to be based on objective criteria. In fact, it seems reasonable to require that the criteria of empirical confirmation, besides being objective in character, should contain no reference to the specific subject matter of the hypothesis or of the evidence in question; it ought to be possible, one feels, to set up purely formal criteria of confirmation in a manner similar to that in which deductive logic provides purely for malcriteria for the validity of deductive inferences.

With this goal in mind, we now turn to a study of the non-quantitative concept of confirmation. We shall begin by examining some current conceptions of confirmation and exhibiting their logical and methodological inadequacies; in the course of this analysis, we shall develop a set of conditions for the adequacy of any proposed definition of confirmation; and finally, we shall construct a definition of confirmation which satisfies those general standards of adequacy.

3. NICOD'S CRITERION OF CONFIRMATION AND ITS SHORTCOMINGS

We consider first a conception of confirmation which underlies many recent studies of induction and of scientific method. A very explicit statement of this conception has been given by Jean Nicod in the following passage: "Consider the formula or the law: *A entails B*. How can a particular proposition, or more briefly, a fact, affect its probability? If this fact consists of the presence of B in a case of A, it is favourable to the law '*A entails B*'; on the contrary, if it consists of the absence of B in a case of A, it is unfavourable to this law. It is conceivable that we have here the only two direct modes in which a fact can influence the probability of a law . . . . Thus, the entire influence of particular truths or facts on the probability of universal propositions or laws would operate by means of these two elementary relations which we shall call *confirmation* and *invalidation*."[15] Note that the applicability of the criterion is restricted to hypotheses of the form "*A entails B*." Any hypothesis H of this kind may be expressed in the notation of symbolic logic[16] by means of a universal conditional sentence, such as, in the simplest case,

$$(x)(P(x) \supset Q(x)),$$

i.e., "For any object x: If x is a P, then x is a Q," or also, "Occurrences of the quality P entails occurrence of the quality Q." According to the above criterion this hypothesis is confirmed by an object a, if a is P and Q; and the hypothesis is disconfirmed by a if a is P, but not Q. In other words, an object confirms a universal conditional hypothesis if and only if it satisfies both the antecedent (here: "$P(x)$") and the consequent (here: "$Q(x)$") of the conditional; it disconfirms the hypothesis if and only if it satisfies the antecedent, but not the consequent of the conditional; and (we add to this Nicod's statement) it is neutral, or irrelevant, with respect to the hypothesis if it does not satisfy the antecedent.

This criterion can readily be extended so as to be applicable also to universal conditionals containing more than one quantifier, such as "Twins always resemble each other," or, in symbolic notation, "$(x)(y)(\text{Twins}(x, y) \supset \text{Rsbl}(x, y))$." In these cases, a confirming instance consists of an ordered couple, or triple, etc., of objects satisfying the antecedent and the consequent of the conditional. (In the case of the last illustration, any two persons who are twins and who resemble each other would confirm the hypothesis; twins who do not resemble each other would disconfirm it; and any two persons not twins—no matter whether they resemble each other or not—would constitute irrelevant evidence.)

We shall refer to this criterion as Nicod's criterion.[17] It states explicitly what is perhaps the most common tacit interpretation of the concept of confirmation. While seemingly quite adequate, it suffers from serious shortcomings, as will now be shown.

(a) First, the applicability of this criterion is restricted to hypotheses of universal conditional form; it provides no standards for existential hypotheses (such as "There exists organic life on other stars," or "Poliomyelitis is caused by some virus") or for hypotheses whose explicit formulation calls for the use of both universal and existential quantifiers (such as "Every human being dies some finite number of years after his birth," or the psychological hypothesis, "You can fool all of the people some of the time and some of the people all of the time, but you cannot fool all of the people all of the time," which may be symbolized by "$(x)(Et)\mathrm{Fl}(x, t) \cdot (Ex)(t)\mathrm{Fl}(x, t) \cdot \sim (x)(t)\mathrm{Fl}(x, t)$," (where "$\mathrm{Fl}(x, t)$" stands for "You can fool (person) x at time t"). We note, therefore, the desideratum of establishing a criterion of confirmation which is applicable to hypotheses of any form.[18]

(b) We now turn to a second shortcoming of Nicod's criterion. Consider the two sentences:

$$S_1: \quad (x)(\mathrm{Raven}(x) \supset \mathrm{Black}(x));$$

$$S_2: \quad (x)(\sim\mathrm{Black}(x) \supset \sim\mathrm{Raven}(x))$$

(i.e. "All ravens are black" and "Whatever is not black is not a raven"), and let a, b, c, d be four objects such that a is a raven and black, b a raven but not black, c not a raven but black, and d neither a raven nor black. Then, according to Nicod's criterion, a would confirm S_1, but be neutral with respect to S_2; b would disconfirm both S_1 and S_2; c would be neutral with respect to both S_1 and S_2, and d would confirm S_2, but be neutral with respect to S_1.

But S_1 and S_2 are logically equivalent; they have the same content, they are different formulations of the same hypothesis. And yet, by Nicod's criterion, either of the objects a and d would be confirming for one of the two sentences, but neutral with respect to the other. This means that Nicod's criterion makes confirmation depend not only on the content of the hypothesis, but also on its formulation.[19]

One remarkable consequence of this situation is that every hypothesis to which the criterion is applicable—i.e. every universal conditional—can be stated in a form for which there cannot possibly exist any confirming instances. Thus, e.g., the sentence:

$$(x)[(\mathrm{Raven}(x) \cdot \sim\mathrm{Black}(x)) \supset (\mathrm{Raven}(x) \cdot \sim\mathrm{Raven}(x)]$$

is readily recognized as equivalent to both S_1 and S_2 above; yet no object whatever can confirm this sentence, i.e., satisfy both its antecedent and its consequent; for the consequent is contradictory. An analogous transformation is, of course, applicable to any other sentence of universal conditional form.

4. THE EQUIVALENCE CONDITION

The results just obtained call attention to a condition which an adequately defined concept of confirmation should satisfy, and in the light of which Nicod's criterion has to be rejected as inadequate: *Equivalence condition*: Whatever confirms (disconfirms) one of two equivalent sentences, also confirms (disconfirms) the other.

Fulfilment of this condition makes the confirmation of a hypothesis independent of the way in which it is formulated; and no doubt it will be conceded that this is a necessary condition for the adequacy of any proposed criterion of confirmation. Otherwise, the question as to whether certain data confirm a given hypothesis would have to be answered by saying: "That depends on which of the different equivalent formulations of the hypothesis is considered" —which appears absurd. Furthermore—and this is a more important point than an appeal to a feeling of absurdity—an adequate definition of confirmation will have to do justice to the way in which empirical hypotheses function in theoretical scientific contexts such as explanations and predictions; but when hypotheses are used for purposes of explanation or prediction,[20] they serve as premisses in a deductive argument whose conclusion is a description of the event to be explained or predicted. The deduction is governed by the principles of formal logic, and according to the latter, a deduction which is valid will remain so if some or all of the premisses are replaced by different, but equivalent statements; and indeed, a scientist will feel free, in any theoretical reasoning involving certain hypotheses, to use the latter in whichever of their equivalent formulations is most convenient for the development of his conclusions. But if we adopted a concept of confirmation which did not satisfy the equivalence condition, then it would be possible, and indeed necessary, to argue in certain cases that it was sound scientific procedure to base a prediction on a given hypothesis if formulated in a sentence S_1, because a good deal of confirming evidence had been found for S_1; but that it was altogether inadmissible to base the prediction (say, for convenience of deduction) on an equivalent formulation S_2, because no confirming evidence for S_2 was available. Thus, the equivalence condition has to be regarded as a necessary condition for the adequacy of any definition of confirmation.

5. THE "PARADOXES" OF CONFIRMATION

Perhaps we seem to have been labouring the obvious in stressing the necessity of satisfying the equivalence condition. This impression is likely to vanish upon consideration of certain consequences which derive from a combination of the equivalence condition with a most natural and plausible assumption concerning a sufficient condition of confirmation.

The essence of the criticism we have levelled so far against Nicod's criterion is that it certainly cannot serve as a necessary condition of confirmation; thus, in the illustration given in the beginning of section 3, the object *a* confirms S_1 and should therefore also be considered as confirming S_2, while according to Nicod's criterion it is not. Satisfaction of the latter is therefore not a necessary condition for confirming evidence.

On the other hand, Nicod's criterion might still be considered as stating a particularly obvious and important sufficient condition of confirmation. And indeed, if we restrict ourselves to universal conditional hypotheses in one variable[21]—such as S_1 and S_2 in the above illustration—then it seems perfectly reasonable to qualify an object as confirming such a hypothesis if it satisfies both its antecedent and its consequent. The plausibility of this view will be further corroborated in the course of our subsequent analyses.

Thus, we shall agree that if a is both a raven and black, then a certainly confirms S_1: "$(x)(\text{Raven}(x) \supset \text{Black}(x))$," and if d is neither black nor a raven, d certainly confirms S_2:

$$(x)(\sim\text{Black}(x) \supset \sim\text{Raven}(x)).$$

Let us now combine this simple stipulation with the equivalence condition: Since S_1 and S_2 are equivalent, d is confirming also for S_1; and thus, we have to recognize as confirming for S_1 any object which is neither black nor a raven. Consequently, any red pencil, any green leaf, any yellow cow, etc., becomes confirming evidence for the hypothesis that all ravens are black. This surprising consequence of two very adequate assumptions (the equivalence condition and the above sufficient condition of confirmation) can be further expanded: The following sentence can readily be shown to be equivalent to S_1: S_3: $(x)[(\text{Raven}(x) \vee \sim\text{Raven}(x)) \supset (\sim\text{Raven}(x) \vee \text{Black}(x))]$, i.e., "Anything which is or is not a raven is either no raven or black." According to the above sufficient condition, S_3 is certainly confirmed by any object, say e, such that (1) e is or is not a raven and, in addition, (2) e is not a raven or also black. Since (1) is analytic, these conditions reduce to (2). By virtue of the equivalence condition, we have therefore to consider as confirming for S_1 any object which is either no raven or also black (in other words: any object which is no raven at all, or a black raven).

Of the four objects characterized in section 3, a, c and d would therefore constitute confirming evidence for S_1, while b would be disconfirming for S_1. This implies that any non-raven represents confirming evidence for the hypothesis that all ravens are black.

We shall refer to these implications of the equivalence criterion and of the above sufficient condition of confirmation as the *paradoxes of confirmation*.

How are these paradoxes to be dealt with? Renouncing the equivalence condition would not represent an acceptable solution, as is shown by the consideration presented in section 4. Nor does it seem possible to dispense with the stipulation that an object satisfying two conditions, C_1 and C_2, should be considered as confirming a general hypothesis to the effect that any object which satisfies C_1, also satisfies C_2.

But the deduction of the above paradoxical results rests on one other assumption which is usually taken for granted, namely, that the meaning of general empirical hypotheses, such as that all ravens are black, or that all sodium salts burn yellow, can be adequately expressed by means of sentences of universal conditional form, such as "$(x)(\text{Raven}(x) \supset \text{Black}(x))$" and "$(x)(\text{Sod. Salt}(x) \supset \text{Burn Yellow}(x))$," etc. Perhaps this customary mode of presentation has to be modified; and perhaps such a modification would automatically remove the paradoxes of confirmation? If this is not so, there seems to be only one alternative left, namely to show that the impression of the paradoxical character of those consequences is due to misunderstanding and can be dispelled, so that no theoretical difficulty remains. We shall now consider these two possibilities in turn: The subsections 5.11 and 5.12 are devoted to a discussion of two different proposals for a modified representation of general hypotheses; in subsection 5.2, we shall discuss the second alternative, i.e. the possibility of tracing the impression of paradoxicality back to a misunderstanding.

5.11. It has often been pointed out that while Aristotelian logic, in agreement with prevalent every day usage, confers "existential import" upon sentences of the form "All *P*'s are *Q*'s," a universal conditional sentence, in the sense of modern logic, has no existential import; thus, the sentence:

$$(x)(\text{Mermaid}(x) \supset \text{Green}(x))$$

does not imply the existence of mermaids; it merely asserts that any object either is not a mermaid at all, or a green mermaid; and it is true simply because of the fact that there are no mermaids. General laws and hypotheses in science, however—so it might be argued—are meant to have existential import; and one might attempt to express the latter by supplementing the customary universal conditional by an existential clause. Thus, the hypothesis that all ravens are black would be expressed by means of the sentence S_1: "$(x)(\text{Raven}(x) \supset \text{Black}(x)) \cdot (Ex)\text{Raven}(x)$"; and the hypothesis that no non-black things are ravens by S_2: "$(x)(\sim\text{Black}(x) \supset \sim\text{Raven}(x)) \cdot (Ex) \sim\text{Black}(x))$." Clearly, these sentences are not equivalent, and of the four objects *a*, *b*, *c*, *d* characterized in section 3, part (*b*), only *a* might reasonably be said to confirm S_1, and only *d* to confirm S_2. Yet this method of avoiding the paradoxes of confirmation is open to serious objections:

(a) First of all, the representation of every general hypothesis by a conjunction of a universal conditional and an existential sentence would invalidate many logical inferences which are generally accepted as permissible in a theoretical argument. Thus, for example, the assertions that all sodium salts burn yellow, and that whatever does not burn yellow is no sodium salt are logically equivalent according to customary understanding and usage; and their representation by universal conditionals preserves this equivalence; but if existential clauses are added, the two assertions are no longer equivalent, as is illustrated above by the analogous case of S_1 and S_2.

(b) Second, the customary formulation of general hypotheses in empirical science clearly does not contain an existential clause, nor does it, as a rule, even indirectly determine such a clause unambiguously. Thus, consider the hypothesis that if a person after receiving an injection of a certain test substance has a positive skin reaction, he has diphtheria. Should we construe the existential clause here as referring to persons, to persons receiving the injection, or to persons who, upon receiving the injection, show a positive skin reaction? A more or less arbitrary decision has to be made; each of the possible decisions gives a different interpretation to the hypothesis, and none of them seems to be really implied by the latter.

(c) Finally, many universal hypotheses cannot be said to imply an existential clause at all. Thus, it may happen that from a certain astrophysical theory a universal hypothesis is deduced concerning the character of the phenomena which would take place under certain specified extreme conditions. A hypothesis of this kind need not (and, as a rule, does not) imply that such extreme conditions ever were or will be realized; it has no existential import. Or consider a biological hypothesis to the effect that whenever man and ape are crossed, the offspring will have such and such characteristics. This is a general hypothesis; it might be contemplated as a mere conjecture, or as a consequence of a broader genetic theory, other implications of which may already have been tested with positive results; but unquestionably the hypothesis does not imply an existential clause asserting that the contemplated kind of crossbreeding referred to will, at some time, actually take place.

While, therefore, the adjunction of an existential clause to the customary symbolization of a general hypothesis cannot be considered as an adequate *general* method of coping with the paradoxes of confirmation, there is a purpose which the use of an existential clause may serve very well, as was pointed out to me by Dr. Paul Oppenheim:[22] If somebody feels that objects of the types c and d mentioned above are irrelevant rather than confirming for the hypothesis in question, and that qualifying them as confirming evidence does violence to the meaning of the hypothesis, then this may indicate that he is consciously or unconsciously construing the latter as having existential import; and this kind of understanding of general hypotheses is in fact very common. In this case, the "paradox" may be removed by pointing out that an adequate symbolization of the intended meaning requires the adjunction of an existential clause. The formulation thus obtained is more restrictive than the universal conditional alone; and while we have as yet set up no criteria of confirmation applicable to hypotheses of this more complex form, it is clear that according to every acceptable definition of confirmation objects of the types c and d will fail to qualify as confirming cases. In this manner, the use of an existential clause may prove helpful in distinguishing and rendering explicit different possible interpretations of a given general hypothesis which is stated in nonsymbolic terms.

5.12. Perhaps the impression of the paradoxical character of the cases discussed in the beginning of section 5 may be said to grow out of the feeling that the hypothesis that all ravens are black is about ravens, and not about nonblack things, nor about all things. The use of an existential clause was one attempt at expressing this presumed peculiarity of the hypothesis. The attempt has failed, and if we wish to reflect the point in question, we shall have to look for a stronger device. The idea suggests itself of representing a general hypothesis by the customary universal conditional, supplemented by the indication of the specific "field of application" of the hypothesis; thus, we might represent the hypothesis that all ravens are black by the sentence "$(x)(\text{Raven}(x) \supset$

Black(x))" (or any one of its equivalents), plus the indication "Class of ravens" characterizing the field of application; and we might then require that every confirming instance should belong to the field of application. This procedure would exclude the objects c and d from those constituting confirming evidence and would thus avoid those undesirable consequences of the existential-clause device which were pointed out in 5.11 (c). But apart from this advantage, the second method is open to objections similar to those which apply to the first: (a) The way in which general hypotheses are used in science never involves the statement of a field of application; and the choice of the latter in a symbolic formulation of a given hypothesis thus introduces again a considerable measure of arbitrariness. In particular, for a scientific hypothesis to the effect that all P's are Q's, the field of application cannot simply be said to be the class of all P's; for a hypothesis such as that all sodium salts burn yellow finds important applications in tests with negative results, i.e., it may be applied to a substance of which it is not known whether it contains sodium salts, nor whether it burns yellow; and if the flame does not turn yellow, the hypothesis serves to establish the absence of sodium salts. The same is true of all other hypotheses used for tests of this type. (b) Again, the consistent use of a domain of application in the formulation of general hypotheses would involve considerable logical complications, and yet would have no counterpart in the theoretical procedure of science, where hypotheses are subjected to various kinds of logical transformation and inference without any consideration that might be regarded as referring to changes in the fields of application. This method of meeting the paradoxes would therefore amount to dodging the problem by means of an *ad hoc* device which cannot be justified by reference to actual scientific procedure.

5.2. We have examined two alternatives to the customary method of representing general hypotheses by means of universal conditionals; neither of them proved an adequate means of precluding the paradoxes of confirmation. We shall now try to show that what is wrong does not lie in the customary way of construing and representing general hypotheses, but rather in our reliance on a misleading intuition in the matter: The impression of a paradoxical situation is not objectively founded; it is a psychological illusion.

(a) One source of misunderstanding is the view, referred to before, that a hypothesis of the simple form "Every P is a Q" such as "All sodium salts burn yellow," asserts something about a certain limited class of objects only, namely, the class of all P's. This idea involves a confusion of logical and practical considerations: Our interest in the hypothesis may be focussed upon its applicability to that particular class of objects, but the hypothesis nevertheless asserts something about, and indeed imposes restrictions upon, *all* objects (within the logical type of the variable occurring in the hypothesis, which in the case of our last illustration might be the class of all physical objects). Indeed, a hypothesis of the form "Every P is a Q" forbids the occurrence of any objects having the property P but lacking the property Q, i.e., it restricts all objects whatsoever to the class of those which either lack the property P or also have the property Q. Now, every object either belongs to this class or falls outside it, and thus, every object—and not only the P's—either conforms to the hypothesis or violates it; there is no object which is not implicitly "referred to" by a hypothesis of this type. In particular, every object which either is no sodium salt or burns yellow conforms to, and thus "bears out" the hypothesis that all sodium salts burn yellow; every other object violates that hypothesis.

The weakness of the idea under consideration is evidenced also by the observation that the class of objects about which a hypothesis is supposed to assert something is in no way clearly determined, and that it changes with the context, as was shown in 5.12 (a).

(b) A second important source of the appearance of paradoxicality in certain cases of confirmation is exhibited by the following consideration.

Suppose that in support of the assertion "All sodium salts burn yellow" somebody were to adduce an experiment in which a piece of pure ice was held into a colourless flame and did not turn the flame yellow. This result would confirm the assertion, "Whatever does not burn yellow is no sodium salt," and consequently, by virtue of the equivalence condition, it would confirm the original formulation. Why does this impress us as paradoxical? The reason becomes clear when we compare the previous situation with the case of an experiment where an object whose chemical constitution is as yet unknown to us is held into a flame and fails to turn it yellow, and where subsequent analysis reveals it to contain no sodium salt. This outcome, we should no doubt agree, is what was to be expected on the basis of the hypothesis that all sodium salts burn yellow—no matter in which of its various equivalent formulations it may be expressed; thus, the data here obtained constitute confirming evidence for the hypothesis. Now the only difference between the two situations here considered is that in the first case we are told beforehand the test substance is ice, and we happen to "know anyhow" that ice contains no sodium salt; this has the consequence that the outcome of the flame-colour test becomes entirely irrelevant for the confirmation of the hypothesis and thus can yield no new evidence for us. Indeed, if the flame should not turn yellow, the hypothesis requires that the substance contain no sodium salt—and we know beforehand that ice does not—and if the flame should turn yellow, the hypothesis would impose no further restrictions on the substance; hence, either of the possible outcomes of the experiment would be in accord with the hypothesis.

The analysis of this example illustrates a general point: In the seemingly paradoxical cases of confirmation, we are often not actually judging the relation of the given evidence, E, alone to the hypothesis H (we fail to observe the "methodological fiction," characteristic of every case of confirmation, that we have no relevant evidence for H other than that included in E); instead, we tacitly introduce a comparison of H with a body of evidence which consists of E in conjunction with an additional amount of information which we happen to have at our disposal; in our illustration, this information includes the knowledge (1) that the substance used in the experiment is ice, and (2) that ice contains no sodium salt. If we assume this additional information as given, then, of course, the outcome of the experiment can add no strength to the hypothesis under consideration. But if we are careful to avoid this tacit reference to additional knowledge (which entirely changes the character of the problem), and if we formulate the question as to the confirming character of the evidence in a manner adequate to the concept of confirmation as used in this paper, we have to ask: Given some object *a* (it happens to be a piece of ice, but this fact is not included in the evidence), and given the fact that *a* does not turn the flame yellow and is no sodium salt—does *a* then constitute confirming evidence for the hypothesis? And now—no matter whether *a* is ice or some other substance—it is clear that the answer has to be in the affirmative; and the paradoxes vanish.

So far, in section (b), we have considered mainly that type of paradoxical case which is illustrated by the assertion that any non-black non-raven constitutes confirming evidence for the hypothesis, "All ravens are black." However, the general idea just outlined applies as well to the even more extreme cases exemplified by the assertion that any non-raven as well as any black object confirms the hypothesis in question. Let us illustrate this by reference to the latter case. If the given evidence *E* (i.e., in the sense of the required methodological fiction, all our data relevant for the hypothesis) consists only of one object which, in addition, is black, then *E* may reasonably be said to support even the hypothesis that all objects are black, and *a fortiori E* supports the weaker assertion that all ravens are black. In this case, again, our factual knowledge that not all objects are black tends to create an impression of paradoxicality which is not justified on logical grounds. Other "paradoxical" cases of confirmation may be dealt with analogously, and it thus turns out that the "paradoxes of confirmation," as fomulated above, are due to a misguided intuition in the matter rather than to a logical flaw in the two stipulations from which the "paradoxes" were derived.[23,24]

. . .

8. CONDITIONS OF ADEQUACY FOR ANY DEFINITION OF CONFIRMATION

The two most customary conceptions of confirmation, which were rendered explicit in Nicod's criterion and in the prediction criterion, have thus been found unsuitable for a general definition of confirmation. Besides this negative result, the preceding analysis has also exhibited certain logical characteristics of scientific prediction, explanation, and testing, and it has led to the establishment of certain standards which an adequate definition of confirmation has to satisfy. These standards include the equivalence condition and the requirement that the definition of confirmation be applicable to hypotheses of any degree of logical complexity, rather than to the simplest type of universal conditional only. An adequate definition of confirmation, however, has to satisfy several further logical requirements, to which we now turn.

First of all, it will be agreed that any sentence which is entailed by, i.e., a logical consequence of, a given observation report has to be considered as confirmed by that report: Entailment is a special case of confirmation. Thus, e.g., we want to say that the observation report "*a* is black" confirms the sentence (hypothesis) "*a* is black or grey"; and—to refer to one of the illustrations given in the preceding section—the observation sentence $R_2(a, b)$ should certainly be confirming evidence for the sentence $(Ez)R_2(a, z)$. We are therefore led to the stipulation that any adequate definition of confirmation must insure the fulfilment of the

(8.1) *Entailment condition.* Any sentence which is entailed by an observation report is confirmed by it.[25]

This condition is suggested by the preceding consideration, but of course not proved by it. To make it a standard of adequacy for the definition of confirmation means to lay down the stipulation that a proposed definition of confirmation will be rejected as logically inadequate if it is not constructed in such a way that (8.1) is unconditionally satisfied. An analogous remark applies to the subsequently proposed further standards of adequacy.

Second, an observation report which confirms certain hypotheses would invariably be qualified as confirming any consequence of those hypotheses. Indeed, any such consequence is but an assertion of all or part of the combined content of the original hypotheses and has therefore to be regarded as confirmed by any evidence which confirms the original hypotheses. This suggests the following condition of adequacy:

(8.2) *Consequence Condition.* If an observation report confirms every one of a class *K* of sentences, then it also confirms any sentence which is a logical consequence of *K*.

If (8.2) is satisfied, then the same is true of the following two more special conditions:

(8.21) *Special Consequence Condition.* If an observation report confirms a hypothesis *H*, then it also confirms every consequence of *H*.

(8.22) *Equivalence Condition.* If an observation report confirms a hypothesis *H*, then it also confirms every hypothesis which is logically equivalent with *H*.

(This follows from (8.21) in view of the fact that equivalent hypotheses are mutual consequences of each other.) Thus, the satisfaction of the consequence condition entails that of our earlier equivalence condition, and the latter loses its status of an independent requirement.

In view of the apparent obviousness of these conditions, it is interesting to note that the definition of confirmation in terms of successful prediction, while satisfying the equivalence condition, would violate the consequence condition. Consider, for example, the formulation of the prediction-criterion given in the earlier part of the preceding section. Clearly, if the observational findings B_2

can be predicted on the basis of the findings B_1 by means of the hypothesis *H*, the same prediction is obtainable by means of any equivalent hypothesis, but not generally by means of a weaker one.

On the other hand, any prediction obtainable by means of *H* can obviously also be established by means of any hypothesis which is stronger than *H*, i.e., which logically entails *H*. Thus, while the consequence condition stipulates in effect that whatever confirms a given hypothesis also confirms any weaker hypothesis, the relation of confirmation defined in terms of successful prediction would satisfy the condition that whatever confirms a given hypothesis also confirms every stronger one.

But is this "converse consequence condition," as it might be called, not reasonable enough, and should it not even be included among our standards of adequacy for the definition of confirmation? The second of these two suggestions can be readily disposed of: The adoption of the new condition, in addition to (8.1) and (8.2), would have the consequence that any observation report *B* would confirm any hypothesis *H* whatsoever. Thus, e.g., if *B* is the report "*a* is a raven" and *H* is Hooke's law, then, according to (8.1), *B* confirms the sentence "*a* is a raven," hence *B* would, according to the converse consequence condition, confirm the stronger sentence "*a* is a raven, and Hooke's law holds"; and finally, by virtue of (8.2), *B* would confirm *H*, which is a consequence of the last sentence. Obviously, the same type of argument can be applied in all other cases.

But is it not true, after all, that very often observational data which confirm a hypothesis *H* are considered also as confirming a stronger hypothesis? Is it not true, for example, that those experimental findings which confirm Galileo's law, or Kepler's laws, are considered also as confirming Newton's law of gravitation?[26] This is indeed the case, but this does not justify the acceptance of the converse entailment condition as a general rule of the logic of confirmation; for in the cases just mentioned, the weaker hypothesis is connected with the stronger one by a logical bond of a particular kind: It is essentially a substitution instance of the stronger one; thus, e.g., while the law of gravitation refers to the force obtaining between any two bodies, Galileo's law is a specialization referring to the case where one of the bodies is the earth, the other an object near its surface. In the preceding case, however, where Hooke's law was shown to be confirmed by the observation report that *a* is a raven, this situation does not prevail; and here, the rule that whatever confirms a given hypothesis also confirms any stronger one becomes an entirely absurd principle. Thus, the converse consequence condition does not provide a sound general condition of adequacy.[27]

A third condition remains to be stated:[28]

(8.3) *Consistency Condition.* Every logically consistent observation report is logically compatible with the class of all the hypotheses which it confirms.

The two most important implications of this requirement are the following:

(8.31) Unless an observation report is self-contradictory,[29] it does not confirm any hypothesis with which it is not logically compatible.

(8.32) Unless an observation report is self-contradictory, it does not confirm any hypotheses which contradict each other.

The first of these corollaries will readily be accepted; the second, however, —and consequently (8.3) itself—will perhaps be felt to embody a too severe restriction. It might be pointed out, for example, that a finite set of measurements concerning the variation of one physical magnitude, x, with another, y, may conform to, and thus be said to confirm, several different hypotheses as to the particular mathematical function in terms of which the relationship of x and y can be expressed; but such hypotheses are incompatible because to at least one value of x, they will assign different values of y.

No doubt it is possible to liberalize the formal standards of adequacy in line with these considerations. This would amount to dropping (8.3) and (8.32) and retaining only (8.31). One of the effects of this measure would be that when a logically consistent observation report B confirms each of two hypotheses, it does not necessarily confirm their conjunction; for the hypotheses might be mutually incompatible, hence their conjunction self-contradictory; consequently, by (8.31), B could not confirm it. This consequence is intuitively rather awkward, and one might therefore feel inclined to suggest that while (8.3) should be dropped and (8.31) retained, (8.32) should be replaced by the requirement (8.33): If an observation sentence confirms each of two hypotheses, then it also confirms their conjunction. But it can readily be shown that by virtue of (8.2) this set of conditions entails the fulfilment of (8.32).

If, therefore, the condition (8.3) appears to be too rigorous, the most obvious alternative would seem to lie in replacing (8.3) and its corollaries by the much weaker condition (8.31) alone; and it is an important problem whether an intuitively adequate definition of confirmation can be constructed which satisfies (8.1), (8.2), and (8.31), but not (8.3). One of the great advantages of a definition which satisfies (8.3) is that it sets a limit, so to speak, to the strength of the hypotheses which can be confirmed by given evidence.[30]

The remainder of the present study, therefore, will be concerned exclusively with the problem of establishing a definition of confirmation which satisfies the more severe formal conditions represented by (8.1), (8.2), and (8.3) together.

The fulfilment of these requirements, which may be regarded as general laws of the logic of confirmation, is of course only a necessary, not a sufficient, condition for the adequacy of any proposed definition of confirmation. Thus, e.g., if "B confirms H" were defined as meaning "B logically entails H," then the above three conditions would clearly be satisfied; but the definition would not be adequate because confirmation has to be a more comprehensive relation than entailment (the latter might be referred to as the special case of *conclusive* confirmation). Thus, a definition of confirmation, to be acceptable, also has to be materially adequate: It has to provide a reasonably close approximation to that conception of confirmation which is implicit in scientific procedure and methodological discussion. That conception is vague and to some extent quite unclear, as I have tried to show in earlier parts of this paper; therefore, it would be too much to expect full agreement as to the material adequacy of a proposed definition of confirmation; on the other hand, there will be rather general agreement on certain points; thus, e.g., the identification of confirmation with entailment, or the Nicod criterion of confirmation as analyzed above, or any definition of confirmation by reference to a "sense of evidence," will probably now be admitted not to be adequate approximations to that concept of confirmation which is relevant for the logic of science.

On the other hand, the soundness of the logical analysis (which, in a clear sense, always involves a logical reconstruction) of a theoretical concept cannot be gauged simply by our feelings of satisfaction at a certain proposed analysis; and if there are, say, two alternative proposals for defining a term on the basis of a logical analysis, and if both appear to come fairly close to the intended meaning, then the choice has to be made largely by reference to such features as the logical properties of the two reconstructions, and the comprehensiveness and simplicity of the theories to which they lead.

9. THE SATISFACTION CRITERION OF CONFIRMATION

As has been mentioned before, a precise definition of confirmation requires reference to some definite "language of science," in which all observation reports and all hypotheses under consideration are assumed to be formulated, and whose logical structure is supposed to be precisely determined. The more complex this language, and the richer its logical means of expression, the more difficult it will be, as a rule, to establish an adequate definition of confirmation for it. However, the problem has been solved at least for certain cases: With respect to languages of a comparatively simple logical structure, it has been possible to construct an explicit definition of confirmation which satisfies all of the above logical requirements, and which appears to be intuitively rather adequate. An exposition of the technical details of this definition has been published elsewhere;[31] in the present study, which is concerned with the general logical and methodological aspects of the problem of confirmation rather than with technical detail, it will be attempted to characterize the definition of confirmation thus obtained as clearly as possible with a minimum of technicalities.

Consider the simple case of the hypothesis H: $(x)(\text{Raven}(x) \supset \text{Black}(x))$, where "Raven" and "Black" are supposed to be terms of our observational vocabulary. Let B be an observation report to the effect that $\text{Raven}(a) \cdot \text{Black}(a) \cdot {\sim}\text{Raven}(c) \cdot \text{Black}(c) \cdot {\sim}\text{Raven}(d) \cdot {\sim}\text{Black}(d)$. Then B may be said to confirm H in the following sense: There are three objects altogether mentioned in B, namely a, c, and d; and as far as these are concerned, B informs us that all those which are ravens (i.e., just the object a) are also black.[32] In other words, from the information contained in B we can infer that the hypothesis H does hold true within the finite class of those objects which are mentioned in B.

Let us apply the same consideration to a hypothesis of a logically more complex structure. Let H be the hypothesis "Everybody likes somebody"; in symbols: $(x)(Ey)\text{Likes}(x, y)$, i.e. for every (person) x, there exists at least one (not necessarily different person) y such that x likes y. (Here again, "Likes" is supposed to be a relation-term which occurs in our observational vocabulary.) Suppose now that we are given an observation report B in which the names of two persons, say e and f, occur. Under what conditions shall we say that B confirms H? The previous illustration suggests the answer: If from B we can infer that H is satisfied within the finite class $\{e, f\}$; i.e. that within $\{e, f\}$ everybody likes somebody. This in turn means that e likes e or f, and f likes e or f. Thus, B would be said to confirm H if B entailed the statement "e likes e or f, and f likes e or f." This latter statement will be called the development of H for the finite class $\{e, f\}$.

The concept of *development of a hypothesis, H, for a finite class of individuals, C*, can be defined in a general fashion; the development of H for C states what H would assert if there existed exclusively those objects which are elements of C. Thus, e.g., the development of the hypothesis $H_1 = (x)(P(x) \vee Q(x))$ (i.e., "Every object has the property P or the property Q") for the class $\{a, b\}$ is $(P(a) \vee Q(a)) \cdot (P(b) \vee Q(b))$ (i.e., "a has the property P or the property Q, and b has the property P or the property Q"); the development of the existential hypothesis H_2 that at least one object has the property P, i.e., $(Ex)P(x)$, for $\{a, b\}$ is $P(a) \vee P(b)$; the development of a hypothesis which contains no quantifiers, such as H_3: $P(c) \vee Q(c)$ is defined as that hypothesis itself, no matter what the reference class of individuals is.

A more detailed formal analysis based on considerations of this type leads to the introduction of a general relation of confirmation in two steps; the first consists in defining a special relation of direct confirmation along the lines just indicated; the second step then defines the general relation of confirmation by reference to direct confirmation.

Omitting minor details, we may summarize the two definitions as follows:

(9.1 Df.) An observation report B directly confirms a hypothesis H if B entails the development of H for the class of those objects which are mentioned in B.

(9.2 Df.) An observation report B confirms a hypothesis H if H is entailed by a class of sentences each of which is directly confirmed by B.

The criterion expressed in these definitions might be called the satisfaction criterion of confirmation because its basic idea consists in construing a hypothesis as confirmed by a given observation report if the hypothesis is satisfed in the finite class of those individuals which are mentioned in the report. Let us now apply the two definitions to our last examples: The observation report B_1: $P(a) \cdot Q(b)$ directly confirms (and therefore also confirms) the hypothesis H_1, because it entails the development of H_1 for the class $\{a, b\}$, which was given above. The hypothesis H_3 is not directly confirmed by B, because its development, i.e., H_3 itself, obviously is not entailed by B_1. However, H_3 is entailed by H_1, which is directly confirmed by B_1; hence, by virtue of (9.2), B_1 confirms H_3.

Similarly, it can readily be seen that B_1 directly confirms H_2.

Finally, to refer to the first illustration given in this section: The observation report $\text{Raven}(a) \cdot \text{Black}(a) \cdot \sim \text{Raven}(c) \cdot \sim \text{Black}(c) \cdot \sim \text{Raven}(d) \cdot \sim \text{Black}(d)$ confirms (even directly) the hypothesis $(x)(\text{Raven}(x) \supset \text{Black}(x))$, for it entails the development of the latter for the class $\{a, c, d\}$, which can be written as follows: $(\text{Raven}(a) \supset \text{Black}(a)) \cdot (\text{Raven}(c) \supset \text{Black}(c)) \cdot (\text{Raven}(d) \supset \text{Black}(d))$.

It is now easy to define disconfirmation and neutrality:

(9.3 Df.) An observation report B disconfirms a hypothesis H if it confirms the denial of H.

(9.4 Df.) An observation report B is neutral with respect to a hypothesis H if B neither confirms nor disconfirms H.

By virtue of the criteria laid down in (9.2), (9.3), (9.4), every consistent observation report, B, divides all possible hypotheses into three mutually exclusive classes: Those confirmed by B, those disconfirmed by B, and those with respect to which B is neutral.

The definition of confirmation here proposed can be shown to satisfy all the formal conditions of adequacy embodies in (8.1), (8.2), and (8.3) and their consequences; for the condition (8.2) this is easy to see; for the other conditions the proof is more complicated.[33]

Furthermore, the application of the above definition of confirmation is not restricted to hypotheses of universal conditional form (as Nicod's criterion is, for example), nor to universal hypotheses in general; it applies, in fact, to any hypothesis which can be expressed by means of property and relation terms of the observational vocabulary of the given language, individual names, the customary connective symbols for "not," "and," "or," "if-then," and any number of universal and existential quantifiers.

Finally, as is suggested by the preceding illustrations as well as by the general considerations which underlie the establishment of the above definition, it seems that we have obtained a definition of confirmation which also is materially adequate in the sense of being a reasonable approximation to the intended meaning of confirmation.

. . .

NOTES

1. The present analysis of confirmation was to a large extent suggested and stimulated by a cooperative study of certain more general problems which were raised by Dr. Paul Oppenheim, and which I have been investigating with him for several years. These problems concern the form and the function of scientific laws and the comparative methodology of the different branches of empirical science. The discussion with Mr. Oppenheim of these issues suggested to me the central problem of the present essay. The more comprehensive problems just referred to will be dealt with by Mr. Oppenheim in a publication which he is now preparing.

In my occupation with the logical aspects of confirmation, I have benefited greatly by discussions with several students of logic, including Professor R. Carnap, Professor A. Tarski, and particularly Dr. Nelson Goodman, to whom I am indebted for several valuable suggestions which will be indicated subsequently.

A detailed exposition of the more technical aspects of the analysis of confirmation presented in this article is included in my article "A Purely Syntactical Definition of Confirmation," *The Journal of Symbolic Logic*, Vol. VIII (1943).

2. This point as well as the possibility of conclusive verification and conclusive falsification will be discussed in some detail in Sec. 10 of the present paper.

3. Or of the "logic of science," as understood by R. Carnap; cf. *The Logical Syntax of Language* (New York and London, 1937), Sec. 72, and the supplementary remarks in *Introduction to Semantics* (Cambridge, Mass., 1942), p. 250.

4. See the lucid presentation of this point in Karl Popper's *Logik der Forschung* (Wien, 1935), esp. Secs. 1, 2, 3, and 25, 26, 27; cf. also Albert Einstein's remarks in his lecture *On the Method of Theoretical Physics* (Oxford, 1933), pp. 11–12. Also of interest in this context is the critical discussion of induction by H. Feigl in "The Logical Character of the Principle of Induction," *Philosophy of Science*, Vol. I (1934).

5. Following R. Carnap's usage in "Testability and Meaning," *Philosophy of Science*, Vols. III (1936) and IV (1937); esp. Sec. 3 (in Vol. III).

6. This group includes the work of such writers as Janina Hosiasson-Lindenbaum (cf. for instance, her article "Induction et analogie: Comparaison de leur fondement," *Mind*, Vol. L (1941); (also see n. 24), H. Jeffreys, J. M. Keynes, B. O. Koopman, J. Nicod (see n. 15), St. Mazurkiewicz, F. Waismann. For a brief discussion of this conception of probability, see Ernest Nagel, *Principles of the Theory of Probability* (Internat. Encyclopedia of Unified Science, Vol. I, No. 6, Chicago, 1939), esp. Secs. 6 and 8.

7. The chief proponent of this view is Hans Reichenbach; cf. especially "Ueber Induktion und Wahrscheinlichkeit," *Erkenntnis*, Vol. V (1935), and *Experience and Prediction* (Chicago, 1938), Chap. v.

8. Cf. Karl Popper, *Logik der Forschung* (Wien, 1935), Sec. 80; Ernest Nagel, *loc. cit.*, Sec. 8, and "Probability and the Theory of Knowledge," *Philosophy of Science*, Vol. VI (1939); C. G. Hempel, "Le problème de la vérité," *Theoria* (Göteborg), Vol. III (1937), Sec. 5, and "On the Logical Form of Probability Statements," *Erkenntnis*, Vol. VII (1937–38), esp. Sec. 5. Cf. also Morton White, "Probability and Confirmation," *The Journal of Philosophy*, Vol. XXXVI (1939).

9. See, for example, J. M. Keynes, *A Treatise on Probability* (London, 1929), esp. Chap. iii; Ernest Nagel, *Principles of the Theory of Probability* (cf. n. 6 above), esp. p. 70. Compare also the somewhat less definitely sceptical statement by Carnap, *loc. cit.* (see n. 5), Sec. 3, p. 427.

10. See especially the survey of such factors given by Ernest Nagel in *Principles of the Theory of Probability* (cf. n. 6), pp. 66–73.

11. Cf. for example, A. J. Ayer, *Language, Truth and Logic* (London and New York, 1936), Chap. i; R. Carnap, "Testability and Meaning" (cf. n. 5), Secs. 1, 2, 3; H. Feigl, "Logical Empiricism" in *Twentieth Century Philosophy*, ed. Dagobert D. Runes (New York, 1943); P. W. Bridgman, *The Logic of Modern Physics* (New York, 1928).

12. It should be noted, however, that in his essay "Testability and Meaning" (cf. n. 5), R. Carnap has constructed definitions of testability and confirmability which avoid reference to the concept of confirming and of disconfirming evidence; in fact, no proposal for the definition of these latter concepts is made in that study.

13. A view of this kind has been expressed, for example, by M. Mandelbaum in "Causal Analyses in History," *Journal of the History of Ideas*, Vol. III (1942); cf. esp. pp. 46–47.

14. See Karl Popper's pertinent statement, *loc. cit.*, Sec. 8.

15. Jean Nicod, *Foundations of Geometry and Induction*, trans. P. P. Wiener (London, 1930), p. 219; cf. also R. M. Eaton's discussion of "Confirmation and Infirmation," which is based on Nicod's views; it is included in Chap. iii of his *General Logic* (New York, 1931).

16. In this paper, only the most elementary devices of this notation are used; the symbolism is essentially that of *Principia Mathematica*, except that parentheses are used instead of dots, and that existential quantification is symbolized by "(E)" instead of by the inverted "E."

17. This term is chosen for convenience, and in view of the above explicit formulation given by Nicod; it is not, of course, intended to imply that this conception of confirmation originated with Nicod.

18. For a rigorous formulation of the problem, it is necessary first to lay down assumptions as to the means of expression and the logical structure of the language in which the hypotheses are supposed to be formulated; the desideratum then calls for a definition of confirmation applicable to any hypotheses which can be expressed in the given language. Generally speaking, the problem becomes increasingly difficult with increasing richness and complexity of the assumed "language of science."

19. This difficulty was pointed out, in substance, in my article "Le problème de la vértité," *Taeoria* (Göteborg), Vol. III (1937), esp. p. 222.

20. For a more detailed account of the logical structure of scientific explanation and prediction, cf. C. G. Hempel, "The Function of General Laws in History," *The Journal of Philosophy*, Vol. XXXIX (1942), esp. Secs. 2, 3, 4. The characterization, given in that paper as well as in the above text, of explanations and predictions as arguments of a deductive logical structure, embodies an oversimplification: as will be shown in Sec. 7 of the present essay, explanations and predictions often involve "quasi-inductive" steps besides deductive ones. This point, however, does not affect the validity of the above argument.

21. This restriction is essential: In its general form, which applies to universal conditionals in any number of variables, Nicod's criterion cannot even be construed as expressing a sufficient condition of confirmation. This is shown by the following rather surprising example: Consider the hypothesis S_1:

$$(x)(y)[\sim(R(x, y) \cdot R(y, x)) \supset (R(x, y) \cdot \sim R(y, x))].$$

Let a, b be two objects such that $R(a, b)$ and $\sim R(b, a)$. Then clearly, the couple (a, b) satisfies both the antecedent and the consequent of the universal conditional S_1; hence, if Nicod's criterion in its general form is accepted as stating a sufficient condition of confirmation, (a, b) constitutes confirming evidence for S_1. However, S_1 can be shown to be equivalent to

$$S_2: (x)(y)R(x, y)$$

Now, by hypothesis, we have $\sim R(b, a)$; and this flatly contradicts S_2 and thus S_1. Thus, the couple (a, b), although satisfying both the antecedent and the consequent of the universal conditional S_1 actually constitutes disconfirming evidence of the strongest kind (conclusively disconfirming evidence, as we shall say later) for that sentence. This illustration reveals a striking and—as far as I am aware—hitherto unnoticed weakness of that conception of confirmation which underlies Nicod's criterion. In order to realize the bearing of our illustration upon Nicod's original formulation, let A and B be $\sim(R(x, y) \cdot R(y, x))$ and $R(x, y) \cdot \sim R(y, x)$ respectively. Then S_1 asserts that A entails B, and the couple (a, b) is a case of the presence of B in the presence of A; this should, according to Nicod, be favourable to S_1.

22. This observation is related to Mr. Oppenheim's methodological studies referred to in n. 1.

23. The basic idea of sect. (b) in the above analysis of the "paradoxes of confirmation" is due to Dr. Nelson Goodman, to whom I wish to reiterate my thanks for the help he rendered me, through many discussions, in clarifying my ideas on this point.

24. The considerations presented in section (b) above are also influenced by, though not identical in content with, the very illuminating discussion of the "paradoxes" by the Polish methodologist and logician Janina Hosiasson-Lindenbaum; cf. her article "On Confirmation," *The Journal of Symbolic Logic*, Vol. V (1940), especially Sec. 4. Dr. Hosiasson's attention had been called to the paradoxes by the article referred to in n. 2, and by discussions with the author. To my knowledge, hers has so far been the only publication which presents an explicit attempt to solve the problem. Her solution is based on a theory of degrees of confirmation, which is developed in the form of an uninterpreted axiomatic system (cf. n. 6 and part (b) in Sec. 1 of the present article), and most of her arguments presuppose that theoretical framework. I have profited, however, by some of Miss Hosiasson's more general observations which proved relevant for the analysis of the paradoxes of the non-gradated relation of confirmation which forms the object of the present study.

One point in those of Miss Hosiasson's comments which rest on her theory of degrees of confirmation is of particular interest, and I should like to discuss it briefly. Stated in reference to the raven-hypothesis, it consists in the suggestion that the finding of one non-black object which is no raven, while constituting confirming evidence for the hypothesis, would increase the degree of confirmation of the hypothesis by a smaller amount than the finding of one raven which is black. This is said to be so because the class of all ravens is much less numerous than that of all non-black objects, so that—to put the idea in suggestive though somewhat misleading terms—the finding of one black raven confirms a larger portion of the total content of the hypothesis than the finding of one non-black non-raven. In fact, from the basic assumptions of her theory, Miss Hosiasson is able to derive a theorem according to which the above statement about the relative increase in degree of confirmation will hold provided that actually the number of all ravens is small compared with the number of all non-black objects. But is this last numerical assumption actually warranted in the present case and analogously in all other "paradoxical" cases? The answer depends in part upon the logical structure of the language of science. If a "coordinate language" is used, in which, say, finite space-time regions figure as individuals, then the raven-hypothesis assumes some such form as "Every space-time region which contains a raven, contains something black"; and even if the total number of ravens ever to exist is finite, the class of space-time regions containing a raven has the power of the continuum, and so does the class of space-time regions containing something non-black; thus, for a coordinate language of the type under consideration, the above numerical assumption is not warranted. Now the use of a coordinate language may appear quite artificial in this particular illustration; but it will seem very appropriate in many other contexts, such as, e.g., that of physical field theories. On the other hand, Miss Hosiasson's numerical assumption may well be justified on the basis of a "thing language," in which physical objects of finite size function as individuals. Of course, even on this basis, it remains an empirical question, for every hypothesis of the form "All *P*'s are *Q*'s," whether actually the class of non-*Q*'s is much more numerous than the class of *P*'s; and in many cases this question will be very difficult to decide.

25. As a consequence of this stipulation, a contradictory observation report, such as {Black(a), ~Black(a)} confirms every sentence, because it has every sentence as a consequence. Of course, it is possible to exclude the possibility of contradictory observation reports altogether by a slight restriction of the definition of "observation report." There is, however, no important reason to do so.

26. Strictly speaking, Galileo's law and Kepler's laws can be deduced from the law of gravitation only if certain additional hypotheses—including the laws of motion—are presupposed; but this does not affect the point under discussion.

27. William Barrett, in a paper entitled "Discussion on Dewey's Logic" (*The Philosophical Review*, Vol. L (1941), pp. 305 ff., esp. p. 312) raises some questions closely related to what we have called above the consequence condition and the converse consequence condition. In fact, he invokes the latter (without stating it explicitly) in an argument which is designed to show that "not every observation which confirms a sentence need also confirm all its con-

sequences," in other words, that the special consequence condition (8.21) need not always be satisfied. He supports his point by reference to "the simplest case: the sentence C is an abbreviation of $A \cdot B$, and the observation 0 confirms A, *and so* C, but is irrelevant to B, which is a consequence of C." (Italics mine.)

For reasons contained in the above discussion of the consequence condition and the converse consequence condition, the application of the latter in the case under consideration seems to us unjustifiable, so that the illustration does not prove the author's point; and indeed, there seems to be every reason to preserve the unrestricted validity of the consequence condition. As a matter of fact, Mr. Barrett himself argues that "the degree of confirmation for the consequence of a sentence cannot be less than that of the sentence itself"; this is indeed quite sound; but it is hard to see how the recognition of this principle can be reconciled with a renunciation of the special consequence condition, since the latter may be considered simply as the correlate, for the non-gradated relation of confirmation, of the former principle which is adapted to the concept of degree of confirmation.

28. For a fourth condition, see n. 33.

29. A contradictory observation report confirms every hypothesis (cf. n. 8) and is, of course, incompatible with every one of the hypotheses it confirms.

30. This was pointed out to me by Dr. Nelson Goodman. The definition later to be outlined in this essay, which satisfies conditions (8.1), (8.2), and (8.3), lends itself, however, to certain generalizations which satisfy only the more liberal conditions of adequacy just considered.

31. In my article referred to in n. 1. The logical structure of the languages to which the definition in question is applicable is that of the lower functional calculus with individual constants, and with predicate constants of any degree. All sentences of the language are assumed to be formed exclusively by means of predicate constants, individual constants, individual variables, universal and existential quantifiers for individual variables, and the connective symbols of denial, conjunction, alternation, and implication. The use of predicate variables or of the identity sign is not permitted.

As to the predicate constants, they are all assumed to belong to the observational vocabulary, i.e. to denote a property or a relation observable by means of the accepted techniques. ("Abstract" predicate terms are supposed to be defined in terms of those of the observational vocabulary and then actually to be replaced by their *definientia*, so that they never occur explicitly.)

As a consequence of these stipulations, an observation report can be characterized simply as a conjunction of sentences of the kind illustrated by $P(a)$, $\sim P(b)$, $R(c, d)$, $\sim R(e, f)$, etc., where P, R, etc., belong to the observational vocabulary, and a, b, c, d, e, f, etc., are individual names, denoting specific objects. It is also possible to define an observation report more liberally as any sentence containing no quantifiers, which means that besides conjunctions also alternations and implication sentences formed out of the above kind of components are included among the observation reports.

32. I am indebted to Dr. Nelson Goodman for having suggested this idea; it initiated all those considerations which finally led to the definition to be outlined below.

33. For these proofs, see the article referred to in Part I, n. 1. I should like to take this opportunity to point out and to remedy a certain defect of the definition of confirmation which was developed in that article, and which has been outlined above: This defect was brought to my attention by a discussion with Dr. Olaf Helmer.

It will be agreed that an acceptable definition of confirmation should satisfy the following further condition which might well have been included among the logical standards of adequacy set up in Sec. 8 above: (8.4). If B_1 and B_2 are logically equivalent observation reports and B_1 confirms (disconfirms, is neutral with respect to) a hypothesis H, then B_2, too, confirms (disconfirms, is neutral with respect to) H. This condition is indeed satisfied if observation reports are construed, as they have been in this article, as classes or conjunctions of observation sentences. As was indicated at the end of n. 14, however, this restriction of observation reports to a conjunctive form is not essential; in fact, it has been adopted here only for greater convenience of exposition, and all the preceding results, including especially the definitions and theorems of the present section, remain applicable without change if observation reports are given the more liberal interpretation characterized at the end of n. 14. (In this case, if P and Q belong to the observational vocabulary, such sentences as $P(a) \vee Q(a)$, $P(a) \vee \sim Q(b)$, etc., would qualify as observation reports.) This broader conception of observation reports was therefore adopted in the article referred to in Part I, n. 1; but it has turned out that in this case, the definition of confirmation summarized above does not generally satisfy the requirement (8.4). Thus, e.g., the observation reports, $B_1 = P(a)$ and $B_2 = P(a) \cdot (Q(b) \vee \sim Q(b))$ are logically equivalent, but while B_1 confirms (and even directly confirms) the hypothesis $H_1 = (x)P(x)$, the second report does not do so, essentially because it does not entail $P(a) \cdot P(b)$, which is the development of H_1 for the class of those objects mentioned in B_2. This deficiency can be remedied as follows: The fact that B_2 fails to confirm H_1 is obviously due to the circumstance that B_2 contains the individual constant b, without asserting anything about b: The object b is mentioned only in an analytic component of B_2. The atomic constituent $Q(b)$ will therefore be said to occur (twice) inessentially in B_2. Generally, an atomic constituent A of a molecular sentence S will be said to occur inessentially in S if by virtue of the rules of the sentential calculus S is equivalent to a molecular sentence in which A does not occur at all. Now an object will be said to be mentioned inessentially in an observation report if it is mentioned only in such components of that report as occur inessentially in it. The sentential calculus clearly provides mechanical procedures for deciding whether a given observation report mentions any object inessentially, and for establishing equivalent formulations of the same report in which no object is mentioned inessentially. Finally, let us say that an object is mentioned essentially in an observation report if it is mentioned, but not only mentioned inessentially, in that report. Now we replace (9.1) by the following definition:

(9.1*a*) An observation report B directly confirms a hypothesis H if B entails the development of H for the class of those objects which are mentioned essentially in B.

The concept of confirmation as defined by (9.1*a*) and (9.2) now satisfies (8.4) in addition to (8.1), (8.2), (8.3) even if observation reports are construed in the broader fashion characterized earlier in this footnote.

B. A. Brody

CONFIRMATION AND EXPLANATION

Carl Hempel's "Studies in the Logic of Confirmation"[1] has given rise to much discussion. This discussion has centered primarily, however, around a side issue, the paradox of the ravens, and little has been said about the primary concern of that paper, viz., the formulation of certain conditions of adequacy for any qualitative confirmation function and the definition of a function that satisfies these conditions. This paper is concerned with this more central issue.

Hempel showed that, if a qualitative confirmation function satisfies the following two conditions, then any evidence that (qualitatively) confirms any statement (qualitatively) confirms every statement:

> *Special-consequence condition:* If a statement confirms (qualitatively) a hypothesis H, it also confirms (qualitatively) every consequence of H.
> *Converse-consequence condition:* If a statement confirms (qualitatively) a hypothesis H, it also confirms (qualitatively) every hypothesis that entails H.

Given this result, and given certain additional arguments which will be discussed below, various authors have suggested dropping one (or both) of these conditions of adequacy.

It is the purpose of this paper to show that one cannot simply drop either of these two conditions, that one can drop them only if one replaces them by some other conditions that often entail the same thing, and that the adoption of these other conditions sheds much light upon the process of qualitative confirmation.

From the *Journal of Philosophy*, LXV, No. 10, May 16, 1968. Reprinted by permission of the editor and the author.

I

What is a qualitative confirmation function? It is a function whose domain contains ordered pairs of statements (the first of which is the hypothesis and the second of which is the evidence) and whose converse domain contains the three relations of confirmation, disconfirmation, and irrelevancy. Roughly speaking, a qualitative confirmation function says *whether* a given body of evidence confirms, disconfirms, or is irrelevant to a given hypothesis. It does not, however, give any information about degree of confirmation or disconfirmation, if any.

The domain of such a function comprises ordered pairs of statements. It might be objected that such a function is not what is needed. After all, the typical case in science is one in which we have a certain body of evidence for a hypothesis and the question before us is whether some additional evidence confirms, disconfirms, or is irrelevant to the hypothesis. What is really needed, therefore, is a function whose domain contains ordered triplets, the first member of which is the hypothesis, the second member of which is the background evidence, and the third member of which is the new evidence.

This objection rests, of course, upon the assumption that whether some new evidence confirms, disconfirms, or is irrelevant to a hypothesis is partially determined by the additional evidence already present. But why should this be the case? One might have in mind the possibility that some evidence is irrelevant by itself but highly significant when added to some previous evidence. But in that case, a proper function should say that the evidence, by itself, is irrelevant, but that new evidence conjoined to some part of the previous evidence is not irrelevant; this is exactly what a function whose domain comprises ordered pairs would say, so this type of case does not force us to have functions whose domain contains ordered triplets. One might also have in mind the possibility of the new evidence's being irrelevant because, given the old evidence, it "tells us nothing new," i.e., it does not make the hypothesis any more likely than it was. This is, however, an introduction of quantitative considerations, and we are not at present concerned with these. It seems,[2] therefore, that nothing will be lost (if we are not concerned with quantitative questions) if we confine ourselves to functions whose domains are ordered pairs.

There seem to be three reasons for adopting something like both of the above conditions for qualitative confirmation functions:

(1) There are many perfectly acceptable scientific inferences that are justified if one's qualitative confirmation function satisfies these conditions. Consider, for example, the following two cases:

(a) Evidence for Boyle's law is, if one's qualitative confirmation function satisfies the converse-consequence condition, evidence for the Boyle-Charles law. Consequently, if one's qualitative confirmation function satisfies the special-consequence condition, it is indirect evidence for Charles's law.

(b) Evidence for the law of definite proportions is, if one's qualitative confirmation function satisfies the converse-consequence condition, evidence for the atomic theory. Consequently, if one's qualitative confirmation function satisfies the special-consequence condition, it is indirect evidence for the law of equivalent proportions. In other words, the validity of the usual inferences whereby evidence for one law is indirect evidence for other laws can plausibly be understood as depending upon qualitative confirmation functions satisfying these two conditions.

(2) There are good reasons for supposing that inferences whose validity can plausibly be understood as depending upon qualitative confirmation functions satisfying these two conditions are essential in science. To begin with, consider the process whereby experimental evidence confirms theoretical statements. This type of confirmation has posed great difficulties even to those who claim to have an account of the process whereby observational data confirm observational laws.

A very plausible way of understanding this process is to suppose that experimental evidence directly confirms observational laws, and, if these laws are entailed by some theory, then, by virtue of the fact that qualitative confirmation functions satisfy the converse-consequence condition, the experimental evidence also confirms the theory.

Israel Scheffler has objected to Hempel's satisfaction criterion of confirmation on the grounds that it does not explain how theoretical statements are confirmed by experimental evidence. We can now see that this objection is not well put. The confirmation of theoretical statements is a problem for Hempel because the function he proposes does not satisfy the converse-consequence condition and not because of the particular function that he proposes. This last point can be put as follows: Scheffler claims that,

> Even if the notion of an instance according with a hypothesis is satisfactorily explained for purely observational languages, then, it seems likely that theoretical languages present problems of a different order, requiring solutions of quite different sorts.[3]

If, however, qualitative confirmation functions satisfy the converse-consequence condition, then a satisfactory account of the confirmation of hypotheses in the "purely observational vocabulary" quickly yields a satisfactory account of the confirmation of theoretical hypotheses.

Let us now consider the case of the application of scientific theory to a new area. It is very common to deduce from a theory (and, possibly, some additional information) what must happen in these new areas, and if the theory is well confirmed, one may act upon its consequences in this new area without testing them. This is one of the things that make the possession of theories so desirable. But the ability to do this seems to depend upon qualitative confirmation functions satisfying the special-consequence condition, since one is assuming that the evidence for the theory is also evidence for the consequences of the theory in this new area.

The following objection might be raised at this point: The arguments I have given so far rest upon the assumption that the relation between theory and particular observation statements is mediated by observational generalizations. Thus, the first argument rests upon the assumption that observational data can directly confirm only observational generalizations; they confirm theoretical hypotheses indirectly by virtue of something like the converse-consequence condition. The second argument rests upon the assumption that the way you apply a theory to a new area is to deduce from it a generalization from which you can deduce particular consequences. But this is a mistaken view of the relation between theories and particular observation statements. Once this assumption is given up, then both of the above arguments collapse. If data can directly confirm theories, then the converse-consequence condition is not needed to account for the confirmation of theories, and if a theory is directly applied to a particular case, then there is no need for the special-consequence condition to account for our being warranted in depending upon the consequence of the theory.

The second half of this objection simply rests upon a confusion. It doesn't make any difference for the purpose of our argument whether one derives the prediction directly from the theory or indirectly via a generalization. In either case, the warrant for the prediction can only be based upon the evidence for the theory, and this seems to presuppose that qualitative confirmation functions do satisfy the special-consequence condition.

The first half of this objection does not rest upon any such confusion. Indeed, it does refute one version of the claim that the confirmation of theoretical statements presupposes something like the converse-consequence condition. That claim might mean that, in order to account for the confirmation of theoretical statements, one would need two confirmation functions:

(1) A function f_1 that would say whether a given evidence statement directly confirms, disconfirms, or is irrelevant to a given observational hypothesis;

(2) A function f_2 that would say whether a given evidence statement confirms, disconfirms, or is irrelevant to a given hypothesis. This function would be governed by the following principles:

(a) If $f_1(H, E) =$ confirms, then $f_2(H, E) =$ confirms. The same thing holds for "disconfirms."

(b) If there exists an H such that $f_1(H, E) =$ confirms, then $f_2(G, E) =$ confirms, for every G such that G entails H. Similarly for "disconfirms."

But if, as the objection would have it, empirical evidence can directly confirm theoretical hypotheses, then the above claim would not be justified, i.e., the above version of the converse-consequence condition is not needed to account for the confirmation of theoretical statements.[4]

This does not mean, however, that the converse-consequence condition is still not needed. Although it need not be part of the definition of a qualitative confirmation function, it still is a condition of adequacy for qualitative confirmation functions. After all, in those cases where $f_2(G, E) =$ confirms because there exists an H such that G entails H and $f_1(H, E) =$ confirms, we expect it to be the case (for any adequate qualitative confirmation function that allows for the direct confirmation of theoretical hypotheses) that the evidence E

directly confirms *G*. So, in either case, we do have, and must have, something like the converse-consequence condition.[5]

(3) Finally, over and above the fact that something like both of these conditions is used and the fact that something like both of them seems necessary, one should also note that these conditions are intuitively very plausible. Hempel has expressed very well the intuitive motivation for the special-consequence condition:

> . . . any such consequence is but an assertion of all or part of the combined content of the original hypotheses and has therefore to be regarded as confirmed by any evidence which confirms all of the latter (31).

But there are also extremely powerful intuitive motivations for the converse-consequence condition. After all, when some observational data confirm a law, it is not the case that the evidence exhausts the content of the law. Rather, given the fact that part of what the law says (the part embodied in the evidence) holds, that fact serves to confirm the hypothesis that the rest of what the law says also holds. But if this is so, why shouldn't the data also confirm a more general law (or a theory)? After all, the evidence does show that part of what the more general law (or theory) says is the case actually is the case.

Of course, these intuitive considerations are not by themselves sufficient to definitely justify the retention of both of the conditions. Indeed, one might look at the proof of the paradox that results from adopting them as a proof that our intuitions have misled us here. But given these intuitive considerations and given our other reasons for retaining the two conditions, we might do better to look for a different way of avoiding the paradoxical result. We shall discuss such alternative ways out in a later section.

II

Before turning to the question of the supposed paradoxical result, let us look at several other arguments that have been raised against the adoption of one or both of these conditions. Brian Skyrms[6] has recently offered the following argument against the adoption of the special-consequence condition:

> If we mean by "confirming instance" an instance on whose sole evidence the probability of the hypothesis is higher than on merely tautological evidence (and there is quite a lot of textual support in Hempel's article for this reading), then the special consequence condition is false. For an instance may confirm a statement p without confirming its consequence $p \vee \sim p$, since the latter presumably has probability 1 on tautological evidence, and thus this probability cannot be raised by adding a statement of the instance to the evidential base (238).

This argument, as it stands, involves essentially quantitative considerations concerning the question of whether or not the evidence raises the degree of confirmation of some statement. As such, it does not apply directly to qualitative confirmation functions.[7] But a parallel argument could be given involving purely qualitative considerations, and the reply to it will also handle the quantitative form of the argument. The purely qualitative argument is: One intuitively thinks of empirical evidence as neither confirming nor disconfirming a tautological statement. But a tautological statement is a consequence of any statement. So, if qualitative confirmation functions satisfy the special-consequence condition, a tautology is confirmed by evidence that confirms any statement.

There is an obvious way to avoid this difficulty. It involves modifying the special-consequence condition slightly so that it now becomes:

> If a statement confirms a hypothesis *H*, then it also confirms every consequence of *H* that is not a tautology.

The rationale for this modification is: (a) it enables us to avoid Skyrm's argument without sacrificing anything, and (b) intuitively, the tautology is not part of the content of the original hypothesis, and the intuitive motivation for adopting the special-consequence condition is that one feels that the consequence is part of the content of the statement that has been confirmed.

This same type of move can also be used to avoid an argument recently given by Hempel against the consequence condition, a generalized form of the special-consequence condition, which says that:

> If an observation report confirms every one of a class *K* of sentences, it also confirms any sentence that is a logical consequence of *K*.

The argument against this condition goes as follows: Imagine some observational evidence which confirms two incompatible hypotheses.[8] Then let the class *K* be the class whose members are just those two hypotheses. *K* is inconsistent, and entails any statement. Consequently, if qualitative confirmation functions satisfy the general-consequence condition, then the original evidence confirms any statement. The way out is, of course, to avoid this type of case by modifying the condition as follows:

> If an observation report confirms every one of a class *K* of sentences and this class is not inconsistent, then it also confirms any sentence that is a logical consequence of *K*.

Let us now consider the more elaborate arguments given by Carnap against both of the conditions. Carnap first constructs the following case, where some evidence confirms a hypothesis *H* and another hypothesis *K* but does not confirm $H \vee K$. Since $H \vee K$ is a consequence both of *H* and of *K*, the special consequence condition should not be adopted:

> The initial evidence is that ten chess players will participate in a tournament. Some are men (M) and others women (W). Some are from New York and others not. Some are junior players and others are senior players. Their distribution is as follows:

	New Yorkers	Strangers
Juniors	M, W, W	M, M
Seniors	M, M	W, W, W

> Moreover, we know that only one will win and that each has an equal chance of winning. Let *H* = "A woman wins," and let *K* = "A stranger wins."

> Then the degree of confirmation of H and the degree of confirmation of K on the initial evidence is .5. Now let us suppose that we are given the additional information that a senior player has won. Then the degree of confirmation of H given this new evidence is .6; so this new evidence confirms H. Similarly, the degree of confirmation of K given this new evidence is .6; so this new evidence confirms K. The degree of confirmation of $H \vee K$ on the initial evidence was, however, .7; but, given this new information, it is only .6. The new evidence therefore disconfirms $H \vee K$ even though it confirms H and confirms K, and either of the latter entails the former.

A similar example is used by Carnap against the converse-consequence condition:

> Let the initial evidence be as before, let $H =$ "A New Yorker wins," and let $K =$ "A junior player wins." The degree of confirmation of H and of K on the initial evidence is .5, and the degree of confirmation of $H \cdot K$ on that evidence is .3. Now let us suppose that we are given the additional information that a man has won. The degree of confirmation of H and of K given this new evidence is .6; the degree of confirmation of $H \cdot K$ given this new evidence is .2. So, although the new evidence confirms both H and K, it disconfirms $H \cdot K$, even though the latter entails either of the former.

One might object to this type of argument against the conditions on the grounds that it presupposes certain numerical assignments of degrees of confirmation. Perhaps these examples show that these assignments are mistaken since they lead to the violation of certain acceptable principles, viz., the special-consequence and the converse-consequence conditions.

Such an objection is not, however, very persuasive. After all, the assignment of degree of confirmation in the above cases rested on the following very plausible principle: If the statistical probability that an event of type A is also an event of type B is r, then the degree of confirmation of the hypothesis that a particular event is an event of type B on the evidence that it is an event of type A and the information about the statistical probability is also r. Besides being very plausible, this principle seems necessary if we are to relate "degree of confirmation" to "fair betting quotient."

Another, and more serious, objection to Carnap's argument is that he is assuming that the proper quantitative analogue to the qualitative notion of "confirmatory evidence" is "evidence that raises the degree of confirmation from what it was on the prior evidence alone." But perhaps the proper quantitative analogue is "evidence such that, on the basis of it, the hypothesis has a degree of confirmation greater than r, where r is a fixed value (perhaps 1/2)."

The adoption of this proposal will do away with Carnap's objection to the special-consequence condition. After all, it is not possible to construct a case where A entails B, $c(A, E) > r$, and $c(B, E) \leq r$; and it is only the fulfillment of all these conditions that would enable Carnap to modify his argument to meet this objection. This objection does not, however, do away with Carnap's objection to the converse-consequence condition. Indeed, if $r = 1/2$, then H (and K) would be confirmed both by the initial evidence and by the new evidence, but $H \cdot K$ would be confirmed by neither. It is easy to see, moreover, that nothing will be accomplished by raising the value of r; one can easily replace this example with another in which the hypotheses have the appropriate degrees of confirmation.

One might conclude, therefore, that we ought to adopt this new proposal as the correct relation between qualitative and quantitative confirmation functions, and conclude that, whereas the special-consequence condition ought to be retained, it is a mistake to retain the converse-consequence condition. It seems to me, however, that this would be a mistake. After all, one would intuitively want to say that a statement reporting the observation of a black raven confirms (qualitatively) the hypothesis "all ravens are black," but it is not at all clear that it, by itself, would give the hypothesis a degree of confirmation greater than some plausible value for r. So it would probably be very rash to adopt this proposal as a way out of Carnap's objections.

There is a much more fundamental objection to both of Carnap's arguments. As the second objection has rightly pointed out, all that Carnap's arguments really show is that neither condition holds for reasonable quantitive confirmation functions. The conditions show anything about qualitative confirmation functions only if certain quantitaive notions are treated as analogous to certain qualitative notions. The second objection claims that Carnap was wrong about what is the proper quantitative analogue of the qualitative notion of confirmatory evidence. It assumed, as Carnap had, that there was such an analogue; perhaps it is this assumption that should be challenged.

These are two ways in which a quantitative notion can be the analogue of a qualitative confirmation function:

(1) If E (qualitatively) confirms H, E raises the degree of confirmation of H.
(2) If E raises the degree of confirmation of H, E (qualitatively) confirms H.

Now let us for the moment grant[9] that there is, in sense (1), a quantitative analogue for a qualitative confirmation function. One might nevertheless deny that it is an analogue in sense (2), on the grounds that there are certain confirmation relations that essentially involve degrees of confirmation and should not therefore be reflected in a qualitative confirmation function. Indeed, Carnap's examples are prime candidates for such confirmation relations.

This point can also be put as follows: There are two ways of looking at a qualitative confirmation function. One way (which Carnap usually, and Hempel sometimes, adopts) is to view it as a function that says whether or not the degree of confirmation has been raised or lowered, but does not say how much. If one has this picture of a qualitative confirmation function, then there should be, in sense (2), an analogous quantitative confirmation function. There is, however, another way of viewing such a function. According to this view, a qualitative confirmation function tells us whether or not certain events, as reported in the evidence, are instances of a general hypothesis. In other words, this views sees a qualitative confirmation function as being concerned with one type of confirmation, a purely qualitative type of confirmation, and does not see it as a dequantified version of a quantitative confirmation function. If one adopts this view, then there is no reason to suppose that an adequate qualitative confirmation function need have, in sense (2), a quantitative analogue.

This suggestion naturally raises the following question: When does confirmatory evidence qualitatively confirm a hypothesis? One might suggest that only evidence that necessarily raises (or, at least, cannot lower) the degree of confirmation of a hypothesis, no matter what the initial evidence was, qualitatively confirms that hypothesis. But evidence, like the new evidence in Carnap's cases, that raises the degree of confirmation of a hypothesis from what it was on certain initial evidence, but which would have lowered it from what it would have been on other initial evidence, does not qualitatively confirm that hypothesis. The trouble with this suggestion is that it would entail that no evidence qualitatively confirms any hypothesis. As I. J. Good has shown,[10] even "*a* is a raven and *a* is black" can lower the degree of confirmation of "all ravens are black" from what it would have been on certain initial evidence.

One might perhaps also suggest that only evidence that raises the degree of confirmation of a hypothesis from what it was on purely tautological evidence qualitatively confirms that hypothesis. But this suggestion allows in too much. If we change Carnap's cases so that the initial evidence is only tautological evidence and the new evidence is his initial evidence together with his new evidence, then those cases will also be cases of purely qualitative confirmation, but those are just the cases we intuitively want to rule out.

The trouble with both of these suggestions is that they attempt to explain the notion of purely qualitative evidence in terms of what this evidence does to the degree of confirmation, and it seems that this cannot be done. But our inability to do this does not destroy the distinction between purely qualitative evidence, i.e., instantial evidence, and evidence that essentially involves quantitative considerations. Having this distinction, we are able to deny the claim that qualitative confirmation functions have to have quantitative analogues in sense (2).

If, however, qualitative confirmation functions do not have quantitative analogues in sense (2), then Carnap's argument, which depends upon the assumption that they do, fails. Indeed, Carnap's cases are good examples of essentially quantitative confirmation relations, and none of the evidence (whether initial or new), in either of his two cases, either confirms or disconfirms (qualitatively) the hypotheses in question. Consequently, although Carnap has shown that neither the special-consequence nor the converse-consequence condition holds for quantitative confirmation functions, he has not shown that they do not hold for qualitative confirmation functions.

III

We have seen so far that there are several good reasons for retaining something like the special-consequence and the converse-consequence conditions for qualitative confirmation functions, and that various arguments (other than the paradox)[11] for rejecting either of them are not very persuasive. Let us now consider Hempel's argument:

(1) Assume that both of these conditions hold and that there is a statement E that confirms a hypothesis H_1.
(2) Let H_2 be any other hypothesis, no matter how irrelevant. Then $H_1 \cdot H_2$ entails H_1.
(3) Therefore, by the converse-consequence condition, E confirms $H_1 \cdot H_2$.
(4) $H_1 \cdot H_2$ entails H_2.
(5) Therefore, by the special-consequence condition, E confirms H_2; and any evidence that confirms a hypothesis confirms any other hypothesis as well.

This paradox, like all good paradoxes, perplexes because its argument mirrors so closely other quite legitimate arguments. The structure of the argument is the same as that of all arguments whereby evidence for one hypothesis indirectly confirms another, e.g., the argument given above for the indirect confirmation of Charles's law by the evidence that directly confirms Boyle's law.

Skyrms (p. 238) has recently shown that the mere adoption of the converse-consequence condition will lead to the same type of result:

(1) H_1 certainly confirms $H_1 \vee H_2$.
(2) H_2 entails $H_1 \vee H_2$.
(3) Therefore, by the converse-consequence condition, H_1 is evidence for H_2; so any statement confirms any other statement.

Something has gone wrong. But what? The argument is certainly valid; so it must be one of the assumptions. The most prominent of these are the special-consequence and converse-consequence conditions; so the standard move has been to drop one or both of these. If, however, our arguments so far have been correct, then it is not at all clear that we can simply drop either of these conditions without replacing it by something else that does the legitimate work that it can do. Nor can we (without absurdity) drop the assumption that there is some statement that confirms some hypothesis, or that there is some other hypothesis, or that H_1 confirms $H_1 \vee H_2$. There remain the assumptions that $H_1 \cdot H_2$ entails both H_1 and H_2, and that H_2 entails $H_1 \vee H_2$, despite the irrelevance of H_1 and H_2 to each other.

But can these assumptions be dropped? If by "entailment" one means "necessity of the material implication," then, of course, this assumption cannot be challenged. Saying this immediately suggests, however, a possible way out of these paradoxes: Perhaps what these paradoxes show us is that we cannot simply transfer our ordinary deductive logic from the context of mathematics, where it is perfectly acceptable, to the context of inductive confirmation. Perhaps what these paradoxes show us is that a different deductive logic, perhaps involving a different notion of "entailment," is needed for use in the logic of confirmation. But before we try this move, let us see if we can find a simpler way of changing our logic so as to avoid the difficulty raised by Hempel and Skyrms.

Perhaps the simplest suggestion for a nonstandard logic that would avoid the paradoxes is to change the formation rules of logic so as to bring "·" and "v" closer to "and" and "or." One is inclined to say that the sentence "John went home and the earth is about 93,000,000 miles from the sun" is deviant despite the fact that it contains two nondeviant declarative sentences conjoined by "and." This reflects an important difference between "and" and "·" (as used in ordinary logic). Whereas "·" can be placed between any two declarative sentences, "and" cannot. P. F. Strawson has put this point well:

> But we do not string together at random any assertions we consider true; we bring them together, in spoken or written sentences or paragraphs, only when there is some further reason for the rapprochement.[12]

What we have said so far applies just as well to the case of "or" and "v." "Either John went home or the earth is about 93,000,000 miles from the sun" is just as deviant as "John went home and the earth is about 93,000,000 miles from the sun." Once more, there must be some reason why the disjuncts are joined together.

Given these facts, it would seem that there is a simple way of avoiding the paradoxes. If H_1 and H_2 are not relevant to each other, then $H_1 \cdot H_2$ cannot, in a correct system for confirmation functions, entail either H_1 or H_2 because $H_1 \cdot H_2$ is not well-formed. Similarly, when H_1 and H_2 are not relevant to each other, H_2 does not entail, in a correct system for confirmation functions, $H_1 \vee H_2$ because the latter is not well-formed.

But what is the appropriate sense of "relevant"? It is not easy to see how this question is to be answered. It must be remembered, moreover, that our proposal will not work if every sentence that is nondeviant in English is well-formed in our logic. "John went home and had dinner" is nondeviant in English, but it cannot be well-formed in our logic (the two sentences are not relevant to each other in the appropriate sense) since that would entail the counterintuitive result that evidence for "John went home" is always evidence for "John had dinner." Once this is realized, however, one begins to suspect that the appropriate sense of "relevant to each other" is simply "evidence for one is evidence for the other," and we aren't going to be able to use the former to explicate the latter. This proposal is initially plausible only because one has in mind some more prosaic sense of "relevant"; once one realizes what is involved here, one feels little confidence in this way out of the paradox.

Are there any other syntactic methods for solving our problems? The following might be suggested: The derivation of the paradoxical results depends upon the assumptions that $A \cdot B$ entails A (and entails B) and that A (or B) entails $A \vee B$. Now it is certainly a theorem that $A \cdot B \supset A(A \cdot B \supset B)$ and that $A(B) \supset A \vee B$, and it is also certainly true that if we use "—entails . . . " as an equivalent for "the material implication whose antecedent is—and whose consequence is . . . is necessarily true," then $A \cdot B$ does entail $A(B)$ and $A(B)$ does entail $A \vee B$. But there are well-known objections to such an interpretation of "entails," and one of them, due to E. Nelson, is certainly relevant here:

> It cannot be asserted that the conjunction of p and q *entails* p, for q may be totally irrelevant to and independent of p, in which case p and q do not entail p, but it is

> only p that entails p . . . Furthermore, "p entails p or q" cannot be asserted on logical grounds, because from an analyses of p we cannot derive the propositional function "p or q" where q is a variable standing for just any other propositional function whatsoever.[13]

If, then, we can find some system that characterizes this different notion of entailment, then perhaps we can use it in our conditions, thereby retaining them in the perfectly acceptable cases but avoiding the paradoxical results that Hempel and Skyrms derived from them.

What condition does such a notion of entailment have to satisfy in order to solve our problem? One might begin by suggesting that it is sufficient that it satisfy the following conditions:

> *Condition 1:* In order for p to entail q, either p is not of the form $r_1 \cdot \ldots \cdot r_n$ $(n \geq 2)$, or p is of the form $r_1 \cdot \ldots \cdot r_n$ $(n \geq 2)$ but there is no $r_a, \ldots, r_i$ such that $(r_a, \ldots, r_i)$ is a proper subset of $(r_1, \ldots, r_n)$ and $r_a \cdot \ldots \cdot r_i$ entails q.
>
> *Condition 2:* In order for p to entail q, either q is not of the form $r_1 \vee \ldots \vee r_n$ $(n \geq 2)$, or q is of the form $r_1, \vee \ldots \vee r_n$ $(n \geq 2)$ but there is no $r_a, \ldots, r_i$ such that $(r_a, \ldots, r_i)$ is a proper subset of $(r_1, \ldots, r_n)$ and p entails $r_a \vee \ldots \vee r_i$.

One can immediately see that, if we replace, in the special-consequence and converse-consequence conditions, the ordinary notion of entailment by some notion of entailment that satisfies conditions 1 and 2, then Hempel's and Skyrms's derivations of the paradoxical results will be blocked.

Such a notion of entailment, however, need not necessarily solve our problem, because it necessarily rules out only entailments whose premises and/or conclusion are of a certain form, and does not necessarily rule out entailments whose premises and/or conclusion are not of that form but are equivalent to statements of that form. To see that this is a problem, consider the following derivations:

(1) Assume that both the special-consequence and the converse-consequence conditions hold and that there is a statement (e) that confirms a hypothesis H_1.
(2) Let H_2 be any other hypothesis, no matter how irrelevant. Then $\sim(\sim H_1 \vee \sim H_2)$ entails H_1.
(3) Therefore, by the converse-consequence condition, e confirms $\sim(\sim H_1 \vee \sim H_2)$.
(4) But $\sim(\sim H_1 \vee \sim H_2)$ also entails H_2.
(5) Therefore, by the special-consequence condition, e confirms H_2; so evidence that confirms any statement confirms every other statement.

(1) H_1 certainly confirms $\sim(\sim H_1 \cdot \sim H_2)$.
(2) H_2 entails $\sim(\sim H_1 \cdot \sim H_2)$.
(3) Therefore, by the converse-consequence condition, H_1 is evidence for H_2; so any statement confirms every other statement.

These derivations depend, of course, upon the assumptions that $\sim(\sim H_1 \vee \sim H_2)$ entails $H_1(H_2)$ and $H_2(H_1)$ entails $\sim(\sim H_1 \cdot \sim H_2)$, and the problem is that a notion of entailment that merely satisfies conditions 1 and 2 might allow these entailments. Consequently, in order to solve our problem, it is not sufficient that we substitute (in the statement of the special-consequence and converse-consequence conditions) a notion of entailment that satisfies conditions 1 and 2.

Once this point has been noted, however, the following suggestion seems almost inevitable: In order to solve our problem, it is sufficient that we replace (in the statement of the special-consequence and converse-consequence conditions) the ordinary notion of entailment by a notion of entailment that satisfies the following conditions:

Condition 3: Let $(r_1{}^1 \cdot \ldots \cdot r^1{}_m, r_1{}^2 \cdot \ldots \cdot r_m{}^2, \ldots)$ be the set of all conjunctive propositions normally equivalent to p (the set of p's coneqs). Then, in order for p to entail q, there must be no $r_a, \ldots, r_i$ $(i \geq 1)$ such that $r_a \cdot \ldots \cdot r_i$ entails q and $(r_a, \ldots, r_i)$ is a proper subset of some $(r_\alpha, \ldots, r_\varepsilon)$ such that $r_\alpha \cdot \ldots \cdot r_\varepsilon$ is a member of the set of p coneqs.

Condition 4: Let $(r_1{}^1 \vee \ldots \vee r_m{}^1, r_1{}^2 \vee \ldots \vee r_m{}^2, \ldots)$ be the set of all disjunctive propostions normally equivalent to q (the set of q's diseqs). Then in order for p to entail q, there must be no $r_a, \ldots, r_i$ $(i \geq 1)$ such that p entails $r_a \vee \ldots \vee r_i$ and $(r_a, \ldots, r_i)$ is a proper subset of some $(r_\alpha, \ldots, r_\varepsilon)$ such that $r_\alpha \vee \ldots \vee r_\varepsilon$ is a member of the set of q's diseqs.

It would appear that if we replace, in the special-consequence and converse-consequence conditions, the ordinary notion of entailment by some notion of entailment that satisfies conditions 3 and 4, then both Hempel's and Skyrms's original derivations of the paradoxical results and our revised derivations of these results will be blocked.

But this proposal is too drastic for two reasons. To begin with,[14] no proposition p entails any proposition q if condition 3 is satisfied. After all, assume that some p entails some q. Then p is normally equivalent to $p \cdot q$, and therefore, if condition 3 is satisfied, p does not entail q. Our assumption has led to a contradiction, and we can conclude that p does not entail q, no matter what propositions p and q are. Secondly, the Boyle-Charles law is, in ordinary logic, equivalent to the conjunction of Boyle's law and Charles's law. Now evidence that qualitatively confirms Boyle's law must qualitatively confirm the Boyle-Charles law. We argued, earlier in this paper, that this was due to the fact that qualitative confirmation functions satisfy the converse-consequence condition. But if we substitute a notion of entailment that satisfies condition 3, this explanation will not work, since the Boyle-Charles law will then not entail Boyle's law.

But even if we could construct a system that would seem to avoid all these difficulties and even if we could have reasons for supposing that similar difficulties would not arise again, this would not constitute a solution to the puzzles that have arisen. Instead, it raises a very deep philosophical issue: Why does the adoption of this strange notion of entailment eliminate the difficulties we have raised? Technical moves of the type we have been discussing shed no light on the process of qualitative confirmation, and we are left with the problem of understanding why they do work.[15]

IV

We have seen so far that the following is true: (1) It would be a mistake to drop either of our conditions of adequacy without substituting something else that would ensure the validity of certain legitimate and necessary inferences; (2) there are serious difficulties with the suggestion that we can avoid Hempel's and Skyrms's paradoxes by modifying the notion of entailment involved in these conditions; (3) even if these technical difficulties could be avoided, such an attempt would not really solve our philosophical puzzle. In this section, we will suggest an alternative solution to our problem, one that will not only resolve it but will also explain why our previous attempts were not successful.

It has been suggested[16] that when, on the basis of inductive evidence, we accept a certain conclusion, the conclusion that we accept is the best explanation of the evidence, and it is accepted precisely because it is the best explanation. Although this suggestion was made for inductive acceptance rules and not for qualitative confirmation functions, it seems to me that something like it could be fruitfully applied to such functions. Indeed, the adoption of this type of move may solve the problems that we have been concerned with until now.

Let us formulate the following condition to replace the converse-consequence condition:

If e confirms H_1 and H_2 explains H_1, then e confirms H_2.

The first thing to note is that if H_2 explains H_1 it is often the case that it will also entail H_1. Consequently, if we adopt this condition, there will be many cases in which evidence for an H_1 is also evidence for an H_2 which entails that H_1, i.e., there will be many cases in which the import of the converse-consequence condition for those cases does hold. On the other hand, there are many cases in which H_2 entails H_1 but does not explain it. If we adopt our new condition, then, in these cases, evidence for H_1 will not be evidence for H_2, i.e., the import of the converse-consequence condition for these cases will not hold in these cases. So this condition allows for only some of the evidential relations that the converse-consequence condition allowed.

The second important thing to note is that this condition allows all the inferences for which the converse-consequence condition was used and was needed, but does not allow those inferences which gave rise to Hempel's and Skyrms's paradoxes. The main type of inference that seemed to require something like the converse-consequence condition was the inference by which evidence that confirmed some observational generalization necessarily confirmed some theoretical hypothesis. But this case will be allowed by our new condition since it is one of the main characteristics of theoretical hypotheses (indeed, it is probably one of the main reasons why such hypotheses are introduced into science) that they serve to explain the observational generalizations that can be deduced from them. Thus, evidence for the law of definite proportions is also evidence for the atomic theory, since the latter explains the former. On the other hand, evidence for A or for B will not in general be evidence for $A \cdot B$, since the latter does not in general explain either of the former. Similarly, evidence for $A \vee B$ will not in general be evidence either for A or for B, since neither of the latter in general explain the former.

A third important point to note is that this proposal explains the failure of our earlier proposals. Substituting a notion of entailment that satisfies conditions 1 and 2 still allows too many cases in which evidence for H_1 would also be evidence for some H_2 that entails (but does not explain) H_1. On the other hand, the Boyle-Charles law suggests that substituting a notion of entailment that satisfies conditions 3 and 4 would rule out (illegitimately) cases where H_2 explains the H_1. The trouble with both of these proposals (and any other like them) is that you cannot find a syntactic equivalent to the relation between explanandum and explanans that is needed for the formulation of our conditions.

Finally, there is a different type of advantage which our new condition has and which has not yet been mentioned. It is well known that there are H_1's and H_2's such that H_2 explains (but does not entail) H_1. One such case (but not necessarily the only such case) occurs when H_2 is a statistical hypothesis which (together with auxiliary hypothesis) makes H_1 highly likely. In such cases, evidence for H_1 should also be evidence for H_2, but neither the original converse-consequence condition nor any of the modifications proposed in section 3 covers such cases. Our new suggestion does, however, cover these cases as well.

Several issues remain. To begin with, should we also replace the special-consequence condition by the following:

If E confirms H_1 and H_1 explains H_2, then E is evidence for H_2.

When we discussed previous modifications of our conditions, we did not raise the question whether we should modify both of the conditions or whether it was sufficient merely to modify one (since Skyrms's paradox involved only the converse-consequence condition, it would have to be that condition). But had we raised that question, it would have been very unclear as to what to say since we had no account of why we were doing what we were doing. With our new proposed change, however, intuitive considerations might indicate that we should modify only the converse-consequence condition. In the case of the consequences of a hypothesis, we are not, after all, giving evidence for an explanation; we are merely pointing out that this evidence for the whole is also evidence for part of that whole. Nevertheless, this matter probably needs further consideration.

The second remaining issue relates to the exact formulation of the condition that is to replace the converse-consequence condition. We have not required that the H_2 be the best explanation of the H_1. In this, our condition differs from proposed acceptance rules where it is required that the hypothesis be the best explanation. There is an obvious reason for this difference. When we are accepting a single hypothesis, there must be some reason why this hypothesis is to be preferred to any other. But in the case of confirmation functions, the evidence need not, and usually does not, indirectly confirm only one hypothesis. It would therefore be a mistake to make our condition exactly like the condition for acceptance rules. What is not clear, however, is whether or not we should strengthen our condition so that it becomes:

> If E is evidence for H_1, and H_2 is just as good an explanation of H_1 as any other hypothesis, then E is also evidence for H_2.

This would still allow evidence for one hypothesis to confirm indirectly more than one additional hypothesis, but it would rule out many confirmation relations that our original proposal allowed.

There remains, finally, the difficult question of defining a qualitative confirmation function that satisfies these new conditions. One possibility is to take a standard syntactic qualitative confirmation function (e.g., Hempel's) and use it to define a function that says that E confirms H_2 either if it does so according to the standard function or if it confirms, according to the standard function, some H_1 that is explained by H_2. Another, and more radical, possibility is to stop attempting to define a qualitative confirmation function syntactically and define it, instead, in terms of the notion of explanation. In a future paper, I hope to show that the adoption of this proposal avoids many of the serious difficulties that have been raised against the various standard qualitative confirmation functions.

For the moment, however, it suffices to conclude that although the final details for our proposal are not entirely clear, it is clear that something along the lines we have proposed would go a long way toward solving some very perplexing philosophical puzzles.

NOTES

1. *Mind*, LIV, 213, 214 (January, April 1945): 1–26, 97–121, reprinted in *Aspects of Scientific Explanation* (New York: Free Press, 1965). All page references will be to the reprinted version.

2. I say "seems" because I have certainly not proved that there could not be an essentially qualitative case where the background evidence should be taken into account. Something more will be said about this issue in the next section.

3. *The Anatomy of Inquiry* (New York: Knopf, 1963) p. 258.

4. It seems to me, on the basis of the work that has been done so far on confirmation functions, highly unlikely that an adequate confirmation function will be discovered that will allow for the direct confirmation by observational data of theoretical hypotheses. But the objection is quite correct in pointing out that such a function is possible.

5. It should be noted that the two conditions were originally suggested as adequacy conditions.

6. "Nomological Necessity and the Paradoxes of Confirmation," *Philosophy of Science*, XXXIV, 3 (September, 1966), 230–49.

7. Skyrms thought that it does because of his attempt to identify the qualitative notion with some quantitative notion. We shall see later in this section that such attempts face serious difficulties. In any case, our modification of the condition does avoid the difficulty.

8. At one time Hempel did not allow for this possibility. He now seems to agree that there are strong reasons for allowing for it: "One and the same observable phenomenon may well be accounted for by each of two incompatible hypotheses, and the observation report describing its occurrence would then normally be regarded as confirmatory for either hypothesis" (49).

9. The existence of cases where confirming instances might not add to the degree of confirmation because one already had too many cases of that type suggests that it might even be a mistake to grant this much.

10. In "The White Shoe Is a Red Herring," *British Journal for the Philosophy of Science*, XVII, No. 4 (February, 1967), 322.

11. We have not considered the following argument against these conditions: Either of these conditions entails the equivalence condition and this condition yields the paradox of the ravens. Our reason for not considering this argument is that it isn't clear whether the argument shows that one must adopt the paradoxical results or that one must drop both of these conditions. Perhaps, as has recently been suggested (subnote 11*), we can avoid some of the paradoxical results while still retaining the equivalence condition. Given strong reasons for keeping both of these conditions, this approach seems most promising, but we shall have to wait and see how it is developed more fully.

* In Richard Grandy, "Some Comments on Confirmation and Selective Confirmation," *Philosophical Studies*, XVII, Nos. 1–2 (January-February, 1967), 19–23.

12. *Introduction to Logical Theory* (New York: Wiley, 1960), p. 81.

13. "Intensional Relations," *Mind*, XXXIX, No. 156 (October, 1930), 447–48.

14. This was pointed out to me by Richard Grandy.

15. I owe this point to David Rosenthal.

16. Most notably by Gilbert Harman; see "The Inference to the Best Explanation," *Philosophical Review*, LXXIV, No. 10 (January, 1965), 88–95. In an unpublished manuscript of his that I have recently seen, he notes that since $A \cdot B$ does not usually explain A, one can perhaps use his theory to avoid the lottery paradox. As far as I know, no author has yet pointed out the desirability of introducing the notion of explanation into the theory of qualitative confirmation.

Nelson Goodman

SELECTIVE CONFIRMATION

. . .

New difficulties promptly appear from other directions, however. One is the infamous paradox of the ravens. The statement that a given object, say this piece of paper, is neither black nor a raven confirms the hypothesis that all non-black things are non-ravens. But this hypothesis is logically equivalent to the hypothesis that all ravens are black. Hence we arrive at the unexpected conclusion that the statement that a given object is neither black nor a raven confirms the hypothesis that all ravens are black. The prospect of being able to investigate ornithological theories without going out in the rain is so attractive that we know there must be a catch in it. The trouble this time, however, lies not in faulty definition, but in tacit and illicit reference to evidence not stated in our example. Taken by itself, the statement that the given object is neither black nor a raven confirms the hypothesis that everything that is not a raven is not black as well as the hypothesis that everything that is not black is not a raven. We tend to ignore the former hypothesis because we know it to be false from abundant other evidence—from all the familiar things that are not ravens but are black. But we are required to assume that no such evidence is available. Under this circumstance, even a much stronger hypothesis is also obviously confirmed: that nothing is either black or a raven. In the light of this confirmation of the hypothesis that there are no ravens, it is no longer surprising that under the artificial restrictions of the example, the hypothesis that all ravens are black is also confirmed. And the prospects for indoor ornithology vanish when we notice that under these same conditions the contrary hypothesis that no ravens are black is equally well confirmed.

Richard E. Grandy

SOME COMMENTS ON CONFIRMATION AND SELECTIVE CONFIRMATION

In order to lay the groundwork for a discussion of alternative solutions to the paradox of confirmation I shall review briefly some of the adequacy conditions which have been proposed for confirmation.[1] The first of these is derived from the Nicod criterion which was originally formulated as follows: "Consider the formula or law: A entails B. How can a particular proposition, or more briefly, a fact, affect its probability? If this fact consists of the presence of B in a case of A, it is favorable to the law "A entails B" . . ."[2] The most plausible way of construing this comment is that an observation report confirms a universal conditional statement if it entails the conjunction of two sentences which are corresponding substitution instances of the antecedent and consequent and does not entail the negation of any substitution instance of the consequent. Corresponding instances would be the antecedent and consequent of a substitution instance of the whole formula.

As Hempel has shown[3] the Nicod criterion is not sufficient in the case of statements containing relations. He has also argued convincingly that the criterion does not provide a necessary condition even for sentences formalizable in monadic functional calculus. This inadequacy follows from a second plausible condition on confirmation, the equivalence condition. This condition requires that if an observation report M confirms S, then M confirms any statement logically equivalent to S. The justification of the principle is obvious—it seems

---

AUTHOR'S NOTE. This paper was written while the author was a National Science Foundation Graduate Fellow. He wishes to express his gratitude to C. G. Hempel, P. Benacerraf, and L. G. Creary, who pointed out inaccuracies in an earlier version.

Reprinted from *Philosophical Studies* (1967) by permission of the editor.

perverse to assert that we have confirmed a statement but not one of its equivalents, since equivalence intuitively means having the same content. Furthermore, to deny the equivalence condition would place severe restrictions on the use which could be made of a statement. In general we are interested in confirming statements which are then to be used to infer other statements which are considered to inherit some of the preferential status.

The same considerations may be advanced in favor of a stronger requirement, the consequence condition. It is desirable that if we are given a set of confirmed statements, then any consequence of the set is also confirmed.[4] It is also plausible to retain the Nicod criterion as a sufficient condition for confirmation in the case of non-relational statements. After all, what better confirmation could we expect for $(x)\cdot Fx \supset Gx$ than $Fa\cdot Ga$?

We are now in a position to state the paradox of confirmation. By Nicod's criterion the observation report that *b* is a non-black non-raven confirms the statement "All non-black things are non-ravens," but by the equivalence condition it confirms the equivalent assertion "All ravens are black." But now our two intuitively plausible conditions entail the non-intuitive result that the observation of a non-black non-raven confirms that all ravens are black. In order to resolve this paradox it will be necessary either to show why our intuitions are misled in accepting the conditions or in rejecting the soundness of the result.

D. Pears[5] has attempted to resolve the paradox by distinguishing the extent to which various observation statements confirm a hypothesis. He suggests that $Fa\cdot Ga$ confirms $(x)\cdot Fx \supset Gx$ more (less) than does $\sim Fa\cdot \sim Ga$ if $\{x: Fx\}$ is numerically smaller (larger) than $\{x: \sim Gx\}$. The intuitive idea behind this is that by examining the smaller class of objects we can more quickly establish (or refute) the hypothesis. Thus the solution of the raven paradox for Pears is that a black raven gives better confirmation than a non-black non-raven because there are fewer ravens than non-black things.

Unfortunately this approach has serious defects. While the criterion does preserve the symmetry between a statement and its contrapositive (for $(x)\cdot Rx \supset Bx$ we compare $\{x: Rx\}$ and $\{x: \sim Bx\}$ and similarly for its contrapositive) it does not in general satisfy the equivalence condition. Consider the statement $(x): Rx\cdot \sim Bx \supset Rx\cdot \sim Rx$, which is equivalent to $(x)\cdot Rx \supset Bx$. The class of non-black ravens is smaller than the universal class, so according to Pears's criterion the best confirmation for this form of the raven hypothesis is a non-black raven. One might feel that this example is unfair or avoidable because of the self-contradictory predicate involved. But $(x): Rx\cdot \sim Bx \supset \sim Rx$ would do equally well.

Most of the discussions of the paradox have centered around the Hempelian definition of confirmation which may be briefly summarized as follows: If M and S are formulas of an observational first-order language without identity and M is a molecular formula, then M directly confirms S if the quantifier-free expansion of S in universe V is entailed by M (where V is the universe consisting of all individuals whose names occur essentially in M). Confirmation can now be defined in terms of direct confirmation and entailment.

If S is entailed by some set of statements each of which is directly confirmed by M, then M confirms S.

It is easily shown that this definition satisfies the conditions previously discussed and thus encounters the paradox. Hempel's suggested resolution of the paradox consists in arguing that it is only a psychological illusion which leads to the feeling of uneasiness with regard to the conclusion. The two sources of the illusion are alleged to be the assumption that universal conditionals only assert something about the antecedent and the illicit and implicit assumption of further information. Thus, Hempel points out, we ignore the fact that $\sim Ra \cdot Ba$ confirms (x) Bx since we know the latter to be false on independent grounds. But once we realize this, it is not difficult to accept that the report confirms the weaker conclusion $(x) \cdot Rx \supset Bx$.

But as Goodman and Scheffler[6] have noted, given that $\sim Ra \cdot Ba$ is the sum of our information, it follows that $(x) \sim Rx$ is confirmed. But this implies both $(x) \cdot Rx \supset Bx$ and $(x) \cdot Rx \supset \sim Bx$, which are contraries.[7] Scheffler concludes that the concept defined by Hempel does not accord with our intuitive notion. The notion of confirmation seems to him to contain an element of preference—that is, if M confirms S, then S is considered to be preferable to its contrary. But the Hempelian definition does not satisfy this.

The difficulty arises partly from the fact that any given hypothesis has several non-equivalent contraries depending on its formulation. For example, A1, A2, and A3 below are equivalent but their respective contraries B, C, and D are not:

A1. $(x) \cdot Rx \supset Bx$	B. $(x) \cdot Rx \supset \sim Bx$
A2. $(x) \cdot \sim Bx \supset \sim Rx$	C. $(x) \cdot \sim Bx \supset Rx$
A3. $(x): Rx \cdot \sim Bx \supset Rx \cdot \sim Rx$	D. $(x): Rx \cdot \sim Bx \supset Rx \vee \sim Rx$

Thus the previous explanation was somewhat misleading. The reason that $Ra \cdot Ba$ seems to be a confirming instance of A1 but not of A2 is that it disconfirms the contrary of A1 but not the contrary of A2. According to our conditions $Ra \cdot Ba$ confirms both A2 and *C*.

At this point we seem to have neither a definition of confirmation nor an explanation of the paradox. The next step then is to see whether we can find a new definition, possibly building on the older one. Scheffler offers some suggestions in this direction.[8] One may define a relation of selective confirmation such that M selectively confirms S if M (Hempel) confirms S and disconfirms the contrary of S.

Two objections arise immediately. First, this definition rejects the equivalence and consequence conditions. Second, it is inapplicable to statements not having the form of a conditional. The latter objection by itself could be met since for any statement there is an equivalent statement which has the form of a conditional. But this way out is not open since the criterion does not permit the interchange of equivalents. Thus selective confirmation is inapplicable to a statement of the form (x) Rx which has no syntactical contrary.

Furthermore, even some perfectly ordinary conditio ıls can never be selectively confirmed. For example, A3, an equivalent of A1, has as its contrary D.

But D is a logical truth and can never be disconfirmed. (This is a comment on the definition of confirmation, not on logical truth.) So A3 can never be selectively confirmed.

There is some motivation to the rejection of the equivalence condition, namely that contraries of different forms of a hypothesis need not be equivalent. But in view of the arguments in favor of the equivalence and consequence conditions, it would seem desirable to find some analysis of selective confirmation which satisfies these.

It is worth noting that there are two distinct aspects to the paradox. The first is that anything which confirms A1 also confirms A2. This follows immediately from the equivalence condition. The second is that the observation reports $Ra \cdot Ba$ and $\sim Ra \cdot \sim Ba$ are both confirmatory evidence for A1. This follows from the conjunction of Nicod's criterion as a sufficient condition and the equivalence condition.

With respect to the first aspect of the paradox Hempel's discussion seems completely sound. Any discomfort on this point seems to be due to a misunderstanding of the meaning of the two sentences. Scheffler's definition of selective confirmation removes both aspects of the paradox, essentially by dropping the equivalence condition. For the reasons presented, this seems too high a price to pay if one can find reasonable alternatives.

In view of these considerations and the fact that scientists are often concerned with choosing between explicitly stated hypotheses, the possibility suggests itself of relativizing confirmation to a set of hypotheses. That is, instead of defining the binary relation "M selectively confirms S" we might try defining the ternary relation "M selectively confirms S relative to K," where K is a specified set of hypotheses, no two of which are equivalent. A definition of this relation is not difficult to find: M selectively confirms S from K if S is a member of K and M confirms S and M disconfirms every other member of K.[9] Scheffler's selective confirmation is the special case where K consists of S and one of its contraries. Hempel's concept of confirmation is the special case where K is the unit set of S.

In order to see that this definition differentiates $Ra \cdot Ba$ and $\sim Ra \cdot \sim Ba$ consider the set $\{(x) \cdot Rx \supset Bx, (x) \cdot Rx \supset \sim Bx\}$. $Ra \cdot Ba$ selectively confirms the raven hypothesis but $\sim Ra \cdot \sim Ba$ does not. On the other hand, $\sim Ra \cdot \sim Ba$ selectively confirms the raven hypothesis from the set $\{(x) \cdot Rx \supset Bx, (x)\, Rx\}$, but $Ra \cdot Ba$ does not.

Our definition retains the equivalence condition in the following form: If S is selectively confirmed from K by M, then M will selectively confirm S′ from K′, where S′ is a statement equivalent to S and where K′ is obtained from K by replacing statements by their equivalents. Our definition, however, satisfies the Nicod criterion only in the weakened form: If M is the conjunction of corresponding substitution instances of the antecedent and consequent of a conditional statement S in K and S contains no relational terms, then S will be selectively confirmed by M from K if anything is. It is because our definition satisfies only the weakened Nicod criterion that we manage to break the symmetry of $Ra \cdot Ba$ and $\sim Ra \cdot \sim Ba$ with respect to the raven hypothesis.

NOTES

1. For a fuller discussion see C. G. Hempel, "Studies in the Logic of Confirmation," *Mind*, LIV (1945), 1–26, 97–121, or "A Purely Syntactic Definition of Confirmation," *Journal of Symbolic Logic*, VIII (1943), 122–43.

2. J. Nicod, *Foundations of Geometry and Induction* (New York: Harcourt, Brace, 1930), p. 219.

3. See "Studies in the Logic of Confirmation," p. 13.

4. Carnap has offered some criticisms of the consequence condition, but it is not clear that his explication of confirmation fits that intended by Hempel. See sections 86–88 of his *Logical Foundations of Probability* (2nd ed.) (Chicago: University of Chicago Press, 1962).

5. D. Pears, "Hypotheticals," *Analysis*, X (1950), 49–63.

6. I. Scheffler, *The Anatomy of Inquiry* (New York: Knopf, 1963), pp. 286–88, following N. Goodman, *Fact, Fiction, and Forecast* (Cambridge, Mass.: Harvard University Press, 1955), pp. 71–72.

7. $(x) \cdot Fx \supset Gx$ and $(x) \cdot Fx \supset \sim Gx$ are not actually contraries unless $(Ex)\ Fx$ is true. The terminology is convenient, however, and we shall continue to use it.

8. *The Anatomy of Inquiry*, pp. 289–91. Actually Scheffler does not offer a definition. He merely states that ". . . in terms of Hempel's definitions of confirmation, disconfirmation and neutrality, requisite notions of *selective confirmation*, of various sorts, can be further explained." The definition given in the text seems the most natural.

9. Another possible definition of selective confirmation would be the following: M selectively confirms S from K if S is the only member of K confirmed by M.

J. W. N. Watkins

THE "PARADOXES OF CONFIRMATION"

Hempel went out of his way to emphasize that his satisfaction criterion has the counterintuitive or "paradoxical" consequence that it makes every non-A constitute "confirming" evidence for a hypothesis of the form "All A's are B." For this hypothesis is "confirmed," according to that criterion, if it is satisfied or instantiated by the evidence; but it is equivalent to "Everything is B or non-A," and so will be satisfied or instantiated not only by any A which is B but also by anything which is B and by anything which is non-A. In short, it will be satisfied by everything except A's which are non-B. Hempel tried to remove the counterintuitiveness of this consequence by pointing out that a hypothesis of the form "All A's are B" asserts something not only about A's, but about *every* thing (namely, that each thing is non-A or B). And since it asserts something about every thing we should not be surprised or perplexed to find that every thing may confirm it ("Studies," pp. 18–19).

What, if anything, is wrong with a criterion of confirmation which involves this "paradoxical" consequence? It is not, of course, paradoxical in the strict sense: It involves no logical inconsistency. It is only rather startlingly counterintuitive. But there is not necessarily anything wrong with that. A counterintuitive philosophical theory which shows up the mistakenness of our intuitions may well be superior to one which accords with them. And, indeed, Hempel claimed to have shown that the counterintuitiveness of the idea that every non-A confirms "All A's are B" is a "psychological illusion" created by "a misleading intuition in the matter" ("Studies," p. 18).

But it *may* also happen that a counterintuitive theory, by provoking us to examine the whole matter afresh, helps us to realize that the counterintuitive theory is itself due to "a misleading intuition in the matter."

Reprinted from J. W. N. Watkins "Confirmation, the Paradoxes, and Positivism," in M. Bunge (ed.) *The Critical Approach to Science and Philosophy*, © 1964, The Free Press, by permission of the publisher.

It is important to remember that the counterintutitive conclusion which Hempel upheld was not just that a non-A may, in certain circumstances, provide confirmation for "All A's are B," but that every non-A in any circumstance automatically does so. A serious objection to this is that it makes the confirming of scientific hypotheses a terribly facile business. But we must remember that Hempel claimed that this conclusion is well-nigh unavoidable. (He conceded that there are certain *ad hoc* dedges for avoiding it, but claimed that to resort to any of these would have very unfortunate consequences—a claim I shall not dispute.) If it is unavoidable it would be no good complaining about facile "confirmation"; we should have to lump it.

I shall now argue, however, that it is not only avoidable but had already been avoided, easily and naturally, by Popper's theory of corroboration.

I shall also argue that Hempel arrived at his counterintuitive conclusion because he stuck to a seemingly self-evident idea which is, however, mistaken. In the course of rejecting various ways in which one might try to avoid the "paradoxes," Hempel stated: "Nor does it seem possible to dispense with the stipulation that an object satisfying two conditions, C_1 and C_2, should be considered as confirming a general hypothesis to the effect that any object which satisfies C_1 also satisfies C_2" ("Studies," p. 14). I shall argue that Hempel's achievement here was to show, despite himself, that this seemingly obvious idea —the idea which governs his whole approach to confirmation—*does* have to be dispensed with.

In this problem area it is particularly desirable to have all the alternatives laid out before us. Call an A which is B an AB. With regard to the confirmation of "All A's are B," the seemingly obvious proposition which Hempel did not contemplate dispensing with was:

(1) An AB always confirms it.

The counter-intuitive or "paradoxical" conclusion to which, Hempel showed, (1) leads is:

(1′) A non-A always confirms it.

To this pair of propositions we can add two alternative pairs:

(2) An AB sometimes confirms it.
(2′) A non-A sometimes confirms it.
(3) An AB never confirms it.
(3′) A non-A never confirms it.

It is not difficult to show that (3′) leads to (3) and that (2) leads to (2′) by reasoning similar to that whereby Hempel showed that (1) leads to (1′).

Many people have felt that the ideal combination would be (1)-cum-(3′)—is it not obvious that all white swans do, whereas no non-swan does, confirm "All swans are white"? Hempel showed that (1)-cum-(3′) is an impossible combination, that *if* we accept (1) we must accept (1′); and he believed that there can be no question of dispensing with (1). But in fact we still have a *choice* here—between (1)-cum-(1′) and (2)-cum-(2′). And if (1)-cum-(1′) generates much too much "confirmation" while (3)-cum-(3′) allows much too little (viz., none at all), we may conjecture that perhaps (2)-cum-(2′) allows the right amount. Hempel's appraisal of (1′) was revolutionary. But this was dictated by his *conservative* retention of (1). Perhaps a moderately reformist approach to both yields better results. Popper's theory, I shall show in a moment, is less revolutionary than Hempel's in endorsing (2′) rather than (1′), and less conservative in endorsing (2) rather than (1). I shall now try to show that (2)-cum-(2′) is indeed the best combination. I shall do this by examining a possible defense of Hempel's (1)-cum-(1′) against Popper's (2)-cum-(2′).

It might be claimed that if Popper's position involves (2)-cum(2′) then it is even more satisfactory than Hempel's. First, it allows some non-A's to confirm "Every A is B" which is *nearly* as bad as making every non-A do so. Second, it has the shocking consequence that it *stops* some AB's doing so.

I shall take these two points in turn.

The first objection presupposes that (3′: a non-A never confirms) is the ideal from which (2′: a non-A sometimes confirms) lapses, though not so badly as does (1′: a non-A always confirms).

But there are various hypotheses which have surely been confirmed but for which we are just unable to obtain direct evidence about instances of their antecedent clauses. For instance, there are hypotheses about live dinosaurs for which there exists today a good deal of confirming evidence, though there are no live dinosaurs. Again, modern physics consists largely of hypotheses about unobservable entities for which confirming evidence is provided by things which are not unobservable entities.

Moreover, even in cases where instances of the antecedent clause of a hypothesis are open to observation it may very well happen that confirming evidence is provided by non-instances of it. Agassi has given a nice example of this. Before reproducing it I will indicate very briefly those features of Popper's theory of corroboration which are relevant to it.

Popper's fundamental idea is that a hypothesis is corroborated whenever it survives a test; the more severe the test the stronger the corroboration. The relative severity of tests is appraised in this way. Suppose we have background information k, an hypothesis h, and evidence e brought to light by an experimental test on h. Then the more improbable was e, given k alone, and the more probable was e, given the amalgamation of k and h, the more severe was the test and the better e corroborates h (*L.S.D.*, appendix *ix).

Here is Agassi's example.[1] He asks us to consider a situation where Galileo's hypothesis that freely falling bodies move with constant acceleration is being tested "by dropping steel balls off an electromagnet by switching off the electric current. These experiments can be carried out in deep mines to allow the steel balls to fall freely a long way." Let us suppose that the same test is being carried out in two mine-shafts.

Assume that background information by no means suggests that a freely falling body falls with constant acceleration, but does suggest that a ball dropped down a vertical shaft normally falls freely. Now introduce Galileo's hypothesis.

This raises sharply the probability that a ball dropped down a vertical shaft will fall with constant acceleration. It *also* raises the probability that *if* such a ball does *not* fall with constant acceleration, it will be found *not to have fallen freely*. Suppose that observations in one of the mine-shafts confirm that the balls dropped there fell with constant acceleration. This evidence corroborates Galileo's hypothesis, since its probability was raised by that hypothesis. In this case a hypothesis of the form "All A's are B" has been corroborated by evidence of the form "These A's were not found to be non-B." Suppose that the balls dropped down the other mine-shaft are definitely observed to fall with inconstant acceleration; and suppose further that, as in Agassi's example, the cause of the aberration were found to be that the shaft passed through a magnetic rock. Then the evidence that these inconstantly accelerating balls were not falling freely *also* corroborates Galileo's hypothesis, since its probability was also raised by that hypothesis. In *this* case a hypothesis of the form "All non-B's are non-A" has been corroborated by evidence of the form "These non-B's were found to be non-A."

Thus it is wrong to suppose that, if Hempel's theory is pregnant with paradox in endorsing (1′), Popper's is "just a little pregnant" with paradox in endorsing (2′).[2] With reference to Agassi's example the relative merits of propositions (1′), (2′), and (3′) could be summarized thus:

Proposition (1′) makes the steel balls which fell with inconstant acceleration because of the magnetic rock confirming evidence for Galileo's law; but it also makes the Tower of Pisa and all other things which are not freely falling bodies confirmatory.

Proposition (2′) allows those steel balls to be confirmatory without requiring all other things which are not freely falling bodies to be confirmatory.

Proposition (3′) makes those steel balls nonconfirmatory.

I conclude that (3′) is as unsatisfactory, though in the opposite direction, as (1′), and that (2′), far from being a little "paradoxical," is entirely satisfactory.

I now turn to the relative merits of Hempel's (1): "An AB always confirms" and Popper's (2): "An AB sometimes confirms."

What I have to say here proceeds from two assumptions. First, confirmation theory bears on the appraisal of scientific hypotheses. Given two hypotheses and a body of evidence, the first hypothesis being more strongly confirmed than the second by that evidence according to a particular theory of confirmation, then my first assumption is that this theory of confirmation suggests that scientists should, other things being equal, *prefer* the first. Hempel accepts this.

My second assumption is that a predictively powerful hypothesis which has withstood severe testing *is* preferable (to put it mildly) to a hypothesis which is completely *ad hoc* relative to known evidence, a hypothesis from which no new predictions can be derived. This assumption is at the core of Popper's philosophy of science.

Verificationist theories of confirmation have an awkward tendency to imply that the *ad hoc* hypothesis is preferable, since it may be *verified*, whereas the predictively powerful hypothesis (especially if it is mathematically exact and postulates theoretical entities) may (as we have seen) turn out to be altogether "unconfirmable."

I shall now argue that proposition (1) is an expression of verificationism and leads precisely to a preference for *ad hoc* hypotheses over predictively powerful ones, whereas (2) enables us to avoid this. I will show this with the help of an artificial example.

Imagine a botanist living in Cornwall. Suppose that part of his background information k consists of the evidence e_1 that all the A's so far investigated in Cornwall have been found to be B, but that k does not include any information about A's in Sumatra, Greenland, or Peru. Now consider two hypotheses which state respectively:

All A's are B
All A's so far investigated in Cornwall are B

I will call these h_2 and h_1 respectively. Hypothesis h_1 is, of course, completely *ad hoc* relative to e_1. Now suppose that our botanist visits Sumatra, Greenland, and Peru in order to investigate A's under varying conditions, and that he finds many A's in each of these countries and that each of them turns out to be B. I will call the evidence he collects in these countries e_2.

By Popper's criterion (see p. 435, above) neither e_1 nor e_2 corroborates h_1. The evidence e_1 is a part of, and therefore entailed by, k so that the introduction of the *ad hoc* hypothesis h_1 cannot *raise* its probability. As for e_2, its probability was not raised by h_1, which predicts nothing about A's in Sumatra, Greenland and Peru.

Hypothesis h_2, on the other hand, is corroborated by e_2. Thus for Popper, h_2 is preferable to h_1.

But this appraisal is not possbile for a theory of confirmation which involves (1) rather than (2). For by (1), an AB is always confirmatory. Thus the AB's investigated so far in Cornwall confirm h_1. Indeed, they *conclusively* confirm it. The AB's observed in Cornwall, Sumatra, Greenland, and Peru confirm h_2, but only *inconclusively*. So h_1 is more strongly confirmed than, and therefore preferable to, h_2.

My last remark to anyone who still hankers after (1) is a reminder that the choice is between (1)-cum-(1′) and (2)-cum-(2′). If he still feels uneasy about (2): "An AB sometimes confirms," I ask him whether he does not feel even more uneasy about (1′): "A non-A always confirms."

Thus Popper's theory, I contend, avoids the "paradoxes" by saying that "Every A is B" is (strongly) corroborated by evidence (whether of AB's or non-A's) if and only if such evidence is rendered (considerably) more probable by the hypothesis than it was by background knowledge alone.

The appeal to background knowledge is essential. Now Hempel deliberately excluded background knowledge, citing in his support "the 'methodological fiction,' characteristic of every case of confirmation, that we have no relevant evidence for H other than that included in E" ("Studies," p. 20). He did not say where this strange fiction comes from or why we should submit to it.

Now genuine testing may still be possible even with very impoverished and sketchy background knowledge. But if we suppose *no background knowledge at all* (a completely unrealistic supposition, of course) the notion of testing becomes empty. Without *any* indications about where to look for likely counterevidence, *all* undirected looking equally "tests" a scientific hypothesis. Anything, anywhere may be an A which is not B, and the discovery that something is, after all, a non-A will always constitute the favorable outcome of a pseudo-test.

Thus even a testability theory will generate the "paradoxes" (viz., the (1)-cum-(1′) combination) if background knowledge is willfully excluded. But why in heaven's name (for I would rather appeal to heaven than to a methodological fiction) *should* it be excluded? Hempel claimed that the "paradoxes" are well-nigh unavoidable. Yet we have only to reintroduce background knowledge and replace a satisfaction theory involving (1)-cum-(1′) by a testability theory involving (2)-cum-(2′), and they disappear.

. . .

NOTES

1. "Corroboration versus Induction," *Brit. Jour. Phil. of Science*, Feb. 1959, especially pp. 313–14.

2. I take the simile from I. Scheffler, *Philosophical Studies*, Jan.-Feb. 1961, p. 19.

Degrees of Confirmation

Rudolf Carnap

STATISTICAL AND INDUCTIVE PROBABILITY

If you ask a scientist whether the term "probability" as used in science has always the same meaning, you will find a curious situation. Practically everyone will say that there is only one scientific meaning; but when you ask that it be stated, two different answers will come forth. The majority will refer to the concept of probability used in mathematical statistics and its scientific applications. However, there is a minority of those who regard a certain nonstatistical concept as the only scientific concept of probability. Since either side holds that its concept is the only correct one, neither seems willing to relinquish the term "probability." Finally, there are a few people—and among them this author—who believe that an unbiased examination must come to the conclusion that both concepts are necessary for science, though in different contexts.

I will now explain both concepts—distinguishing them as "statistical probability" and "inductive probability"—and indicate their different functions in science. We shall see, incidentally, that the inductive concept, now advocated by a heretic minority, is not a new invention of the twentieth century, but was the prevailing one in an earlier period and only forgotten later on.

The *statistical concept of probability* is well known to all those who apply in their scientific work the customary methods of mathematical statistics. In this field, exact methods for calculations employing statistical probability are developed and rules for its application are given. In the simplest cases, probability in this sense means the relative frequency with which a certain kind of event occurs within a given reference class, customarily called the "population." Thus, the statement "The probability that an inhabitant of the United States belongs to blood group A is p" means that a fraction p of the inhabitants belongs to this group. Sometimes a statement of statistical probability refers, not to an

Published by The Galois Institute of Mathematics and Art, 1955. Reprinted by permission of the author and the publisher.

actually existing or observed frequency, but to a potential one, i.e., to a frequency that would occur under certain specifiable circumstances. Suppose, for example, a physicist carefully examines a newly made die and finds it is a geometrically perfect and materially homogeneous cube. He may then assert that the probability of obtaining an ace by a throw of this die is 1/6. This means that *if* a sufficiently long series of throws with this die were made, the relative frequency of aces would be 1/6. Thus, the probability statement here refers to a potential frequency rather than to an actual one. Indeed, if the die were destroyed before any throws were made, the assertion would still be valid. Exactly speaking, the statement refers to the physical microstate of the die; without specifying its details (which presumably are not known), it is characterized as being such that certain results would be obtained if the die were subjected to certain experimental procedures. Thus the statistical concept of probability is not essentially different from other disposition concepts which characterize the objective state of a thing by describing reactions to experimental conditions, as, for example, the I.Q. of a person, the elasticity of a material object, etc.

Inductive probability occurs in contexts of another kind; it is ascribed to a hypothesis with respect to a body of evidence. The hypothesis may be any statement concerning unknown facts, say, a prediction of a future event, e.g., tomorrow's weather or the outcome of a planned experiment or of a presidential election, or a presumption concerning the unobserved cause of an observed event. Any set of known or assumed facts may serve as evidence; it consists usually in results of observations which have been made. To say that the hypothesis h has the probability p (say, 3/5) with respect to the evidence e, means that for anyone to whom this evidence but no other relevant knowledge is available, it would be reasonable to believe in h to the degree p or, more exactly, it would be unreasonable for him to bet on h at odds higher than p:(1 − p) (in the example, 3:2). Thus inductive probability measures the strength of support given to h by e or the *degree of confirmation* of h on the basis of e. In most cases in ordinary discourse, even among scientists, inductive probability is not specified by a numerical value but merely as being high or low or, in a comparative judgment, as being higher than another probability. It is important to recognize that every inductive probability judgment is relative to some evidence. In many cases no explicit reference to evidence is made; it is then to be understood that the totality of relevant information available to the speaker is meant as evidence. If a member of a jury says that the defendant is very probably innocent or that, of two witnesses A and B who have made contradictory statements, it is more probable that A lied than that B did, he means it with respect to the evidence that was presented in the trial plus any psychological or other relevant knowledge of a general nature he may possess. Probability as understood in contexts of this kind is not frequency. Thus, in our example, the evidence concerning the defendant, which was presented in the trial, may be such that it cannot be ascribed to any other person; and if it could be ascribed to several people, the juror would not know the relative frequency of innocent persons among them. Thus the probability concept used here cannot be the statistical one. While a statement of statistical probability asserts a matter of fact, a statement of inductive probability is of a purely logical nature. If hypothesis and evidence are given, the probability can be determined by logical analysis and mathematical calculation.

One of the basic principles of the theory of inductive probability is the *principle of indifference*. It says that, if the evidence does not contain anything that would favor either of two or more possible events, in other words, if our knowledge situation is symmetrical with respect to these events, then they have equal probabilities relative to the evidence. For example, if the evidence e_1 available to an observer X_1 contains nothing else about a given die than the information that it is a regular cube, then the symmetry condition is fulfilled and therefore each of the six faces has the same probability 1/6 to appear uppermost at the next throw. This means that it would be unreasonable for X_1 to bet more than one to five on any one face. If X_2 is in possession of the evidence e_2 which, in addition to e_1, contains the knowledge that the die is heavily loaded in favor of one of the faces without specifying which one, the probabilities for X_2 are the same as for X_1. If, on the other hand, X_3 knows e_3 to the effect that the load favors the ace, then the probability of the ace on the basis of e_3 is higher than 1/6. Thus, inductive probability, in contradistinction to statistical probability, cannot be ascribed to a material object by itself, irrespective of an observer. This is obvious in our example; the die is the same for all three observers and hence cannot have different properties for them. Inductive probability characterizes a hypothesis relative to available information; this information may differ from person to person and vary for any person in the course of time.

A brief look at the historical development of the concept of probability will give us a better understanding of the present controversy. The mathematical study of problems of probability began when some mathematicians of the sixteenth and seventeenth centuries were asked by their gambler friends about the odds in various games of chance. They wished to learn about probabilities as a guidance for their betting decisions. In the beginning of its scientific career, the concept of probability appeared in the form of inductive probability. This is clearly reflected in the title of the first major treatise on probability, written by Jacob Bernoulli and published posthumously in 1713; it was called *Ars Conjectandi*, the art of conjecture, in other words, the art of judging hypotheses on the basis of evidence. This book may be regarded as marking the beginning of the so-called classical period of the theory of probability. This period culminated in the great systematic work by Laplace, *Theorie analytique des probabilités* (1812). According to Laplace, the purpose of the theory of probability is to guide our judgments and to protect us from illusions. His explanations show clearly that he is mostly concerned, not with actual frequencies, but with methods for judging the acceptability of assumptions, in other words, with inductive probability.

In the second half of the last century and still more in our century, the application of statistical methods gained more and more ground in science. Thus attention was increasingly focussed on the statistical concept of probability. However, there was no clear awareness of the fact that this development constituted a transition to a fundamentally different meaning of the word "probability." In the 1920's the first probability theories based on the frequency interpretation were proposed by men like the statistician R. A. Fisher, the mathematician R. von Mises, and the physicist-philosopher H. Reichenbach. These authors and their followers did not explicitly suggest to abandon that concept of probability which had prevailed since the classical period and to replace it by a new one. They rather believed that their concept was essentially the same as that of all earlier authors. They merely claimed that they had given a more exact definition for it and had developed more comprehensive theories on this improved foundation. Thus, they interpreted Laplace's word "probability" not in his inductive sense, but in their own statistical sense. Since there is a strong, though by far not complete analogy between the two concepts, many mathematical theorems hold in both interpretations, but others do not. Therefore these authors could accept many of the classical theorems but had to reject others. In particular, they objected strongly to the principle of indifference. In the frequency interpretation, this principle is indeed absurd. In our earlier example with the observer X_1, who knows merely that the die has the form of a cube, it would be rather incautious for him to assert that the six faces will appear with equal frequency. And if the same assertion were made by X_2, who has information that the die is biased, although he does not know the direction of the bias, he would contradict his own knowledge. In the inductive interpretation, on the other hand, the principle is valid even in the case of X_2, since in this sense it does not predict frequencies but merely says, in effect, that it would be arbitrary for X_2 to have more confidence in the appearance of one face than in that of any other face and therefore it would be unreasonable for him to let his betting decisions be guided by such arbitrary expectations. Therefore it seems much more plausible to assume that Laplace meant the principle of indifference in the inductive sense rather than to assume that one of the greatest minds of the eighteenth century in mathematics, theoretical physics, astronomy, and philosophy chose an obvious absurdity as a basic principle.

The great economist John Maynard Keynes made the first attempt in our century to revive the old but almost forgotten inductive concept of probability. In his *Treatise on Probability* (1921) he made clear that the inductive concept is implicitly used in all our thinking on unknown events both in everyday life and in science. He showed that the classical theory of probability in its application to concrete problems was understandable only if it was interpreted in the inductive sense. However, he modified and restricted the classical theory in several important points. He rejected the principle of indifference in its classical form. And he did not share the view of the classical authors that it should be possible in principle to assign a numerical value to the probability of any hypothesis whatsoever. He believed that this could be done only under very special, rarely fulfilled conditions, as in games of chance where there is a well determined number of possible cases, all of them alike in their basic features, e.g., the six possible results of a throw of a die, the possible distributions of cards among the players, the possible final positions of the ball on a roulette table, and the like. He thought that in all other cases at best only comparative judgments of probability could be made, and even these only for hypotheses which belong, so to speak, to the same dimension. Thus one might come to the result that, on the basis of available knowledge, it is more probable that the next child of a specified couple will be male rather than female; but no comparison could be made between the probability of the birth of a male child and the probability of the stocks of General Electric going up tomorrow.

A much more comprehensive theory of inductive probability was constructed by the geophysicist Harold Jeffreys (*Theory of Probability*, 1939). He agreed with the classical view that probability can be expressed numerically in all cases. Furthermore, in view of the fact that science replaces statements in qualitative terms (e.g., "the child to be born will be very heavy") more and more by those in terms of measurable quantities ("the weight of the child will be more than eight pounds"), Jeffreys wished to apply probability also to hypotheses of quantitative form. For this reason, he set up an axiom system for probability much stronger than that of Keynes. In spite of Keynes's warning, he accepted the principle of indifference in a form quite similar to the classical one: "If there is no reason to believe one hypothesis rather than another, the probabilities are equal." However, it can easily be seen that the principle in this strong form leads to contradictions. Suppose, for example, that it is known that every ball in an urn is either blue or red or yellow but that nothing is known either of the color of any particular ball or of the numbers of blue, red, or yellow balls in the urn. Let B be the hypothesis that the first ball to be drawn from the urn will be blue, R, that it will be red, and Y, that it will be yellow. Now consider the hypotheses B and non-B. According to the principle of indifference as used by Laplace and again by Jeffreys, since nothing is known concerning B and non-B, these two hypotheses have equal probabilities, i.e., one half. Non-B means that the first ball is not blue, hence either red or yellow. Thus "R or Y" has probability one half. Since nothing is known concerning R and Y, their probabilities are equal and hence must be one fourth each. On the other hand, if we start with the consideration of R and non-R, we obtain the result that the probability of R is one half and that of B one fourth, which is incompatible with the previous result. Thus Jeffreys's system as it stands is inconsistent. This defect cannot be eliminated by simply omitting the principle of indifference. It plays an essential role in the system; without it, many important results can no longer be derived. In spite of this defect, Jeffreys's book remains valuable for the new light it throws on many statistical problems by discussing them for the first time in terms of inductive probability.

Both Keynes and Jeffreys discussed also the statistical concept of probability, and both rejected it. They believed that all probability statements could be formulated in terms of inductive probability and that therefore there was no need for any probability concept interpreted in terms of frequency. I think that in this point they went too far. Today an increasing number of those who study both sides of the controversy which has been going on for thirty years are coming to the conclusion that here, as often before in the history of scientific thinking, both sides are right in their positive theses, but wrong in their polemic remarks about the other side. The statistical concept, for which a very elaborate mathematical theory exists, and which has been fruitfully applied in many fields in science and industry, need not at all be abandoned in order to make room for the inductive concept. Both concepts are needed for science, but they fulfill quite different functions. Statistical probability characterizes an objective situation, e.g., a state of a physical, biological, or social system. Therefore it is this concept which is used in statements concerning concrete situations or in laws expressing general regularities of such situations. On the other hand, inductive probability, as I see it, does not occur *in* scientific statements, concrete or general, but only in judgments *about* such statements; in particular, in judgments about the strength of support given by one statement, the evidence, to another, the hypothesis, and hence about the acceptability of the latter on the basis of the former. Thus, strictly speaking, inductive probability belongs not to science itself but to the methodology of science, i.e., the analysis of concepts, statements, theories, and methods of science.

The theories of both probability concepts must be further developed. Although a great deal of work has been done on statistical probability, even here some problems of its exact interpretation and its application, e.g., in methods of estimation, are still controversial. On inductive probability, on the other hand, most of the work remains still to be done. Utilizing results of Keynes and Jeffreys and employing the exact tools of modern symbolic logic, I have constructed the fundamental parts of a mathematical theory of inductive probability or inductive logic (*Logical Foundations of Probability*, 1950). The methods developed make it possible to calculate numerical values of inductive probability ("degree of confirmation") for hypotheses concerning either single events or frequencies of properties and to determine estimates of frequencies in a population on the basis of evidence about a sample of the population. A few steps have been made towards extending the theory to hypotheses involving measurable quantities such as mass, temperature, etc.

It is not possible to outline here the mathematical system itself. But I will explain some of the general problems that had to be solved before the system could be constructed and some of the basic conceptions underlying the construction. One of the fundamental questions to be decided by any theory of induction is whether to accept a principle of indifference and, if so, in what form. It should be strong enough to allow the derivation of the desired theorems, but at the same time sufficiently restricted to avoid the contradictions resulting from the classical form.

The problem will become clearer if we use a few elementary concepts of inductive logic. They will now be explained with the help of the first two columns of the accompanying diagram. We consider a set of four individuals, say four balls drawn from an urn. The individuals are described with respect to a given division of mutually exclusive properties; in our example, the two properties black (B) and white (W). An *individual distribution* is specified by ascribing to each individual one property. In our example, there are sixteen individual distributions; they are pictured in the second column (e.g., in the individual distribution No. 3, the first, second, and fourth ball are black, the third is white). A *statistical distribution*, on the other hand, is characterized by merely stating the number of individuals for each property. In the example, we have five statistical distributions, listed in the first column (e.g., the statistical distribution No. 2 is described by saying that there are three B and one W, without specifying *which* individuals are B and which W).

By the *initial probability* of a hypothesis ("probability a priori" in traditional terminology) we understand its probability before any factual knowledge concerning the individuals is available. Now we shall see that, if any initial probabilities which sum up to one are assigned to the individual distributions, all other probability values are thereby fixed. To see how the procedure works, put a slip of paper on the diagram alongside the list of individual distributions and write down opposite each distribution a fraction as its initial probability; the sum of the sixteen fractions must be one, but otherwise you may choose them just as you like. We shall soon consider the question whether some choices might be preferable to others. But for the moment we are only concerned with the fact that any arbitrary choice constitutes one and only one *inductive method* in the sense that it leads to one and only one system of probability values which contain an initial probability for any hypothesis (concerning the given individuals and the given properties) and a relative probability for any hypothesis with respect to any evidence. The procedure is as follows. For any given statement we can, by perusing the list of individual distributions, determine those in which it holds (e.g., the statement "among the first three balls there is exactly one W" holds in distributions Nos. 3, 4, 5, 6, 7, 9). Then we assign to it as initial probability the sum of the initial probabilities of the individual distributions in which it holds. Suppose that an evidence statement e (e.g., "The first ball is B, the second W, the third B") and a hypothesis h (e.g., "The fourth ball is B") are given. We ascertain first the individual distributions in which e holds (in the example, Nos. 4 and 7), and then those among them in which also h holds (only No. 4). The former ones determine the initial probability of e; the latter ones determine that of e and h together. Since the latter are among the former, the latter initial probability is a part (or the whole) of the former. We now divide the latter initial probability by the former and assign the resulting fraction to h as its relative probability with respect to e. (In our example, let us take the values of the initial probabilities of individual distributions given in the diagram for methods I and II, which will soon be explained. In method I the values for

	STATISTICAL DISTRIBUTIONS		INDIVIDUAL DISTRIBUTIONS	METHOD I	METHOD II	
	Number of Blue	Number of White		Initial Probability of Individual Distributions	Initial Probability of Statistical Distributions	Initial Probability of Individual Distributions
1.	4	0	1. ● ● ● ●	1/16	1/5	1/5 = 12/60
2.	3	1	2. ● ● ● ○	1/16	1/5	1/20 = 3/60
			3. ● ● ○ ●	1/16		1/20 = 3/60
			4. ● ○ ● ●	1/16		1/20 = 3/60
			5. ○ ● ● ●	1/16		1/20 = 3/60
3.	2	2	6. ● ● ○ ○	1/16	1/5	1/30 = 2/60
			7. ● ○ ● ○	1/16		1/30 = 2/60
			8. ● ○ ○ ●	1/16		1/30 = 2/60
			9. ○ ● ● ○	1/16		1/30 = 2/60
			10. ○ ● ○ ●	1/16		1/30 = 2/60
			11. ○ ○ ● ●	1/16		1/30 = 2/60
4.	1	3	12. ● ○ ○ ○	1/16	1/5	1/20 = 3/60
			13. ○ ● ○ ○	1/16		1/20 = 3/60
			14. ○ ○ ● ○	1/16		1/20 = 3/60
			15. ○ ○ ○ ●	1/16		1/20 = 3/60
5.	0	4	16. ○ ○ ○ ○	1/16	1/5	1/5 = 12/60

Inductive Probability Methods. (From Rudolf Carnap, "What is Probability?" *Scientific American*, September, 1953.)

Nos. 4 and 7—as for all other individual distributions—are 1/16; hence the initial probability of e is 2/16. That of e and h together is the value of No. 4 alone, hence 1/16. Dividing this by 2/16, we obtain 1/2 as the probability of h with respect to e. In method II, we find for Nos. 4 and 7 in the last column the values 3/60 and 2/60 respectively. Therefore the initial probability of e is here 5/60, that of e and h together 3/60; hence the probability of h with respect to e is 3/5.)

The problem of choosing an inductive method is closely connected with the problem of the principle of indifference. Most authors since the classical period have accepted some form of the principle and have thereby avoided the otherwise unlimited arbitrariness in the choice of a method. On the other hand, practically all authors in our century agree that the principle should be restricted to some well-defined class of hypotheses. But there is no agreement as to the class to be chosen. Many authors advocate either method I or method II, which are exemplified in our diagram. Method I consists in applying the principle of indifference to individual distributions, in other words, in assigning equal initial probabilities to individual distributions. In method II the principle is first applied to the statistical distributions and then, for each statistical distribution, to the corresponding individual distributions. Thus, in our example, equal initial probabilities are assigned in method II to the five statistical distributions, hence 1/5 to each; then this value 1/5 or 12/60 is distributed in equal parts among the corresponding individual distributions, as indicated in the last column.

If we examine more carefully the two ways of using the principle of indifference, we find that either of them leads to contradictions if applied without restriction to all divisions of properties. (The reader can easily check the following results by himself. We consider, as in the diagram, four individuals and a division D_2 into two properties; blue (instead of black) and white. Let h be the statement that all four individuals are white. We consider, on the other hand, a division D_3 into three properties: dark blue, light blue, and white. For division D_2, as used in the diagram, we see that h is an individual distribution (No. 16) and also a statistical distribution (No. 5). The same holds for division D_3. By setting up the complete diagram for the latter division, one finds that there are fifteen statistical distributions, of which h is one, and 81 individual distributions (viz., $3 \times 3 \times 3 \times 3$), of which h is also one. Applying method I to division D_2, we found as the initial probability of h 1/16; if we apply it to D_3, we find 1/81; these two results are incompatible. Method II applied to D_2 led to the value 1/5; but applied to D_3 it yields 1/15. Thus this method likewise furnishes incompatible results.) We therefore restrict the use of either method to one division, viz. The one consisting of all properties which can be distinguished in the given universe of discourse (or which we wish to distinguish within a given context of investigation). If modified in this way, either method is consistent. We may still regard the examples in the diagram as representing the modified methods I and II, if we assume that the difference between black and white is the only difference among the given individuals, or the only difference relevant to a certain investigation.

How shall we decide which of the two methods to choose? Each of them is regarded as *the* reasonable method by prominent scholars. However, in my view, the chief mistake of the earlier authors was their failure to specify explicitly the main characteristic of a reasonable inductive method. It is due to this failure that some of them chose the wrong method. This characteristic is not difficult to find. Inductive thinking is a way of judging hypotheses concerning unknown events. In order to be reasonable, this judging must be guided by our knowledge of observed events. More specifically, other things being equal, a future event is to be regarded as the more probable, the greater the relative frequency of similar events observed so far under similar circumstances. This *principle of learning from experience* guides, or rather ought to guide, all inductive thinking in everyday affairs and in science. Our confidence that a certain drug will help in a present case of a certain disease is the higher the more frequently it has helped in past cases. We would regard a man's behavior as unreasonable if his expectation of a future event were the higher the less frequently he saw it happen in the past, and also if he formed his expectations for the future without any regard to what he had observed in the past. The principle of learning from experience seems indeed so obvious that it might appear superfluous to emphasize it explicitly. In fact, however, even some authors of high rank have advocated an inductive method that violates the principle.

Let us now examine the methods I and II from the point of view of the principle of learning from experience. In our earlier example we considered the evidence e saying that of the four balls drawn the first was B, the second W, the third B; in other words, that two B and one W were so far observed. According to the principle, the prediction h that the fourth ball will be black should be taken as more probable than its negation, non-h. We found, however, that method I assigns probability 1/2 to h, and therefore likewise 1/2 to non-h. And we see easily that it assigns to h this value 1/2 also on any other evidence concerning the first three balls. Thus method I violates the principle. A man following this method sticks to the initial probability value for a prediction, irrespective of all observations he makes. In spite of this character of method I, it was proposed as the valid method of induction by prominent philosophers, among them Charles Sanders Peirce (in 1883) and Ludwig Wittgenstein (in 1921), and even by Keynes in one chapter of his book, although in other chapters he emphasizes eloquently the necessity of learning from experience.

We saw earlier that method II assigns, on the evidence specified, to h the probability 3/5, hence to non-h 2/5. Thus the principle of learning from experience is satisfied in this case, and it can be shown that the same holds in any other case. (The reader can easily verify, for example, that with respect to the evidence that the first three balls are black, the probability of h is 4/5 and therefore that of non-h 1/5.) Method II in its modified, consistent form was proposed by the author in 1945. Although it was often emphasized throughout the historical development that induction must be based on experience, nobody as far as I am aware, succeeded in specifying a consistent inductive method satisfying the principle of learning from experience. (The method proposed by Thomas Bayes (1763) and developed by Laplace—sometimes called "Bayes's rule" or "Laplace's rule of succession"—fulfills the principle. It is essentially method II, but in its unrestricted form; therefore it is inconsistent.) I found later that there are infinitely many consistent inductive methods which satisfy the principle (*The Continuum of Inductive Methods*, 1952). None of them seems to be as simple in its definition as method II, but some of them have other advantages.

Once a consistent and suitable inductive method is developed, it supplies the basis for a *general method of estimation*, i.e., a method for calculating, on the basis of given evidence, an estimate of an unknown value of any magnitude. Suppose that, on the basis of the evidence, there are n possibilities for the value of a certain magnitude at a given time, e.g., the amount of rain tomorrow, the number of persons coming to a meeting, the price of wheat after the next harvest. Let the possible values be $x_1, x_2, \ldots, x_n$, and their inductive probabilities with respect to the given evidence $p_1, p_2, \ldots, p_n$, respectively. Then we take the product p_1x_1 as the expectation value of the first case at the present moment. Thus, if the occurrence of the first case is certain and hence $p_1 = 1$, its expectation value is the full value x_1; if it is just as probable that it will occur as that it will not, and hence $p_1 = 1/2$, its expectation value is half its full value ($p_1x_1 = x_1/2$), etc. We proceed similarly with the other possible values. As estimated or total expectation value of the magnitude on the given evidence we take the sum of the expectation values for the possible cases, that is, $p_1x_1 + p_2x_2 + \ldots + p_nx_n$. (For example, suppose someone considers buying a ticket for a lottery and, on the basis of his knowledge of the lottery procedure, there is a probability of 0.01 that the ticket will win the first prize of $200 and a probability of 0.03 that it will win $50; since there are no other prizes, the probability that it will win nothing is 0.96. Hence the estimate of the gain in dollars is $0.01 \times 200 + 0.03 \times 50 + 0.96 \times 0 = 3.50$. This is the value of the ticket for him and it would be irrational for him to pay more for it.) The same method may be used in order to make a rational decision in a situation where one among various possible actions is to be chosen. For example, a man considers several possible ways for investing a certain amount of money. Then he can—in principle, at least—calculate the estimate of his gain for each possible way. To act rationally, he should then choose that way for which the estimated gain is highest.

Bernoulli and Laplace and many of their followers envisaged the idea of a theory of inductive probability which, when fully developed, would supply the means for evaluating the acceptability of hypothetical assumptions in any field of theoretical research and at the same time methods for determining a rational decision in the affairs of practical life. In the more sober cultural atmosphere of the late nineteenth century and still more in the first half of the twentieth, this idea was usually regarded as a utopian dream. It is certainly true that those audacious thinkers were not as near to their aim as they believed. But a few men dare to think today that the pioneers were not mere dreamers and that it will be possible in the future to make far-reaching progress in essentially that direction in which they saw their vision.

Rudolf Carnap

ON INDUCTIVE LOGIC

§1. INDUCTIVE LOGIC

Among the various meanings in which the word "probability" is used in everyday language, in the discussion of scientists, and in the theories of probability, there are especially two which must be clearly distinguished. We shall use for them the terms "probability$_1$" and "probability$_2$." Probability$_1$ is a logical concept, a certain logical relation between two sentences (or, alternatively, between two propositions); it is the same as the concept of degree of confirmation. I shall write briefly "*c*" for "degree of confirmation," and "$c(h, e)$" for "the degree of confirmation of the hypothesis *h* on the evidence *e*"; the evidence is usually a report on the results of our observations. On the other hand, probability$_2$ is an empirical concept; it is the relative frequency in the long run of one property with respect to another. The controversy between the so-called logical conception of probability, as represented, e.g., by Keynes,[1] and Jeffreys,[2] and others, and the frequency conception, maintained, e.g., by Von Mises[3] and Reichenbach,[4] seems to me futile. These two theories deal with two different probability concepts which are both of great importance for science. Therefore, the theories are not incompatible, but rather supplement each other.[5]

In a certain sense we might regard deductive logic as the theory of L-implication (logical implication, entailment). And inductive logic may be construed as the theory of degree of confirmation, which is, so to speak, partial L-implication. *e* L-implies *h* says that *h* is implicitly given with *e*, in other words, that the whole logical content of *h* is contained in *e*. On the other hand, $c(h, e) = 3/4$ says that *h* is not entirely given with *e* but that the assumption of *h* is supported to the degree 3/4 by the observational evidence expressed in *e*.

In the course of the last years, I have constructed a new system of inductive logic by laying down a definition for degree of confirmation and developing a theory based on this definition. A book containing this theory is in preparation.[6] The purpose of the present paper is to indicate briefly and informally the definition and a few of the results found; for lack of space, the reasons for the choice of this definition and the proofs for the results cannot be given here. The book will, of course, provide a better basis than the present informal summary for a critical evaluation of the theory and of the fundamental conception on which it is based.[7]

§2. SOME SEMANTICAL CONCEPTS

Inductive logic is, like deductive logic, in my conception a branch of semantics. However, I shall try to formulate the present outline in such a way that it does not presuppose knowledge of semantics.

Let us begin with explanations of some semantical concepts which are important both for deductive logic and for inductive logic.[8]

The system of inductive logic to be outlined applies to an infinite sequence of finite language systems L_N ($N = 1, 2, 3$, etc.) and an infinite language system L_∞. L_∞ refers to an infinite universe of individuals, designated by the individual constants a_1, a_2, etc. (or a, b, etc.), while L_N refers to a finite universe containing only N individuals designated by $a_1, a_2, \ldots a_N$. Individual variables x_1, x_2, etc. (or x, y, etc.), are the only variables occurring in these languages. The languages contain a finite number of predicates of any degree (number of arguments), designating properties of the individuals or relations between them. There are, furthermore, the customary connectives of negation ($\sim$, corresponding to not), disjunction (V, or), conjunction ($\cdot$, and); universal and existential quantifiers (for every x, there is an x); the sign of identity between individuals $=$, and t as an abbreviation for an arbitrarily chosen tautological sentence. (Thus the languages are certain forms of what is technically known as the lower functional logic with identity.) (The connectives will be used in this paper in three ways, as is customary: (1) between sentences, (2) between predicates (§ 8), (3) between names (or variables) of sentences (so that, if i and j refer to two sentences, iVj is meant to refer to their disjunction).)

A sentence consisting of a predicate of degree n with n individual constants is called an *atomic sentence* (e.g., Pa_1, i.e., a_1 has the property P, or Ra_3a_5, i.e., the relation R holds between a_3 and a_5). The conjunction of all atomic sentences in a finite language L_N describes one of the possible states of the domain of the N individuals with respect to the properties and relations expressible in the language L_N. If we replace in this conjunction some of the atomic sentences by their negations, we obtain the description of another possible state. All the conjunctions which we can form in this way, including the original one, are called *state-descriptions* in L_N. Analogously, a state-description in L_∞ is a class containing some atomic sentences and the negations of the remaining atomic sentences; since this class is infinite, it cannot be transformed into a conjunction.

In the actual construction of the language systems, which cannot be given here, semantical rules are laid down determining for any given sentence j and any state-description i whether j holds in i, that is to say whether j would be true if i described the actual state among all possible states. The class of those state-descriptions in a language system L (either one of the systems L_N or L_∞) in which j holds is called the *range* of j in L.

The concept of range is fundamental both for deductive and for inductive logic; this has already been pointed out by Wittgenstein. If the range of a sentence j in the language system L is universal, i.e., if j holds in every state-description (in L), j must necessarily be true independently of the facts; therefore we call j (in L) in this case *L-true* (logically true, analytic). (The prefix L- stands for "logical"; it is not meant to refer to the system L.) Analogously, if the range of j is null, we call j *L-false* (logically false, self-contradictory). If j is neither L-true nor L-false, we call it *factual* (synthetic, contingent). Suppose that the range of e is included in that of h. Then in every possible case in which e would be true, h would likewise be true. Therefore we say in this case that e *L-implies* (logically implies, entails) h. If two sentences have the same range, we call them *L-equivalent;* in this case, they are merely different formulations for the same content.

The L-concepts just explained are fundamental for deductive logic and therefore also for inductive logic. Inductive logic is constructed out of deductive logic by the introduction of the concept of degree of confirmation. This introduction will here be carried out in three steps: (1) the definition of regular c-functions (§ 3), (2) the definition of symmetrical c-functions (§ 5), (3) the definition of the degree of confirmation c^* (§ 6).

§3. REGULAR *C*-FUNCTIONS

A numerical function m ascribing real numbers of the interval 0 to 1 to the sentences of a finite language L_N is called a regular m-function if it is constructed according to the following rules:

(1) We assign to the state-descriptions in L_N as values of m any positive real numbers whose sum is 1.

(2) For every other sentence j in L_N, the value $m(j)$ is determined as follows:
 (a) If j is not L-false, $m(j)$ is the sum of the m-values of those state-descriptions which belong to the range of j.
 (b) If j is L-false and hence its range is null, $m(j) = 0$.

(The choice of the rule (2)(*a*) is motivated by the fact that j is L-equivalent to the disjunction of those state-descriptions which belong to the range of j and that these state-descriptions logically exclude each other.)

If any regular m-function m is given, we define a corresponding function c as follows:

(3) For any pair of sentences e, h in L_N, where e is not L-false, $c(h, e) = \dfrac{m(e \cdot h)}{m(e)}$.

$m(j)$ may be regarded as a measure ascribed to the range of j; thus the function m constitutes a metric for the ranges. Since the range of the conjunction

$e \cdot h$ is the common part of the ranges of e and of h, the quotient in (3) indicates, so to speak, how large a part of the range of e is included in the range of h. The numerical value of this ratio, however, depends on what particular m-function has been chosen. We saw earlier that a statement in deductive logic of the form e *L*-implies h says that the range of e is entirely included in that of h. Now we see that a statement in inductive logic of the form $c(h, e) = 3/4$ says that a certain part—in the example, three fourths—of the range of e is included in the range of h.[9] Here, in order to express the partial inclusion numerically, it is necessary to choose a regular m-function for measuring the ranges. Any m chosen leads to a particular c as defined above. All functions c obtained in this way are called *regular c-functions*.

One might perhaps have the feeling that the metric m should not be chosen once for all but should rather be changed according to the accumulating experiences.[10] This feeling is correct in a certain sense. However, it is to be satisfied not by the function m used in the definition (3) but by another function m dependent upon e and leading to an alternative definition (5) for the corresponding c. If a regular m is chosen according to (1) and (2), then a corresponding function m_e is defined for the state-descriptions in L_N as follows:

(4) Let i be a state-description in L_N, and e a non-L-false sentence in L_N.
(a) If e does not hold in i, $m_e(i) = 0$.
(b) If e holds in i, $m_e(i) = \frac{m(i)}{m(e)}$.

Thus m_e represents a metric for the state-descriptions which changes with the changing evidence e. Now $m_e(j)$ for any other sentence j in L_N is defined in analogy to (2)(a) and (b). Then we define the function c corresponding to m as follows:

(5) For any pair of sentences e, h in L_N, where e is not L-false, $c(h, e) = m_e(h)$.

It can easily be shown that this alternative definition (5) yields the same values as the original definition (3).

Suppose that a sequence of regular m-functions is given, one for each of the finite languages L_N ($N = 1, 2$, etc.). Then we define a corresponding m-function for the infinite language as follows:

(6) $m(j)$ in L_∞ is the limit of the values $m(j)$ in L_N for $N \to \infty$.

c-functions for the finite languages are based on the given m-functions according to (3). We define a corresponding c-function for the infinite language as follows:

(7) $c(h, e)$ in L_∞ is the limit of the values $c(h, e)$ in L_N for $N \to \infty$.

The definitions (6) and (7) are applicable only in those cases where the specified limits exist.

We shall later see how to select a particular sub-class of regular c-functions (§ 5) and finally one particular c-function c^* as the basis of a complete system of inductive logic (§ 6). For the moment, let us pause at our first step, the definition of regular c-functions just given, in order to see what results this definition alone can yield, before we add further definitions. The theory of regular c-functions, i.e., the totality of those theorems which are founded on the definition stated, is the first and fundamental part of inductive logic. It turns out that we find here many of the fundamental theorems of the classical theory of probability, e.g., those known as the theorem (or principle) of multiplication, the general and the special theorems of addition, the theorem of division, and, based upon it, Bayes's theorem.

One of the cornerstones of the classical theory of probability is the principle of indifference (or principle of insufficient reason). It says that, if our evidence e does not give us any sufficient reason for regarding one of two hypotheses h and h' as more probable than the other, then we must take their probabilities$_1$ as equal: $c(h, e) = c(h', e)$. Modern authors, especially Keynes, have correctly pointed out that this principle has often been used beyond the limits of its original meaning and has then led to quite absurd results. Moreover, it can easily be shown that, even in its original meaning, the principle is by far too general and leads to contradictions. Therefore the principle must be abandoned. If it is and we consider only those theorems of the classical theory which are provable without the help of this principle, then we find that these theorems hold for all regular c-functions. The same is true for those modern theories of probability$_1$ (e.g., that by Jeffreys, *op. cit.*) which make use of the principle of indifference. Most authors of modern axiom systems of probability$_1$ (e.g., Keynes, *op. cit.*, Waismann, *op. cit.*, Mazurkiewicz[11], Hosiasson[12], von Wright[13]) are cautious enough not to accept that principle. An examination of these systems shows that their axioms and hence their theorems hold for all regular c-functions. Thus these systems restrict themselves to the first part of inductive logic, which, although fundamental and important, constitutes only a very small and weak section of the whole of inductive logic. The weakness of this part shows itself in the fact that it does not determine the value of c for any pair h, e except in some special cases where the value is 0 or 1. The theorems of this part tell us merely how to calculate further values of c if some values are given. Thus it is clear that this part alone is quite useless for application and must be supplemented by additional rules. (It may be remarked incidentally, that this point marks a fundamental difference between the theories of probability$_1$ and of probability$_2$ which otherwise are analogous in many respects. The theorems concerning probability$_2$ which are analogous to the theorems concerning regular c-functions constitute not only the first part but the whole of the logico-mathematical theory of probability$_2$. The task of determining the value of probability$_2$ for a given case is—in contradistinction to the corresponding task for probability$_1$—an empirical one and hence lies outside the scope of the logical theory of probability$_2$.)

§4. THE COMPARATIVE CONCEPT OF CONFIRMATION

Some authors believe that a metrical (or quantitative) concept of degree of confirmation, that is, one with numerical values, can be applied, if at all, only in certain cases of a special kind and that in general we can make only a comparison in terms of higher or lower confirmation without ascribing numerical values.

Whether these authors are right or not, the introduction of a merely comparative (or topological) concept of confirmation not presupposing a metrical concept is, in any case, of interest. We shall now discuss a way of defining a concept of this kind.

For technical reasons, we do not take the concept "more confirmed" but "more or equally confirmed." The following discussion refers to the sentences of any finite language L_N. We write, for brevity, $MC(h, e, h', e')$ for h is confirmed on the evidence e more highly or just as highly as h' on the evidence e'.

Although the definition of the comparative concept MC at which we aim will not make use of any metrical concept of degree of confirmation, let us now consider, for heuristic purposes, the relation between MC and the metrical concepts, i.e., the regular c-functions. Suppose we have chosen some concept of degree of confirmation, in other words, a regular c-function c, and further a comparative relation MC; then we shall say that MC is in accord with c if the following holds:

(1) For any sentences h, e, h', e', if $MC(h, e, h', e')$ then $c(h, e) \geqq c(h', e')$.

However, we shall not proceed by selecting one c-function and then choosing a relation MC which is in accord with it. This would not fulfill our intention. Our aim is to find a comparative relation MC which grasps those logical relations between sentences which are, so to speak, prior to the introduction of any particular m-metric for the ranges and of any particular c-function; in other words, those logical relations with respect to which all the various regular c-functions agree. Therefore we lay down the following requirement:

(2) The relation MC is to be defined in such a way that it is in accord with *all* regular c-functions; in other words, if $MC(h, e, h', e')$, then for every regular c, $c(h, e) \geqq c(h', e')$.

It is not difficult to find relations which fulfill this requirement (2). First let us see whether we can find quadruples of sentences h, e, h', e' which satisfy the following condition occurring in (2):

(3) For every regular c, $c(h, e) \geqq c(h', e')$.

It is easy to find various kinds of such quadruples. (For instance, if e and e' are any non-L-false sentences, then the condition (3) is satisfied in all cases where e L-implies h, because here $c(h, e) = 1$; further in all cases where e' L-implies $\sim h'$, because here $c(h', e') = 0$; and in many other cases.) We could, of course, define a relation MC by taking some cases where we know that the condition (3) is satisfied and restricting the relation to these cases. Then the relation would fulfill the requirement (2); however, as long as there are cases which satisfy the condition (3) but which we have not included in the relation, the relation is unnecessarily restricted. Therefore we lay down the following as a second requirement for MC:

(4) MC is to be defined in such a way that it holds in all cases which satisfy the condition (3); in such a way, in other words, that it is the most comprehensive relation which fulfills the first requirement (2).

These two requirements (2) and (4) together stipulate that $MC(h, e, h', e')$ is to hold if and only if the condition (3) is satisfied; thus the requirements determine uniquely one relation MC. However, because they refer to the c-functions, we do not take these requirements as a definition for MC, for we intend to give a purely comparative definition for MC, a definition which does not make use of any metrical concepts but which leads nevertheless to a relation MC which fulfills the requirements (2) and (4) referring to c-functions. This aim is reached by the following definition (where $=_{Df}$ is used as sign of definition).

(5) $MC(h, e, h', e') =_{Df}$ the sentences h, e, h', e' (in L_N) are such that e and e' are not L-false and at least one of the following three conditions is fulfilled:
(a) e L-implies h,
(b) e' L-implies $\sim h'$,
(c) $e' \cdot h'$ L-implies $e \cdot h$ and simultaneously e L-implies $h \vee e'$.

((a) and (b) are the two kinds of rather trivial cases earlier mentioned; (c) comprehends the interesting cases; an explanation and discussion of them cannot be given here.)

The following theorem can then be proved concerning the relation MC defined by (5). It shows that this relation fulfills the two requirements (2) and (4).

(6) For any sentences h, e, h', e' in L_N the following holds:
(a) If $MC(h, e, h' e')$, then, for every regular c, $c(h, e) \geqq c(h', e')$.
(b) If, for every regular c, $c(h, e) \geqq c(h', e')$, then $MC(h, e, h', e')$.

(With respect to L_∞, the analogue of (6)(a) holds for all sentences, and that of (6)(b) for all sentences without variables.)

§5. SYMMETRICAL *C*-FUNCTIONS

The next step in the construction of our system of inductive logic consists in selecting a narrow sub-class of the comprehensive class of all regular c-functions. The guiding idea for this step will be the principle that inductive logic should treat all individuals on a par. The same principle holds for deductive logic; for instance, if $..a..b..$ L-implies $--b--c--$ (where the first expression is meant to indicate some sentence containing a and b, and the second another sentence containing b and c), then L-implication holds likewise between corresponding sentences with other individual constants, e.g., between $..d..c..$ and $--c--a--$. Now we require that this should hold also for inductive logic, e.g., that $c(--b--c--, ..a..b..) = c(--c--a--, ..d..c..)$. It seems that all authors on probability$_1$ have assumed this principle—although it has seldom, if ever, been stated explicitly—by formulating theorems in the following or similar terms: "On the basis of observations of s things of which s_1 were found to have the property M and s_2 not to have this property, the probability that another thing has this property is such and such." The fact that these theorems refer only to the number of things observed and do not mention particular things shows implicitly that it does not matter which things are involved; thus it is assumed, e.g., that $c(Pd, Pa \cdot Pb \cdot \sim Pc) = c(Pc, Pa \cdot Pd \cdot \sim Pb)$.

The principle could also be formulated as follows. Inductive logic should, like deductive logic, make no discrimination among individuals. In other words, the value of *c* should be influenced only by those differences between individuals which are expressed in the two sentences involved; no differences between particular individuals should be stipulated by the rules of either deductive or inductive logic.

It can be shown that this principle of non-discrimination is fulfilled if *c* belongs to the class of symmetrical *c*-functions which will now be defined. Two state-descriptions in a language L_N are said to be *isomorphic* or to have the same structure if one is formed from the other by replacements of the following kind: We take any one-one relation *R* such that both its domain and its converse domain is the class of all individual constants in L_N, and then replace every individual constant in the given state-description by the one correlated with it by *R*. If a regular *m*-function (for L_N) assigns to any two isomorphic state-descriptions (in L_N) equal values, it is called a symmetrical *m*-function; and a *c*-function based upon such an *m*-function in the way explained earlier (see (3) in § 3) is then called a *symmetrical c-function.*

§6. THE DEGREE OF CONFIRMATION C^*

Let *i* be a state-description in L_N. Suppose there are n_i state-descriptions in L_N isomorphic to *i* (including *i* itself), say i, i', i'', etc. These n_i state-descriptions exhibit one and the same structure of the universe of L_N with respect to all the properties and relations designated by the primitive predicates in L_N. This concept of structure is an extension of the concept of structure or relation-number (Russell) usually applied to one dyadic relation. The common structure of the isomorphic state-descriptions i, i', i'', etc., can be described by their disjunction $i \vee i' \vee i'' \vee \cdots$. Therefore we call this disjunction, say *j*, a *structure-description* in L_N. It can be shown that the range of *j* contains only the isomorphic state-descriptions i, i', i'', etc. Therefore (see (2)(a) in § 3) $m(j)$ is the sum of the *m*-values for these state-descriptions. If *m* is symmetrical, then these values are equal, and hence

$$(1) \qquad m(j) = n_i \times m(i).$$

And, conversely, if $m(j)$ is known to be *q*, then

$$(2) \qquad m(i) = m(i') = m(i'') = \cdots = q/n_i.$$

This shows that what remains to be decided is merely the distribution of *m*-values among the structure-descriptions in L_N. We decide to give them equal *m*-values. This decision constitutes the third step in the construction of our inductive logic. This step leads to one particular *m*-function m^* and to the *c*-function c^* based upon m^*. According to the preceding discussion, m^* is characterized by the following two stipulations:

(3) (a) m^* is a symmetrical m-function;
(b) m^* has the same value for all structure-descriptions (in L_N).

We shall see that these two stipulations characterize just one function. Every state-description (in L_N) belongs to the range of just one structure-description. Therefore, the sum of the m^*-values for all structure-descriptions in L_N must be the same as for all state-descriptions, hence 1 (according to (1) in § 3). Thus, if the number of structure-descriptions in L_N is m, then, according to (3)(b),

(4) for every structure-description j in L_N, $m^*(j) = \frac{1}{m}$.

Therefore, if i is any state-description in L_N and n_i is the number of state descriptions isomorphic to i, then, according to (3)(a) and (2),

(5) $$m^*(i) = \frac{1}{mn_i}.$$

(5) constitutes a definition of m^* as applied to the state-descriptions in L_N. On this basis, further definitions are laid down as explained above (see (2) and (3) in § 3): first a definition of m^* as applied to all sentences in L_N, and then a definition of c^* on the basis of m^*. Our inductive logic is the theory of this particular function c^* as our concept of degree of confirmation.

It seems to me that there are good and even compelling reasons for the stipulation (3)(a), i.e., the choice of a symmetrical function. The proposal of any non-symmetrical c-function as degree of confirmation could hardly be regarded as acceptable. The same can not be said, however, for the stipulation (3)(b). No doubt, to the way of thinking which was customary in the classical period of the theory of probability, (3)(b) would appear as validated, like (3)(a), by the principle of indifference. However, to modern, more critical thought, this mode of reasoning appears as invalid because the structure-descriptions (in contradistinction to the individual constants) are by no means alike in their logical features but show very conspicuous differences. The definition of c^* shows a great simplicity in comparison with other concepts which may be taken into consideration. Although this fact may influence our decision to choose c^*, it cannot, of course, be regarded as a sufficient reason for this choice. It seems to me that the choice of c^* cannot be justified by any features of the definition which are immediately recognizable, but only by the consequences to which the definition leads.

There is another c-function c_{w} which at the first glance appears not less plausible than c^*. The choice of this function may be suggested by the following consideration. Prior to experience, there seems to be no reason to regard one state-description as less probable than another. Accordingly, it might seem natural to assign equal m-values to all state-descriptions. Hence, if the number of the state-descriptions in L_N is n, we define for any state-description i

(6) $$m_{\mathrm{w}}(i) = \frac{1}{n}.$$

This definition (6) for m_W is even simpler than the definition (5) for m^*. The measure ascribed to the ranges is here simply taken as proportional to the cardinal numbers of the ranges. On the basis of the m_W-values for the state-descriptions defined by (6), the values for the sentences are determined as before (see (2) in § 3), and then c_W is defined on the basis of m_W (see (3) in § 3).[14]

In spite of its apparent plausibility, the function c_W can easily be seen to be entirely inadequate as a concept of degree of confirmation. As an example, consider the language L_{101} with P as the only primitive predicate. Let the number of state-descriptions in this language be n (it is 2^{101}). Then for any state-description, $m_W = 1/n$. Let e be the conjunction $Pa_1 \cdot Pa_2 \cdot Pa_3 \ldots Pa_{100}$ and let h be Pa_{101}. Then $e \cdot h$ is a state-description and hence $m_W(e \cdot h) = 1/n$. e holds only in the two state-descriptions $e \cdot h$ and $e \cdot \sim h$; hence $m_W(e) = 2/n$. Therefore $c_W(h, e) = 1/2$. If e' is formed from e by replacing some or even all of the atomic sentences with their negations, we obtain likewise $c_W(h, e') = 1/2$. Thus the c_W-value for the prediction that a_{101} is P is always the same, no matter whether among the hundred observed individuals the number of those which we have found to be P is 100 or 50 or 0 or any other number. Thus the choice of c_W as the degree of confirmation would be tantamount to the principle never to let our past experiences influence our expectations for the future. This would obviously be in striking contradiction to the basic principle of all inductive reasoning.

§7. LANGUAGES WITH ONE-PLACE PREDICATES ONLY

The discussions in the rest of this paper concern only those language systems whose primitive predicates are one-place predicates and hence designate properties, not relations. It seems that all theories of probability constructed so far have restricted themselves, or at least all of their important theorems, to properties. Although the definition of c^* in the preceding section has been stated in a general way so as to apply also to languages with relations, the greater part of our inductive logic will be restricted to properties. An extension of this part of inductive logic to relations would require certain results in the deductive logic of relations, results which this discipline, although widely developed in other respects, has not yet reached (e.g., an answer to the apparently simple question as to the number of structures in a given finite language system).

Let $L_N{}^p$ be a language containing N individual constants $a_1, \ldots a_N$, and p one-place primitive predicates $P_1, \ldots P_p$. Let us consider the following expressions (sentential matrices). We start with $P_1x \cdot P_2x \ldots P_px$; from this expression we form others by negating some of the conjunctive components, until we come to $\sim P_1x \cdot \sim P_2x \ldots \sim P_px$, where all components are negated. The number of these expressions is $k = 2^p$; we abbreviate them by $Q_1x, \ldots Q_kx$. We call the k properties expressed by those k expressions in conjunctive form and now designated by the k new Q-predicates the *Q-properties* with respect to the given language $L_N{}^p$. We see easily that these Q-properties are the strongest properties expressible in this language (except for the L-empty, i.e., logically self-contradictory, property); and further, that they constitute an exhaustive and non-overlapping classification, that is to say, every individual has one and only one of the Q-properties. Thus, if we state for each individual which of the Q-properties it has, then we have described the individuals completely. Every state-description can be brought into the form of such a statement, i.e., a conjunction of N Q-sentences, one for each of the N individuals. Suppose that in a given state-description i the number of individuals having the property Q_1 is N_1, the number for Q_2 is $N_2, \ldots$ that for Q_k is N_k. Then we call the numbers $N_1, N_2, \ldots N_k$ the *Q-numbers* of the state-description i; their sum is N. Two state-descriptions are isomorphic if and only if they have the same Q-numbers. Thus here a structure-description is a statistical description giving the Q-numbers N_1, N_2, etc., without specifying which individuals have the properties Q_1, Q_2, etc.

Here—in contradistinction to languages with relations—it is easy to find an explicit function for the number m of structure-descriptions and, for any given state-description i with the Q-numbers $N_1, \ldots N_k$, an explicit function for the number n_i of state-descriptions isomorphic to i, and hence also a function for $m^*(i)$.[15]

Let j be a non-general sentence (i.e., one without variables) in $L_N{}^p$. Since there are effective procedures (that is, sets of fixed rules furnishing results in a finite number of steps) for constructing all state-descriptions in which j holds and for computing m^* for any given state-description, these procedures together yield an effective procedure for computing $m^*(j)$ (according to (2) in §3). However, the number of state-descriptions becomes very large even for small language systems (it is k^N, hence, e.g., in $L_7{}^3$ it is more than two million). Therefore, while the procedure indicated for the computation of $m^*(j)$ is effective, nevertheless in most ordinary cases it is impracticable; that is to say, the number of steps to be taken, although finite, is so large that nobody will have the time to carry them out to the end. I have developed another procedure for the computation of $m^*(j)$ which is not only effective but also practicable if the number of individual constants occurring in j is not too large.

The value of m^* for a sentence j in the infinite language has been defined (see (6) in § 3) as the limit of its values for the same sentence j in the finite languages. The question arises whether and under what conditions this limit exists. Here we have to distinguish two cases. (i) Suppose that j contains no variable. Here the situation is simple; it can be shown that in this case $m^*(j)$ is the same in all finite languages in which j occurs; hence it has the same value also in the infinite language. (ii) Let j be general, i.e., contain variables. Here the situation is quite different. For a given finite language with N individuals, j can of course easily be transformed into an L-equivalent sentence j'_N without variables, because in this language a universal sentence is L-equivalent to a conjunction of N components. The values of $m^*(j'_N)$ are in general different for each N; and although the simplified procedure mentioned above is available for the computation of these values, this procedure becomes impracticable even for moderate N. Thus for general sentences the problem of the existence and the practical computability of the limit becomes serious. It can be shown that for every general sentence the limit exists; hence m^* has a value for all sentences in the infinite language. Moreover, an effective procedure for the computation of $m^*(j)$ for any sentence j in the infinite language has been constructed. This is based on a procedure for transforming any given general sentence j into a non-general sentence j' such that j and j', although not necessarily L-equivalent, have the same m^*-value in the infinite language and j' does not contain more individual constants than j; this procedure is not only effective but also practicable for sentences of customary length. Thus, the computation of $m^*(j)$ for a general sentence j is in fact much simpler for the infinite language than for a finite language with a large N.

With the help of the procedure mentioned, the following theorem is obtained:

If j is a purely general sentence (i.e., one without individual constants) in the infinite language, then $m^*(j)$ is either 0 or 1.

§8. INDUCTIVE INFERENCES

One of the chief tasks of inductive logic is to furnish general theorems concerning inductive inferences. We keep the traditional term "inference"; however, we do not mean by it merely a transition from one sentence to another (viz., from the evidence or premiss e to the hypothesis or conclusion h) but the determination of the degree of confirmation $c(h, e)$. In deductive logic it is sufficient to state that h follows with necessity from e; in inductive logic, on the other hand, it would not be sufficient to state that h follows—not with necessity but to some degree or other—from e. It must be specified to what degree h follows from e; in other words, the value of $c(h, e)$ must be given. We shall now indicate some results with respect to the most important kinds of inductive inference. These inferences are of special importance when the evidence or the hypothesis or both give statistical information, e.g., concerning the absolute or relative frequencies of given properties.

If a property can be expressed by primitive predicates together with the ordinary connectives of negation, disjunction, and conjunction (without the use of individual constants, quantifiers, or the identity sign), it is called an *elementary property*. We shall use M, M', M_1, M_2, etc., for elementary properties. If a property is empty by logical necessity (e.g., the property designated by $P \cdot \sim P$) we call it L-empty; if it is universal by logical necessity (e.g., $P \vee \sim P$), we call it L-universal. If it is neither L-empty nor L-universal (e.g., P_1, $P_1 \cdot \sim P_2$), we call it a *factual property*; in this case it may still happen to be universal or empty, but if so, then contingently, not necessarily. It can be shown that every elementary property which is not L-empty is uniquely analyzable into a disjunction (i.e., or-connection) of Q-properties. If M is a disjunction of n Q-properties ($n \geqq 1$), we say that the (logical) *width* of M is n; to an L-empty property we ascribe the width 0. If the width of M is w ($\geqq 0$), we call w/k its *relative width* (k is the number of Q-properties).

The concepts of width and relative width are very important for inductive logic. Their neglect seems to me one of the decisive defects in the classical theory of probability which formulates its theorems "for any property" without qualification. For instance, Laplace takes the probability a priori that a given thing has a given property, no matter of what kind, to be 1/2. However, it seems clear that this probability cannot be the same for a very strong property (e.g., $P_1 \cdot P_2 \cdot P_3$) and for a very weak property (e.g., $P_1 \vee P_2 \vee P_3$). According to our definition, the first of the two properties just mentioned has the relative width 1/8, and the second 7/8. In this and in many other cases the probability or degree of confirmation must depend upon the widths of the properties involved. This will be seen in some of the theorems to be mentioned later.

§9. THE DIRECT INFERENCE

Inductive inferences often concern a situation where we investigate a whole population (of persons, things, atoms, or whatever else) and one or several samples picked out of the population. An inductive inference from the whole population to a sample is called a direct inductive inference. For the sake of simplicity, we shall discuss here and in most of the subsequent sections only the case of one property M, hence a classification of all individuals into M and $\sim M$. The theorems for classifications with more properties are analogous but more complicated. In the present case, the evidence e says that in a whole population of n individuals there are n_1 with the property M and $n_2 = n - n_1$ with $\sim M$; hence the relative frequency of M is $r = n_1/n$. The hypothesis h says that a sample of s individuals taken from the whole population will contain s_1 individuals with the property M and $s_2 = s - s_1$ with $\sim M$. Our theory yields in this case the same values as the classical theory.[16]

If we vary s_1, then c^* has its maximum in the case where the relative frequency s_1/s in the sample is equal or close to that in the whole population.

If the sample consists of only one individual c, and h says that c is M, then $c^*(h, e) = r$.

As an approximation in the case that n is very large in relation to s, Newton's theorem holds.[17] If furthermore the sample is sufficiently large, we obtain as an approximation Bernoulli's theorem in its various forms.

It is worthwhile to note two characteristics which distinguish the direct inductive inference from the other inductive inferences and make it, in a sense, more closely related to deductive inferences:

(i) The results just mentioned hold not only for c^* but likewise for all symmetrical c-functions; in other words, the results are independent of the particular m-metric chosen provided only that it takes all individuals on a par.

(ii) The results are independent of the width of M. This is the reason for the agreement between our theory and the classical theory at this point.

§10. THE PREDICTIVE INFERENCE

We call the inference from one sample to another the predictive inference. In this case, the evidence e says that in a first sample of s individuals, there are s_1 with the property M, and $s_2 = s - s_1$ with $\sim M$. The hypothesis h says that in a second sample of s' other individuals, there will be s'_1 with M, and $s'_2 = s' - s'_1$ with $\sim M$. Let the width of M be w_1; hence the width of $\sim M$ is $w_2 = k - w_1$.[18]

The most important special case is that where h refers to one individual c only and says that c is M. In this case,

$$c^*(h, e) = \frac{s_1 + w_1}{s + k}. \tag{1}$$

Laplace's much debated rule of succession gives in this case simply the value $(s_1 + 1)/(s + 2)$ for any property whatever; this, however, if applied to different properties, leads to contradictions. Other authors state the value s_1/s, that is, they take simply the observed relative frequency as the probability for the prediction that an unobserved individual has the property in question. This rule, however, leads to quite implausible results. If $s_1 = s$, e.g., if three individuals have been observed and all of them have been found to be M, the last-mentioned rule gives the probability for the next individual being M as 1, which seems hardly acceptable. According to (1), c^* is influenced by the following two factors (though not uniquely determined by them):

(i) w_1/k, the relative width of M;
(ii) s_1/s, the relative frequency of M in the observed sample.

The factor (i) is purely logical; it is determined by the semantical rules. (ii) is empirical; it is determined by observing and counting the individuals in the sample. The value of c^* always lies between those of (i) and (ii). Before any individual has been observed, c^* is equal to the logical factor (i). As we first begin to observe a sample, c^* is influenced more by this factor than by (ii). As the sample is increased by observing more and more individuals (but not including the one mentioned in h), the empirical factor (ii) gains more and more influence upon c^* which approaches closer and closer to (ii); and when the sample is sufficiently large, c^* is practically equal to the relative frequency (ii). These results seem quite plausible.[19]

The predictive inference is the most important inductive inference. The kinds of inference discussed in the subsequent sections may be construed as special cases of the predictive inference.

§11. THE INFERENCE BY ANALOGY

The inference by analogy applies to the following situation. The evidence known to us is the fact that individuals *b* and *c* agree in certain properties and, in addition, that *b* has a further property; thereupon we consider the hypothesis that *c* too has this property. Logicians have always felt that a peculiar difficulty is here involved. It seems plausible to assume that the probability of the hypothesis is the higher the more properties *b* and *c* are known to have in common; on the other hand, it is felt that these common properties should not simply be counted but weighed in some way. This becomes possible with the help of the concept of width. Let M_1 be the conjunction of all properties which *b* and *c* are known to have in common. The known similarity between *b* and *c* is the greater the stronger the property M_1, hence the smaller its width. Let M_2 be the conjunction of all properties which *b* is known to have. Let the width of M_1 be w_1, and that of M_2, w_2. According to the above description of the situation, we presuppose that M_2 L-implies M_1 but is not L-equivalent to M_1; hence $w_1 > w_2$. Now we take as evidence the conjunction $e{\cdot}j$; *e* says that *b* is M_2, and *j* says that *c* is M_1. The hypothesis *h* says that *c* has not only the properties ascribed to it in the evidence but also the one (or several) ascribed in the evidence to *b* only, in other words, that *c* has all known properties of *b*, or briefly that *c* is M_2. Then

$$(1) \qquad c^*(h, e{\cdot}j) = \frac{w_2 + 1}{w_1 + 1}.$$

j and *h* speak only about *c*; *e* introduces the other individual *b* which serves to connect the known properties of *c* expressed by *j* with its unknown properties expressed by *h*. The chief question is whether the degree of confirmation of *h* is increased by the analogy between *c* and *b*, in other words, by the addition of *e* to our knowledge. A theorem[20] is found which gives an affirmative answer to this question. However, the increase of c^* is under ordinary conditions rather small; this is in agreement with the general conception according to which reasoning by analogy, although admissible, can usually yield only rather weak results.

Hosiasson[21] has raised the question mentioned above and discussed it in detail. She says that an affirmative answer, a proof for the increase of the degree of confirmation in the situation described, would justify the universally accepted reasoning by analogy. However, she finally admits that she does not find such a proof on the basis of her axioms. I think it is not astonishing that neither the classical theory nor modern theories of probability have been able to give a satisfactory account of and justification for the inference by analogy. For, as the theorems mentioned show, the degree of confirmation and its increase depend here not on relative frequencies but entirely on the logical widths of the properties involved, thus on magnitudes neglected by both classical and modern theories.

The case discussed above is that of simple analogy. For the case of multiple analogy, based on the similarity of c not only with one other individual but with a number n of them, similar theorems hold. They show that c^* increases with increasing n and approaches 1 asymptotically. Thus, multiple analogy is shown to be much more effective than simple analogy, as seems plausible.

§12. THE INVERSE INFERENCE

The inference from a sample to the whole population is called the inverse inductive inference. This inference can be regarded as a special case of the predictive inference with the second sample covering the whole remainder of the population. This inference is of much greater importance for practical statistical work than the direct inference, because we usually have statistical information only for some samples and not for the whole population.

Let the evidence e say that in an observed sample of s individuals there are s_1 individuals with the property M and $s_2 = s - s_1$ with $\sim M$. The hypothesis h says that in the whole population of n individuals, of which the sample is a part, there are n_1 individuals with M and n_2 with $\sim M$ ($n_1 \geqq s_1, n_2 \geqq s_2$). Let the width of M be w_1, and that of $\sim M$ be $w_2 = k - w_1$. Here, in distinction to the direct inference, $c^*(h, e)$ is dependent not only upon the frequencies but also upon the widths of the two properties.[22]

§13. THE UNIVERSAL INFERENCE

The universal inductive inference is the inference from a report on an observed sample to a hypothesis of universal form. Sometimes the term "induction" has been applied to this kind of inference alone, while we use it in a much wider sense for all non-deductive kinds of inference. The universal inference is not even the most important one; it seems to me now that the role of universal sentences in the inductive procedures of science has generally been overestimated. This will be explained in the next section.

Let us consider a simple law l, i.e., a factual universal sentence of the form "all M are M'" or, more exactly, "for every x, if x is M, then x is M'," where M and M' are elementary properties. As an example, take "all swans are white." Let us abbreviate $M \cdot \sim M'$ ("non-white swan") by M_1 and let the width of M_1 be w_1. Then l can be formulated thus: "M_1 is empty," i.e. "there is no individual (in the domain of individuals of the language in question) with the property M_1" ("there are no non-white swans"). Since l is a factual sentence, M_1 is a factual property; hence $w_1 > 0$. To take an example, let w_1 be 3; hence M_1 is a disjunction of three Q-properties, say $Q \vee Q' \vee Q''$. Therefore, l can be transformed into: "Q is empty, and Q' is empty, and Q'' is empty." The weakest factual laws in a language are those which say that a certain Q-property is empty;

we call them Q-laws. Thus we see that l can be transformed into a conjunction of w_1 Q-laws. Obviously l asserts more if w_1 is larger; therefore we say that the law l has the strength w_1.

Let the evidence e be a report about an observed sample of s individuals such that we see from e that none of these s individuals violates the law l; that is to say, e ascribes to each of the s individuals either simply the property $\sim M_1$ or some other property L-implying $\sim M_1$. Let l, as above, be a simple law which says that M_1 is empty, and w_1 be the width of M_1; hence the width of $\sim M_1$ is $w_2 = k - w_1$. For finite languages with N individuals, $c^*(l, e)$ is found to decrease with increasing N, as seems plausible.[23] If N is very large, c^* becomes very small; and for an infinite universe it becomes 0. The latter result may seem astonishing at first sight; it seems not in accordance with the fact that scientists often speak of "well-confirmed" laws. The problem involved here will be discussed later.

So far we have considered the case in which only positive instances of the law l have been observed. Inductive logic must, however, deal also with the case of negative instances. Therefore let us now examine another evidence e' which says that in the observed sample of s individuals there are s_1 which have the property M_1 (non-white swans) and hence violate the law l, and that $s_2 = s - s_1$ have $\sim M_1$ and hence satisfy the law l. Obviously, in this case there is no point in taking as hypothesis the law l in its original forms, because l is logically incompatible with the present evidence e', and hence $c^*(l, e') = 0$. That all individuals satisfy l is excluded by e'; the question remains whether at least all unobserved individuals satisfy l. Therefore we take here as hypothesis the restricted law l' corresponding to the original unrestricted law l; l' says that all individuals not belonging to the sample of s individuals described in e' have the property $\sim M_1$. w_1 and w_2 are, as previously, the widths of M_1 and $\sim M_1$ respectively. It is found that $c^*(l', e')$ decreases with an increase of N and even more with an increase in the number s_1 of violating cases.[24] It can be shown that, under ordinary circumstances with large N, c^* increases moderately when a new individual is observed which satisfies the original law l. On the other hand, if the new individual violates l, c^* decreases very much, its value becoming a small fraction of its previous value. This seems in good agreement with the general conception.

For the infinite universe, c^* is again 0, as in the previous case. This result will be discussed in the next section.

§14. THE INSTANCE CONFIRMATION OF A LAW

Suppose we ask an engineer who is building a bridge why he has chosen the building materials he is using, the arrangement and dimensions of the supports, etc. He will refer to certain physical laws, among them some general laws of mechanics and some specific laws concerning the strength of the materials. On further inquiry as to his confidence in these laws he may apply to them phrases like "very reliable," "well founded," "amply confirmed by numerous experiences." What do these phrases mean? It is clear that they are intended to say something about probability$_1$ or degree of confirmation. Hence, what is meant could be formulated more explicitly in a statement of the form "$c(h, e)$ is high" or the like. Here the evidence e is obviously the relevant observational knowledge of the engineer or of all physicists together at the present time. But what is to serve as the hypothesis h? One might perhaps think at first that h is the law in question, hence a universal sentence l of the form: "For every space-time point x, if such and such conditions are fulfilled at x, then such and such is the case at x." I think, however, that the engineer is chiefly interested not in this sentence l, which speaks about an immense number, perhaps an infinite number, of instances dispersed through all time and space, but rather in one instance of l or a relatively small number of instances. When he says that the law is very reliable, he does not mean to say that he is willing to bet that among the billion of billions, or an infinite number, of instances to which the law applies there is not one counter-instance, but merely that this bridge will not be a counter-instance, or that among all bridges which he will construct during his lifetime, or among those which all engineers will construct during the next one thousand years, there will be no counter-instance. Thus h is not the law l itself but only a prediction concerning one instance or a relatively small number of instances. Therefore, what is vaguely called the reliability of a law is measured not by the degree of confirmation of the law itself but by that of one or several instances. This suggests the subsequent definitions. They refer, for the sake of simplicity, to just one instance; the case of several, say one hundred, instances can then easily be judged likewise. Let e be any non-L-false sentence without variables. Let l be a simple law of the form earlier described (§ 13). Then we understand by the *instance confirmation* of l on the evidence e, in symbols $c^*_i(l, e)$, the degree of confirmation, on the evidence e, of the hypothesis that a new individual not mentioned in e fulfills the law l.[25]

The second concept, now to be defined, seems in many cases to represent still more accurately what is vaguely meant by the reliability of a law l. We suppose here that l has the frequently used conditional form mentioned earlier: "For every x, if x is M, then x is M'" (e.g., "all swans are white"). By the *qualified-instance confirmation* of the law that all swans are white we mean the degree of confirmation for the hypothesis h' that the next swan to be observed will likewise be white. The difference between the hypothesis h used previously for the instance confirmation and the hypothesis h' just described consists in the fact that the latter concerns an individual which is already qualified as fulfilling the condition M. That is the reason why we speak here of the qualified-instance confirmation, in symbols c^*_{qi}.[26] The results obtained concerning instance confirmation and qualified-instance confirmation[27] show that the values of these two functions are independent of N and hence hold for all finite and infinite universes. It has been found that, if the number s_1 of observed counter-instances is a fixed small number, then, with the increase of the sample s, both c^*_i and c^*_{qi} grow close to 1, in contradistinction to c^* for the law itself. This justifies the customary manner of speaking of "very reliable" or "well-founded" or "well-confirmed" laws, provided we interpret these phrases as referring to a high value of either of our two concepts just introduced. Understood in this sense, the phrases are not in contradiction to our previous results that the degree of confirmation of a law is very small in a large domain of individuals and 0 in the infinite domain (§ 13).

These concepts will also be of help in situations of the following kind. Suppose a scientist has observed certain events, which are not sufficiently explained by the known physical laws. Therefore he looks for a new law as an explanation. Suppose he finds two incompatible laws l and l', each of which would explain the observed events satisfactorily. Which of them should he prefer? If the domain of individuals in question is finite, he may take the law with the higher degree of confirmation. In the infinite domain, however, this method of comparison fails, because the degree of confirmation is 0 for either law. Here the concept of instance confirmation (or that of qualified-instance confirmation) will help. If it has a higher value for one of the two laws, then this law will be preferable, if no reasons of another nature are against it.

It is clear that for any deliberate activity predictions are needed, and that these predictions must be "founded upon" or "(inductively) inferred from" past experiences, in some sense of those phrases. Let us examine the situation with the help of the following simplified schema. Suppose a man X wants to make a plan for his actions and, therefore, is interested in the prediction h that c is M'. Suppose further, X has observed (1) that many other things were M and that all of them were also M', let this be formulated in the sentence e; (2) that c is M, let this be j. Thus he knows e and j by observation. The problem is, how does he go from these premisses to the desired conclusion h? It is clear that this cannot be done by deduction; an inductive procedure must be applied. What is this inductive procedure? It is usually explained in the following way. From the evidence e, X infers inductively the law l which says that all M are M'; this inference is supposed to be inductively valid because e contains many positive and no negative instances of the law l; then he infers h ("c is white") from l ("all swans are white") and j ("c is a swan") deductively. Now let us see what the procedure looks like from the point of view of our inductive logic. One might perhaps be tempted to transcribe the usual description of the procedure just given into technical terms as follows. X infers l from e inductively because $c^*(l, e)$ is high; since $l \cdot j$ L-implies h, $c^*(h, e \cdot j)$ is likewise high; thus h may be inferred inductively from $e \cdot j$. However, this way of reasoning would not be correct, because, under ordinary conditions, $c^*(l, e)$ is not high but very low, and even 0 if the domain of individuals is infinite. The difficulty disappears when we realize on the basis of our previous discussions that X does not need a high c^* for l in order to obtain the desired high c^* for h; all he needs is a high c^*_{qi} for l; and this he has by knowing e and j. To put it in another way, X need not take the roundabout way through the law l at all, as is usually believed; he can instead go from his observational knowledge $e \cdot j$ directly to the prediction h. That is to say, our inductive logic makes it possible to determine $c^*(h, e \cdot j)$

directly and to find that it has a high value, without making use of any law. Customary thinking in every-day life likewise often takes this short-cut, which is now justified by inductive logic. For instance, suppose somebody asks Mr. *X* what color he expects the next swan he will see to have. Then *X* may reason like this: He has seen many white swans and no non-white swans; therefore he presumes, admittedly not with certainty, that the next swan will likewise be white; and he is willing to bet on it. He does perhaps not even consider the question whether all swans in the universe without a single exception are white; and if he did, he would not be willing to bet on the affirmative answer.

We see that the use of laws is not indispensable for making predictions. Nevertheless it is expedient of course to state universal laws in books on physics, biology, psychology, etc. Although these laws stated by scientists do not have a high degree of confirmation, they have a high qualified-instance confirmation and thus serve us as efficient instruments for finding those highly confirmed singular predictions which we need for guiding our actions.

§15. THE VARIETY OF INSTANCES

A generally accepted and applied rule of scientific method says that for testing a given law we should choose a variety of specimens as great as possible. For instance, in order to test the law that all metals expand by heat, we should examine not only specimens of iron, but of many different metals. It seems clear that a greater variety of instances allows a more effective examination of the law. Suppose three physicists examine the law mentioned; each of them makes one hundred experiments by heating one hundred metal pieces and observing their expansion; the first physicist neglects the rule of variety and takes only pieces of iron; the second follows the rule to a small extent by examining iron and copper pieces; the third satisfies the rule more thoroughly by taking his one hundred specimens from six different metals. Then we should say that the third physicist has confirmed the law by a more thoroughgoing examination than the two other physicists; therefore he has better reasons to declare the law well founded and to expect that future instances will likewise be found to be in accordance with the law; and in the same way the second physicist has more reasons than the first. Accordingly, if there is at all an adequate concept of degree of confirmation with numerical values, then its value for the law, or for the prediction that a certain number of future instances will fulfill the law, should be higher on the evidence of the report of the third physicist about the positive results of his experiments than for the second physicist, and higher for the second than for the first. Generally speaking, the degree of confirmation of a law on the evidence of a number of confirming experiments should depend not only on the total number of (positive) instances found but also on their variety, i.e., on the way they are distributed among various kinds.

Ernest Nagel[28] has disscussed this problem in detail. He explains the difficulties involved in finding a quantitative concept of degree of confirmation that would satisfy the requirement we have just discussed, and he therefore expresses his doubt whether such a concept can be found at all. He says (pp. 69ff.): "It follows, however, that the degree of confirmation for a theory seems to be a function not only of the absolute number of positive instances but also of the kinds of instances and of the relative number in each kind. It is not in general possible, therefore, to order degrees of confirmation in a linear order, because the evidence for theories may not be comparable in accordance with a simple linear schema; and a fortiori degrees of confirmation cannot, in general, be quantized." He illustrates his point by a numerical example. A theory T is examined by a number E of experiments all of which yield positive instances; the specimens tested are taken from two non-overlapping kinds K_1 and K_2. Nine possibilities $P_1, \ldots P_9$ are discussed with different numbers of instances in K_1 and in K_2. The total number E increases from 50 in P_1 to 200 in P_9. In P_1, 50 instances are taken from K_1 and none from K_2; in P_9, 198 from K_1 and 2 from K_2. It does indeed seem difficult to find a concept of degree of confirmation that takes into account in an adequate way not only the absolute number E of instances but also their distribution among the two kinds in the different cases. And I agree with Nagel that this requirement is important. However, I do not think it impossible to satisfy the requirement; in fact, it is satisfied by our concept c^*.

This is shown by a theorem in our system of inductive logic, which states the ratio in which the c^* of a law l is increased if s new positive instances of one or several different kinds are added by new observations to some former positive instances. The theorem, which is too complicated to be given here, shows that c^* is greater under the following conditions: (1) if the total number s of the new instances is greater, *ceteris paribus;* (2) if, with equal numbers s, the number of different kinds from which the instances are taken is greater; (3) if the instances are distributed more evenly among the kinds. Suppose a physicist has made experiments for testing the law l with specimens of various kinds and he wishes to make one more experiment with a new specimen. Then it follows from (2), that the new specimen is best taken from one of those kinds from which so far no specimen has been examined; if there are no such kinds, then we see from (3) that the new specimen should best be taken from one of those kinds which contain the minimum number of instances tested so far. This seems in good agreement with scientific practice. (The above formulations of (2) and (3) hold in the case where all the kinds considered have equal width; in the general and more exact formulation, the increase of c^* is shown to be dependent also upon the various widths of the kinds of instances.) The theorem shows further that c^* is much more influenced by (2) and (3) than by (1); that is to say, it is much more important to improve the variety of instances than to increase merely their number.

The situation is best illustrated by a numerical example. The computation of the increase of c^*, for the nine possible cases discussed by Nagel, under certain plausible assumptions concerning the form of the law l and the widths of the properties involved, leads to the following results. If we arrange the nine possibilities in the order of ascending values of c^*, we obtain this: P_1, P_3, P_7, P_9; P_2, P_4, P_5, P_6, P_8. In this order we find first the four possibilities with a bad distribution among the two kinds, i.e., those where none or only very few (two) of the instances are taken from one of the two kinds, and these four possibilities occur in the order in which they are listed by Nagel; then the five possibilities with a good or fairly good distribution follow, again in the same order as Nagel's. Even for the smallest sample with a good distribution (viz., P_2, with 100 instances, 50 from each of the two kinds) c^* is considerably higher—under the assumptions made, more than four times as high—than for the largest sample with a bad distribution (viz., P_9, with 200 instances, divided into 198 and 2). This shows that a good distribution of the instances is much more important than a mere increase in the total number of instances. This is in accordance with Nagel's remark (p. 69): "A large increase in the number of positive instances of one kind may therefore count for less, in the judgment of skilled experimenters, than a small increase in the number of positive instances of another kind."

Thus we see that the concept c^* is in satisfactory accordance with the principle of the variety of instances.

§16. THE PROBLEM OF THE JUSTIFICATION OF INDUCTION

Suppose that a theory is offered as a more exact formulation—sometimes called a "rational reconstruction"—of a body of generally accepted but more or less vague beliefs. Then the demand for a justification of this theory may be understood in two different ways. (1) The first, more modest task is to validate the claim that the new theory is a satisfactory reconstruction of the beliefs in question. It must be shown that the statements of the theory are in sufficient agreement with those beliefs; this comparison is possible only on those points where the beliefs are sufficiently precise. The question whether the given beliefs are true or false is here not even raised. (2) The second task is to show the validity of the new theory and thereby of the given beliefs. This is a much deeper going and often much more difficult problem.

For example, Euclid's axiom system of geometry was a rational reconstruction of the beliefs concerning spatial relations which were generally held, based on experience and intuition, and applied in the practices of measuring, surveying, building, etc. Euclid's axiom system was accepted because it was in sufficient agreement with those beliefs and gave a more exact and consistent formulation for them. A critical investigation of the validity, the factual truth, of the axioms and the beliefs was only made more than two thousand years later by Gauss.

Our system of inductive logic, that is, the theory of c^* based on the definition of this concept, is intended as a rational reconstruction, restricted to a simple language form, of inductive thinking as customarily applied in everyday life and in science. Since the implicit rules of customary inductive thinking are rather vague, any rational reconstruction contains statements which are neither supported nor rejected by the ways of customary thinking. Therefore, a comparison is possible only on those points where the procedures of customary inductive thinking are precise enough. It seems to me that on these points sufficient agreement is found to show that our theory is an adequate reconstruction; this agreement is seen in many theorems, of which a few have been mentioned in this paper.

An entirely different question is the problem of the validity of our or any other proposed system of inductive logic, and thereby of the customary methods of inductive thinking. This is the genuinely philosophical problem of induction. The construction of a systematic inductive logic is an important step towards the solution of the problem, but still only a preliminary step. It is important because without an exact formulation of rules of induction, i.e., theorems on degree of confirmation, it is not clear what exactly is meant by "inductive procedures," and therefore the problem of the validity of these procedures cannot even be raised in precise terms. On the other hand, a construction of inductive logic, although it prepares the way towards a solution of the probelm of induction, still does not by itself give a solution.

Older attempts at a justification of induction tried to transform it into a kind of deduction, by adding to the premisses a general assumption of universal form, e.g., the principle of the uniformity of nature. I think there is fairly general agreement today among scientists and philosophers that neither this nor any other way of reducing induction to deduction with the help of a general principle is possible. It is generally acknowledged that induction is fundamentally different from deduction, and that any prediction of a future event reached inductively on the basis of observed events can never have the certainty of a deductive conclusion; and, conversely, the fact that a prediction reached by certain inductive procedures turns out to be false does not show that those inductive procedures were incorrect.

The situation just described has sometimes been characterized by saying that a theoretical justification of induction is not possible, and, hence, that there is no problem of induction. However, it would be better to say merely that a justification in the old sense is not possible. Reichenbach[29] was the first to raise the problem of the justification of induction in a new sense and to take the first step towards a positive solution. Although I do not agree with certain other features of Reichenbach's theory of induction, I think it has the merit of having first emphasized these important points with respect to the problem of justification: (1) The decisive justification of an inductive procedure does not consist in its plausibility, i.e., its accordance with customary ways of inductive reasoning, but must refer to its success in some sense; (2) the fact that the truth of the predictions reached by induction cannot be guaranteed does not preclude a justification in a weaker sense; (3) it can be proved (as a purely logical result) that induction leads in the long run to success in a certain sense, provided the world is "predictable" at all, i.e., such that success in that respect is possible. Reichenbach shows that his rule of induction R leads to success in the following sense: R yields in the long run an approximate estimate of the relative frequency in the whole of any given property. Thus suppose that we observe the relative frequencies of a property M in an increasing series of samples, and that we determine on the basis of each sample with the help of the rule R the probability q that an unobserved thing has the property M, then the values q thus found approach in the long run the relative frequency of M in the whole. (This is, of course, merely a logical consequence of Reichenbach's definition or rule of induction, not a factual feature of the world.)

I think that the way in which Reichenbach examines and justifies his rule of induction is an important step in the right direction, but only a first step. What remains to be done is to find a procedure for the examination of any given rule of induction in a more thoroughgoing way. To be more specific, Reichenbach is right in the assertion that any procedure which does not possess the characteristic described above (viz., approximation to the relative frequency in the whole) is inferior to his rule of induction. However, his rule, which he calls "the" rule of induction, is far from being the only one possessing that characteristic. The same holds for an infinite number of other rules of induction, e.g., for Laplace's rule of succession (see above, § 10; here restricted in a suitable way so as to avoid contradictions), and likewise for the corresponding rule of our theory of c^* (as formulated in theorem (1), § 10). Thus our inductive logic is justified to the same extent as Reichenbach's rule of induction, as far as the only criterion of justification so far developed goes. (In other respects, our inductive logic covers a much more extensive field than Reichenbach's rule; this can be seen by the theorems on various kinds of inductive inference mentioned in this paper.) However, Reichenbach's rule and the other two rules mentioned yield different numerical values for the probability under discussion, although these values converge for an increasing sample towards the same limit. Therefore we need a more general and stronger method for examining and comparing any two given rules of induction in order to find out which of them has more chance of success. I think we have to measure the success of any given rule of induction by the total balance with respect to a comprehensive system of wagers made according to the given rule. For this task, here formulated in vague terms, there is so far not even an exact formulation; and much further investigation will be needed before a solution can be found.

NOTES

1. J. M. Keynes, *A Treatise on Probability*, 1921.
2. H. Jeffreys, *Theory of Probability*, 1939.
3. R. von. Mises, *Probability, Statistics, and Truth* (orig. 1928), 1939.
4. H. Reichenbach, *Wahrscheinlichkeitslehre*, 1935.
5. The distinction briefly indicated here is discussed more in detail in my paper "The Two Concepts of Probability," which appears in *Philos. and Phenom. Research*, V, No. 4, 1945.

6. The reader is referred to the later works of Carnap, especially *Logical Foundations of Probability* (University of Chicago Press, 1950) (2nd ed., 1962), and "The Aim of Inductive Logic" in *Logic, Methodology and Philosophy of Science*, eds. E. Nagel, P. Suppes, and A. Tarski (Stanford University Press, 1962.)

7. In an article by C. G. Hempel and Paul Oppenheim in the present issue of this journal, a new concept of degree of confirmation is proposed, which was developed by the two authors and Olaf Helmer in research independent of my own.

8. For more detailed explanations of some of these concepts see my *Introduction to Semantics*, 1942.

9. See F. Waismann, "Logische Analyse des Wahrscheinlichkeitsbegriffs," *Erkenntnis*, I (1930), 228–48.

10. See Waismann, op. cit., p. 242.

11. St. Mazurkeiwicz, "Zur Axiomatik der Wahrscheinlichkeitsrechnung," *C. R. Soc. Science Varsovie*, Cl. III, Vol. 25, 1932, 1–4.

12. Janina Hosiasson-Lindenbaum, "On Confirmation," *Journal of Symbolic Logic*, V, (1940), 133–48.

13. G. H. von Wright, *The Logical Problem of Induction* (Acta Phil. Fennica, 1941, Fasc. III). See also C. D. Broad, *Mind*, LIII, 1944.

14. It seems that Wittgenstein meant this function c_w in his definition of probability, which he indicates briefly without examining its consequences. In his *Tractatus Logico-Philosophicus*, he says: "A proposition is the expression of agreement and disagreement with the truth-possibilities of the elementary [i.e., atomic] propositions" (*4.4); "The world is completely described by the specification of all elementary propositions plus the specification, which of them are true and which false" (*4.26). The truth-possibilities specified in this way correspond to our state-descriptions. Those truth-possibilities which verify a given proposition (in our terminology, those state-descriptions in which a given sentence holds) are called the truth-grounds of that proposition (*5.101). "If T_r is the number of the truth-grounds of the proposition r, T_{rs} the number of those truth-grounds of the proposition s which are at the same time truth-grounds of r, then we call the ratio T_{rs}: T_r the measure of the *probability* which the proposition r gives to the proposition s" (*5.15). It seems that the concept of probability thus defined coincides with the function c_w.

15. The results are as follows.

$$m = \frac{(N + k - 1)!}{N!(k - 1)!} \tag{1}$$

$$n_i = \frac{N!}{N_1!\,N_2! \cdots N_k!} \tag{2}$$

Therefore (according to (5) in §6):

$$m^*(i) = \frac{N_1!\,N_2! \cdots N_k!\,(k - 1)!}{(N + k - 1)!} \tag{3}$$

16. The general theorem is as follows: $c^*(h, e) = \dfrac{\binom{n_1}{s_1}\binom{n_2}{s_2}}{\binom{n}{s}}$

17. $(c)^*h, e) = \binom{s}{s_1} r^{s_1}(1 - r)^{s_2}$.

18. The general theorem is as follows:

$$c^*(h, e) = \frac{\binom{s_1 + s_1' + w_1 - 1}{s_1'}\binom{s_2 + s_2' + w_2 - 1}{s_2'}}{\binom{s + s' + k - 1}{s'}}.$$

19. Another theorem may be mentioned which deals with the case where, in distinction to the case just discussed, the evidence already gives some information about the individual c mentioned in h. Let M_1 be a factual elementary property with the width w_1 ($w_1 \geqq 2$); thus M_1 is a disjunction of w_1 Q-properties. Let M_2 be the disjunction of w_2 among those w_1 Q-properties ($1 \leqq w_2 < w_1$); hence M_2 L-implies M_1 and has the width w_2. e specifies first how the s individuals of an observed sample are distributed among certain properties, and, in particular, it says that s_1 of them have the property M_1 and s_2 of these s_1 individuals have also the property M_2; in addition, e says that c is M_1; and h says that c is also M_2. Then,

$$c^*(h, e) = \frac{s_2 + w_2}{s_1 + w_1}.$$

This is analogous to (1); but in the place of the whole sample we have here that part of it which shows the property M_1.

20. $$\frac{c^*(h, e \cdot j)}{c^*(h, j)} = 1 + \frac{w_1 - w_2}{w_2(w_1 + 1)}$$

This theorem shows that the ratio of the increase of c^* is greater than 1, since $w_1 > w_2$.

21. Janina Lindenbaum-Hosiasson, "Induction et analogie: Comparaison de leur fondement," *Mind*, L, (1941), 351–65; see especially pp. 361–65.

22. The general theorem is as follows:

$$c^*(h, e) = \frac{\binom{n_1 + w_1 - 1}{s_1 + w_1 - 1}\binom{n_2 + w_2 - 1}{s_2 + w_2 - 1}}{\binom{n + k - 1}{n - s}}$$

Other theorems, which cannot be stated here, concern the case where more than two properties are involved, or give approximations for the frequent case where the whole population is very large in relation to the sample.

23. The general theorem is as follows:

(1) $$c^*(l, e) = \frac{\binom{s + k - 1}{w_1}}{\binom{N + k - 1}{w_1}}.$$

In the special case of a language containing M_1 as the only primitive predicate, we have $w_1 = 1$ and $k = 2$, and hence $c^*(l, e) = \frac{s + 1}{N + 1}$. The latter value is given by some authors as holding generally (see Jeffreys, op. cit., p. 106 (16)). However, it seems plausible that the degree of confirmation must be smaller for a stronger law and hence depend upon w_1.

If s, and hence N, too, is very large in relation to k, the following holds as an approximation:

(2) $$c^*(l, e) = \left(\frac{s}{N}\right)^{\frac{w_1}{k}}.$$

For the infinite language L_∞ we obtain, according to definition (7) in §3:

(3) $$c^*(l, e) = 0.$$

24. The theorem is as follows:

$$c^*(l', e') = \frac{\binom{s + k - 1}{s_1 + w_1}}{\binom{N + k - 1}{s_1 + w_1}}.$$

25. In technical terms, the definition is as follows: $c^*_i(l, e) =_{\text{Df}} c^*(h, e)$, where h is an instance of l formed by the substitution of an individual constant not occurring in e.

26. The technical definition will be given here. Let l be "for every x, if x is M, then x is M'."

Let l be non-L-false and without variables. Let c be any individual constant not occurring in e; let j say that c is M, and h' that c is M'. Then the qualified-instance confirmation of l with respect to M and M' on the evidence e is defined as follows: $c^*_{qi}(M, M', e) =_{\text{Df}} c^*(h', e \cdot j)$.

27. Some of the theorems may here be given. Let the law l say, as above, that all M are M'. Let M_1 be defined, as earlier, by $M \cdot \sim M'$ ("non-white swan") and M_2 by $M \cdot M'$ ("white swan"). Let the widths of M_1 and M_2 be w_1 and w_2 respectively. Let e be a report about s observed individuals saying that s_1 of them are M_1 and s_2 are M_2, while the remaining ones are $\sim M$ and hence neither M_1 nor M_2. Then the following holds:

(1) $$c^*_i(l, e) = 1 - \frac{s_1 + w_1}{s + k}.$$

(2) $$c^*_{qi}(M, M', e) = 1 - \frac{s_1 + w_1}{s_1 + w_1 + s_2 + w_2}.$$

The values of c^*_i and c^*_{qi} for the case that the observed sample does not contain any individuals violating the law l can easily be obtained from the values stated in (1) and (2) by taking $s_1 = 0$.

28. E. Nagel, *Principles of the Theory of Probability*. Int. Encycl. of Unified Science, I, No. 6, 1939; see pp. 68–71.

29. Hans Reichenbach, *Experience and Prediction*, 1938, §§38 ff., and earlier publications.

Ernest Nagel

CARNAP'S THEORY OF INDUCTION

. . .

II

Carnap bases his quantitative inductive logic on a definition of the notion of the degree of confirmation (or probability$_1$) of a hypothesis h relative to (noncontradictory) evidence e–written for short as $c(h, e)$, and even more briefly when no confusion arises as c. A fundamental condition Carnap imposes upon c is that it satisfies a set of postulates, essentially the postulates usually assumed for the mathematical calculus of probability. These postulates require, among other things, that c be associated with a real number in the interval from 0 to 1 inclusive, for every pair of statements h and e–provided only that e is not self-contradictory. On the other hand, these postulates define c only implicitly, so that there is a nondenumerable infinity of ways in which c can be *explicitly* defined so as to conform with the postulates. Carnap therefore indicates how, for a certain class of specially constructed languages, explicit definitions for the c can be given, each definition corresponding to what he calls an "inductive method." These languages possess a relatively simple syntactical structure, adequate for formulating certain parts of scientific discourse, though not the whole of it. Carnap's problem then reduces to that of selecting from these infinitely numerous inductive methods, just those (possibly just one) which promise to be adequate for actual inductive practice and which are in reasonably good agreement with our habitual (or "intuitive") notions concerning the assessment of inductive evidence.

It turns out, however, that the infinitely numerous possible definitions fall into one or the other of two classes. The c's falling into the first class are functions of the number of primitive predicates in the language for which they are

---

From *The Philosophy of Rudolf Carnap*, ed. Paul A. Schilpp (The Library of Living Philosophers; La Salle, Ill.: The Open Court Publishing Company, La Salle, Illinois, 1963). Reprinted by permission of the publisher.

defined; the c's belonging to the second class are not functions of this number, but depend on a parameter whose value is assigned in some other way. Carnap appears to believe, though whether he really does so is not quite certain on the basis of his published statements, that a certain c belonging to the former class and designated as c^* is particularly appropriate as the foundation for an inductive logic which can serve to clarify, systematize and extend actual inductive practice. I shall therefore first discuss c^*, and postpone comment on other definitions in Carnap's repertory of inductive methods.

Every language for which c (and therefore c^*) is defined has a finite number of primitive predicates, and a finite or denumerably infinite number of individual constants. Although the predicates may be of any degree, Carnap develops his system in detail mainly for the case that the predicates are all monadic. Moreover, although the individuals named by the individual constants may be of any sort (e.g., physical objects, events, etc.), he suggests that for technical reasons it is preferable to take them to be spatio-temporal positions.[1] In any case, the only characteristics that are to be ascribed to the individuals mentioned in a given language are those expressible in terms of its primitive predicates and the explicit definitions constructed out of these. Accordingly, there is one indispensable condition which the primitive predicates must satisfy, if the language in which they occur and the inductive logic based on c^* are to be adequate for the aims of science: The set of primitives must be *complete*, in the sense that they must suffice to express every "qualitative attribute" we may ever have the occasion to predicate of the individuals in our universe.

The reason for this requirement of completeness is that c^* is so defined that its numerical value for a given h and e is in general a function of the number of primitive predicates in the language. Thus, suppose a language is adopted containing π independent monadic primitive predicates. Then there will be k $(= 2^{\pi})$ "narrowest classes" specifiable with the help of these predicates and their negations. Suppose, moreover, that M is a predicate which is expressible as the disjunct of w of these narrowest classes, w being the "logical width" of M. If now the evidence e asserts that in a sample of s individuals, s_1 have the property M, and h is the hypothesis that an individual not mentioned in e also has the property M, then

$$c^*(h, e) = \frac{s_1 + w}{s + k}.$$

However, if the language is not complete, and if new primitive predicates must be added to express some feature of the universe, the logical width of M in the new language will be increased, even though the *relative* logical width w/k of M will be unaltered. It follows immediately that $c^*(h, e)$ in the first language will differ from $c^*(h, e)$ in the second enlarged language. To be sure, as Carnap has explicitly noted, the values of the c^*'s will remain in the interval with the end-points s_1/s and w/k, where s_1/s is the observed relative frequency of M in the sample of size s and w/k is the constant relative width of M. He has also pointed out that if the sample size s is increased but the relative frequency s_1/s of M

remains the same, then even though the number of primitive predicates is augmented, the relative frequency s_1/s will swamp the influence of π (the number of primitive predicates) upon the value of c^*, and that as s increases without limit c^* (h, e) will approach s_1/s as the limit. Nevertheless, the fact that c^* varies at all with the number of primitive predicates in the language appears to be strongly counterintuitive. Certainly no biologist, for example, would be inclined to alter his estimate of the support given by the available evidence to the hypothesis that the next crow to be hatched will be black, merely because the language of science becomes enriched through the introduction in some branch of sociology of a new primitive predicate. Nor is there any prima facie good reason why such an altered estimate should be made. On the other hand, if the set of primitive predicates is complete, their number cannot be augmented, and the difficulty disappears.

But is the proposed cure an improvement on the disease? Unless we do have good reasons for fixing the number of primitive predicates in a complete set, we cannot, even in principle, calculate the value of c^* for nontrivial cases, so that the inductive logic based on c^* is simply inapplicable. But the assumption that a complete set of primitives contains a given number π of predicates is not a truth of logic; it is at best a logically contingent hypothesis which can be accepted only on the basis of empirical evidence. The assumption is not a logical truth, for it in effect asserts that the universe exhibits exactly π elementary and irreducible qualitative traits, into which all other traits found in nature are analyzable without remainder. It is an assumption which would be contradicted by the discovery of some hitherto unnoted property of things (e.g., an odor or distinct type of physical force) that is not explicitly analyzable in terms of the assumed set of basic traits. Since the assumption must therefore be evaluated in the light of available empirical evidence, the obvious question arises as to how the weight of this evidence is to be estimated. It cannot be measured by way of c^* defined for the language with π primitive predicates. For in that language the assumption is an analytic truth, and its c^* has the maximum value of 1, contrary to the supposition that the assumption is a contingent hypothesis. Nor can the weight of the evidence for the assumption be measured in terms of a c^* defined for some different language. For this latter language would then have to have a complete set of primitive predicates, and we would thus be faced with an infinite regress. Perhaps a c, different from c^*, is needed, one for which the condition of completeness is not essential? But if so, there are no clues as to which alternative to c^* is to be employed; and in any event, if a c different from c^* is required in order to select a language in which c^* is to be defined, then c^* would not be the *uniquely* and *universally* adequate measure of evidential support—contrary to the supposition underlying the present discussion that c^* is such a measure.

However this may be, it is difficult to avoid the conclusion that the assumption that we have, or some day shall have, a complete set of primitive predicates is thoroughly unrealistic, and that in consequence an inductive logic based on that assumption is a form of science fiction. Although in certain areas of experience we are fairly confident that all the directly observable traits have already been noted, there are no good reasons for believing that we have already catalogued such traits occurring in all parts of the universe. All possible experiments upon all individuals spread through time have not been, and are not likely to be, performed; and the ancient discovery of the previously unknown magnetic property of loadstones has had its analogue frequently repeated in the past and may continue to be repeated in the future. Moreover, though this point perhaps bears only on eventual developments of Carnap's system so as to make it potentially applicable to the whole of the language of science and not only to a fragment of it as is the case at present, even a presumptively complete catalogue of predicates referring to *directly observable* traits would not exhaust the primitive predicates actually required in *theoretical science*. The theoretical predicates which enter into modern systems of natural science (e.g., such predicates as "entropy," "gene," or "electron") are not explicitly definable in terms of directly observable things, though without them scientific research as we know it would be impossible. Such theoretical predicates are usually the products of great feats of scientific imagination; and the introduction of new theoretical predicates into a branch of science is often accompanied by the elimination of older ones—this has been the fate of such terms as "phlogiston" and "caloric." The theoretical parts of the language of science, at any rate, undergo frequent changes, and the direction of change does not appear to be converging towards a limit. The supposition that some day we shall have a complete list of theoretical predicates is thus tantamount to the assumption that after a certain date, no further intellectual revolutions in science will occur. But this is an assumption that is incredible on the available evidence.

As Carnap recognizes, the requirement of completeness is related to John Maynard Keynes's principle of limited variety (and incidentally, also to Francis Bacon's doctrine of "forms"), according to which the amount of variety in the universe is so limited that no one object possesses an infinite number of independent properties. Carnap does not find this principle implausible, and cites in its support the success of modern physics in "reducing" the great variety of phenomena to a small number of fundamental theoretical magnitudes.[2] But this evidence does not seem to be compelling, if only because it is at least debatable whether the phenomenal qualities of things are *explicitly definable* in terms of the theoretical concepts of physics; and if they are not so definable, the total number of primitive predicates has not in fact been diminished. Moreover, though no a priori limits can be set to the scope of physical theory, and it may well be that the physics of the future will account for larger areas of our phenomenal experience than it does at present, two points should be noted. In the first place, current physical theory does not in fact embrace all that experience, and it may never do so. In the second place, the evidence of history seems to show that as the scope of physics is enlarged, the number of its primitive theoretical predicates does not converge to any fixed value, and no plausible upper bound can be assigned to such a number, if indeed there is one. But without a reliable estimate of the value of such an upper bound (to say nothing of offering a reasonably based conjecture as to what will be the actual primitive predicates that a possibly complete physical theory of the future will require), a fully satisfactory inductive logic based on c^* cannot be constructed. The fulfillment of the requirement of completeness depends on our possessing more knowledge than we possess at present, or are likely to possess in the foreseeable future. And if the requirement should ever be fulfilled, we would, by hypothesis, have acquired so much knowledge about the universe that much of our present need for an inductive logic will no longer be actual.

III

There is a further difficulty (which may, however, be only an apparent one) that faces an inductive logic based on c^*—and more generally, on a c that is a function of the number of primitive predicates in the language. It is a familiar fact that two deductive systems may be logically equivalent, so that statements in one are translatable into statements in the other and conversely, even though each system is based on a distinctive set of primitive predicates and a distinct set of axioms. Thus, Euclidean geometry can be developed in the manner of Veblen (who employs, among others, the terms "point" and "between" as primitives), or in the fashion of Huntington (who uses "sphere" and "includes" as primitive predicates); and there is no statement in the Veblen system which cannot be matched in the Huntington codification, and vice versa. If two languages, each containing only monadic primitive predicates, are intertranslatable, then it can be shown that the number of primitives in one must be equal to the number in the other. But in general, if at least one of two intertranslatable languages has polyadic primitives, then it seems that the number of primitives in one may be different from the number in the other. But if this is so, the consequences are serious. For suppose that a hypothesis h and the evidence e for it can be formulated in two intertranslatable languages L_1 and L_2, where the number of polyadic primitives in the former is unequal to the number of primitive in the latter. It then follows that since c^* is a function of the number of primitive predicates in the language for which it is defined, the value of $c^*(h, e)$ calculated for L_1 will be unequal to the value of $c^*(h, e)$ calculated for L_2. Accordingly, the degree of support which the same evidence provides for a given hypothesis will depend on which of two equivalent languages is used for codifying the evidence and the hypothesis. This result is strongly counterintuitive. If the premise of this argument is sound (and I frankly do not know whether it is or not), then the degree of support which a hypothesis receives from given evidence on the basis of Carnap's approach is contingent on the arbitrary choice of one among several logically equivalent languages. But such a conception of evidential weight seems of dubious value as the basis for the practice of induction.

Two considerations occur to me, however, which may make this difficulty only a spurious one. One of them is Carnap's suggestion that in addition to satisfying the requirement of completeness, the primitive predicates of a language for which c^* is defined must also satisfy the requirement of simplicity. As Carnap once formulated this requirement, "the qualities and relations designated by the primitive predicates must not be analyzable into simpler components."[3] The required simplicity of primitive predicates must, on this stipulation, be an "absolute" one, and not merely relative to some given language or mode of analysis. If this notion of simplicity could be assumed to be sufficiently clear, the difficulty under discussion would presumably vanish. For if two intertranslatable languages are constructed on the basis of two sets of unequally numerous primitive predicates, it might be possible to show in general that one of the sets of primitives is simpler than the other, and that therefore the value of c^* must be calculated for the simpler of the two languages. Nevertheless, the notion of absolute simplicity is far from clear. If we do not employ psychological criteria such as familiarity, what rules are to be used in deciding whether, for example, the Veblen set of primitives for geometry are simpler than the Huntington set? When Carnap first proposed the requirement of simplicity he himself admitted his inability to give an exact explication of the notion; and the obscurity of the notion perhaps accounts for the fact that he has not mentioned this requirement in more recent publications. But in any event, the use of the notion of absolute simplicity for outflanking the above difficulty in c^* generates difficulties that are no less grave.

The second consideration mentioned above is of a more technical sort. The values of c^* is in fact an explicit function of the number of *state-descriptions* constructable in the language adopted, and only indirectly of the number of primitive predicates in it. Now a state-description states for every individual, and for every property or relation designated by the primitives, whether or not the individual has the property or relation. Accordingly, a state-description is a noncontradictory conjunction of atomic statements or of their negations (or, in case of languages with an infinity of individual constants, it is an infinite class of such statements)—where an atomic statement ascribes a property (or relation) designated by a primitive predicate to an individual (or individuals) named by an individual constant (or constants). It follows that if the primitive predicates of a language are not totally and logically independent of each other, not every conjunction of atomic statements or of their negations will be a state-description—since in that case some of the conjunctions will be self-contradictory in virtue of the logical relations between the predicates, and must therefore be omitted from the count of all possible state-descriptions. But if two intertranslatable languages L_1 and L_2 are based on two unequally numerous sets of primitive predicates, there will presumably be relations of logical dependence between the primitives in each set, so that the number of state-descriptions in L_1 will be the same as the numer of state-descriptions in L_2. It will then follow that for given h and e, the values of $c^*(h, e)$ will also be the same for the two languages, so that the objection to c^* under present discussion loses its point.

There are, however, two comments on this solution of the difficulty which seem to me in order. The solution assumes that it is possible to give, for each of two intertranslatable languages, an exhaustive catalogue of the rules or postulates which specify the relations of logical dependence between its primitives. Such a catalogue can of course be offered for "artificial" languages, since artificial languages are actually constructed by explicitly stipulating what are the relations of logical dependence between the primitives and what are the logically contingent connections between them. But this is not so readily accomplished for the so-called "natural" languages (including much of the language of science), for in such languages it is not always clear which statements are logically necessary and which have the status of logically contingent hypotheses. Indeed, as is well-known, the same *sentence* may alter its status in this respect with the progress of inquiry or with alternate codifications of a scientific system. (For example, the sentence expressing the ostensibly contingent second law of motion in Newton's system of mechanics, appears as a statement of a logical truth in Mach's reformulation of the system.) On the other hand, though this problem of codifying a natural language is in practice often a difficult one, it is not a problem that is distinctive to Carnap's system of inductive logic.

The second comment is this. Carnap has thus far defined the notion of logical width only for languages with monadic predicates. But it seems plausible to assume that when he does develop his system for more complex languages, he will require an analogous notion for the latter. I have no idea how he will define the notion for the general case. However, it seems to me a reasonable conjecture that for languages with polyadic predicates, as for languages with exclusively monadic ones, the logical width of a predicate must also be some function of the number of primitive predicates in the system. But if this conjecture is sound, an important question immediately arises, one which bears directly on the adequacy of the suggested solution to the difficulty under discussion. Given two intertranslatable languages based on two sets of unequally numerous primitives, and granted that the number of state-descriptions in each is the same, will corresponding predicates in the two languages (i.e., predicates that designate the same property) also have the same logical width? If not, and since the logical width presumably enters into the value of c^*, then for given h and e the value of $c^*(h, e)$ in the two languages will not be the same. In that eventuality, however, the difficulty under discussion will not have been put to final rest.

IV

I wish next to raise an issue that concerns not only c^* but also the whole continuum of inductive methods Carnap regards as possible candidates for explicating the notion of evidential support. Among the condition he lays down which any reasonable c must satisfy, there are two that bear considerable resemblance to the notorious Principle of Indifference, often regarded as the Achilles' heel of the classical theory of probability. The first of these stipulates that all the individuals are to be treated on par, the second introduces a similar requirement for the primitive predicates. According to the first, for example, if the evidence e asserts that the individuals a_1 and a_2 have the property M, while the hypothesis h declares that the individual a_3 has M, then $c(h, e)$ must be equal to $c(h', e')$, where e' asserts that the individuals a_4 and a_5 have M and h' declares that the individual a_6 has M. According the second requirement, if P_1 and P_2 are primitive predicates, e asserts that a_1 and a_2 have the property P_1, and h asserts that a_3 has P_1, then $c(h, e)$ must equal $c(h', e')$—where e' declares that a_1 and a_2 have the property P_2 and h' declares that a_3 has P_2. In short, $c(h, e)$ must be invariant with respect to any permutation of individual constants as well as with respect to any permutation of the primitive predicates.

These requirements are initially plausible, and as Carnap points out assumptions very much like them are tacitly employed in deductive logic. Taken in context, they formulate a feature of actual inductive practice; and in generalizing sciences like physics they are unavoidable, on pain of putting an end to the use of repeated experiments for establishing universal hypotheses. For example, it obviously makes no difference to the evaluation of the evidence for the generalization that water expands on freezing, whether the evidence is obtained from one sample lot of water rather than another sample—*provided* that the samples are taken from a reservoir of the substance that is homogeneous in certain respects. Similarly, it makes no difference to the credibility of a generalization, whether the generalization under inquiry is that copper expands on heating or whether the generalization is that copper is a good electrical conductor —*provided* again that the instances used as evidence are the same in both cases, and *provided* also that the hypothetical relations between the properties under investigation are assumed to be dependent only on the traits of things explicitly mentioned.

On the other hand, as the examples just mentioned suggest, such judgments of indifference are made within contexts controlled by *empirical assumptions* as to what are the relevant properties of individuals that must be noted in using the individuals for evidential purposes, and as to the relevant factors that must be introduced into general statements concerning the concomitances of properties. Thus, different samples of water must be sufficiently homogeneous in their chemical composition, though not in their historical origins, if they are to be on par as evidence for the generalization concerning the expansion of water when cooled. Again, if given instantial evidence is to carry the same weight for the generalization that copper expands when heated as it does for the generalization that copper is a good conductor, the concomitances asserted must be assumed to be independent of variations in other properties exhibited by the instances—for example, of differences in the shapes or the weights of the individuals upon which observation is being made. It is clear, however, that these judgments of relevance and irrelevance are based on *prior experience*, and cannot be justified by purely a priori reasoning.

Within the framework of Carnap's construction, however, the status of the requirements concerning the indicated invariance of c is different. For in his system, the invariance is absolute, not relative to special contexts involving special empirical assumptions. Indeed, the invariance is postulated antecedently to any empirical evidence which might make the postulation a reasonably plausible one. It is not easy to see, therefore, what grounds—other than purely arbitrary and a priori ones—can be adduced for such a requirement of absolute invariance. Carnap defends the requirement by arguing in effect that since the primitive predicates are stipulated to be logically independent, there is no reason for assigning unequal *c*'s to two hypotheses relative to the evidence for them, when the respective hypotheses and evidential statements are isomorphic under a permutation of individual constants or primitive predicates. But though there is no a priori reason for assigning *unequal* *c*'s in such a case, neither is there a compelling a priori reason for assigning *equal* ones. There is surely the alternative, suggested by actual scientific practice, that the matter is not to be decided once and for all by fiat, but settled differently for different classes of cases in the light of available empirical knowledge. In any event, the value of the Principle of Indifferences is as debatable when it is used, as Carnap uses it, to specify in inductive logic which pairs of hypotheses and evidential statements are significantly isomorphic, as when the principle is employed in the classical manner to determine the magnitudes of empirical probabilities.

V

Some of the consequences which follow from the adoption of c^* as the measure of evidential support must now be examined. One of these consequences is that for a language with an infinity of individual constants (i.e., in a universe with a non-finite number of individuals), the value of c^* for any universal empirical hypothesis, relative to any finite number of confirming instances for it, is always zero. For example, despite the great number of known corroborative instances for the generalization that water expands on freezing, the evidential support provided by those instances for this ostensible law is zero when measured by c^* in an infinite universe, and is no better than the evidential support given by those instances to the contrary hypothesis that water contracts on freezing. Moreover, even if the number of individuals in the universe is assumed to be finite but very large, the c^* for the generalization relative to the available instantial evidence will normally differ from zero only by a negligible amount.

Accordingly, if c^* were a proper measure of what ought to be our degree of reasonable belief in hypotheses, none of the generalizations proclaimed by various sciences as laws of nature merits our rational confidence. The search by scientists for critical evidence to support such claims is then pointless, for however much evidence is accumulated in favor of universal laws, increments in the degree of that support remain at best inappreciable. This outcome of adopting c^*, however, is patently in disharmony with our customary way of judging such matters.

There are several ways, nevertheless, in which this apparently fatal consequence entailed by c^* may be made more palatable. It might be argued, in the first place, that it is gratuitous to assume the universe to contain an infinity of individuals, so that the theorem concerning the value of c^* for universal hypotheses in infinite languages simply does not apply to the actual world. It must of course be admitted that we have no certain knowledge that our universe does indeed contain an infinity of empirically specifiable individuals, even if the universe is taken to be extended in time without limit. On the other hand, neither do we know with certainty that the individuals in the universe are only finite in number. If the use of c^* is defensible only on condition that this number really is finite, its use must be postponed indefinitely until that fact is established; and we shall have to carry on our inductive studies (including the inquiry into the number of individuals in the universe) without the help of an inductive logic based on c^*. Moreover, as has already been noted, even if the universe contains only a very large finite number of individuals—and there surely is competent evidence that this number is very large indeed—for all practical purposes such a number entails the same undesirable consequences as if it were nonfinite.

It might be claimed, in the second place, that it is just a mistake to raise questions about the "probability" of universal hypotheses, and thereby to view them as statements on par with instantial ones, concerning which it is significant to ask what measure of support they receive from given evidence. For universal hypotheses, so it is often said, function as guides to the conduct of inquiry, as instruments for predicting concrete events, or as means for organizing systematically the outcome of previous investigations. Universal hypotheses, on this view, are intellectual devices concerning which it is appropriate to ask whether they are adequate for achieving the ends for which they have been designed, but not whether they are true or false. Accordingly, the circumstance that for universal hypotheses the value of c^* is uniformly zero, simply calls attention to the absurdity of treating them as factual statements for which evidence is to be assessed. However, whatever the merits or limitations of this intellectual gambit may be, it is not one which Carnap can employ. For on his approach, universal hypotheses are considered to be on par with instantial ones in respect to their status as empirical statements. It is indeed a central feature of his system that for *any* hypothesis h and (noncontradictory) evidence e, $c^*(h, e)$ must have a determinate value.

Carnap's own proposed resolution of the difficulty bears a certain resemblance to the one just mentioned. But he offers a technically different answer, by way of the notion of the "instance confirmation" of universal laws. He introduces his discussion (though a full account by him is still not available) with the following general explanation:

> Suppose we ask an engineer who is building a bridge why he has chosen the particular design. He will refer to certain physical laws and tell us that he regards them as "very reliable," "well founded," "amply confirmed by numerous experiences." What do these phrases mean? It is clear that they are intended to say something about probability$_1$ or degree of confirmation. Hence, what is meant could be formulated more explicitly in a statement of the form "$c(h, e)$ is high" or the like. Here the evidence e is obviously the relevant observational knowledge. But what

> is to serve as the hypothesis h? One might perhaps think at first that h is the law in question, hence a universal sentence l of the form: "For every space-time point x, if such and such conditions are fulfilled at x, then such and such is the case at x." I think, however, that the engineer is chiefly interested not in this sentence l, which speaks about an immense number, perhaps an infinite number, of instances dispersed through all time and space, but rather in one instance of l or a relatively small number of instances. When he says that the law is very reliable, he does not mean to say that he is willing to bet that among the billion of billions, or an infinite number, of instances to which the law applies there is not one counterinstance, but merely that this bridge will not be a counterinstance, or that among all bridges which he will construct during his lifetime there will be no counterinstance. Thus h is not the law l itself but only a prediction concerning one instance or a relatively small number of instances. Therefore, what is vaguely called the reliability of a law is measured not by the degree of confirmation of the law itself but by that of one or several instances.[4]

Carnap thereupon defines the instance confirmation of a law l on evidence e as the c^* value of the support given by e for the hypothesis that an individual not mentioned in e fulfills l. Furthermore, he defines the qualified-instance confirmation of the law l as the c^* value of the support given by e to the hypothesis that an individual not mentioned in e, but possessing the property mentioned in the antecedent clause of the universal conditional l, also has the property mentioned in the consequent clause of l.

Carnap then argues that contrary to usual opinion, the use of laws is not essential for making predictions, since the inference to a new case can be made *directly* from the instantial evidence, rather than through the mediating office of the law. Thus, suppose the hypothesis h under discussion is whether some given individual a has the property B, and that the evidence e asserts that all the many other individuals which have been observed to possess the property A also possess B. Suppose further that j is the instantial datum that a has A. The usual account, as Carnap formulates it, is that from e we first inductively infer the law l: All A's are B's; and from l together with j we deductively infer h. However, since $c^*(l, e)$ is zero or very close to it, this argument is unsatisfactory. But according to Carnap we really do not need a high value for $c^*(l, e)$ in order to obtain a high value for $c^*(h, e \cdot j)$—that is, for a qualified-instance confirmation of the law. In his view, on the contrary, the person X conducting the inquiry

> need not take the roundabout way through the law l at all, as is usually believed; he can instead go from his observational knowledge $e \cdot j$ directly to the singular prediction h. That is to say, our inductive logic makes it possible to determine $c^*(h, e \cdot j)$ directly and to find that it has a high value, without making use of any law. Customary thinking in everyday life likewise often takes this short cut, which is now justified by inductive logic. For instance, suppose somebody asks X what he expects to be the color of the next swan he will see. Then X may reason like this: He has seen many white swans and no nonwhite swans; therefore he presumes, admittedly not with certainty, that the next swan will likewise be white; and he is willing to bet on it. Perhaps he does not even consider the question

> whether all swans in the universe without a single exception are white; and, if he did, he would not be willing to bet on the affirmative answer.
>
> We see that the use of laws is not indispensable for making predictions. Nevertheless it is expedient, of course, to state universal laws in books on physics, biology, psychology, etc. Although these laws stated by scientists do not have a high degree of confirmation, they have a high qualified-instance confirmation and thus serve as efficient instruments for finding those highly confirmed singular predictions which are needed in practical life.[5]

In short, Carnap appears to be in substantial agreement with J. S. Mill's view that the fundamental type of inductive reasoning is "from particulars to particulars."

Carnap's proposed solution of the difficulty is brilliantly ingenious. But is it satisfactory? Several considerations make this doubtful to me. (1) His solution is predicated on the assumption that the evidence in the qualified-instance confirmation of a law can in general be identified and established without even the tacit acceptance and use of laws, since otherwise a regress would be generated that would defeat the objective of his analysis. The assumption is illustrated by his own example, in which the instantial statement "*a* is a swan," constituting part of the evidence for the hypothesis that *a* is white, can presumably be affirmed on the strength of a direct observation of the individual *a* without the implicit use of any universal laws. I shall not dispute this particular claim, even though legitimate doubts may be expressed as to whether the assertion that *a* is a swan does not "go beyond" what is directly present to observation, and does not carry with it implicit assumptions about invariable connections between anatomical structure, physiological function, and other biological properties—connections which are assumed when an organism is characterized as a swan. But I do dispute the ostensible claim that this example is typical of the way laws are in general confirmed, and that the instances which confirm many scientific theories are quite so simply obtained. To be sure, most of these theories cannot be formulated in the restricted languages for which Carnap has thus far constructed his system of inductive logic; but I do not believe the point under discussion is affected by this fact. Consider, therefore, some of the confirming instances for the Newtonian theories of mechanics and gravitation. One of them is the obloid shape of the earth. The fact that it is obloid, however, can be established only through the use of a system of geometry and of optical instruments for making geodetic measurements—all of which involve at least the tacit acceptance of universal laws as well-founded. Could we inductively infer this fact from the instantial evidence alone, without including in the evidence for it any general laws? It would not advance the solution of the problem, were we to construct an inductive argument, to parallel the schema suggested by Carnap, to read as follows: The qualified-instance confirmation of the Newtonian laws, where the instance is the obloid shape of the earth, is high relative to the instantial evidence that all of the many rotating solids which have been observed in the past have an equatorial bulge, supplemented by the additional evidence that the earth is a rotating solid. For how can the fact that the earth is a rotating solid be established, except by way of assuming an astronomical theory? But unless this fact (or an analogous one) is granted, it is difficult to see how Newtonian theory is relevant to ascertaining the earth's shape, or to understand why the earth's obloid shape is to be counted as a confirming instance for that theory. I do not believe, therefore, that Carnap has successfully defended c^* against the objection that it leads to grave difficulties when it is applied to universal hypotheses.

(2) There is a further point bearing on the present issue, which is suggested by Carnap's discussion of Laplace's Rule of Succession. As is well known, Laplace derived a theorem from the assumptions of his theory of probability, which asserts that if a property is known to be present in each member of a sample of s events, the probability that the next event will also have this property is $(s + 1)/(s + 2)$. Using the evidence available to him concerning the past risings of the sun, Laplace then calculated the probability of another sunrise to be $\frac{1,826,214}{1,826,215}$. This result has been severely criticized by many authors for a variety of reasons. Carnap also finds Laplace's conclusion unsatisfactory, because Laplace allegedly violated the "requirement of total evidence." According to this requirement, "the total evidence available must be taken as basis for determining the degree of confirmation" [or probability$_1$] in the application of the theorems of inductive logic to actual situations.[6] Carnap points out that Laplace assumed the available evidence to consist merely of the known past sunrises, and that he thereby neglected other evidence for the hypothesis that the sun would rise again—in particular, the evidence involved in his knowledge of mechanics. As Carnap puts the matter,

> the requirement of total evidence is here violated because there are many other known facts which are relevant for the probability of the sun's rising tomorrow. Among them are all those facts which function as confirming instances for the laws of mechanics. They are relevant because the prediction of the sunrise for tomorrow is a prediction of an instance of these laws.[7]

Although I do not believe, as Carnap does, that there is no analogue to the requirement of total evidence in deductive logic,[8] I shall not pursue this side issue, and will assume that Carnap's diagnosis of Laplace's error is well taken. The question I do wish to raise is whether it is the *confirming instances* of the laws of mechanics, or the *laws of mechanics* themselves, which are to be included in the evidence when tomorrow's sunrise is predicted. Carnap appears to adopt the former alternative. It is not clear, however, why in that case most of the evidence—taken simply as so many *independent* instantial statements—is *relevant* to the prediction of another sunrise, and why it should raise the c^* for the predictive hypothesis. For example, the confirming instances of the laws of mechanics include observations on tidal behavior, on the motions of double stars, on the rise of liquids in thin tubes, on the shapes of rotating liquids, and much else, in addition to observations on the rising of the sun. There is, however, no purely logical dependence between instantial statements about the height of the tides or instantial statements about phenomena of capillarity, on the one hand, and instantial statements on the rising of the sun on the other hand. Apart from the laws of mechanics, these statements express just so many disparate facts, no more related to each other than they are related to other statements which do *not* formulate confirming instances of these laws—for example, statements about the magnetic properties of a given metal bar, or about the color of a man's eyes. Why should the inclusion of the former instantial data increase the evidential support for the prediction of another sunrise, but the inclusion of the latter ones not do so? To make the point more fully, consider a language with three logically independent monadic primitive predicates, P_1, P_2 and P_3. Then according to the definition of c^*,

$$c^*(P_1a_1, P_1a_2) = \frac{5}{9}$$

and if the evidence is enlarged to include P_1a_3,

$$c^*(P_1a_1, P_1a_2 \cdot P_1a_3) = \frac{6}{10}$$

so that the c^* for the hypothesis is increased. But if the evidence is further enlarged by including P_2a_4, then

$$c^*(P_1a_1, P_1a_2 \cdot P_1a_3 \cdot P_2a_4) = \frac{6}{10},$$

so that this additional evidence is irrelevant; and the situation remains the same when the evidence is further augmented by adding to it P_3a_5, P_2a_6, or in fact any number of instantial statements which ascribe properties to individuals other than the property designated by the predicate P_1. Now most of the predicates occurring in the formulation of confirming instances for the laws of mechanics are *prima facie* quite analogous to the predictive predicates in this example in respect to their being logically independent of each other. Indeed, there are cases in the history of science when, on the basis of some well established theory, an event has been predicted which had rarely if ever been observed previously. In such cases, though the predictive hypothesis receives a considerable measure of support from the theory, the predicates in the instantial evidence for the theory are for the most part different from, and logically independent of, the predicates occurring in the hypothesis. To cite a notorious example, William R. Hamilton predicted the phenomenon of conical refraction from theoretical considerations, though this phenomenon had not been previously observed, so that instances of the phenomenon did not constitute a part of the evidence for the theory Hamilton employed. In consequence of all this, it does not appear plausible that, in conformity with the requirement of total evidence, it is the *instantial evidence* for the laws of mechanics, but rather the *laws of mechanics* themselves, which must be included in the evidence for the hypothesis of another sunrise.

But if this is so, I am also compelled to conclude that the use of general laws in inductive inference is not eliminable, in the manner proposed by Carnap. Accordingly, the notion of instance confirmation (or qualified-instance confirmation) of a law as a measure of the law's "reliability" does not achieve what he thinks this notion can accomplish. In short, I do not believe he has succeeded in outflanking the difficulty which arises from the counter-intuitive consequences of adopting c^* as a measure of evidential support in infinite languages.

VI

I must now raise an issue that affects not only an inductive logic based on c^*, but nearly all of the inductive methods Carnap has outlined.

It is commonly assumed that the evidential support for a hypothesis (whether singular or universal) is generally increased by increasing the sheer number of its confirming instances. For example, it is usually claimed that the hypothesis that the next marble to be drawn from an urn will be white is better supported by evidence consisting of 100 previous drawings each of which yielded a white marble, than by evidence consisting of only 50 such drawings; and many accounts of inductive logic attribute this difference in evidential weight entirely to the difference in the relative size of the two samples. Again, it is often supposed that simply by repeating an experiment on the period of a pendulum, where each experiment shows this period to be proportional to the square-root of the pendulum's length, the weight of the evidence for the generalization that the period of any pendulum follows this law is augmented. In any event, this assumption is implicit in most of the inductive methods (including the one based on c^*) which Carnap discusses. But although the assumption appears to be eminently plausible, I think it is a reasonable one only when it is employed under certain conditions, so that the assumption is acceptable only in a qualified form. I want to show, however, that most of Carnap's inductive methods in effect adopt it without such qualifications.

Consider first a language with two monadic primitive predicates R and S and N individual constants, so that if Q is defined as $R \cdot S$, Q specifies one of the four narrowest classes of individuals which is formulable in this language. The relative logical width of Q is then $w/k = 1/4$. Suppose now that a sample of size s is drawn from the population, that just s_i individuals in the sample have the property Q, and that h is the hypothesis asserting that an individual a not contained in the sample also has Q. Carnap shows that for all the measures c satisfying the conditions he regards as minimal for a measure of evidential support,

$$c(h, Qa_1 \ldots Qa_{si} \cdot -Qa_{si+1} \ldots -Qa_s) = \frac{s_i + \lambda/4}{s + \lambda}, \text{ with } 0 \leq \lambda \leq \infty,$$

where the value of the parameter λ depends on the inductive method adopted and thus fixes the measure c of evidential support. For c^*, this parameter is equal to k (the total number of narrowest classes specifiable in the language), and in the present example is equal to 4. When all the individuals in the sample have the property Q,

$$s_t = s \text{ and } c(h, Qa_1 \ldots Qa_s = \frac{s + \lambda/4}{s + \lambda} = \frac{4s + \lambda}{4(s + \lambda)}.$$

Suppose now the size of the sample is increased by n, so that it contains $s + n$ individuals, and that every member of the sample has Q. Then

$$c(h, Qa_1 \ldots Qa_{s+n}) = \frac{4(s + n) + \lambda}{4(s + n + \lambda)}.$$

Since for $\lambda > 0$ and $n > 0$

$$\frac{4(s + n) + \lambda}{4(s + n + \lambda)} > \frac{4s + \lambda}{4(s + \lambda)},$$

it follows that

$$c(h, Qa_1 \ldots Qa_{s+n}) > c(h, Qa_1 \ldots Qa_s).$$

Accordingly, when all the individuals in a sample belong to the class Q (so that, since Q determines one of the narrowest classes specifiable in the language, the individuals are indistinguishable in respect to the properties they exhibit) and the sample size is increased, the measure of evidential support for the hypothesis is also increased. Indeed, in an infinite language if all the individuals in progressively more inclusive samples belong to Q and if the sample size is increased without limit, the degree of confirmation for the hypothesis approaches the maximum value of 1.

Suppose, next, that "All A is B" is the formulation of a law in a language having only monadic predicates and N individual constants, where A and B are any molecular predicates defined in terms of the primitives, and where the predicate $A \cdot -B$ has the logical width w. (It is clear that the law is logically equivalent to "Nothing is $A \cdot -B$".) Suppose, further, that the evidence e for the law asserts that s distinct individuals do not have the property $A \cdot -B$ and that all the individuals fall into *one* of the k narrowest classes specifiable in the language. (e is then the conjunction of s instantial statements, each of which asserts that some individual has the property determining this class—the property in question being incompatible with $A \cdot -B$.) Carnap then shows that when s is very large in relation to k, c^* (All A is B, e) is approximately equal to $(s/N)^w$.[9] Accordingly, if the evidence for the law is increased by the addition of further instances all of which continue to fall in to the same narrowest class, the degree of confirmation for the law is also increased. For infinite languages (i.e., when $N = \infty$) this degree of confirmation is of course zero, as has already been noted. On the other hand, if the logical width of the predicate $A \cdot B$ is w', and if the evidence consists of a sample of s individuals all of which have the property $A \cdot B$ and all of which, moreover, fall into the same narrowest class, Carnap proves that for the measure c^* the qualified-instance confirmation of the law is equal to

$$1 - \frac{w}{s + w + w'}.$$

This latter is the value of the degree of confirmation of the hypothesis that an individual, known to have the property A, also has the property B, on the evidence that s other individuals all falling into one of the narrowest classes have both A and B. Since the value of the qualified-instance confirmation of the law is independent of the number of individual constants in the language, it will differ from zero even for infinite languages, and it will be close to 1 when s is made sufficiently large.

In my judgment, however, these results are incongruous with the normal practice of scientific induction, as well as with any plausible rationale of controlled experimentation. For according to the formulas Carnap obtains for his system, the degree of confirmation for a hypothesis is in general increased if the confirming instances for the hypothesis are multiplied—*even when the individuals mentioned in the evidence cannot be distinguished from each other by any property expressible in the language for which the inductive logic is constructed.* But it seems to me most doubtful whether under these conditions we would in fact regard the evidential support for a hypothesis to be strengthened. Suppose we undertook to test a proposed law, say the law that all crows are black, by making a number of observations or experiments on individuals; and suppose further that the individuals we examined were *known to be completely alike* in respect to all the properties which we can formulate. Would there be any virtue in repeating the observations or experiments in such a case? Would we not be inclined to say that under the imagined circumstances *one* observation carries as much weight as an *indefinite number* of observations?

We do, of course, repeat observations and experiments intended to test proposed laws. But apart from our desire to make allowances for and to correct personal carelessness and "random" errors of observation, we do so only when we have some grounds for believing that the individuals are *not* completely alike in the properties they possess. In fact, we generally *select* the individuals upon which tests are to be performed so that they are *unlike* in as large a variety of features as possible, compatible with the requirement that the individuals exhibit the properties mentioned in the antecedent clause of the law we are testing. The rationale for this standard procedure is to show that the connections between properties asserted by the proposed law do hold in just the way the proposed law asserts them to hold, and that the hypothetical connections are not contingent upon the occurrence of some other property not mentioned by the proposed law. Accordingly, test-cases for the law that all crows are black will be drawn from a wide assortment of geographic regions, climatic conditions, and other variable circumstances under which crows may be found, in the hope that despite variations in these circumstances the color of the plumage is indeed uniformly associated with the anatomical structure that identifies crows, and in the desire to show that the color does not depend on the occurrence of some other properties which crows may have. In short, the sheer repetition of confirming instances does not, by itself, appear to carry much weight in the support given by the evidence to a hypothesis. But if this point is well taken, all those inductive methods considered by Carnap (including c^*) in which a contrary result is obtained (the only method for which such a contrary result does not hold is the one for which $\lambda = \infty$) are inadequate rational reconstructions of generally accepted canons of scientific inquiry.

The point just argued also has some bearing on the notion of the instance confirmation of a law. For if, as is required by Carnap's analysis, increasing the number of otherwise indistinguishable confirming instances for a law augments the degree of confirmation for a still unobserved additional instance of the law, why should not the degree of confirmation also be augmented by an equal amount for an *indefinite* number of further instances of the law—or even for the law itself? Under the conditions supposed, is there really a better reason for expecting that a *single* hitherto untested individual will conform to the law than for the hypothesis that many such individuals will do so? Thus, if all observed instances of crows are black, and if these instances are *known* to be alike in all respects formulable in the (hypothetically complete) language we employ (e.g., the crows observed come from the same locality, they have the same genetic constitution, their diet is the same, etc.), why should this evidence give stronger support to the prediction that the next crow to be observed will be black, than for the hypothesis that the next ten crows to be observed will be black, or for the hypothesis that all crows are black? The contrary view seems to me to reflect sound inductive practice.

To see the point more clearly, consider the following schematic example constructed in conformity with Carnap's procedure. Assume a language with four monadic primitive predicates P_1, P_2, P_3, and P_4. The law "Anything that is both P_1 and P_2 is also P_3" is proposed for testing. A sample consisting of $2s$ individuals is now examined, and each individual is found to possess the property $P_1 \cdot P_2 \cdot P_3$. Two possible cases will be considered: (1) All the $2s$ individuals belong to the narrowest class determined by $P_1 \cdot P_2 \cdot P_3 \cdot P_4$, and are therefore otherwise indistinguishable; (2) only s individuals in the sample belong to this class, while the remaining half belong to the narrowest class determined by $P_1 \cdot P_2 \cdot P_3 {\sim} P_4$. Now the evidence in the first case leaves it unsettled whether P_3 is always present when the property $P_1 \cdot P_2$ alone is present, or whether the occurrence of P_3 is contingent not only on the presence of this property but also on the presence of P_4 as well. In the second case, however, the evidence shows that P_3 is dependent only on $P_1 \cdot P_2$, irrespective of the presence or absence of P_4. It therefore appears reasonable to maintain that the evidence in the second case is better than in the first as a support for the hypothesis that an individual not included in the sample, but known to possess $P_1 \cdot P_2$, also possesses P_3—and if I judge the matter aright, such a claim is in agreement with standard scientific practice. Now this point is also recognized by Carnap, since the value of c^* in the second case is higher than the value in the first case. It is clear, therefore that the variety of instances contributes to the strength of evidence for a hypothesis. On Carnap's analysis, however, complete absence of variety in the instances is compatible with a high degree of confirmation, provided that the sheer number of instances is large; and this seems to me a defect in his system. Accordingly, if my argument holds water, his system fails to take into consideration an essential feature of sound inductive reasoning.

Moreover, I have not been able to persuade myself that the evidence in the first case supports to a lower degree the hypothesis that an *indefinite* number of further individuals possessing $P_1 \cdot P_2$ also possess P_3, than it supports the hypothesis that just one further individual is so characterized. It might be retorted that to deny this is counterintuitive—since we normally do say, for example, that on the evidence of having drawn 100 white marbles from an urn, the "probability" of getting a white marble on the next trial is greater than the "probability" of getting two white marbles on the next two trials. But I do not find this rejoinder convincing. If we *know* that the 100 white marbles constituting the evidence are fully alike *in all relevant respects* upon which obtaining a white marble from the urn depends, then it seems to me that the evidence supports the hypothesis that any further marbles, resembling in those respects the marbles already drawn, will also be white—irrespective of how many further marbles will be drawn from the urn. It is because we generally do *not* know what are the complete set of properties in respect to which marbles may differ, and because we therefore do *not* know whether the marbles in the sample lot are alike in all relevant respects, that the evidence offers better support for the hypothesis concerning a single additional marble than it does for the hypothesis concerning two or more additional marbles. I cannot therefore evade the conclusion that because of the consequences noted, c^* as the measure of evidential support runs counter to sound inductive practice.

. . .

NOTES

1. Rudolf Carnap, *Logical Foundations of Probability* (Chicago, 1950), p. 62. This work will be cited in the sequel as LFP. Carnap's *The Continuum of Inductive Methods* (Chicago, 1952) will be cited as CIM.

2. LFP, p. 75.

3. Rudolf Carnap, "On the Application of Inductive Logic," *Philosophy and Phenomenological Research*, VIII (1947), 137.

4. LFP, pp. 571ff.

5. LFP, pp. 574ff.

6. LFP, p. 211.

7. LFP, pp. 212ff.

8. Thus, if we assume that only gravitational forces are present, we can deduce from Newtonian theory certain conclusions about the orbit of a given body. But if there are also magnetic forces in operation which enter into the determination of the orbit and which we have unwittingly ignored, our original conclusions are clearly wrong.

9. LFP, p. 571.

Jaakko Hintikka

TOWARDS A THEORY OF INDUCTIVE GENERALIZATION

In discussing the relation of the notions of probability and rational belief, one of the main problems concerns the status of the inductive generalizations of sciences and of everyday life. A great number of such generalizations are accepted and believed in by scientists and by men of practical affairs. In many cases, their attitude is obviously rational. Can this rationality be interpreted in probabilistic terms? Can the rationality of our acceptance of the generalizations we in fact accept be explained in terms of the high probability of these generalizations, in some interesting sense of probability?

In recent literature a slightly different question often comes to the fore. This literature is characterized by a contrast of two main schools of thought, led by Rudolf Carnap and Karl Popper, respectively.[1] What Popper is primarily concerned with are not our beliefs in various generalizations, but rather our choice of the best possible generalization, best not in the sense of the safest one but rather best in the sense of the most informative and most thoroughly testable (and tested) one.[2] Popper claims that this choice of the best possible generalization cannot be explained in terms of its high probability.[3] On the contrary, Popper maintains that the most informative generalization is typically one with the lowest probability.[4] His answer to our second initial question is thus negative.

Carnap denigrates the importance of (strict, nonstatistical) generalizations for our inductive thinking and inductive practice.[5] According to his attempted "rational reconstruction" of our inductive practice we should not strictly speaking believe in the general laws of science and of everyday life, for their probability on evidence (a posteriori probability, Carnap's "degree of confirmation") is according to Carnap normally very small. Carnap writes: "Although . . . laws stated by scientists do not have a high degree of confirmation, they have a high

From *Logic, Methodology, and the Philosophy of Science*. Reprinted by permission of the publishers, the North-Holland Publishing Company.

qualified-instance confirmation and thus serve as efficient instruments for finding those highly confirmed singular predictions which are needed in practical life."[6] However, this provokes the question whether the acceptance of the general laws which the scientists in fact accept can be justified on Carnap's grounds; more generally, whether accepting generalizations because of their high instance confirmation leads us to choose the generalizations which we in fact choose on sound rational grounds. Carnap does not by any means demonstrate that this is the case, and in this paper I shall argue that in some typical cases he is wrong.

Thus instead of discussing directly one's beliefs in inductive generalizations I am led to discuss the question: Why do we believe that one generalization is better than another? Can our rational preferences in this matter be interpreted in terms of a high probability of the preferred generalizations? The latter question might be called the basic question of this paper. Popper's answer to it is negative. Carnap also gives a negative answer to it in the sense that he does not think that the generalizations themselves are highly probable a posteriori. However, he thinks that the preferable generalizations are characterized by a high degree of instance confirmation, and that our choice of a generalization is therefore in a sense guided by probabilistic considerations. This is firmly denied by Popper.

In this paper I shall try to offer some suggestions concerning the basic question just formulated, and thereby to put this conflict of views into a new perspective. In order to fix our ideas, let us start by considering a situation which may look like a rather special case but which will subsequently turn out to be quite representative. Let us assume that we are given k primitive monadic predicates $P_j(j = 1, 2, \ldots, k)$. By means of these predicates and propositional connectives, a partition of our domain of individuals into exactly $K = 2^k$ different kinds of individuals can be defined.[7] Let us assume that these kinds of individuals are defined by the expressions (complex predicates)

$$Ct_1(x), Ct_2(x), \ldots, Ct_K(x). \tag{1}$$

(These are of course simply Carnap's Q-predicates in a new guise.[8]) Let us also assume that we have observed exactly n individuals in a domain of individuals (universe of discourse) which contains a totality of N individuals. (In the sequel, it will normally be assumed that $N \gg n \gg K$, i.e., that N is large as compared with n and n large compared with $K = 2^k$.) Let us finally assume that among the n observed individuals exactly c different kinds of individuals are exemplified, say the ones specified by the expressions

$$Ct_{i_1}(x), Ct_{i_2}(x), \ldots, Ct_{i_c}(x) \tag{2}$$

which constitute a subset of all the expressions (1).

Apart from unavoidable oversimplification, this situation does not appear to be entirely unrepresentative of what one is likely to encounter in actual applications.

The question we want to ask is: What generalization concerning the whole universe of discourse ought we to prefer in this situation?

It seems to me that from an intuitive point of view the answer is obvious, provided that $n \gg K$. If n is large in relation to K, there have been plenty of opportunities for the remaining kinds of individuals to prove that they are not empty. Their failure to do so therefore strongly suggests that they are in fact empty. Hence the only rational thing to do in these circumstances is to assume that they are empty, i.e., to prefer the following sentence to all the other comparable generalizations:

$$(3) \qquad ((Ex)Ct_{i_1}(x) \,\&\, (Ex)Ct_{i_2}(x) \,\&\, \ldots \,\&\, (Ex)Ct_{i_c}(x)) \\ \&\, (x)(Ct_{i_1}(x) \vee Ct_{i_2}(x) \vee \ldots \vee Ct_{i_c}(x)).$$

This is then the desirable hypothesis independently of what $N \geqq n$ is. It also appears to be preferable generalization on Popper's principles.[9]

It is not at all clear, however, how this preference could be justified in terms of inductive logic. The problem is to compare (3) with a competing generalization, say with

$$(4) \qquad ((Ex)Ct_{i_1}(x) \,\&\, \ldots \,\&\, (Ex)Ct_{i_c}(x) \,\&\, \ldots \,\&\, (Ex)Ct_{i_w}(x)) \\ \&\, (x)(Ct_{i_1}(x) \vee \ldots \vee Ct_{i_c}(x) \vee \ldots \vee Ct_{i_w}(x))$$

where $w \geqq c$. (The possibility that $w < c$ is excluded by the evidence.) Carnap's theories offer us two different ways of comparing (3) with (4). We may compare the degrees of confirmation of the two on the evidence in question, or we may compare their respective degrees of instance confirmation.[10] These comparisons are facilitated by the observation that for Carnap the degree of confirmation of the existential part of (4) relative to its universal part (last member of conjunction (4)) is very close to one if n is large in relation to K. Hence we may consider the degree of confirmation of the universal part only, and also apply to it the notion of instance confirmation which does not otherwise apply to (4). According to Carnap's results,[11] the degree of confirmation of the universal part of (4) is approximately

$$(5) \qquad \left(\frac{n}{N}\right)^{K-w}$$

when n is large in relation to K, and its degree of instance confirmation is

$$(6) \qquad \frac{n+w}{n+K}.$$

Both (5) and (6) have the greater value the greater w is. Thus the recomendation which Carnap's theory gives us is not to assume that (3) is true, that is to say, not to assume that as few kinds of individuals are exemplified in the world as is compatible with evidence. Instead, we ought to assume that as many of them as possible are exemplified—ideally, all of them. It is obvious that this recommendation is counterintuitive. It is more like a counsel of despair than a rational direction for choosing one's generalization.

It is perhaps worth pointing out that this difficulty of Carnap's theory is largely independent of the possibility that N might be infinite. It appears as soon as $N \gg n \gg K$. It is obvious that this possibility cannot be ruled out in many important types of applications. And it is obvious that this difficulty cannot be eliminated by means of the notion of instance confirmation.

Our example thus seems to add grist to the mill of Carnap's critics. It certainly shows that, however much value Carnap's theory has in other respects, there is at least one task it does not perform in a satisfactory manner, to wit, the task of guiding us in our choice of a generalization.

Our example might seem to accomplish even more than this. It might seem to point to the (fallacious) conclusion that there is *no* way of explaining our rational preference of (3) as compared to (4) ($w > c$) in terms of the higher probability of (3). For it might seem that since (4) allows more possibilities for the unexamined individuals of our universe of discourse than (3), (4) must be more probable a priori than (3) on any reasonable notion of probability. And since there are usually more unexamined than examined individuals the same would seem to be the case also with a posteriori probability. Conversely, since the preferable generalization (3) allows fewer possibilities for the unexamined individuals than any of the competing generalizations (4), it may seem to be the most *im*probable hypothesis in any natural sense of probability. It might thus seem not only that the special form of quantitative inductive logic which Carnap has developed does not account for our reasons for preferring one generalization to another in terms of the higher probability (on evidence) of the former, but also that *no* system of quantitative inductive logic can accomplish this.

It thus seems to me that our example brings out some of the reasons which have led the critics of inductive logic to give a negative answer to our basic question, although their reasons have been derived from somewhat different examples.

Their conclusion is a much too hasty one, however. It is the main purpose of this paper to outline a way of constructing a quantitative inductive logic (for the same kinds of language systems as Carnap) which avoids some of the main disadvantages of Carnap's theory in so far as the problem of dealing with strict (non-statistical) inductive generalizations is concerned.

Such an inductive logic can even be built along lines closely reminiscent of Carnap's. In developing an inductive logic along Carnapian lines the fundamental problem concerns the choice of the underlying measure function. This choice may be thought of as a choice of a method of assigning a weight (or an a priori probability, if you prefer) to each of the different state-descriptions that can be formulated in the language we are considering. (These weights are here assumed to be between zero and one and their sum is assumed to equal one.)

Now one way in which one comes to choose the particular measure function m^* which Carnap favours may perhaps be described as follows:[12] What we are primarily interested in are not the different "possible worlds" described by state-descriptions but rather the different *kinds of* possible worlds that can be described in our language. These different kinds of worlds are interpreted by Carnap as structurally different worlds. Accordingly, Carnap assigns an equal weight not to all the state-descriptions but rather to what he calls structure-descriptions.[13] Here each structure-description is the disjunction of all the different state-descriptions that can be transformed into each other by permuting free singular terms (names of individuals). The weight which a structure-description receives is then divided evenly among its members.

The main drawback of this procedure seems to me to be its dependence on the domain of individuals which the language in question presupposes. In order to know the weight of a structure-description, one has to know the number of all the structure-descriptions; and this depends on the number of individuals in the domain. In order to apply Carnap's procedure, one therefore has to know the whole universe at least to the extent of knowing its size. In most applications, however, this domain is largely unknown. In general it seems to me perverse to start one's inductive logic from the assumption that one is in some important respect already familiar with the whole of one's world, for this logic is largely designed to reconstruct some of the procedures we use in coming to know it. In view of this perversity, it is small wonder that Carnap's theory in fact leads to great difficulties in connection with generalizations.[14]

Hence we cannot use structure-descriptions as an explication of the notion of the description of a possible kind of world. These descriptions must be independent of one's list of individuals. They must be described by general and not by singular sentences.[15]

Now the question becomes: Can we find, for each finite set of predicates (each with one or more argument-places), such descriptions of possible kinds of worlds in some natural sense of the word? The answer is easily seen to be affirmative if we limit the number of quantifiers whose scopes may all overlap in our sentences. For a given sentence, we shall call the maximum number of quantifiers whose scopes have a common part in it the *depth* of the sentence in question. More loosely expressed, the depth of a sentence is the number of layers of quantifiers which it contains at its thickest. The restriction just mentioned is a restriction on this parameter. Let it be restricted to d_0 at most.

With this restriction, how can one arrive at the different descriptions of possible kinds of worlds that we are looking for? An answer is not very difficult to give. Suppose first that we have already arrived at a list of all the possible kinds of individuals that can be described by means of the given set of predicates which we are presupposing and by means of at most $d_0 - 1$ layers of quantifiers, and let these kinds of individuals be described by sentences (1). Then we can arrive at a description of a possible kind of world simply by running through the list and indicating, for each kind of individual, whether individuals of that kind exist or not. Each description of a possible world will in other words be of the form

(7) $$(\pm)(Ex)Ct_1(x) \mathbin{\&} (\pm)(Ex)Ct_2(x) \mathbin{\&} \ldots \mathbin{\&} (\pm)(Ex)Ct_K(x)$$

where in the place of each symbol $(\pm)$ there is either a negation-sign or nothing at all. It is obvious that each sentence of form (7) can be rewritten: Instead of specifying, for each possible kind of individuals, whether individuals of that kind exist or not, it suffices to list all the kinds of individuals that are exemplified, and then to add that they are *all* the kinds that are not empty. But this means that (7) can be written in form (3) where the sentences $Ct_{i_1}(x)$, $Ct_{i_2}(x)$, . . ., $Ct_{i_c}(x)$ form a subset (proper or improper) of the set of all sentences (1). Each sentence of form (3) will then be called a *constituent*.

But how can we obtain the descriptions (1) of all the possible kinds of individuals? In the monadic case the answer is obvious: The relevant kinds of individuals are specified by Carnap's Q-predicates, i.e., by our sentences (1). Hence in the monadic case our constituents are just the sentences of form (3) considered in our example. Part of the general significance of the example is due to this fact.

Thus it remains to deal with the polyadic (general) case. This case can be handled inductively. Suppose that the following sentences describe all the possible ways in which the value of the variable x_j can be related to the individuals specified by $x_1\ x_2 \ldots x_{j-1}$:

(8) $$Ct_1^d(x_1, x_2, \cdots, x_{j-1}, x), Ct_2^d(x_1, x_2, \cdots, x_{j-1}, x_j), \cdots.$$

By "all the possible ways" we here mean all the ways which can be specified by using only the given predicates, propositional connectives, and at most d layers of quantifiers. We might also say that (8) is a list of all kinds of individuals x_j that can be specified by these means plus the "reference point" individuals specified by $x_1, \cdots, x_{j-1}$. Then we may obtain a similar list for the preceding variable x_{j-1} and for $d + 1$ layers of quantifiers in the same way in which (7) (or (3)) was obtained from (1). This list is given by the sentences

(9) $$(\pm)(Ex_j)Ct_1^d(x_1, x_2, \cdots, x_{j-1}, x_j) \mathbin{\&} (\pm)(Ex_j)Ct_2^d(x_1, x_2, \cdots, x_{j-1}x_j) \\ \mathbin{\&} \ldots \mathbin{\&} (\pm)A_1(x_1, x_2, \cdots, x_{j-1}) \mathbin{\&} (\pm)A_2(x_1, x_2, \cdots, x_{j-1}) \mathbin{\&} \cdots$$

where at the place of each $(\pm)$ there again may or may not be a negation-sign and where $A_1(x_1, x_2, \cdots, x_{j-1})$, $A_2(x_1, x_2, \cdots, x_{j-1})$, $\cdots$ are all the atomic sentences which can be formed from the given predictes and from the variables $x_1, \cdots, x_{j-1}$, and which contain at least one occurrence of x_{j-1}.

What happens in (9) can be given an intuitive interpretation. Again we run through the list (8) and indicate, for each possible kind of an individual x_j, whether individuals of that kind exist or not. This adds a new layer of quantifiers, but it also transforms the result to something we can assert of the referent of x_{j-1} in relation to the referents of $x_1, x_2, \ldots, x_{j-2}$. Finally, we specify how the referent of x_{j-1} is related to the referents of $x_1, x_2, \ldots, x_{j-2}$.

If $d = 0$, sentences (8) are simply conjunctions of the form

$$(10) \qquad (\pm)A_1(x_1, x_2, \cdots, x_j) \;\&\; (\pm)A_2(x_1, x_2, \cdots, x_j) \;\&\; \cdots.$$

This gives us a basis for induction.

Obviously we can rewrite each (9) in the same way in which (7) was rewritten as (3). In the sequel, we shall assume that this has always been done. The resulting sentences will be called attributive constituents with depth $d + 1$ and with the free variables $x_1, x_2, \ldots, x_{j-1}$. From attributive constituents of depth $d_0 - 1$ with one free variable we may obtain constituents of depth d_0 in the way described above.

Every consistent general sentence has a normal form (its *distributive normal form*) which is a disjunction of constituents with the same predicates and with the same depth (or any fixed greater depth). For instance, every consistent constituent of depth d is equivalent to a disjunction of constituents of depth $d + e$, for each $e = 1, 2, 3, \ldots$. These are called its *subordinate* constituents. Similarly, attributive constituents may be split into disjunctions of deeper attributive constituents (their subordinates) with the same free individual variables and constants.

The theory of constituents, attributive constituents, and of distributive normal forms has been developed in greater detail elsewhere.[16] Here I shall only explain briefly how they might be used for the purposes of inductive logic.

Given a finite set of predicates, it may be stipulated that an equal weight be given to all the consistent constituents of some fixed depth d_0 which must be larger than the greatest number of argument-places among our predicates.[17] If constituents of different depths are considered, the weight of each constituent of depth d_0 or more (say of depth d) is then divided evenly among its consistent subordinate constituents of depth $d + 1$. There remains an important open problem as to how the different systems obtained by choosing d_0 differently are to be compared with each other.

If we are given, in addition to a finite number of predicates, also a domain of individuals, then weights will also be assigned to the different state-descriptions as equally as is compatible with the previous assignments of weights to constituents. In the monadic case, the weight of each constituent is divided evenly among all the state-descriptions which make it true.[18]

It is easily seen that the resulting assignment of a priori probabilities to all our general sentences satisfies all the usual axioms of probability calculus (including Kolmogorov's axiom of continuity).[19] In order to make the requisite set-theoretical concepts applicable, each sentence may be considered as standing for the set of all its models.[20]

On this basis, a quantitative inductive logic may be built.[21] Here I can only indicate some of its main features, restricting most of my remarks to the monadic case. I shall try to make the conceptual situation clear rather than to develop the theory itself very far in any particular direction.[22]

For most purposes it suffices to consider only such generalizations as are formulated by a constituent. These generalizations will be called *strong* ones. Since all consistent general sentences are disjunctions of constituents, their probability can be obtained as sums of the probabilities of the constituents occurring in their respective distributive normal forms. In our initial example we already restricted ourselves tacitly to strong generalizations.

What can be said of this example from the point of view of my inductive logic? It is easily seen that in this example our evidence admits of exactly

$$(11)\quad \begin{aligned} m(w) &= w^{N-n} - (w-c)(w-1)^{N-n} + \frac{(w-c)(w-c-1)}{2!}(w-2)^{N-n} \\ &\quad - \cdots + (-1)^{i}\binom{w-c}{i}(w-i)^{N-n} + \cdots \end{aligned}$$

state-descriptions which would make (4) true. In the absence of all evidence there would be $M(w)$ such state-descriptions, where $M(w)$ is obtained from (11) by putting $n = 0$, $c = 0$. In particular we obtain $m(c) = c^{N-n}$.

According to Bayes's formula, the probability of (4) on the evidence we have is

$$(12)\quad \frac{\dfrac{m(w)}{M(w)}}{\dfrac{m(c)}{M(c)} + (K-c)\dfrac{m(c+1)}{M(c+1)} + \dfrac{(K-c)(K-c-1)}{2!}\dfrac{m(c+2)}{M(c+2)} + \cdots}$$

A good idea of the behaviour of this expression is obtained by letting N become infinite (infinite domain). Then $m(w)/M(w)$ approaches $1/w^n$ as a limit and (12) becomes

$$(13)\quad \frac{\dfrac{1}{w^n}}{\dfrac{1}{c^n} + (K-c)\dfrac{1}{(c+1)^n} + \dfrac{(K-c)(K-c-1)}{2!}\dfrac{1}{(c+2)^n} + \cdots}$$

It is seen that (13) has the greater value the smaller w is, and that (13) assumes its greatest value for $w = c$. Thus the recommendation our inductive logic yields is the intuitively acceptable one: Generalization (3) is to be preferred to (4) whenever c is smaller than w.

There is a stronger reason for preferring (3) to (4), however. If we let $n \to \infty$ in (13), the value of (13) is seen to approach zero, except in the case $w = c$ where it approaches one. The more evidence compatible with (3) we thus have, the more probable (3) therefore becomes and the less probable the competing hypotheses (4) become.

In the case $w = c$, (13) assumes the form

$$\frac{1}{1 + (K - c)\left(\frac{c}{c+1}\right)^n + \frac{(K-c)(K-c-1)}{2!}\left(\frac{c}{c+2}\right)^n + \cdots} \tag{14}$$

This expression shows how the inductive probability of (3) grows when more and more confirmation is obtained (in an infinite domain). It approaches one when $n \to \infty$. In the case of an ordinary general implication $(x)(P_1(x) \subset P_2(x))$, with two primitive monadic predicates P_1 and P_2 ($k = 2$, $K = 4$, $c = 3$, presupposing that all the combinations of predicates compatible with the implication have been observed), (14) becomes

$$\frac{1}{1 + \left(\frac{3}{4}\right)^n} \tag{15}$$

The "inductive behaviour" suggested by this function is not completely unreasonable qualitatively, although it is clearly far too overoptimistic for small values of n.[23]

Preferring (3) to (4) in the situation envisaged in our example may thus be explained in terms of the higher degree of confirmation (a posteriori probability) of the preferred generalization. The basic question of this paper thus receives an affirmative answer in the monadic case, which may be expected to be representative of most of the cases for which an inductive logic has so far been developed. In fact, the situation appears essentially similar in the polyadic case.

It is worth pointing out that no difficulties are caused by the possibility that our domain may be infinite; on the contrary, this is usually the most clear-cut case of all. Another disadvantage of Carnap's theory is thus avoided. In general, we may say that the power of a quantitative inductive logic to deal with strict generalizations has been vindicated.

I do not want to suggest, however, that Carnap's critics have been altogether mistaken. On the contrary, the system of inductive logic which I have sketched may be thought of as a partial formalization of ideas closely related to some of their ideas. For instance, it has been maintained by Popper that the best generalizations are typically the *simplest* ones. A closely related idea can now be seen to be incorporated in my system; the fact that our methods prefer (3) to (4) can likewise be related to an interesting notion of simplicity.

Distributive normal forms yield in fact more than one way of measuring the simplicity of a sentence. Among sentences with the same predicates and the same depth, a sentence is naturally said to be the simpler the fewer constituents there occur in its distributive normal form. This criterion of simplicity is closely related to Popper's use of the greater content of a sentence as an indication of its greater simplicity.[24]

This does not yet enable us to compare two constituents (with the same predicates and the same depth) for simplicity. A basis for such comparisons is nevertheless seen from (4). The smaller the number w is, the fewer kinds of individuals are allowed to exist by (4), and the simpler the possible world described by (4) therefore is. This number w may therefore serve as a measure of the complexity of (4) in the monadic case.

The polyadic case is in need of further study. It is obvious, in any case, that similar criteria of simplicity are given to us there by the distributive normal forms. For instance, one natural procedure would be to give a weight to each attributive constituent of depth $d - 1$ occurring in a constituent of depth d, say in (3). The complexity of (3) is then the sum of these weights. The weight of each of these attributive constituents is in turn obtained by adding the weights of all the attributive constituents of depth $d - 2$ occurring in it; and so on, until we reach attributive constituents of depth zero, which all have the same weight.

Although these criteria do not assign a unique measure of simplicity to each closed sentence, they enable us to make several interesting observations. For one thing, our second criterion of simplicity is related rather closely to (although it is certainly different from) Popper's use of what he calls the dimension of a theory as an indication of its complexity.[25] The dimension of a theory is defined by him essentially as the least number of atomic sentences needed to refute it. In the monadic case, the least number of (negated or unnegated) atomic sentences needed to falsify a strong generalization like (4) does not depend on w only. It usually depends also on the order of the predicates which may enter into our atomic sentences.[26] It is easily seen, however, that his number is in no circumstances larger than $1 + \log_2 w$, which therefore seems to be the best approximation to Popper's notion of dimension that we can define for the whole of the monadic case. Since $1 + \log_2 w$ grows together with w, we obtain by means of it the same simplicity ordering of constituents as was already proposed above.

More important than this is the connection which there exists between our notion of simplicity and the comparison between different generalizations which we have made in the monadic case. Since the complexity of (4) is the greater the larger w is, in preferring (3) to (4) our inductive logic recommends to us the simplest generalization compatible with evidence.[27] Contrary to a frequent suggestion, it is therefore not inevitable that simplicity should always go together with low probability. We have just seen that on a suitable assignment of weights (a priori probabilities) to constituents it goes together with high probability on evidence. A simpler constituent does not have a higher a priori probability (it would in fact be more natural to give it a lower one),[28] but when more and more evidence comes in, the simplest constituent consistent with it will become more probable than others. In the long run, this simplest constituent is the only one which is confirmed by the evidence.[29]

Interesting further results obtainable in our inductive logic could be mentioned. From a theoretical point of view it is almost as interesting, however, to see what it does *not* enable us to do.

Although it seems to offer a reasonable account of our rational preferences concerning *strict* (inductive) generalizations, it is powerless to cope with inductive generalizations concerning relative number of individuals of different kinds. For the sake of a short name, let us call the latter *statistical* generalizations. A simple way of seeing this is to calculate the probability that the next individual observed in our example (in a situation where (3) is the simplest strong generalization compatible with the evidence we already have) should exemplify one fixed kind of individuals, say the one specified by $Ct_{i_1}(x)$. It is easily seen[30] that for an infinite N and for an n large in comparison with K this probability is approximately $(1/c)$. This means that the statistical generalizations which our inductive logic yields take into account the fact which possible kinds of individuals are exemplified in our experience and which ones are not. However, they do not take into account the observed relative frequencies of individuals of the different kinds; as far as the distribution of the individuals among the observed kinds are concerned, our inductive logic adheres simply to the a priori recommendation of even distribution.

This suggests that there are two essentially different problems we have to solve independently of each other in developing a quantitative inductive logic, viz., to formulate the principles that underlie our strict inductive generalizations, and to formulate the principles that underlie our statistical generalizations.[31] Carnap's methods yield undesirable results when applied to the former problem, while our approach does not give much better results in the case of the latter than a strictly a priori method. To what extent and in what way the solutions that can be given to these two problems may be combined, remains to be investigated.

If these two problems really turn out to be distinct, we will have to say that some confusion has ensued from the current practice of discussing them as if they were one and the same problem. For instance, we will then have to say that Carnap and Popper have unwittingly addressed themselves to somewhat different problems. Most of the work the former has in fact done pertains to the different methods of statistical generalizations, while the latter appears to be more interested in strict generalizations than in statistical ones.

NOTES

1. See Rudolf Carnap, *Logical Foundations of Probability* (Chicago: The University of Chicago Press, 1950) (2nd ed.), Chicago, 1963; and Karl R. Popper, *The Logic of Scientific Discovery* (London: Hutchinson & Co., 1959). These two books also supply further references. In discussing Carnap's views in the present paper, I am confining my attention to the views put forward in the first edition of *Logical Foundations of Probability*, disregarding whatever subsequent changes there have been in Carnap's point of view.

2. Popper, *op. cit.*, p. 399: "Science does not aim, primarily, at high probabilities. It aims at a high informative content, well backed by experience."

3. According to Popper, *op. cit.*, p. 387, his aim has been "to show that degree of corroboration was not a probability; that is to say, that it was not one of the possible interpretations of the probability calculus." "Degree of corroboration" is Popper's measure "of the severity of tests to which a theory has been subjected, and of the manner in which it has passed these tests, or failed them. . . " (*loc. cit.*).

4. More accurately, one with what Popper calls the lowest "absolute logical probability." Cf. *op. cit.*, Sec. 35.

5. See *op. cit.*, pp. 570–77.

6. *Op. cit.*, p. 575.

7. They are defined by the expressions $(\pm)P_1(x)$ & $(\pm)P_2(x)$ & . . . &$(\pm)P_k(x)$, where the symbols $(\pm)$ are replaced by negation-signs or by nothing at all in all the different combinations. Some of these kinds of individuals may of course be empty.

8. For the *Q*-predicates, see *op. cit.*, pp. 124–26.

9. Some remarks on the connection between this preference and Popper's ideas will be made later. It is also obvious that (3) is the preferable generalization on any reasonable method of maximum likelihood or of "straight rule."

10. For the notion of instance confirmation, see Carnap, *op. cit.*, pp. 571–77.

11. For (5), see Carnap, *op. cit.*, p. 571, formula (11); for (6), see *op. cit.*, p. 573, formula (16), or p. 568, formula (7).

12. This is not intended as a reconstruction of Carnap's reasons for preferring m^* (some of these reasons are mentioned briefly in *op. cit.*, pp. 562–66), but rather as an intuitively persuasive line of thought which leads to the same result.

13. For the notion of structure-description, see Carnap, *op. cit.*, pp. 114–17.

14. The worst difficulty is probably the fact that in Carnap's approach a universally quantified sentence (which is not logically true) always has zero as its degree of confirmation when the domain of individuals is infinite.

15. A *general* sentence means in this paper a sentence of applied first-order logic (no predicate variables) without individual constants and without free individual variables. A sentence which is not general is called *singular*.

16. See my papers, "Distributive Normal Forms in First-Order Logic," in *Formal Systems and Recursive Functions, Proceedings of the 1963 Logic Colloquium in Oxford*, ed. John N. Crossley, *Studies in Logic and the Foundations of Mathematics* (Amsterdam: North-Holland Publishing Company, 1965), and "Distributive Normal Forms and Deductive Interpolation," *Zeitschrift für mathematische Logik und Grundlagen der Mathematik*, X (1964), 185–91, as well as the older monograph "Distributive Normal Forms in the Calculus of Predicates," *Acta Philosophica Fennica*, VI (1953).

17. It would also be possible, and for many purposes more natural, to assign to them *different* constant weights, the larger the more attributive constituents they contain. This would make little difference to my arguments in the sequel.

18. In the polyadic case, it may turn out to be desirable to modify this procedure somewhat. It would take us too far, however, to discuss the matter here.

19. These a priori probabilities are independent of the domain of individuals. Similar a priori probabilities can be assigned to singular sentences only in some given fixed universe of discourse.

20. Suppose that we are given a sequence of general sentence S_1, S_2, . . . , the models of each of which are all also models of its predecessors and which do not possess any single

model in common. There will then exist (by the compactness theorem for first-order logic) a member of the sequence which is contradictory and whose probability is therefore zero. This proves the continuity axiom. (For the axiom, see A. N. Kolmogorov, *Grundbergriffe der Wahrscheinlichkeitsrechnung* (Berlin, 1933), Chap. ii, §1.)

21. This quantitative inductive logic is in the case of each finite universe based on the use of a regular and symmetrical measure function. Hence all the results Carnap establishes (*op. cit.*, Chap. viii) for regular and symmetrical *c*-functions (concerning direct inference, binomial theorem, and Bernoulli's laws) are valid in our inductive logic for each finite system. However, it is easily seen that Carnap's requirement of fitting together (*op. cit.*, pp. 290–92) is not satisfied. Hence the infinite case may have to be dealt with in a way different from Carnap's.

22. One interesting feature of this conceptual situation is the following: In order to calculate the a priori and therefore also the a posteriori probability of a general sentence S we have to know which constituents (with the same predicates and the same depth as S) are consistent and which ones are inconsistent. However, the general problem of deciding which constituents are consistent is easily seen to be equivalent to the decision problem for first-order logic (predicate calculus), and hence recursively unsolvable. Hence the general problem of calculating the degree of confirmation of an arbitrary sentence on given evidence is recursively unsolvable. This gives in a sense an answer to the much-debated question: How difficult might induction be? It turns out to be exactly as difficult as the decision problem for first-order logic. The unsolvability of the problem of determining the degree of confirmation of an arbitrary generalization is no argument against my approach, however, but rather for it. Hilary Putnam has shown (roughly speaking) that an optimal inductive strategy, if such a strategy should exist, cannot be computable. [See "Degree of Confirmation and Inductive Logic," *The Philosophy of Rudolf Carnap*, ed. P. A. Schilpp (La Salle, Illinois: Open Court, 1963.)] In other words, any optimal inductive strategy must exhibit recursive undecidability similar to ours.

23. The reasons for this overoptimism are not difficult to diagnose although it would be out of place to discuss them here. It can also be corrected very easily along the lines of n. 17, e.g., by making the weight of (4) proportional to K^w or perhaps K^{2w}. Then we would have instead of (15) an expression which appears much less overoptimistic.

24. Popper, *op. cit.*, Secs. 43, 33–35.

25. Popper, *op. cit.*, Sec. 38, appendices i, *viii.

26. In other words, the answer to the question: How many of the sentences $(\pm)P_1(a)$, $(\pm)P_2(a)$, . . . , *taken in this order*, are needed to refute (4)? often changes when these sentences are permuted.

27. This recommendation may be thought of as a direct application of Occam's Razor: One should not assume the existence of more kinds of individuals than is made necessary by evidence.

28. Cf. notes 17 and 23.

29. When $w > c$, (3) is also easier to falsify than (4) in the sense that fewer (negated or unnegated) atomic sentences are needed to contradict it. This is reminiscent of what Popper says of the connection between simplicity and falsifiability. Perhaps one should not overemphasize the role of falsification here, however, for a certain symmetry obtains here between verification and falsification. Let us write (4) as a conjunction C_j & D_j of its existential and universal part. Then in an infinite domain C_j can be finitely verified, but not finitely falsified, whereas D_j can be finitely falsified but not finitely verified. Moreover, C_j is the easier to verify the smaller w is in exactly the same sense in which D_j is the easier to falsify the smaller w is. Hence in a sense a simpler constituent is not only easier to falsify but also easier to verify, namely, easier to verify to the extent it can be finitely verified at all.

30. Let our evidence be e, and let the event that the next individual exemplifies $Ct_{i_1}(x)$ be s. Then we have $p(s/e)$ = the probability of s on the evidence $e = p(s \,\&\, e)/p(e)$. Now $p(e)$ =

$$(*) \qquad \frac{1}{2^K}\left(\frac{m(c)}{M(c)} + (K - c)\frac{m(c+1)}{M(c+1)} + \frac{(K-c)(K-c-1)}{2!}\frac{m(c+2)}{M(c+2)} + \cdots\right).$$

Here the function m is defined as in (11). Clearly, $p(s \,\&\, e)$ is obtained from (*) by replacing n by $n + 1$. Since (11) is independent of the distribution of the observed n individuals among the c exemplified kinds, the same holds of $p(s/e)$. In fact, an easy computation shows that in the case of an infinite domain it has approximately the value $1/c$ when n is large.

On this basis, it is easily seen that our inductive method does not fall within Carnap's "continuum of inductive methods." The reason is that our method does not satisfy Carnap's condition C9 [see Rudolf Carnap, *The Continuum of Inductive Methods* (Chicago: The University of Chicago Press, 1952), p. 13].

31. One question that is likely to be affected by a distinction between these two problems concerns Carnap's requirement of "fitting together" (cf. n. 21). This requirement may be formulated by saying that the degree of confirmation of a sentence without quantifiers must not change when a new individual is introduced into our language. If the distinction is made, it perhaps suffices to require only that no such change takes place when the new individual does not disturb any of the general laws that hold in our universe.

Which Statements are Confirmable

Nelson Goodman

THE NEW RIDDLE OF INDUCTION

Confirmation of a hypothesis by an instance depends rather heavily upon features of the hypothesis other than its syntactical form. That a given piece of copper conducts electricity increases the credibility of statements asserting that other pieces of copper conduct electricity, and thus confirms the hypothesis that all copper conducts electricity. But the fact that a given man now in this room is a third son does not increase the credibility of statements asserting that other men now in this room are third sons, and so does not confirm the hypothesis that all men now in this room are third sons. Yet in both cases our hypothesis is a generalization of the evidence statement. The difference is that in the former case the hypothesis is a *lawlike* statement; while in the latter case, the hypothesis is a merely contingent or accidental generality. Only a statement that is *lawlike*—regardlesss of its truth or falsity or its scientific importance—is capable of receiving confirmation from an instance of it; accidental statements are not. Plainly, then, we must look for a way of distinguishing lawlike from accidental statements.

So long as what seems to be needed is merely a way of excluding a few odd and unwanted cases that are inadvertently admitted by our definition of confirmation, the problem may not seem very hard or very pressing. We fully expect that minor defects will be found in our definition and that the necessary refinements will have to be worked out patiently one after another. But some further examples will show that our present difficulty is of a much graver kind.

Suppose that all emeralds examined before a certain time *t* are green. At time *t*, then, our observations support the hypothesis that all emeralds are green; and this is in accord with our definition of confirmation. Our evidence statements assert that emerald *a* is green, that emerald *b* is green, and so on;

---

and each confirms the general hypothesis that all emeralds are green. So far, so good.

Now let me introduce another predicate less familiar than "green." It is the predicate "grue" and it applies to all things examined before *t* just in case they are green but to other things just in case they are blue. Then at time *t* we have, for each evidence statement asserting that a given emerald is green, a parallel evidence statement asserting that that emerald is grue. And the statements that emerald *a* is grue, that emerald *b* is grue, and so on, will each confirm the general hypothesis that all emeralds are grue. Thus according to our definition, the prediction that all emeralds subsequently examined will be green and the prediction that all will be grue are alike confirmed by evidence statements describing the same observations. But if an emerald subsequently examined is grue, it is blue and hence not green. Thus although we are well aware which of the two incompatible predictions is genuinely confirmed, they are equally well confirmed according to our present definition. Moreover, it is clear that if we simply choose an appropriate predicate, then on the basis of these same observations we shall have equal confirmation, by our definition, for any prediction whatever about other emeralds—or indeed about anything else. As in our earlier example, only the predictions subsumed under lawlike hypotheses are genuinely confirmed; but we have no criterion as yet for determining lawlikeness. And now we see that without some such criterion, our definition not merely includes a few unwanted cases, but is so completely ineffectual that it virtually excludes nothing. We are left once again with the intolerable result that anything confirms anything. This difficulty cannot be set aside as an annoying detail to be taken care of in due course. It has to be met before our definition will work at all.

Nevertheless, the difficulty is often slighted because on the surface there seem to be easy ways of dealing with it. Sometimes, for example, the problem is thought to be much like the paradox of the ravens. We are here again, it is pointed out, making tacit and illegitimate use of information outside the stated evidence: the information, for example, that different samples of one material are usually alike in conductivity, and the information that different men in a lecture audience are usually not alike in the number of their older brothers. But while it is true that such information is being smuggled in, this does not by itself settle the matter as it settles the matter of the ravens. There the point was that when the smuggled information is forthrightly declared, its effect upon the confirmation of the hypothesis in question is immediately and properly registered by the definition we are using. On the other hand, if to our initial evidence we add statements concerning the conductivity of pieces of other materials or concerning the number of older brothers of members of other lecture audiences, this will not in the least affect the confirmation, according to our definition, of the hypothesis concerning copper or of that concerning other lecture audiences. Since our definition is insensitive to the bearing upon hypotheses of evidence so related to them, even when the evidence is fully declared, the difficulty about accidental hypotheses cannot be explained away on the ground that such evidence is being surreptitiously taken into account.

A more promising suggestion is to explain the matter in terms of the effect of this other evidence not directly upon the hypothesis in question but *in*directly through other hypotheses that *are* confirmed, according to our definition, by such evidence. Our information about other materials does by our definition confirm such hypotheses as that all pieces of iron conduct electricity, that no pieces of rubber do, and so on; and these hypotheses, the explanation runs, impart to the hypothesis that all pieces of copper conduct electricity (and also to the hypothesis that none do) the character of lawlikeness—that is, amenability to confirmation by direct positive instances when found. On the other hand, our information about other lecture audiences *dis*confirms many hypotheses to the effect that all the men in one audience are third sons, or that none are; and this strips any character of lawlikeness from the hypothesis that all (or the hypothesis that none) of the men in *this* audience are third sons. But clearly if this course is to be followed, the circumstances under which hypotheses are thus related to one another will have to be precisely articulated.

The problem, then, is to define the relevant way in which such hypotheses must be alike. Evidence for the hypothesis that all iron conducts electricity enhances the lawlikeness of the hypothesis that all zirconium conducts electricity, but does not similarly affect the hypothesis that all the objects on my desk conduct electricity. Wherein lies the difference? The first two hypotheses fall under the broader hypothesis—call it *H*—that every class of things of the same material is uniform in conductivity; the first and third fall only under some such hypothesis as—call it *K*—that every class of things that are either all of the same material or all on a desk is uniform in conductivity. Clearly the important difference here is that evidence for a statement affirming that one of the classes covered by *H* has the property in question increases the credibility of any statement affirming that another such class has this property; while nothing of the sort holds true with respect to *K*. But this is only to say that *H* is lawlike and *K* is not. We are faced anew with the very problem we are trying to solve: the problem of distinguishing between lawlike and accidental hypotheses.

The most popular way of attacking the problem takes its cue from the fact that accidental hypotheses seem typically to involve some spatial or temporal restriction, or reference to some particular individual. They seem to concern the people in some particular room, or the objects on some particular person's desk; while lawlike hypotheses characteristically concern all ravens or all pieces of copper whatsoever. Complete generality is thus very often supposed to be a sufficient condition of lawlikeness; but to define this complete generality is by no means easy. Merely to require that the hypothesis contain no term naming, describing, or indicating a particular thing or location will obviously not be enough. The troublesome hypothesis that all emeralds are grue contains no such term; and where such a term does occur, as in hypotheses about men in *this room*, it can be suppressed in favor of some predicate (short or long, new or old) that contains no such term but applies only to exactly the same things. One might think, then, of excluding not only hypotheses that actually contain terms for specific individuals but also all hypotheses that are equivalent to others that do contain such terms. But, as we have just seen, to exclude only hypotheses of which *all* equivalents are free of such terms is to exclude nothing. On the other hand, to exclude all hypotheses that have *some* equivalent containing such a term is to exclude everything; for even the hypothesis

All grass is green

has as an equivalent

All grass in London or elsewhere is green.

The next step, therefore, has been to consider ruling out predicates of certain kinds. A syntactically universal hypothesis is lawlike, the proposal runs, if its predicates are "purely qualitative" or "nonpositional." This will obviously accomplish nothing if a purely qualitative predicate is then conceived either as one that is equivalent to some expression free of terms for specific individuals, or as one that is equivalent to no expression that contains such a term; for this only raises again the difficulties just pointed out. The claim appears to be rather that at least in the case of a simple enough predicate we can readily determine by direct inspection of its meaning whether or not it is purely qualitative. But even aside from obscurities in the notion of "the meaning" of a predicate, this claim seems to me wrong. I simply do not know how to tell whether a predicate is qualitative or positional, except perhaps by completely begging the question at issue and asking whether the predicate is "well-behaved"—that is, whether simple syntactically universal hypotheses applying it are lawlike.

This statement will not go unprotested. "Consider," it will be argued, "the predicates 'blue' and 'green' and the predicate 'grue' introduced earlier, and also the predicate 'bleen' that applies to emeralds examined before time *t* just in case they are blue and to other emeralds just in case they are green. Surely it is clear," the argument runs, "that the first two are purely qualitative and the second two are not; for the meaning of each of the latter two plainly involves reference to a specific temporal position." To this I reply that indeed I do recognize the first two as well-behaved predicates admissible in lawlike hypotheses, and the second two as ill-behaved predicates. But the argument that the former but not the latter are purely qualitative seems to me quite unsound. True enough, if we start with "blue" and "green," then "grue" and "bleen" will be explained in terms of "blue" and "green" and a temporal term. But equally truly, if we start with "grue" and "bleen," then "blue" and "green" will be explained in terms of "grue" and "bleen" and a temporal term; "green," for example, applies to emeralds examined before time *t* just in case they are grue, and to other emeralds just in case they are bleen. Thus qualitativeness is an entirely relative matter and does not by itself establish any dichotomy of predicates. This relativity seems to be completely overlooked by those who contend that the qualitative character of a predicate is a criterion for its good behavior.

Of course, one may ask why we need worry about such unfamiliar predicates as "grue" or about accidental hypotheses in general, since we are unlikely to use them in making predictions. If our definition works for such hypotheses as are normally employed, isn't that all we need? In a sense, yes; but only in the sense that we need no definition, no theory of induction, and no philosophy of knowledge at all. We get along well enough without them in daily life and in scientific research. But if we seek a theory at all, we cannot excuse gross anomalies resulting from a proposed theory by pleading that we can avoid them in practice. The odd cases we have been considering are the clinically pure cases that, though seldom encountered in practice, nevertheless display to best advantage the symptoms of a widespread and destructive malady.

We have so far neither any answer nor any promising clue to an answer to the question what distinguishes lawlike or confirmable hypotheses from accidental or nonconfirmable ones; and what may at first have seemed a minor technical difficulty has taken on the stature of a major obstacle to the development of a satisfactory theory of confirmation. It is this problem that I call the new riddle of induction.

Steven Barker and
Peter Achinstein

ON THE NEW RIDDLE OF INDUCTION

I

In his *Fact, Fiction, and Forecast* Nelson Goodman formulates a striking puzzle which he calls "the new riddle of induction."[1] He reminds us of how we are willing to "project" a predicate like "green": Since all the emeralds we have so far observed have been green, we adopt the hypothesis that all future emeralds probably are going to be green also. But then he asks us to consider a predicate such as "grue," which is to be understood as applying to a thing at a given time if and only if either the thing is then green and the time is prior to time *t*, or the thing is then blue and the time not prior to *t*.[2] Let us take *t* as a time in the future, say, the year 2000 A.D. Now, all emeralds observed in the past have been green, and therefore they have been grue. Thus the hypothesis that all emeralds are grue has as much inductive support as the hypothesis that all emeralds are green, for we have equal numbers of positive instances in favor of each hypothesis and no negative instances. But the two hypotheses yield incompatible predictions regarding emeralds in the future, since a grue emerald after the year 2000 A.D. will be blue. We cannot accept both these conflicting hypotheses, yet they seem equally to have good inductive support. Why ought one predicate to be "projected" rather than the other? This is the "new riddle of induction."

One answer which has been suggested to the riddle is this: Predicates not involving any sort of reference to time ought to be projected in preference to ones that do involve such reference. "Grue" and "bleen" (where "bleen" is a predicate which is to be understood as applying to a thing at a given time if and only if either the thing is then blue and the time is prior to *t*, or the thing is then green and the time is not prior to *t*) are predicates which do involve reference

Reprinted from the *Philosophical Review* (1960) by permission of the editor and the authors.

to time, whereas "green" and "blue" are not. For this reason "blue" and "green" ought to be projected in preference to "grue" and "bleen."[3]

But this answer will not satisfy Goodman. In his most scathing tone he will ask, "And what, pray, is it for a predicate to be a predicate involving reference to time?" It is true that a person brought up to speak the language of green and blue will not understand "grue" and "bleen" unless these latter are defined for him in terms of green and blue and time. But, Goodman will insist, a person brought up to speak the language of grue and bleen will define "green" and "blue" in an exactly symmetrical manner: That is, he will not understand "green" and "blue" until their uses are explained for him in terms of "grue" and "bleen" and time; he will employ the definition that "green" applies to all and only those things which are grue prior to the year 2000 or bleen thereafter, and the definition that "blue" applies to all and only those things which are bleen prior to the year 2000 or grue thereafter. Goodman will maintain that "grue" is not intrinsically a more nontemporal predicate than is "green"; the situation is perfectly symmetrical, he will say. For the speaker of the green-blue language, "grue" and "bleen" appear to be temporal predicates; but for the speaker of the grue-bleen language, "green" and "blue" appear to be temporal predicates. Which predicate seems temporal to you and which one not will all depend on which predicate happens to have got more deeply entrenched in the language that you are accustomed to speaking.

Goodman's way of rejecting this proposed solution to his riddle of induction is acute. But does he not go too far? Does he not exaggerate the extent to which the grue-bleen language and the green-blue language are symmetrical in nature? To see whether he does, we must consider just what claims he makes.

In maintaining that there exists so much symmetry between these two languages, Goodman implicitly is making two especially important claims. (1) He is claiming that people brought up to use "grue" understand their predicate in such a way that it applies to all and only those objects of all kinds to which speakers of the green-blue language would apply the phrase "green if the date is prior to the year 2000 or blue if the date is thereafter." And this surely means that such people must understand not only what it is for actual objects to be grue but also what it means for fictional objects, imaginary objects, or objects in pictures to be grue. A grue-bleen speaker, on this assumption, when confronted with pictures of Robin Hood in Sherwood Forest, would be able to tell that the foliage of the forest in the picture is grue, not bleen. And (2) Goodman is claiming that the predicate "grue" is for the speaker of the grue-bleen language a nontemporal predicate, in that such a speaker does not need to ascertain the date of an object before being able to tell whether the object is grue or not. On the other hand, if we, who speak the green-blue language, were asked to determine whether an object in a painting satisfies the condition of being either green if the date is before 2000 A.D. or blue if the date is thereafter, we would have to ascertain the date of the scene in the picture. If the picture contained people or buildings, for example, we might determine the date of the scene by noticing the style of the people's costumes or the style of the architecture, and dating this style through historical research if it is a past style, or through careful prediction if it is a future style. Our use of these procedures indicates that the predicate "green if the date is prior to 2000 or blue if the date is thereafter" is for us a temporal predicate, that is, one which we know how to apply to a thing only after we have ascertained the date of the thing. But Goodman's speaker of the grue-bleen language (whom we shall call Mr. Grue) must not need to go through such procedures in determining whether or not something is grue; for if he did, "grue" would be for him a temporal predicate. Nor is it enough to say that Mr. Grue must be able to use "grue" without explicitly consulting history books and looking at calendars; he must be able to use "grue" without even implicitly basing his judgment upon observable temporal clues, such as styles of costumes or of architecture.

Goodman must take for granted these two suppositions about speakers of the grue-bleen language. But what would it be like for there to be such people? Is there not some difficulty latent in this conception?

II

Mr. Grue and Mr. Green (a speaker of the green-blue language) both will often be using their color predicates in discussions of other things besides actual objects. For instance, they will sometimes want to talk about objects in pictures. Let us consider their respective conceptions of what it is for ordinary, straightforwardly representational pictures to picture colored obejcts.

With regard to this favorite color, Mr. Green of course will have the notion that (excluding special subtleties of painting) an object in a picture is green if and only if the pigment used in representing the object is green pigment. This rule bleongs to his conception of what it is for an object in a picture to be colored.

But what of Mr. Grue? His case is a bit more complicated. With regard to his favorite color, there would seem to be two different possibilities about what his notion of pictorial representation will be. (A) Perhaps he calls an object in a picture "grue" when and only when the pigment used in representing it is grue pigment. Or (B) perhaps he calls an object in a picture "grue" when and only when either the pigment used in representing it is green and the date in the picture is prior to 2000, or the pigment used is blue and the date in the picture 2000 or later.

But we see that (A) and (B) are not equally possible, if we look back at Goodman's assumption (1) that Mr. Grue's term "grue" is to be applicable to all and only those objects to which Mr. Green's phrase "green if the date is prior to 2000 or blue if the date is thereafter" correctly applies. Mr. Grue, if he is to satisfy assumption (1), cannot classify in accordance with notion (A); for (A) and (1) directly conflict. To see that this is so, consider a painting done this year, which depicts a pastoral scene of the year 2001; in this painting green pigment is used in representing the grass. What color is the grass in the picture? If (A) were the case then Mr. Grue would have to regard the grass in the picture as grue, since it is represented by means of grue pigment. But Mr. Green will hold that the grass in the picture is green and that therefore the phrase "green if the date is prior to 2000 or blue if the date is thereafter" does not properly apply to it, since the date in the picture is after 2000. Of course Mr. Green is willing to apply the phrase "green if the date is prior to 2000 or blue if the date is thereafter" to the pigment used to represent the grass; but he will not apply it to the grass itself. Thus if (A) were the case, assumption (1) would be violated. Since we have assumed that Mr. Grue does satisfy assumption (1), we must suppose that his notion of pictorial representation is in accordance with (B) rather than (A).

In the light of this, let us proceed to see whether we can detect any difficulties latent in Goodman's formulation of his new riddle of induction.

III

A person who spoke the grue-bleen language, who was able to tell whether a thing is grue at a given time without first needing to ascertain what the date is at that time, and who applied the predicate "grue" to all and only those objects to which speakers of the green-blue language apply the phrase "green if the date is prior to the year 2000 or blue if the date is thereafter" would be a person of remarkable powers.

Suppose that we present Mr. Grue with two black-and-white drawings of plots of grass, drawings which we carefully construct so that he cannot tell them apart. Now we prepare two larger drawings of Commencement in the Harvard Yard, into one of which we incorporate one of our drawings of grass, into the other of which we incorporate the other of our drawings of grass. These two larger drawings of the Harvard Yard are not alike, however; one represents the Yard as it is nowadays, the other shows the Yard not as it is but as it will be. In the first drawing the buildings all have their familiar present-day aspect, and an alumni procession is straggling past, those at the very rear bearing a placard announcing themselves as the Class of '58. The second drawing has an altered aspect, for the fifty-story Boylston Street Apartments tower in the distance, the air is filled with flying saucers, and the rear of the procession is brought up by a group whose placard proclaims them to be the Class of '00. Suppose that we now confront Mr. Grue with these two large drawings; we do not, however, let him see either of them entire, for we cleverly screen off most of the area of each drawing from the gaze of Mr. Grue, forcing him to peer at each drawing through a small opening in a screen, so that in each case all that he is allowed to see is the area of grass which he saw previously, before it was incorporated into the larger drawing. We now provide him with a palette and put to him the question, "What paint should be used in order to color these two pictures in such a way that the grass in the pictures will be grue?"

Now, according to Goodman's supposition, Mr. Grue is a person who is able to tell whether a thing is grue just by looking at it, without first needing to ascertain its date. So presumably in this case he will be able to tell, just by looking at the grass, what sort of look it needs to be given in order to be grue. Presumably he will correctly respond to our query and will answer that one kind of paint (which we call green) needs to be used for the first picture while another kind of paint (which we call blue) needs to be used for the second. But if Mr. Grue chooses the appropriate paints, green for the pre-2000 picture and blue for the other, then he must be recognizing a difference between the pictures, a difference which makes necessary this difference in pigments. And he must recognize that the two pictures are different even though what he sensuously perceives is the same in both cases, even though the light reaching his eye and stimulating his retina is just the same in both cases. How should we describe Mr. Grue's power? It is clear that the power which he possesses is a type of extrasensory perception.

Thus, one curious consequence of Goodman's set of suppositions is that the speaker of the grue-bleen language must possess a faculty of extrasensory perception not possessed by people such as ourselves who are speakers of the green-blue language.

IV

Suppose we were to grant that there is nothing logically absurd about the idea that Mr. Grue possesses this queer faculty of extrasensory perception. Still, is it really the case that the predicate "grue" as he uses it is a nontemporal predicate, whose applicability he can determine without needing to ascertain the date?

Let us suppose that Mr. Grue knows how to apply temporal terms such as "now," "before 2000 A.D.," and "after 2000 A.D."; or, if he does not know how to do this, let us teach him how. Now we present him with a single black-and-white sketch of the Harvard Yard, just showing the Holden Chapel with its tympanum which is so well maintained that its color never changes. We first say to Mr. Grue, "This sketch is a picture of the Holden Chapel in 1959. But black-and-white pictures are dull, aren't they? The tympanum in the picture is bleen, of course. Just color it in for us, won't you?" And we give Mr. Grue a palette of paints. He will reach for the paint that we call blue. But next we tell Mr. Grue that we wish to have a picture of the Holden Chapel in the year 2001; and we explain to him that the Holden Chapel is so carefully maintained that it never changes in appearance. We now ask Mr. Grue to give us his advice; we say, "Please show us what paint should be used if we are now to paint the tympanum in a picture which will show the Holden Chapel in the year 2001." If Mr. Grue is the type of person Goodman supposes him to be, he will of course be able to answer this question. He will point out the paint which we call green.

In this case we see that Mr. Grue uses two different paints in order to color the two pictures bleen; but the difference between the two pictures is solely a temporal difference. Thus his knowledge of the temporal difference is here sufficient to affect his choice of the paint needed in order to make the object in the picture look bleen. Moreover, knowledge of the temporal difference is necessary in order to enable him to decide what paint is needed. For if we were simply to say to him, "What paint should be used now in order to color a picture of the bleen tympanum of the Holden Chapel? We are not saying what the date of the scene in the picture is to be," then he will be unable under these circumstances to tell us which paint is the correct one.

Thus it is clear here that the predicates "grue" and "bleen" cannot be used successfully by Mr. Grue unless he is somehow apprised of the date of the object to which the predicates are to be applied. In this sense, whether he himself is aware of it or not, these predicates are for Mr. Grue temporal predicates.

V

What about the predicates "green" and "blue," on the other hand? Mr. Grue might try to argue that as he understands the term "green" it is for him a temporal predicate. He may try to argue that he cannot tell whether an object in a picture is green or not, unless he knows whether the object is grue before 2000 A.D. or bleen thereafter; and he may claim that he cannot ascertain this unless he knows the date of the object in the picture. He may say that it is only when there is some indication of the date of the scene in the picture, as, for example, the presence of Robin Hood and his men, that he can tell that the foliage in the picture is green. If we presented him with two black-and-white pictures of the Harvard Yard of no definite date which he could not tell apart, and if we now used grue paint to paint the grass in one and bleen paint to paint the grass in the other, he might claim that this would not be enough to enable him to tell the pictured scenes apart as regards their greenness. According to Mr. Grue, then, "green" and "blue" are temporal predicates.

His claim is misleading, however. We should point out to Mr. Grue how even he is really capable of detecting whether an object in a picture is green or not without having to ascertain what date the picture represents. First, we ask Mr. Grue always to notice whether or not the paint on the canvas is grue in color. This, of course, Mr. Grue can do. We then tell him that nowadays whenever the paint on a certain area of the canvas is grue, then the object being portrayed at that area of the canvas is green, and whenever the paint is bleen, the object represented is blue. These instructions are sufficient for the present, and if he uses the terms "green" and "blue" according to these instructions he will be able to tell whether the object in the picture is green or blue without having to ascertain its date; for recognizing whether the paint used at that spot on the canvas is grue or not does not entail ascertaining the date of the object which the paint is being used to portray.

Mr. Grue cannot point out for Mr. Green's benefit an analogous device for recognizing grue objects in pictures without noticting their dates. Mr. Grue cannot tell Mr. Green, for example, that whenever the paint on the canvas is green then the object in that area of the canvas is grue, nowadays; for a picture of a post-2000 scene painted with green paint is bleen nowadays and not grue. And this reinforces the claim that the predicates "green" and "grue" are asymmetrical in their logical character.

VI

So far, we have seen that there are reasons for believing that Goodman's set of suppositions about Mr. Grue contains latent difficulties. Not only would Mr. Grue have to possess extrasensory perception, but what is worse, his use of the predicate "grue" would logically involve prior ascertaining of the date, and in this sense "grue" could not be for him a nontemporal predicate. On the other hand, "green" is a nontemporal predicate for him.

It is possible, however, that Goodman might seek to reply in the following vein. He might maintain that the situation is nevertheless symmetrical as between the grue-bleen language and the green-blue language; that is, granting the aptness of the description which we have so far given of Mr. Grue, Goodman might nevertheless wish to argue that if we were to frame a description of Mr. Green it would be necessary from the point of view of Mr. Grue to describe Mr. Green as possessing a faculty of extrasensory perception and as using his predicate "green" in a way which involves essential reference to the date. But would this counterargument be sound?

What kind of experiment might we consider in hope of showing that Mr. Green possesses a faculty of extrasensory perception? Suppose we were to conduct with Mr. Green an experiment analogous to the one which demonstrated Mr. Grue's occult power. Here we would show Mr. Green two black-and-white drawings of plots of grass carefully constructed so that he cannot tell them apart. Then we prepare two larger drawings of the Harvard Yard into which we incorporate these smaller drawings of grass. One of the larger drawings represents the Commencement of 1959, the other that of 2001. We confront Mr. Green with these two large drawings, preventing him however from seeing them entire; he is allowed only to view the portions of grass which he has previously seen. We now ask him which paint should be used in order now to color these two drawings in such a way that the grass in the pictures will be green. What will be his response? Here of course Mr. Green will use the same paint in both cases. Thus the outcome of this experiment gives no reason whatever for imputing to Mr. Green any faculty of extrasensory perception, not even from Mr. Grue's point of view.

We might try to alter the experiment, however, so that Mr. Green will have to give different reactions instead of the same one. Instead of telling Mr. Green to point the grass green in both pictures, we tell him that we would like to have the grass in the first picture green and that in the second picture blue. In this case, Mr. Green will reach for different paints. Mr. Grue might be inclined to say that Mr. Green is recognizing a difference between the two pictures, a difference which requires that he use one kind of paint on the first picture and another kind on the second. This difference, Mr. Grue may be inclined to say, is a difference that is not sensuously perceptible; hence, Mr. Green possesses a faculty of extrasensory perception. But of course this inference on the part of Mr. Grue would be mistaken. Even in this experiment no evidence has been given of any occult power being possessed by Mr. Green. The fact that Mr. Green reaches for two different kinds of paint reflects only the fact that he was given different instructions with respect to the different pictures.

Thus it seems impossible to demonstrate that Mr. Green possesses any faculty of extrasensory perception. But are predicates such as "green" and "blue" temporal predicates for Mr. Green?

Suppose again that we were to say to Mr. Green that we want to color a sketch of the Holden Chapel in the Harvard Yard; we tell him that we want the tympanum of the Chapel in the picture to be blue, but we do not tell him what the date of the pictured scene is to be. Unlike Mr. Grue in the analogous experiment, Mr. Green here will be able to decide which paint to reach for in order to carry out these instructions, even if we tell him nothing about the date of the scene represented in the picture. And if we do tell him that the picture represents the Chapel in 1959 rather than in 2001, this information will not at all affect Mr. Green as he decides what paint to use in order to color the picture. Thus there is no basis here for saying that Mr. Green is using "blue" as a temporal predicate; knowledge of the date is irrelevant for him to the application of "blue."

Thus this possible counterargument fails.

VII

Another kind of objection which might be raised is this. Someone might suppose that Mr. Grue can never tell what kind of paint to use to paint an object in a picture grue or bleen; he might suppose that Mr. Grue's peculiar ability lies only in the fact that, once it is painted, he can tell whether or not the object in the picture is grue, but not until then. So Mr. Grue, if presented with two black-and-white sketches of the Harvard Yard, one pre-2000 and the other post-2000, and if allowed to see only a plot of grass in each, might not be able to tell us how he would paint the pictures so that they would both be pictures of grue grass. He might have to experiment with all the paints on the palette before the grue color would emerge for him, as a gestalt quality, in each picture. How can we say, then, that because Mr. Grue reaches for different paints to paint both sketches grue he therefore recognizes a difference in the two black-and-white sketches? For, according to this supposition, the fact that he does not know what paints to use would be a good indication that he recognzies no difference between the two sketches.

This type of objection is directed against all the earlier examples which were intended to show that Mr. Grue must have a kind of extrasensory perception in order to be able to use "grue" and "bleen" and which were intended to show that these predicates are temporal predicates. According to this objection, there is no sense in asking Mr. Grue what paints he will use in painting a certain picture grue (as we did in our examples), for he will not know.

This supposition is very strange indeed, for it implies that "grue" is hardly a color predicate at all, so peculiar is its status. However, even if Mr. Grue were this type of person, the earlier points can still be made, if the examples are altered. Suppose we show Mr. Grue a picture in color of the Harvard Yard at Commencement time, but this is to be an idealized picture representing Harvard at no definite time; for example, we see a procession with placards of classes ranging from 1640 to 2500, and there are some buildings of the eighteenth century and some of the twenty-sixth century (the picture is entitled "Harvard through the Ages"). But we screen off most of this picture, allowing Mr. Grue to see only one small plot of grass. If we now ask him to tell us the color of the grass in the picture, then, of course, he cannot do so, for it is not grass of any given date. Now, without his being aware of it, we cleverly change the concealed portion of the picture so that it will be a picture of the Harvard Yard at Commencement time in 1959. Now we ask him the color of the grass. This time, on the assumption that he needs no observable temporal clues from the object itself in order to be able to identify grue objects, he will tell us that the grass in the picture is grue. But he is receiving no more sensory stimulation now than he was before, and yet he is cognizant of a change. For this reason, we attribute to him a faculty of extrasensory perception.

In order to show that this Mr. Grue is using "grue" as a temporal predicate, we can change the previous experiment in the following way. We allow Mr. Grue to see entire both pictures of the Commencement scenes. In the first case, he cannot tell us the color of the grass because it is not part of a scene of Harvard at any definite time. When we now change the picture so that it represents the Commencement of 1959, he can tell us that the grass in the picture is indeed grue. Here Mr. Grue is perceptually aware of the change which, in effect, is merely a temporal one. In the first case, when he was aware of no definite date of the objects in the picture, he could not tell whether to apply the term "grue." But when the picture is changed so that he becomes aware of a definite date for the scene in question, then he can correctly identify the grass in the scene as grue. Thus Mr. Grue is using "grue" as a temporal predicate, for he must be cognizant of the date of the object in the picture in order to be able to tell whether it is grue.

VIII

The suggestion that nontemporal predicates are more legitimately projectable than temporal ones was put forward as an answer to the new riddle of induction. The conclusion which seems to emerge here is that one ought not to dismiss this suggestion in the brusque way in which Goodman has been inclined to do. It is not the case that there is no difference (except as regards their entrenchment in our language) between predicates such as "green" and "grue"; it is not the case that there exists a thorough symmetry between the grue-bleen language and the green-blue language. On the contrary, if the examples discussed above are appropriate, they seem to show that a logically important difference does exist between "grue" and "green," a difference sufficiently important to help us to see why it is that we rightly regard "green" as more projectable than "grue."

In formulating his "new riddle of induction" Goodman of course had in mind a wide range of puzzling predicates in addition to "grue" and "bleen." For instance, one might consider the predicate "condulates electricity," a predicate true of a thing at a given time if and only if the thing then conducts electricity and the time is prior to 2000 A.D. or the thing is then an electrical insulator and the time is not prior to 2000 A.D.; and one might ask why we judge that things made of copper are going to conduct electricity rather than judge that they are going to condulate electricity. An unlimited array of similarly puzzling predicates could be constructed. But whatever these new predicates might be, it would seem that what has been said with regard to "grue" suffices to provide a clue for dealing with each of them.

Thus, for example, we could bring out that "condulates electricity," like "grue," is a temporal predicate in a sense in which "green" and "conducts electricity" are not. Suppose we were to make a picture clearly showing an object properly connected to a battery and ammeter, with the ammeter giving a high reading; but suppose that the background of the picture is concealed from the observer's view, so that nothing is visible to indicate whether the date in the picture is prior to 2000 A.D. or not. An observer then can readily tell that the object in the picture is a conductor of electricity; for that, he does not require clues as to the date in the picture. But a person not possessing extrasensory perception of the background will be wholly unable to tell whether the object in the picture is a condulator of electricity. The point is that it is the temporal clues contained in the picture which are essential to telling whether the object in the picture is a condulator or not. For if the background in the picture is such as to show that the date in the picture is prior to 2000 A.D., then the object in the picture is a condulator of electricity; while if the background is such as to show that the date in the picture is not prior to 2000 A.D., then the object in the picture is not a condulator of electricity. And if even the background of the picture contains no clues as to the date in the picture, then the object in the picture neither is nor is not a condulator of electricity, although it is a conductor. This brings out the sense in which determining whether an object condulates electricity at a given time essentially involves prior ascertaining of what the date is at that time.

Hence it would seem that we may be able in any case we meet to explain the difference between the nontemporal and legitimately projectable predicate and its temporal and less legitimately projectable competitors.

NOTES

1. Nelson Goodman, *Fact, Fiction, and Forecast* (Cambridge, Mass., 1955), Chap. iii. The authors are indebted to Professor Goodman for valuable criticism; he does not accept their conclusions, however.

2. This appears to be what Goodman means when he explains "grue" as applying to "all things examined before *t* just in case they are green but to other things just in case they are blue" (*ibid.*, p. 74).

3. This claim was advanced by Carnap, "On the Application of Inductive Logic," *Philosophy and Phenomenlogical Research*, VIII (1947), 133–47; he put it in a stronger form, implying that it is a necessary condition for projectability of a predicate that it be nontemporal.

Haskell Fain

THE VERY THOUGHT OF GRUE*

With *Fact, Fiction and Forecast,*[1] Nelson Goodman bequeathed to us a predicate which, as he correctly put it at the time, was less familiar than "green" —namely, "grue." Since then, the volume of literature on "grue" has made the predicate all too horrifyingly familiar. It threatens to becomes as well entrenched as "green." Should that happen, we would really be in serious inductive trouble. The purpose of this paper is to prevent it from happening.

Which objects are grue? As Goodman originally defined it, the predicate "grue" applies to "all things examined before t just in case they are green but to other things just in case they are blue" (p.74). Stated less elliptically, grue objects are those which are either green and examined prior to t, or blue and not examined prior to t.[2] For making philosophical capital of "grue" in the way that Goodman does, one must assume that nothing can be both green all over and blue all over at the same time. "Grue" can be redefined, however, in a way that makes the kind of concept introduced by the predicate much easier to grasp, while still preserving the uses to which Goodman puts the predicate. Let us define "grue" as "either green and examined prior to t, or not green and not examined prior to t." This way of defining "grue" amounts to the following:

(1) x is grue $=_{df} x$ is green $\equiv$ x is examined prior to t

Imagine a bag of marbles from which we are drawing a sample. Suppose ten and only ten marbles, all of them green, have been drawn prior to some particular time t. Let us try to list the information we have. We could begin with the following itemization:

---

*I am indebted to Anatole Beck, Professor of Mathematics at the University of Wisconsin, for some valuable criticisms of an earlier draft of this paper.

Reprinted from the *Philosophical Review* (1967) by permission of the editor and the author.

Marble 1 is drawn from the bag and marble 1 is green and the time at which marble 1 is examined is prior to t.
Marble 2 is drawn from the bag and marble 2 is green and the time at which marble 2 is examined is prior to t.
And so forth.
Marble 10 is drawn from the bag and marble 10 is green and the time at which marble 10 is examined is prior to t.

Suppose the tenth marble is drawn by time t and there is no time left to draw another. Query: Do we have confirmatory evidence, in the Nicodian sense, that the next marble to be drawn, the eleventh, is green?

Consider the following argument. We have confirmatory evidence that

(2) All the marbles in the bag are green.

Since the eleventh marble is in the bag, we have confirmatory evidence that it too is green. Nothing could be more simple.

Now consider this argument. The same body of evidence confirms equally that

(3) All the marbles in the bag are grue.

The sentence describing the first draw logically implies:

Marble 1 is drawn from the bag and (marble 1 is green if and only if the time at which marble 1 is examined is prior to t).

If we continue in this way, we have a second list of ten sentences which constitute confirmatory evidence, in the Nicodian sense, that

(4) (x) (x is a marble in the bag $\supset$ [x is green $\equiv$ x is examined prior to t]).

Well, perhaps not quite. One could make a Moore-ish point that we have not yet said whether all the marbles *drawn from* the bag were *in* the bag. They are (or were). By the definition given in (1), (4) and (3) are extensionally equivalent. So we have confirmatory evidence that all the marbles in the bag are grue. The "riddle of induction," as Goodman terms it, is this: If (4) is as equally well confirmed as (2), then given that the eleventh marble is *not* examined prior to t, we have confirmatory evidence that the eleventh marble in the bag is *not* green. But the same evidence is confirmatory, to the same degree, that the eleventh marble *is* green. And on the basis of the evidence so far stated, there appears to be no reason for "projecting" the predicate "green" rather than the predicate "grue" to the eleventh marble.

Note first that we have not yet said whether we are sampling with or without replacement (nor does Goodman in *Fact, Fiction and Forecast*). Sampling with replacement allows for the possibility that the eleventh marble to be examined after t is identical with, say, the second marble examined prior to t. But then we might not know, when we examined it after t, whether we had examined it prior to t, or what its color was. Suppose it had changed color. We might never be the wiser. Thus, if we examined it after t and discovered it to be red, we still could not say whether it was grue, because if it had been examined prior to t and had been green, it would still be grue. Only we could not know it. Another possibility is that the first marble examined prior to t is identical with the fourth marble examined prior to t. Suppose this were the case, and the marble turned color from green to red between the time it was first examined and replaced in the bag and the time it was next examined. Given the definition of "grue," we would be obliged to say that the marble was both grue and not grue, assuming in this case that we could recognize it the second time around. To block this possibility, we could attempt to define a new predicate "x is grue at t" by means of the predicate "x is green at t if and only if x is examined prior to t." I don't know what trouble this would entail.

If one believes that objects at different temporal locations cannot be identical with each other, then some of, but not all, the above complications disappear. I myself prefer objects which stick around for a while, but let us try to avoid ontological persiflage. The shortest detour is to conceptualize the procedure as one of sampling without replacement. Of course, if one holds that objects at different temporal locations cannot be identical with each other, then one is always sampling without replacement. There are then, however, surprising numbers of marbles in the bag(s).

The sampling procedure is without replacement. Do we, on the basis of the total evidence, have the same reason for believing the eleventh marble to be green as we have for believing the eleventh marble to be non-green? On the basis of the evidence stated in the first and second lists, we do. The first and second lists, however, do not contain all the evidence. We have not yet stated our evidence that *there are marbles left in the bag*, marbles that can be examined after t. Those slippery lumps, those pleasant clicks as we shake the bag, are all evidence that we have not yet examined all the marbles prior to t. Let us draw up a new list of sentences which state this evidence. Notice now that sentence (4), (x) (x is a marble in the bag $\supset$ [x is green $\equiv$ x is examined prior to t]), logically implies the sentence "All the marbles in the bag are green if and only if all the marbles in the bag are examined prior to t"; that is:

(5) [(x) (x is a marble in the bag $\supset$ x is green)] $\equiv$ [(x) (x is a marble in the bag $\supset$ x is examined prior to t)].

The first list we compiled contains the confirmatory evidence that the left-hand side of biconditional (5) is true. The list just mentioned contains the evidence that the right-hand side of biconditional (5) is false. Because biconditionals are false when their opposite sides have opposite truth values, combining the two lists produces disconfirmatory evidence for (5). Since (4) implies (5) and we have evidence that (5) is false, we also have evidence that (4) is false. But (4) is extensionally equivalent with sentence (3), "all the marbles in the bag are grue." The evidence apparently favors the hypothesis that all the marbles in the bag are green over the hypothesis that all the marbles in the bag are grue.

This argument will not go unchallenged. With "grue," there is always a parallel argument. Let us take a look at one. Suppose we start with the logically true sentence

(6) $(x)(x \text{ is grue} \equiv x \text{ is grue})$.

Universal instantiation and application of the definition stated in (1) yields:

(7) $x \text{ is grue} \equiv (x \text{ is green} \equiv x \text{ is examined prior to } t)$.

A judicious number of commutations and associations yields:

(8) $x \text{ is green} \equiv (x \text{ is grue} \equiv x \text{ is examined prior to } t)$.

(8) truth-functionally implies:

(9) $(x \text{ is a marble in the bag} \supset x \text{ is green}) \equiv (x \text{ is a marble in the bag} \supset [x \text{ is grue} \equiv x \text{ is examined prior to } t])$.

Universal generalization yields a sentence which, in turn, logically implies:

(10) $[(x)(x \text{ is a marble in the bag} \supset x \text{ is green})] \equiv [(x)(x \text{ is a marble in the bag} \supset [x \text{ is grue} \equiv x \text{ is examined prior to } t])]$.

Since (10) is deducible from (6), and (6) is a logical truth, (10) states a logical equivalence between "all marbles in the bag are green" and the sentence "$(x)(x \text{ is a marble in the bag} \supset [x \text{ is grue} \equiv x \text{ is examined prior to } t])$." In particular, according to (10), the sentence "all the marbles in the bag are green" logically implies sentence

(11) $(x)(x \text{ is a marble in the bag} \supset [x \text{ is grue} \equiv x \text{ is examined prior to } t])$.

But (11) implies

(12) All the marbles in the bag are grue $\equiv$ all the marbles in the bag are examined prior to t.

The parallel argument will go something like this. The argument to show that there is disconfirmatory evidence for "all marbles are grue" begs the question. In that argument, we had to assume that we *had* confirmatory evidence that all marbles in the bag are green. This is just the point at issue, though, because we also had confirmatory evidence that all the marbles in the bag are grue. For since "all the marbles in the bag are green" implies (12), and we have some evidence that the left-hand side of biconditional (12) is true and some evidence that the right-hand side of the biconditional is false, we have some evidence that the biconditional itself is false. Inasmuch as "all the marbles in the bag are green" implies (12), and we have evidence that (12) is false, we have evidence that counts against "all marbles in the bag are green." We are apparently back where we started. With a difference, though. If the parallel argument succeeds, and I shall argue that it does, the riddle of induction is transformed into what might be called "the paradox of induction." The riddle of induction was that the same body of evidence apparently confirms, to an equal degree, two incompatible hypotheses. "The paradox of induction," as I shall call it, is that the *same* body of evidence both supports and undermines the *same hypothesis*.

Suppose the body of evidence consists of three lists of true sentences. List A contains ten true sentences of the kind: x is a marble and x is green and x is examined prior to t. List B contains ten true sentences of the form: x is a marble and (x is green $\equiv x$ is examined prior to t). It is harder to state the form of the sentences contained in List C. List C presumably contains a set of true sentences which confer a probability greater than one-half that the bag contains some marbles to be examined after *t*, marbles that were not examined prior to *t*. To make things simple, we will assume that List C contains one and only one sentence which we know is true. The form of the sentence will be: The bag contains *n* number of marbles to be examined after *t*. As we are sampling without replacement, we will assume that none of the 0, 1, 2, 3, . . . *n* marbles left in the bag have been examined prior to *t*.

We can forget the case where the bag contains no marbles to be examined after *t*, for then the sentence "All the marbles in the bag are grue" is extensionally equivalent with "All the marbles in the bag are green." Matters become interesting only when there are marbles yet to be examined. Suppose, then, that there are *n* marbles left in the bag, where $n \neq 0$. Suppose also that the evidence favors both "All the marbles in the bag are green" and "All the marbles in the bag are grue," to the same degree. Then, by the main argument, the same evidence counts against "All the marbles in the bag are grue." And by the parallel argument, the same evidence counts against "All marbles in the bag are green." So *if* the body of evidence is favorable, and in the same degree, for both "All the marbles in the bag are green" *and* "All the marbles in the bag are grue," then the same evidence counts against "All the marbles in the bag are green." It also counts against "All the marbles in the bag are grue."

This result is not merely a riddle; it is a genuine paradox.[3] Accordingly we know that the hypothesis must be false—that the evidence is favorable, and in the same degree, to both "All the marbles in the bag are green" and "All the marbles in the bag are grue." This leaves only two alternatives. One alternative is that the total evidence contained in the three lists is neither favorable nor unfavorable to "All the marbles in the bag are green" as well as neither favorable nor unfavorable to "All the marbles in the bag are grue." The second alternative is that the total evidence discriminates between "All the marbles in the bag are green" and "All the marbles in the bag are grue," conferring different degrees of probability upon them.

The inductive skeptic, of course, chooses the first alternative. "What all this shows," he will say, "is that positive instances are never confirmatory in any degree; I knew that all along." The second alternative is chosen by Goodman. Positive evidence, he thinks, is confirmatory of sentences of the form "All *A*'s are green" and is not confirmatory of sentences of the form "All *A*'s are grue" because the predicate "green" is better entrenched in our language. "Plainly 'green,'" he writes, "as a veteran of earlier and many more projections than 'grue,' has the more impressive biography. The predicate 'green,' we may say, is much better *entrenched*, than the predicate 'grue'" (p.95).

What Goodman is suggesting, I think, is that it would be possible for "grue" to become entrenched in a language if there were many, many occasions when it was successfully projected. Suppose there were a universe in which draws from bags of marbles very often gave first a run of green marbles and, after a certain specified time *t*, nothing but non-green marbles. Suppose also there were some other bags of marbles, but many fewer in number than the former kind, from which one could continue to draw green marbles no matter what the time. Imagine now a society of marble-counters whose sole preoccupation and entire experience consisted in drawing marbles from bags. The language spoken by members of the society would, no doubt, consist of very few words, but "grue" might be among them. Perhaps, in such a society, there would be a universal disposition to project "grue" to the marbles examined after a specified time *t*, rather than the predicate "green." A disposition to project, however, seems to be one thing, confirmatory evidence something else.

Another possibility is that the paradox of induction depends upon the introduction of the definition of "grue" in the parallel argument (p.64). If the parallel argument depends upon such introduction, then blocking it would remove the paradox. If one could not deduce sentence (12), "All the marbles in the bag are grue if and only if all the marbles in the bag are examined prior to *t*," from "All the marbles in the bag are green," then the lesson perhaps is that one cannot introduce *ad hoc* definitions without expecting trouble. If the introduction of the definition led to the contradiction that the same body of evidence counts both for and against the same hypothesis, then we have a very good reason for rejecting the definition.

It is true, of course, that sentence (12) cannot be deduced from "All the marbles in the bag are green" without the aid of the definition of "grue" because (12) contains the expression "grue" in a strategic way. Nonetheless, a sentence extensionally equivalent with sentence (12) can be deduced from "All the marbles in the bag are green" alone. The sentence "All the marbles in the bag are green" logically implies

$$(13)\quad [(x)\,(x \text{ is a marble in the bag} \supset [x \text{ is green} \equiv x \text{ is examined prior to } t])] \equiv [(x)\,(x \text{ is a marble in the bag} \supset x \text{ is examined prior to } t)].$$

Inasmuch as (12) is extensionally equivalent with (13), the paradox of induction makes an unwelcome entry without waiting for its proper introduction by means of the predicate "grue."

This sobering consideration eliminates the theory that the paradox depends upon definitional promiscuity. It also casts doubt on Goodman's theory that "All the marbles in the bag are green" receives positive confirmation over "All the marbles in the bag are grue" because "green" is more entrenched in our language than "grue."

The definition of "grue" is logically innocuous. It seems, though, to have generated all sorts of unnecessary philosophical difficulties. To forestall the objection that the riddle of induction arises only because the definition of "grue" contains the temporal predicate "examined prior to *t*," Goodman quite correctly pointed out that the general problem does not depend upon this fact. He stated the point, however, in a peculiar way. To paraphrase him: If we start with "green," then "grue" will be explained in terms of "green" and a temporal term. But, he went on, if we start with "grue," then "green" will be explained in terms of "grue" and a temporal term. "Thus qualitativeness," he concluded, "is an entirely relative matter and does not by itself establish any dichotomy of predicates. This relativity seems to be completely overlooked by those who contend that the qualitative character of a predicate is a criterion for its good behavior" (p.79).

Is qualitativeness a relative matter because "x is green" can be defined by means of the predicate "x is grue $\equiv x$ is examined prior to t," given that "x is grue" is extensionally equivalent with "x is green $\equiv x$ is examined prior to t"? Many people apparently think so; as a result, there has been a running controversy over whether grue things look alike to speakers of a grue language, whether "green" is really a temporal predicate, and so on. I am not averse to going into blue and brown studies about the nature of the relationship of language to one's perceptions of reality. In the case of "grue," though, the response is disproportionate to the stimulus. To show this, let me introduce a new concept —the Concept of Dred—by means of the following definition: x is dred $=_{df}$(x is red $\equiv x$ is not red). The concept of dred is obviously a self-contradictory one. Equally obvious, the predicate "red" is extensionally equivalent with "x is dred $\equiv x$ is not red." So, I shall argue, contradictoriness is a relative matter because if we start with "dred" instead of with "red," then. . . .

Perhaps a Hegelian would welcome a short proof that even one of the simplest of our concepts contains within it the seeds of a contradiction. Why should we maintain that it is better to be red than dred?

Unfortunately, the puzzle has a prosaic solution. Though the predicate "x is dred $\equiv x$ is not red" contains a self-contradictory predicate, the predicate itself is not self-contradictory. For it is, by the original definition, extensionally equivalent with "(x is red $\equiv x$ is not red) $\equiv x$ is not red," which in turn is equivalent with "x is red." Analogous considerations apply to the question of whether "green" is a temporal predicate and related things. Though "x is green" is extensionally equivalent with "x is grue if and only if x is examined prior to t," and the latter contains not just one but two "temporal" expressions, the expression itself is not a temporal one any more than is "x is dred if and only if x is not red" a self-contradictory expression.

Matters stand thus. If we assume that the initial probability of the sentence "All the marbles in the bag are green" is equal to the initial probability of "All the marbles in the bag are (green $\equiv$ they are examined prior to t)," and we also assume that the total evidence contained in lists A, B, and C favors both hypotheses in the same degree, then we can conclude, by means of the main and parallel argument, that the very same evidence both favors and counts against "All the marbles in the bag are green." It also favors and counts against (4). Hence, either the evidence favors neither hypothesis, or the total evidence does discriminate between them. Let us try one more time to effect the discrimination.

The bag contains marbles. Ten and only ten are examined prior to t. All the marbles examined are green. None has been replaced in the bag. Furthermore, what is most important, we must assume that *before* we began the draw, the hypothesis that all the marbles in the bag are green was just as likely as the hypothesis that (4) is ture.

Note that sentence (4) is logically equivalent with:

> All the *green* marbles are examined prior to *t and* all the marbles in the bag that are examined prior to *t* are green.

Now why should only the green marbles have been drawn prior to *t*? Given that there are eleven marbles in the bag, only ten of which can be drawn prior to *t*, a person who at the beginning of the experiment is betting even money on the hypothesis that all the marbles are grue is committing himself to the proposition that the bag contains ten green marbles and one non-green marble. After ten draws, he might begin to wonder why the non-green marble has not presented itself, for it could have been the first marble drawn, or the second, and so on. In fact, there are eleven ways in which the non-green marble could have been placed. We are given that it was not among the first ten. If, then, there are ten green marbles and one non-green marble in the bag, then an extremely unlikely sample has been drawn. The likelihood of that kind of sample is 1/11. The probability that the eleventh marble is non-green, after ten green marbles have been drawn, is 1/11. But the probability that the eleventh marble is green, given that the first ten were green, is the probability that all the marbles in the bag are green. And the probability that the eleventh marble is non-green, given that ten green marbles have been drawn, is the probability that all the marbles are grue. The total evidence confers a probability of 10/11 on the sentence "All the marbles are green." The total evidence confers a probability of 1/11 on the sentence "All the marbles are grue," if we assume that initially both hypotheses were equally likely. Thus it appears after all as if the total evidence does discriminate between "All the marbles in the bag are green" and "All the marbles in the bag are grue."

The argument just stated may seem plausible, but it is unsound. It is true that *if* the bag contains eleven marbles, then there are eleven different ways of drawing a sample containing ten marbles; and *if* the bag contains one and only one marble that is not green, then ten out of eleven possible samples contain that marble. Affirming the consequent, however, is as fallacious in inductive logic as elsewhere. We cannot argue that because we have drawn a sample containing ten green marbles, it is one of eleven possible samples, ten of which contain a marble that is not green. Suppose, on the other hand, we know that the bag of eleven marbles is itself a random sample drawn from a large population in which the ratio of green to non-green marbles is eleven to one. Under these circumstances, we have a basis for assuming that the initial likelihood that all eleven are green is equal to the initial likelihood that ten of eleven are green. Then, if we continued to draw nothing but green marbles, we would with each draw increase harmonically the likelihood that all are green. At the same time, we would be harmonically decreasing the likelihood that the bag contains ten green marbles and one non-green marble. Thus, before beginning to draw, the probability that all are green would be equal to the probability that ten and only ten are green. If ten green marbles are drawn, the probability that all eleven are green is then 11/12, while the probability that ten and only ten are green is at that point only 1/12.

The above argument is instructive because it presents a case in which a body of evidence does discriminate between two incompatible hypotheses, provided we have information about initial probabilities. If the probability of drawing a green marble is, for example, eleven times greater than the probability of drawing a marble that is not green, then as we continue to draw green marbles from a sample containing eleven marbles, we increase with each draw the probability that all eleven are green, and decrease the probability that ten and only ten are green. This is the way induction works. We could never, however, increase the probability that all eleven are green beyond the probability that any one of the marbles is green. Hence everything depends upon what estimates we make of the initial likelihoods of two hypotheses *before* we begin to collect the evidence that is supposed to decide between them.

Now, the crucial question is whether the above specimen of inductive reasoning can show the kinds of circumstances in which one could increase the likelihood that all are green while, at the same time, *decreasing the likelihood* that all are grue.

The answer, unfortunately, is "no," for in the above example the hypothesis that all the marbles in the bag are grue has been surreptitiously replaced by the hypothesis that the bag of eleven marbles contains ten and only ten green marbles. The grue hypothesis is that *the first ten to be drawn* (prior to t) are green *and* the eleventh marble (drawn after t) is not green, whereas the hypothesis we have been considering is that ten of eleven are green. These are not equivalent, though the former implies the latter. If the probability of drawing a green marble, at any draw, equals 11/12, then the initial likelihood of drawing eleven in a row is $(11/12)^{11}$. The initial likelihood of drawing ten green marbles and one non-green marble, where the order of draw is of course disregarded, is $11 \times (11/12)^{10} \times (1/12)$. These likelihoods are the same. But the initial likelihood that all the marbles are grue is $(11/12)^{10} \times (1/12)$, for according to the grue hypothesis the order of draw is crucial—green marbles prior to t, and non-green marbles after t. Thus the initial likelihood that all are grue is one-eleventh the initial likelihood that all are green. If ten green marbles are drawn, the probability that all are green at that point is, as noted above, 11/12. The probability that all are grue, however, is then 1/12. Thus, though we have increased the probability that all are green by drawing ten green marbles in a row, we have also increased the probability that all are grue. The probability that all are grue remains, at the end of the experiment, one-eleventh of the probability that all are green.

By now, one should be thoroughly convinced that all exits are blocked. There is no way to collect a body of evidence which will discriminate between "All are green" and "All are grue." Therefore, there is only one remaining alternative: namely, that no matter how many green marbles or emeralds we observe prior to some time t, the evidence alone does not favor the green hypothesis over the grue hypothesis. This result has nothing at all to do with the relative "entrenchment" of either "green" or "grue." To see what is at stake here, let us suppose that the hypothesis that all the marbles in the bag are green is initially as likely as the hypothesis that all the marbles in the bag are grue. Let the probability that any one marble is green $= m/n$. Accordingly, the probability that any one marble is not green $= (1 - m/n)$. Then the probability of drawing eleven green marbles in a row is $(m/n)^{10} \times (m/n)$. The pobability of drawing ten green marbles and *then* drawing one that is not green (the grue hypothesis) is $(m/n)^{10} \times (1 - m/n)$. If the green hypothesis and the grue hypothesis are initially equally likely, then $(m/n)^{10} \times (m/n) = (m/n)^{10} \times (1 - m/n)$. In particular, $(m/n) = (1 - m/n) = 1/2$.

Thus, if it is as equally likely to begin with that all marbles or emeralds or whatever are green as that they are grue, it is no surprise that accumulation of positive instances does not discriminate between the two hypotheses. If one thought otherwise, one would be committing a sophisticated version of the old gambler's fallacy.

We can now see the basis on which the paradox of induction rests. A man, born of a sudden and warily approaching his first bag of marbles, would have no reason to suppose it any more likely that all the marbles were green than he would have for supposing that, say, the first fifty marbles to be drawn were green and the remainder nongreen. If he should then proceed to draw nothing but green marbles in the first fifty draws, he would, as a matter of fact, revise upward his original expectation that all are green. If, for example, he were initially disposed to wager even money on either hypothesis, then after observing fifty green marbles drawn from the bag, he would no longer be so disposed. The fundamental question is whether he could justify the revision of his expectations by appealing to what he had observed. I do not think there is much doubt that, in such situations, people will revise their initial estimates—this could be tested empirically. But can anyone justify it by a simple appeal to what has been observed? If it is just as likely initially that all are green as it is that the first n drawn are green and the remaining non-green, then it can be proved that the probability of drawing a green marble on any draw $= 1/2$. But if the probability of drawing a green marble on any one draw $= 1/2$, then it is not reasonable to appeal to what is actually observed for supporting a change in one's expectations, for this would be nothing but the gambler's fallacy. The grue situation is one way of dramatizing this, and it is always possible to construct a grue-like predicate. For the above example, the predicate would be: x is green if and only if x is among the first fifty marbles drawn from the bag. One could then proceed to show that the evidence could not provide the basis for discriminating between the hypothesis that all the marbles are green, and the hypothesis that for all x, if x is a marble in the bag, then x is green if and only if x is among the first fifty drawn. The evidence cannot possibly provide a basis for effecting the discrimination, though this would not pose a problem if we had reason to believe that the hypotheses were not initially equally likely. What comes to the same thing, one must have some reason for believing that in the universe as a whole the ratio of green to non-green marbles is not one to one.

But how could we estimate this ratio without first observing the contents of many bags of marbles? We are in the position of a dog chasing its tail. It seems obvious that the justification of induction must rest upon the reasons we might have, if we have them, for supposing that not all hypotheses are, to begin with, equally likely. Thus the justification of each induction must lie outside of the particular induction itself.

NOTES

1. Cambridge, Mass., 1955.

2. I say "less elliptically" somewhat hesitantly because a number of "less elliptic" paraphrases of Goodman's casual definition of "grue" in *Fact, Fiction and Forecast* have appeared in the literature. On one interpretation, for example, grue objects are those which are green prior to some time t and blue after t. The particular paraphrase used here is, I think, the closest to the text, although it may not adequately represent the concept Goodman actually had in mind. My analysis of the riddle of induction depends upon defining "grue" exactly the way in which I do.

3. Perhaps it would be wise, at this point, to pre-empt a cavil. It might be argued that there is nothing paradoxical in the same body of evidence being both confirmatory and disconfirmatory of the same hypothesis. Consider, for example, a generalization of the form $(x)\ (P[x] \supset Q[x])$, and two true singular sentences of the form $P(a)\ \&\ Q(a)$; $P(b)\ \&\ {-Q(b)}$. The two singular sentences taken together constitute a body of evidence both confirmatory and disconfirmatory of the generalization. In this case, however, the "body of evidence" can be divided into two nonintersecting subsets, one of which contains all the confirmatory evidence and the other of which contains all the disconfirmatory evidence. Suppose, on the other hand, that no such subdivision is possible: e.g., the case of the grue and green hypotheses.

The Acceptance of Scientific Theories

Richard Rudner

THE SCIENTIST *QUA* SCIENTIST MAKES VALUE JUDGMENTS*

The question of the relationship of the making of value judgments in a typically ethical sense to the methods and procedures of science has been discussed in the literature at least to that point which e. e. cummings somewhere refers to as "The Mystical Moment of Dullness." Nevertheless, albeit with some trepidation, I feel that something more may fruitfully be said on the subject.

In particular the problem has once more been raised in an interesting and poignant fashion by recently published discussions between Carnap[1] and Quine[2] on the question of the ontological commitments which one may make in the choosing of language systems.

I shall refer to this discussion in more detail in the sequel; for the present, however, let us briefly examine the current status of what is somewhat loosely called the "fact-value dichotomy."

I have not found the arguments which are usually offered, by those who believe that scientists do essentially make value judgments, satisfactory. On the other hand the rebuttals of some of those with opposing viewpoints seem to have had a least at *prima facie* cogency although they too may in the final analysis prove to have been subtly perverse.

Those who contend that scientists do essentially make value judgments generally support their contentions by either

(a) pointing to the fact that our having a science at all somehow "involves" a value judgment, or

---

*The opinions or assertions contained herein are the private ones of the writer and are not to be construed as official or reflecting the views of the Navy Department or the Naval Establishments at large.

(b) by pointing out that in order to select, say among alternative problems, the scientist must make a value judgment; or (perhaps most frequently)
(c) by pointing to the fact that the scientist cannot escape his quite human self—he is a "mass of predilections," and these predilections must inevitably influence all of his activities not excepting his scientific ones.

To such arguments, a great many empirically oriented philosophers and scientists have responded that the value judgments involved in our decisions to have a science, or to select problem (a) for attention rather than problem (b) are, *of course*, extra-scientific. If (they say) it is necessary to make a decision to have a science before we can have one, then this decision is literally pre-scientific and the act has thereby certainly not been shown to be any part of the *procedures* of science. Similarly the decision to focus attention on one problem rather than another is extra-problematic and forms no part of the procedures involved in dealing with the problem *decided* upon. Since it is *these* procedures which constitute the method of science, value judgments, so they respond, have not been shown to be involved in the scientific method as such. Again, with respect to the inevitable presence of our predilections in the laboratory, most empirically oriented philosophers and scientists agree that this is "unfortunately" the case; but, they hasten to add, if science is to progress toward objectivity the influence of our personal feelings or biases on experimental results must be minimized. We must try not to let our personal idiosyncrasies affect our scientific work. The perfect scientist—the scientist *qua* scientist does not allow this kind of value judgment to influence his work. However much he may find doing so unavoidable *qua* father, *qua* lover, *qua* member of society, *qua* grouch, *when* he does so he is not behaving *qua* scientist.

As I indicated at the outset, the arguments of neither of the protagonists in this issue appear quite satisfactory to me. The empiricists' rebuttals, telling prima facie as they may against the specific arguments that evoke them, nonetheless do not appear ultimately to stand up, but perhaps even more importantly, *the original arguments* seem utterly too frail.

I belive that a much stronger case may be made for the contention that value judgments are essentially involved in the procedures of science. And what I now propose to show is that scientists as scientists *do* make value judgments.

Now I take it that no analysis of what constitutes the method of science would be satisfactory unless it comprised some assertion to the effect that the scientist as scientist accepts or rejects hypotheses.

But if this is so then clearly the scientist as scientist does make value judgments. For, since no scientific hypothesis is ever completely verified, in accepting a hypothesis the scientist must make the decision that the evidence is *sufficiently* strong or that the probability is *sufficiently* high to warrant the acceptance of the hypothesis. Obviously our decision regarding the evidence and respecting how strong is "strong enough," is going to be a function of the *importance*, in the typically ethical sense, of making a mistake in accepting or rejecting the hypothesis. Thus, to take a crude but easily managable example, if the hypothesis under consideration were to the effect that a toxic ingredient of a drug was not present in lethal quantity, we would require a relatively high degree of confirmation or confidence before accepting the hypothesis—for the consequences of making a mistake here are exceedingly grave by our moral standards. On the other hand, if, say, our hypothesis stated that, on the basis of a sample, a certain lot of machine stamped belt buckles was not defective, the degree of confidence we should require would be relatively not so high. *How sure we need to be before we accept a hypothesis will depend on how serious a mistake would be.*

The examples I have chosen are from scientific inferences in industrial quality control. But the point is clearly quite general in application. It would be interesting and instructive, for example, to know just how high a degree of probability the Manhattan Project scientists demanded for the hypothesis that no uncontrollable pervasive chain reaction would occur, before they proceeded with the first atomic bomb detonation or first activated the Chicago pile above a critical level. It would be equally interesting and instructive to know why they decided that *that* probability value (if one was decided upon) was high enough rather than one which was higher; and perhaps most interesting of all to learn whether the problem in this form was brought to consciousness at all.

In general then, before we can accept any hypothesis, the value decision must be made in the light of the seriousness of a mistake, that the probability is *high enough* or that, the evidence is *strong enough*, to warrant its acceptance.

Before going further, it will perhaps be well to clear up two points which might otherwise prove troublesome below. First I have obviously used the term "probability" up to this point in a quite loose and pre-analytic sense. But my point can be given a more rigorous formulation in terms of a description of the process of making statistical inference and of the acceptance or rejection of hypotheses in statistics. As is well known, the acceptance or rejection of such a hypothesis presupposes that a certain level of significance or level of confidence or critical region be selected.[3]

It is with respect at least to the *necessary* selection of a confidence level or interval that the necessary value judgment in the inquiry occurs. For, "the size of the critical region (one selects) is related to *the risk one wants to accept* in testing a statistical hypothesis."[3*, p. 435.]

And clearly how great a risk one is willing to take of being wrong in accepting or rejecting the hypothesis will depend upon how seriously in the typically ethical sense one views the consequences of making a mistake.

I believe, of course, that an adequate rational reconstruction of the procedures of science would show that every scientific inference is properly construable as a statistical inference (i.e., as an inference from a set of characteristics of a sample of a population to a set of characteristics of the total population) and that such an inference would be scientifically in control only in so far as it is statistically in control. But it is not necessary to argue this point, for even if one believes that what is involved in some scientific inferences is not statistical probability but rather a concept like strength of evidence or degree of confirmation, one would still be concerned with making the decision that the evidence was *strong enough* or the degree of confirmation *high enough* to warrant acceptance of the hypothesis. Now, many empiricists who reflect on the foregoing considerations agree that acceptances or rejections of hypotheses do essentially involve value judgments, but they are nonetheless loathe to accept the conclusion. And one objection which has been raised against this line of argument by those of them who are suspicious of the intrusion of value questions into the "objective realm of science," is that actually the scientist's task is only to *determine* the degree of confirmation or the strength of the evidence which *exists* for an hypothesis. In short, they object that while it may be a function of the scientist *qua member of society* to decide whether a degree of probability associated with the hypothesis is high enough to warrant its acceptance, *still* the task of the scientist *qua* scientist is *just the determination* of the degree of probability or the strength of the evidence for a hypothesis and not the acceptance or rejection of that hypothesis.

But a little reflection will show that the plausibility of this objection is merely apparent. For the determination that the degree of confirmation is say, *p*, or that the strength of evidence is such and such, which is on this view being held to be the indispensable task of the scientist *qua* scientist, is clearly nothing more than *the acceptance by the scientist of the hypothesis that the degree of confidence is p or that the strength of the evidence is such and such;* and as these men have conceded, acceptance of hypotheses does require value decisions. The second point which it may be well to consider before finally turning our attention to the Quine-Carnap discussion has to do with the nature of the suggestions which have thus far been made in this essay. In this connection, it is important to point out that the preceeding remarks do *not* have as their import that an empirical description of every present day scientist ostensibly going about his business would include the statement that he made a value judgment at such and such a juncture. This is no doubt the case; but it is a hypothesis which can only be confirmed by a discipline which cannot be said to have gotten extremely far along as yet; namely, the Sociology and Psychology of Science, whether such an empirical description is warranted, cannot be settled from the armchair.

My remarks have, rather, amounted to this: Any adequate analysis or (if I may use the term) rational reconstruction of the method of science must comprise the statement that the scientist *qua* scientist accepts or rejects hypotheses; and further that an analysis of that statement would reveal it to entail that the scientist *qua* scientist makes value judgments.

I think that it is in the light of the foregoing arguments, the substance of which has, in one form or another, been alluded to in past years by a number of inquirers (notably C. W. Churchman, R. L. Ackoff, and A.Wald), that the Quine-Carnap discussion takes on heightened interest. For, if I understand that discussion and its outcome correctly, although it apparently begins a good distance away from any consideration of the fact-value dichotomy, and although all the way through it both men touch on the matter in a way which indicates that they believe that questions concerning the dichotomy are, if anything, merely tangential to their main issue, yet it eventuates with Quine by an independent argument apparently in agreement with at least the conclusion here reached and also apparently having forced Carnap to that conclusion. (Carnap, however, is expected to reply to Quine's article and I may be too sanguine here.)

The issue of ontological commitment between Carnap and Quine has been one of relatively long standing. In this recent article,[1] Carnap maintains that we are concerned with two kinds of questions of existence relative to a given language system. One is what *kinds* of entities it would be permissable to speak about as existing when that language system is used, i.e., what kind of *framework* for speaking of entities should our system comprise. This, according to Carnap, is an *external* question. It is the *practical* question of what sort of linguistic system we want to choose. Such questions as "are there abstract entities?" or "are there physical entities?" thus are held to belong to the category of external questions. On the other hand, having made the decision regarding which linguistic framework to adopt, we can then raise questions like "are there any black swans?" "What are the factors of 544?" etc. Such questions are *internal* questions.

For our present purposes, the important thing about all of this is that while for Carnap *internal* questions are theoretical ones, i.e., ones whose answers have cognitive content, external questions are not theoretical at all. They are *practical questions*—they concern our decisions to employ one language structure or another. They are of the kind that face us when for example we have to decide whether we ought to have a Democratic or a Republican administration for the next four years. In short, though neither Carnap nor Quine employ the epithet, they are *value questions.*

Now if this dichotomy of existence questions is accepted Carnap can still deny the essential involvement of the making of value judgments in the procedures of science by insisting that concern with *external* questions, admittedly necessary and admittedly axiological, is nevertheless in some sense a pre-scientific concern. But most interestingly, what Quine then proceeds to do is to show that the dichotomy, as Carnap holds it, is untenable. This is not the appropriate place to repeat Quine's arguments which are brilliantly presented in the article referred to. They are in line with the views he has expressed in his "Two Dogmas of Empiricism" essay and especially with his introduction to his recent book, *Methods of Logic.* Nonetheless the final paragraph of the Quine article I'm presently considering sums up his conclusions neatly:

> Within natural science there is a continuum of gradations, from the statements which report observations to those which reflect basic features say of quantum theory or the theory of relativity. The view which I end up with, in the paper last cited, is that statements of ontology or even of mathematics and logic form a continuation of this continuum, a continuation which is perhaps yet more remote from observation than are the central principles of quantum theory or relativity. The differences here are in my view differences only in degree and not in kind. Science is a unified structure, and in principle it is the structure as a whole, and not its component statements one by one, that experience confirms or shows to be imperfect. Carnap maintains that ontological questions, and likewise questions of logical or mathematical principle, are questions not of fact but of choosing a

> convenient conceptual scheme or framework for science; and with this I agree only if the same be conceded for every scientific hypothesis. (n. 2, pp. 71–72.)

In the light of all of this I think that the statement that *Scientists qua Scientists* make value judgments is also a consequence of Quine's position.

Now, if the major point I have here undertaken to establish is correct, then clearly we are confronted with a first order crisis in science and methodology. The positive horror which most scientists and philosophers of science have of the intrusion of value considerations into science is wholly understandable. Memories of the (now diminished but a certain extent still continuing) conflict between science and, e.g., the dominant religions over the intrusion of religious value considerations into the domain of scientific inquiry, are strong in many reflective scientists. The traditional search for objectivity exemplifies science's pursuit of one of its most precious ideals. But for the scientist to close his eyes to the fact that scientific method *intrinsically* requires the making of value decisions, for him to push out of his consciousness the fact that he does make them, can in no way bring him closer to the ideal of objectivity. To refuse to pay attention to the value decisions which *must* be made, to make them intuitively, unconsciously, haphazardly, is to leave an essential aspect of scientific method scientifically out of control.

What seems called for (and here no more than the sketchiest indications of the problem can be given) is nothing less than a radical reworking of the ideal of scientific objectivity. The slightly juvenile conception of the coldblooded, emotionless, impersonal, passive scientist mirroring the world perfectly in the highly polished lenses of his steel rimmed glasses—this stereotype—is no longer, if it ever was, adequate.

What is being proposed here is that objectivity for science lies at least in becoming precise about what value judgments are being and might have been made in a given inquiry—and even, to put it in its most challenging form, what value decisions ought to be made; in short that a science of ethics is a necessary requirement if science's progress toward objectivity is to be continuous.

Of course the establishment of such a science of ethics is a task of stupendous magnitude and it will probably not even be well launched for many generations. But a first step is surely comprised of the reflective self awareness of the scientist in making the value judgments he must make.

NOTES

1. R. Carnap, "Empiricism, Semantics, and Ontology," *Revue Internationale de Philosophie*, XI, 1950, pp. 20–40.

2. W. V. Quine, "On Carnap's Views on Ontology," *Philosophical Studies*, II, No. 5, 1951.

3. "In practice three levels are commonly used: 1 per cent, 5 per cent and 0.3 of one per cent. There is nothing sacred about these three values; *they have become established in practice*

without any rigid theoretical justification." (my italics) (subnote 3*, p. 435). To establish significance at the 5 per cent level means that one is willing to take the risk of accepting a hypothesis as true when one will be thus making a mistake, one time in twenty. Or in other words, that one will be wrong (over the long run) once every twenty times if one employed an .05 level of significance. See also (subnote 3† Chap. v) for such statements as "which of these two errors is most *important* to avoid (it being necessary to make such a decision in order to accept or reject the given hypothesis) is a *subjective matter* . . ." (my italics) (subnote 3†, p. 262).

* A. C. Rosander, *Elementary Principles of Statistics* (New York: D. Van Nostrand Co., 1951).

† J. Neyman, *First Course in Probability and Statistics* (New York: Henry Holt & Co., 1950).

*Richard C. Jeffrey**

VALUATION AND ACCEPTANCE OF SCIENTIFIC HYPOTHESES

1. INTRODUCTION

Churchman,[1] Braithwaite,[2] and Rudner[3] have recently argued from premises acceptable to many empiricists to the conclusion that ethical judgments are essentially involved in decisions as to which hypotheses should be included in the body of scientifically accepted propositions[4]. Rudner summarizes the argument:

> Now I take it that no analysis of what constitutes the method of science would be satisfactory unless it comprised some assertion to the effect that the scientist as scientist accepts or rejects hypotheses.
>
> But if this is so, then clearly the scientist as scientist does make value judgments. For, since no scientific hypothesis is ever completely verified, in accepting a hypothesis the scientist must make the decision that the evidence is *sufficiently* strong or that the probability is *sufficiently* high to warrant the acceptance of the hypothesis. Obviously our decision regarding the evidence and respecting how strong is "strong enough" is going to be a function of the *importance*, in the typically ethical sense, of making a mistake in accepting or rejecting the hypothesis (n. 3, p. 2).

The form of this reasoning is hypothetical: *If* it is the job of the scientist to accept and reject hypotheses, *then* he must make value judgments. Now I shall argue

*The author wishes to express his thanks to Prof. C. G. Hempel, at whose suggestion this paper was written, and to Dr. Abner Shimony, for their helpful criticism.

Reprinted from *Philosophy of Science*, XXIII, 237–46.

(in effect) that if the scientist makes value judgements, then he neither accepts nor rejects hypotheses. These two statements together form a *reductio ad absurdum* of the widely held view which our authors presuppose, that science consists of a body of hypotheses which, pending further evidence, have been *accepted* as highly enough confirmed for practical purposes ("practical" in Aristotle's sense).

In place of that picture of science I shall suggest that the activity proper to the scientist is the assignment of probabilities (with respect to currently available evidence) to the hypotheses which, on the usual view, he simply accepts or rejects. This is not presented as a fully satisfactory position, but rather as the standpoint to which we are led when we set the Churchman-Braithwaite-Rudner arguments free from the presupposition that it is the job of the scientist as such to accept and reject hypotheses.

In the following pages we shall frequently have to speak of probabilities in connection with rational choice, and this opens us to the danger of greatly complicating our task through involvement in the dispute between conflicting theories of probability.[5] To avoid this I shall make use of the notion that these theories are conflicting explications[6] of the concept, *reasonable degree of belief*. If this is so we shall usually be able to avoid the controversy by using the subjectivistic language ("rational degree of belief" or of "confidence") which is appropriate to the *explicandum*. This device is justified if the reader finds the relevant statements acceptable after he has freely translated them into the terminology of whichever *explicatum* he prefers.

2. BETTING AND CHOOSING

It is commonly held that although we have no certain knowledge we must often act as if probable hypotheses were known to be true. For example it might be said that when we decide to inoculate a child against polio we are accepting as certain the hypothesis that the vaccine is free from virulent polio virus. Proponents of this view speak of "accepting a hypothesis" as a sort of inductive jump from high probability to certainty.

On this account, betting is an exceptional situation. Let H be the hypothesis that the ice on Lake Carnegie is thick enough to skate on. Now if one is willing to give odds of 4:1 to anyone who will bet that not-H, and if these are the longest odds one is willing to give one has pretty well expressed by a sort of action a degree of belief in H four times as great as one's belief in not-H. Here one is risking a good—money—which admits of degrees, so that in the bet one is able to adjust the degree of risk to the degree of belief in H. But in actually attempting to skate, degree of commitment cannot be nicely tailored to suit degree of belief: One cannot arrange to fall in only part way in case the ice breaks.

Part of the discrepancy between betting and other choosing can be disposed of immediately. Thus far we have stressed the case where the bettor himself proposes the stakes, and indeed there is nothing comparable to this in most other choosing. When you "bet" by trusting the ice with your own weight the stakes are, say, a dunking if you lose and an afternoon's skating if you win. These stakes are fixed by nature, not by the skater. But the same arrangement is common in actual betting, e.g., at a race tack, where the bettor's problem is not to propose the stakes but rather to decide whether odds offered by someone else are fair or advantageous to him. In such cases the bettor cannot pick the exact degree of his commitment any more than the skater can.

In betting as in other choosing the rational agent acts so as to maximize his expectation of value. In betting, the values at stake seem especially easy to measure: The value or utility of winning is measured by the amount of money won, and the value of losing by the amount lost. But it is well known that this identification is vague and approximate. In the bet about the ice, for example, the identification of utility with money would lead one to accept as advantageous odds of 1:1 when the ratio of degrees of belief is 4:1. But clearly it makes a difference whether the 1:1 in question is $1:$1 or $1000:$1000. The former would be a good bet for a man of moderate means, but the latter would not.

The usual way out of this difficulty is to specify that the stakes be small compared with the bettor's fortune, but not so small as to bore him. The importance of finding a way out is that the ratio of stakes which a man finds acceptable is a convenient measure of degree of belief. But we have seen that it is not always a reliable measure. Therefore it seems appropriate to interpret the relationship between odds and utilities in the same way we interpret the relationship between the height of a column of mercury and temperature; the one is a reliable sign of the other within a certain range, but is unreliable outside that range, where we accordingly seek other signs (e.g., alcohol thermometers below and gas thermometers above the range of reliability of mercury).

3. RATIONAL DECISION: THE BAYES CRITERION

The decision whether or not to accept a bet separates naturally into four stages; we illustrate in the case of the bet about the hypothesis (H) that the ice on Lake Carnegie is thick enough for skating. First we draw up a *table of stakes* indicating what will be won or lost in each of the four situations which can arise depending on whether the hypothesis is true or not, and whether the bet is accepted or not.

Choice	Actual state of the ice: H	not-H
A: accept the bet.	Win $1.	Lose $1.
not-A: don't accept the bet.	Neither win nor lose.	

Table of Stakes

In this case we may suppose the utilities of the stakes to be proportional to the stakes themselves, so that the *table of utilities* looks like this:

	H	not-*H*
A	1	−1
not-*A*	0	0

Here we are concerned not with the numbers themselves, but rather with their ratios; the same information about the utilities is contained in any table which is the same as the one above except that all entries are mutlitplied by some positive number, e.g., $\begin{bmatrix} 5 & -5 \\ 0 & 0 \end{bmatrix}$

If somehow we know that the bettor's degrees of belief in *H* and not-*H* stand in the ratio 4 : 1 we can construct a *table of expectations*:

	H	not-*H*
A	4	−1
not-*A*	0	0

where the expectation in each of the four situations is the product of belief in and utility of that situation.

By adding the two numbers in the top row of this table we get the *total expectation* from choice *A* (accepting the bet): 3. The sum of the numbers in the bottom row is the total expectation from not-*A*. The Bayes criterion defines the rational choice to be the one with the greater expectation. Here, then, the rational decision would be to accept the bet.

The decision about actually trying the ice is exactly parallel. Here the table of stakes is

	H	not-*H*
A: try to skate.	Skate	Get wet
not-*A*: don't try.	Neither skate nor get wet.	

If skating and getting wet are equal and opposite goods, the utility table and the rest of the calculation is identical with that for the betting decision, and the recommendation is: Try the ice.

The Bayes criterion has generally been accepted as a satisfactory explication of "rational choice" *relative to a set of numerical utilities and degrees of belief.* If the criterion is accepted, then the rationality of a decision which conforms to it can be attacked only on grounds that the degrees of belief and utilities involved are themselves unreasonable. The most influential school of thought in statistics today holds that in many cases there are no reasonable grounds for assigning probabilities to sets of hypotheses. This does not mean that in such cases the reasonable degree of belief in each hypothesis is zero, or that it is the same for all hypotheses, but simply that no numerical assignment whatever can be justified. Accordingly, statisticians have developed alternatives to Bayes's criterion, one of which (the *minimax criterion*) we shall consider in section 5.

The question of how and whether it is possible to justify the assignment of numerical utilities to situations is even more difficult, and we do not propose to consider it here. But it should be noted that the use made of utilities by the Bayes criterion is not very exacting. As observed earlier, we are concerned not with the utilities themselves, but with their ratios. Further, it is easy to show that often not even that much is required. For example, in applying the Bayes criterion to a choice between two actions it is sufficient to know ratios of certain *differences* between the utilities.[7]

4. CHOICE BETWEEN HYPOTHESES: BAYES'S METHOD

Meno, in the dialogue bearing his name, makes a strong objection to the Socratic concept of inquiry:

> And how will you enquire, Socrates, into that which you do know? What will you put forth as the subject of enquiry? And if you find what you want, how will you ever know that this is the thing which you did not know? (n. 8, Steph. 80).

In reply, Socrates undertakes the famous demonstration of how geometrical ideas can be "recollected" by an ignorant boy. But then he goes on, and apparently weakens the force of the demonstration by admitting

> . . . Some things I have said of which I am not altogether confident. But that we shall be better and braver and less helpless if we think that we ought to enquire, than we should have been if we indulged in the idle fancy that there was no knowledge and no use in seeking to know what we do not know;—that is a theme upon which I am ready to fight, in word and deed, to the utmost of my power (n. 8, Steph. 86).

This is not mere wishful thinking, but rather part of a rational argument, in the Bayes sense of "rational." Using our previous notation, the hypothesis under consideration is (H): "Knowledge is obtainable through inquiry," and the choice is between A, the decision to inquire, and not-A. Meno had made it plausible that H is very improbable; Socrates's first reply (the demonstration of recollection) was an attempt to undermine Meno's argument directly, by showing that in fact H is more probable than Meno would have us believe. Socrates's second reply, quoted above, concedes that Meno may be right, but goes on to say that even if H *is* improbable, the utility of knowledge is so great that even when it is multiplied by a small probability, *the total expectation from inquiry* (*A*) *exceeds that from not-A*.

	H	not-*H*
A	Possibility of obtaining knowledge.	Waste of effort.
not-*A*	No knowledge obtained and no effort wasted in seeking it.	

Table of Stakes

For amusement, we might assign numerical utilities to the table of stakes: $\begin{bmatrix} 1000 & -1 \\ 0 & 0 \end{bmatrix}$. A little calculation shows that with these utilities it is rational to inquire even if the probability of H is as low as .001.

This pattern of argument is fairly common as a justification for faith—in God (Pascal's wager), in inquiry (Socrates), in the unity of the laws of nature (Einstein). But is should be noted that in all three cases what is meant by "faith" is not verbalized intellectual acceptance of the truth of a thesis, but rather commitment to a line of action which would be useless or even damaging if the thesis in question were false. Typically, in these cases, the thesis itself is extremely vague; but it is meaningful to the person who accepts it in the sense that it partly determines his activity.

To take a more precise thesis as an example, consider the problem of quality control in the manufacture of polio vaccine. A sample of the vaccine in a certain lot is tested and found to be free from active polio virus. Let us suppose that this imparts a definite probability to the hypothesis that the entire lot is good. Is this probability high enough for us rationally to accept the hypothesis?

Contrast this with a similar problem about roller skate ball bearings. Imagine that here, too, a sample has been taken and that all the bearings tested have proved satisfactory; and suppose that this evidence imparts to the hypothesis that all the bearings in the lot are good the same probability that we encountered before in the case of the vaccine. As Rudner points out, we might accept the ball bearings and yet reject the vaccine because although the probabilities are the same in the two cases, the utilities are different. If the probability were just enough to lead us to accept the bearings, we should reject the vaccine because of the graver consequences of being wrong about it.

But what determines these consequences? There is nothing in the hypothesis, "This vaccine is free from active polio virus," to tell us what the vaccine is *for*, or what would happen if the statement were accepted when false. One naturally assumes that the vaccine is intended for inoculating children, but for all we know from the hypothesis it might be intended for inoculating pet monkeys. One's confidence in the hypothesis might well be high enough to warrant inoculation of monkeys, but not of children.

The trouble is that implicitly we have been discussing a utility table with these headings

	H	not-H
Accept H		
Reject H		

but there is no way to decide what numbers should be written in the blank spaces unless we know what actions depend on the acceptance or rejection of H. Bruno DeFinetti sums up the case:

> I do not deem the usual expression "to accept hypothesis H_r" to be proper. The "decision" does not really consist of this "acceptance" but in *the choice of a definite action A_r*. The connection between the action A_r and the hypothesis H_r may be very strong, say "the action A_r is that which we would choose if we knew that H_r was the true hypothesis." Nevertheless, this connection cannot turn into an identification (subnote 5**, p. 219).

This fact is obscured when we consider very specialized hypotheses of the sort encountered in industrial quality control, where it is clear from the contexts, although not expressly stated in the hypotheses, what actions are in view. But the vaccine example shows that even in these cases it may be necessary to make the distinction that DeFinetti urges. In the case of lawlike scientific hypotheses the distinction seems to be invariably necessary; there it is certainly meaningless to speak of *the* cost of mistaken acceptance or rejection, for by its nature a putative scientific law will be relevant in a great diversity of choice situations among which the cost of a mistake will vary greatly.

In arguing for his position Rudner concedes, "The examples I have chosen are from scientific inferences in industrial quality control. But the point is clearly general in application" (n. 3, p. 2). Rudner seems to give his reason for this last statement later on:

> I believe, of course, that an adequate rational reconstruction of the procedures of science would show that every scientific inference is properly construable as a statistical inference (i.e., as an inference from a set of characteristics of a sample of a population to a set of characteristics of the whole population (n. 3, p. 3).

But even if analysis should show that lawlike hypotheses are like the examples from industrial quality control in being inferences from characteristics of a sample to characteristics of an entire population, they are different in the respect which is of importance here, namely their generality of application. Braithwaite and Churchman are more cautious here; they confine their remarks to statistical inferences of the ordinary sort. But we have seen that even in statistics the feasibility of blurring the distinction between accepting a hypothesis and acting upon it depends on features of the statement of the problem which are not present in every inference.

5. CHOICE BETWEEN HYPOTHESES: MINIMAX METHOD

In applying the Bayes criterion to quality control we assumed that on the basis of the relative frequency of some property in a sample of a population definite probabilities can be assigned to the various conflicting hypotheses about the relative frequency of that property in the whole population. In general such "inverse inference" presupposes a knowledge of the *prior probabilities* (or of an a priori *probability distribution*) for the hypotheses in question. On the other hand, "direct inference" from relative frequencies in an entire population to relative frequencies in samples involves no such difficulty. Wald writes,

> In many statistical problems the existence of an a priori distribution cannot be postulated, and, in those cases where the existence of an a priori distribution can be assumed, it is usually unknown to the experimenter and therefore the Bayes solution cannot be determined (subnote 4*, p. 16).

For these cases Wald proposes a criterion which makes no use of inverse inference.

For simplicity we consider the case where somehow it is known that one or the other of two hypotheses, H_1 or H_2, must be true. The hypotheses assign different relative frequencies of a property P to a population. A sample consisting of only one member is drawn from the population and will be inspected for this property. Wald's problem is to choose, in advance of the inspection, an *inductive rule* which tells him which hypothesis to accept under every possible assumption as to the outcome of the inspection. In this case the choice is between four rules, since the relative frequency of P in the sample can only be 0 or 100%.

Rule 1. Accept H_1 in either case.
Rule 2. Accept H_1 in case the relative frequency of P in the sample is 0, H_2 if it is 100%.
Rule 3. Accept H_2 in case the relative frequency of P in the sample is 0, H_1 if it is 100%.
Rule 4. Accept H_2 in either case.

By direct inference one can find the conditional probabilities that each of the inductive rules will lead to the right or the wrong hypothesis on the assumption that H_1 is true, and separately on the assumption that H_2 is true. From these eight probabilities together with a knowledge of the losses (negative utilities) that would result from accepting one of the H's when in fact the other is true, one can calculate a table of risks, e.g.:

Inductive rule	Risk in using this rule in case H_1 is true	Risk in using this rule in case H_2 is true
1	7	0
2	1/2	5
3	4	2
4	10	18

Wald's *minimax criterion* is: Minimize the maximum risk. Here this directs us to choose the rule—3—for which the larger of the two risks is least.

The minimax criterion is the counsel of extreme conservatism or pessimism. Wald proves this in two ways. (1) He shows that "a minimax solution is, under some weak restrictions, a Bayes solution relative to a least favorable a priori distribution"[4*, p. 18]. (2) He shows that the situation in which an experimenter uses the minimax criterion to make a decision is formally identical with the situation in which the experimenter is playing a competitive "game" with a personalized Nature in the sense that the experimenter's losses are Nature's gains. Nature plays her hand by selecting a set of prior probabilities for the hypotheses between which the experimenter must choose; being intelligent and malevolent, Nature chooses a set of probabilities which are as unfavorable as possible when viewed in the light of the negative utilities which the experimenter attaches to the acceptance of false hypotheses.

Since different experimenters make different value judgments, it would seem that in applying the minimax criterion each experimenter implicitly assumes that this is the worst of all possible worlds *for him*. We might look at the matter in this way: The minimax criterion is at the pessimistic end of a continuum of criteria. At the other end of this continuum is the "minimin" criterion, which advises each experimenter to minimize his minimum risk. Here each experimenter acts as if this were the *best* of all possible worlds *for him*. The rules at both extremes of the continuum share the same defect: They presuppose a great sensitivity on the part of Nature to human likes and dislikes and are therefore at odds with a basic attitude which we all share, in our lucid moments.

Wald was aware of the sort of objection we have been making:

> The analogy between the decision problem and a two-person game seems to be complete, except for one point. Whereas the experimenter wishes to minimize the risk . . . , we can hardly say that Nature wishes to maximize [the risk]. Nevertheless, since Nature's choice is unknown to the experimenter, it is perhaps not unreasonable for the experimenter to behave as if Nature wanted to maximize the risk (subnote 4*, p. 27).

This suggests that we have overstated our case. As a general inductive rule, the minimax criterion represents an unprofitable extreme of caution; nevertheless we feel that there are conditions under which a man would do well to act with a maximum of caution, even though it would be unwise to follow that policy for all decisions. What we lack is an account of the conditions under which it is appropriate to use the minimax criterion.[9]

Apart from this, our previous objections to the notion of "accepting" a hypothesis apply to the minimax as well as to the Bayes criterion. Wald's procedure leads us to accept an inductive rule which, once the experiment has been made, determines one of the competing hypotheses as the "best." But this means best for making the specific choice in question, e.g., whether to inject a child with polio vaccine from a certain lot. Among the *same* hypotheses, a different one might be best with respect to a different choice, e.g., inoculating a pet monkey. Hence both the Bayes and minimax criteria permit choice between hypotheses only with respect to a set of utilities which in turn are relative to the intended applications of the hypotheses.

6. CONCLUSION

On the Churchman-Braithwaite-Rudner view it is the task of the scientist as such to accept and reject hypotheses in such a way as to maximize the expectation of good for, say, a community for which he is acting. On the other hand, our conclusion is that if the scientist is to maximize good he should refrain from accepting or rejecting hypotheses, since he cannot possibly do so in such a way as to optimize every decision which may be made on the basis of those hypotheses. We note that this difficulty cannot be avoided by making acceptance relative to the most stringent possible set of utilities (even if there were some way of determining what that is) because then the choice would be wrong for all less stringent sets. One cannot, by accepting or rejecting the hypothesis about the polio vaccine, do justice both to the problem of the physician who is trying to decide whether to inoculate a child, and the veterinarian who has a similar problem about a monkey. To accept or reject that hypotheses once and for all is to introduce an unnecessary conflict between the interests of the physician and the veterinarian. The conflict can be resolved if the scientist either contents himself with providing them both with a single probability for the hypothesis (whereupon each makes his own decision based on the utilities peculiar to his problem), or if the scientist takes on the job of making a separate decision as to the acceptability of the hypothesis in each case. In any event, we conclude that it is not the business of the scientist as such, least of all of the scientist who works with lawlike hypotheses, to accept or reject hypotheses.

We seem to have been driven to the conclusion that the scientist's proper role is to provide the rational agents in the society which he represents with probabilities for the hypotheses which on the other account he simply accepts or rejects. There are great difficulties with this view. (1) It presupposes a satisfactory theory of probability in the sense of *degree of confirmation* for hypotheses on given evidence. (2) Even if such a theory were available there would be great practical difficulties in using it in the way we have indicated. (3) This account bears no resemblance to our ordinary conception of science. Books on electrodynamics, for example, simply list Maxwell's equations as laws; they do not add a degree of confirmation. These are only some of the difficulties with the probabilistic view of science.

To these, Rudner adds a very basic objection.

> . . . the determination that the degree of confirmation is say, *p*, . . . which is on this view being held to be the indispensable task of the scientist, is clearly nothing more than *the acceptance by the scientist of the hypothesis that the degree of confidence is p*. . . (n. 3, p. 4.)

But of course we must reply that it is no more the business of the scientist to "accept" hypotheses about degrees of confidence than it is to accept hypotheses of any other sort, and for the same reasons.[10] Rudner's objection must be included as one of the weightiest under heading (1) above as a difficulty of the probabilistic view of science. These difficulties may be fatal for that theory; but they cannot save the view that the scientist, *qua* scientist, accepts hypotheses.

NOTES

1. C. West Churchman, *Theory of Experimental Inference* (The Macmillan Co., New York, 1948).

2. R. B. Braithwaite, *Scientific Explanation* (Cambridge University Press, Cambridge, 1953).

3. Richard Rudner, "The Scientist *Qua* Scientist Makes Value Judgements," *Philosophy of Science*, Vol. 20 (1953), pp. 1–6.

4. Rudner (n. 3) has the most explicit and unqualified statement of the point of view in question. These views stem largely from recent developments in statistics, especially from the work of Abraham Wald; see subnote 4* and references there to earlier writings.

* Abraham Wald, *Statistical Decision Functions* (John Wiley and Sons, Inc., New York 1950).

5. For accounts of this dispute, see subnotes 5*, Chap. ii; 5†, Sec. I; and 5**.

* Rudolf Carnap, *Logical Foundations of Probability* (University of Chicago Press, Chicago, 1950).

**Bruno DeFinetti, pp. 217–225 of *Proceedings of the Second Berkeley Symposium on Mathematical Statistics and Probability*, ed. Jerzy Neyman (University of California Press, Berkeley, 1951).

† Rudolf Carnap, *The Continuum of Inductive Methods* (University of Chicago Press, Chicago, 1952).

6. The view that a theory of probability is an explication of a vague concept in common use (the *explicandum*) by a precise concept (the *explicatum*) is due to Carnap; cf. subnote 5*, Chap. i.

7. Let $H_1, \ldots, H_n$ be mutually exclusive, collectively exhaustive hypotheses for which

$\beta_1, \ldots, \beta_n$ are the corresponding degrees of belief. A decision is to be made between acts A_1 and A_2. Let Δ_i be the difference: (utility of choosing A_1 if H_i is true) – (utility of choosing A_2 if H_i is true). Then in this case the Bayes criterion reduces to: Choose A_1 if $\Delta_1\beta_1 + \ldots + \Delta_n\beta_n > 0$, and choose A_2 if the inequality goes the other way.

8. Plato, "Meno," *The Dialogues of Plato*, tr. Benjamin Jowett (3rd ed., Random House, Inc., New York), I, 349–80.

9. Savage (subnote 9*, Chap. xiii) discusses a number of other objections to the minimax criterion.

* Leonard J. Savage, *The Foundations of Statistics* (John Wiley and Sons, Inc., New York, 1954).

10. In Carnap's confirmation theory there at first seems to be no difficulty since it is a logical rather than a factual question, what the degree of confirmation of a given hypothesis is, with respect to certain evidence. But the difficulty may appear at a deeper level in choosing a particular *c*-function; this Carnap describes as a practical decision. See subnote 5†, sec. 18.

Isaac Levi

MUST THE SCIENTIST MAKE VALUE JUDGMENTS?*

Two assumptions implicit in Pearson's characterization of "the scientific man" have been called into question in recent years: (a) At least one major goal of the scientist *qua* scientist is to make judgments—i.e., to accept or reject hypotheses—and to justify his judgments. (b) The scientific inquirer is prohibited by the canons of scientific inference from taking his attitudes, preferences, temperament, and values into account when assessing the correctness of his inferences.

One currently held view affirms (a) but denies (b). This position maintains that the scientist does and, indeed, must make value judgments when choosing between hypotheses. The other position upholds the value-neutrality thesis (b) at the expense of the claim that scientific inference issues in the acceptance and rejection of hypotheses (a). According to this view, a scientific inquiry does not terminate with the replacement of doubt by belief but with the assignment of probabilities or degrees of confirmation to hypotheses relative to the available evidence.

In this paper, a critical examination of these conflicting conceptions of scientific inference will be undertaken; the *prima facie* tenability of the claim that scientists can, do, and ought to accept and reject hypotheses in accordance with the value-neutrality thesis will be defended; and some indication will be given of the kind of question that must be answered before this plausible view can be converted into a coherent and adequate theory of the relation of values to scientific inference.

---

*I wish to acknowledge my debt to Sidney Morgenbesser, whose critical comments in conversation have greatly influenced my thinking on this question, and to Mortimer Kadish and John McLellan, whose reactions to earlier drafts of this paper have helped shape the final result.

From *The Journal of Philosophy*, LVII, No. 11, May 26, 1960. Reprinted by permission of the author and the editor.

I

The tenability of the value-neutrality thesis has been questioned by C. W. Churchman[1] and R. B. Braithwaite,[2] at least insofar as it applies to statistical inference. However, the most explicit and sweeping attack against the value-neutrality thesis is to be found in an article by Richard Rudner, who argues that the scientist must make value judgments in drawing any kind of non-deductive inference.[3]

> Now I take it that no analysis of what constitutes the method of science would be satisfactory unless it comprised some assertion to the effect that the scientist as scientist accepts or rejects hypotheses.
>
> But if this is so then clearly the scientist as scientist does make value judgments. For, since no scientific hypothesis is ever completely verified, in accepting a hypothesis the scientist must make the decision that the evidence is *sufficiently* strong or that the probability is *sufficiently* high to warrant the acceptance of the hypothesis. Obviously our decision regarding the evidence and respecting how strong is "strong enough" is going to be a function of the *importance*, in the typically ethical sense, of making a mistake in accepting or rejecting the hypothesis. Thus, to take a crude but easily manageable example, if the hypothesis under consideration were to the effect that a toxic ingredient of a drug was not present in lethal quantity, we would require a relatively high degree of confirmation or confidence before accepting the hypothesis—for the consequences of making a mistake here are exceedingly grave by our moral standards. On the other hand, if, say, our hypothesis stated that, on the basis of a sample, a certain lot of machine stamped belt buckles was not defective, the degree of confidence we should require would be relatively not so high. *How sure we need to be before we accept a hypothesis will depend on how serious a mistake would be.*[4]

Rudner's claim is not that values play a role in the scientist's selection of research problems, nor is he arguing that scientists often let their attitudes, values, and temperaments influence their conclusions. These points are relevant to the psychology and sociology of inquiry but not to its logic. Rudner is making an assertion about the requirements imposed upon the inquirer who embraces the goals and the canons of scientific inference.[5] He contends that the scientist in his capacity as a scientist *must* make value judgments even if it is psychologically possible for him to avoid doing so. His argument for this conclusion can be summarized in the following series of statements:

(1) The scientist *qua* scientist accepts or rejects hypotheses.

(2) No amount of evidence ever completely confirms or disconfirms any (empirical) hypothesis but only renders it more or less probable.

(3) As a consequence of (1) and (2), the scientist must decide how high the

probability of a hypothesis relative to the evidence must be before he is warranted in accepting it.

(4) The decision required in (3) is a function of how important it will be if a mistake is made in accepting or rejecting a hypothesis.

The need for assigning minimum probabilities for accepting and rejecting hypotheses (3) is a deductive consequence of the claim that scientists accept and reject hypotheses (1) and the corrigibility of empirical hypotheses (2). Since (2) is a cardinal tenet of an empiricist philosophy of science and will not be questioned in this paper, the first part of Rudner's argument reduces to the correct claim that if (1) is true (3) is true.

Rudner's rejection of the value-neutrality thesis cannot be justified, however, on the basis of (3) alone. He must show that the assignment of minimum probabilities is a function of the importance of making mistakes (4). But (4) cannot be obtained from (3) without further argument.[6] Rudner attempts to fill the gap by citing illustrations from quality control and appealing to current theories of statistical inference.[7] He believes that the problem of choosing how to act in the face of uncertainty, which is the fundamental problem of quality control, is typical of all scientific inquiry and concludes from this that the importance of making mistakes must be taken into account in all scientific inference.

This argument seems to rest upon certain assumptions adopted more or less explicitly by Rudner and Churchman.[8] These assumptions involve the notion of acting on the basis of a hypothesis relative to an objective. To say "*X* acts on the basis of *H* relative to some objective *P*" is to assert that *X* carries out action *A* where *A* is the best proccedure[9] to follow relative to *P*, given that *H* is true. The Rudner-Churchman assumptions can now be stated as follows:

(5) To choose to accept a hypothesis *H* as true (or to believe that *H* is true) is equivalent to choosing to act on the basis of *H* relative to some specific objective *P*.

(6) The degree of confirmation that a hypothesis *H* must have before one is warranted in choosing to act on the basis of *H* relative to an objective *P* is a function of the seriousness of the error relative to *P* resulting from basing the action on the wrong hypothesis.

Assumption (6) is a version of a principle adopted by Pearson, Neyman, and Wald in their theories of statistical inference. The plausibility of Rudner's argument from quality control (where the problem is how to act on the basis of hypotheses) to (4) is due largely to the reasonableness of this presupposition. However, (6) without (5) will not yield (4).

Unlike (6), (5) cannot be justified by an appeal to the authority of the statisticians. Not only are these authorities fallible, but some of them have been noncommittal regarding the acceptability of (5).[10] Substantial grounds can be offered for praising this exercise of caution.

II

An interesting case against the tenability of (5) has been made by Richard Jeffrey. Jeffrey considers the problem of deciding whether a given batch of polio vaccine is free from active polio virus. The seriousness of the consequences of mistakenly accepting the hypothesis would seem to demand that we confirm the hypothesis to a far higher degree before accepting it than would be the case if we were interested in the quality of a batch of roller skate bearings.

> But what determines these consequences? There is nothing in the hypothesis, "This vaccine is free from active polio virus," to tell us what the vaccine is *for*, or what would happen if the statement were accepted when false. One naturally assumes that the vaccine is intended for inoculating children, but for all we know from the hypothesis it might be intended for inoculating pet monkeys. One's confidence in the hypothesis might well be high enough to warrant inoculation of monkeys but not of children.[11]

Jeffrey's point can be reformulated as follows: Action on the basis of a hypothesis H is always relative to an objective P. Consequently if accepting H is identical with acting on the basis of H (5), accepting H in an "open-ended" situation[12] where there is no specific objective is impossible. But accepting H is possible in open-ended situations, for it is compatible with different and even conflicting objectives. Hence, (5) must be rejected.

In a reply to Jeffrey's paper, Churchman compares Jeffrey's open-ended decision problems to situations that occur in production. Suppose that a manufacturer wishes to place on the market a certain product (rope) that has many different uses. Churchman points out that procedures are available to the manufacturer in terms of which he can single out needs that his product should be designed to meet. He contends that similar procedures must be employed if we are to accept and reject hypotheses intelligibly.

> In this sense, it is certainly meaningless to talk of *the* acceptance of the hypothesis about the freedom of a vaccine from active polio virus, provided the information has a number of different uses. Even within one business organization one can readily point out that the many uses of information imply many different criteria for the "acceptance" or "rejection" of hypotheses.[13]

Churchman's argument seems to be this: *If* "accepting a hypothesis H" is understood in a sense that makes (5) true, then open-ended decision problems involving the acceptance or rejection of hypotheses can be treated like open-ended production problems. The solvability and, hence, the intelligibility of such problems requires the elimination of the open-endedness.

This true observation does not meet, however, the major point of Jeffrey's objection. Jeffrey's argument attempts to show that in *one* sense of "accepting a hypothesis" to accept a hypothesis in an open-ended situation is perfectly meaningful and consistent. Consequently, in *that* sense, (5) does not hold.

An easy but cheap victory might be gained at Jeffrey's expense by pointing out that wherever a scientist does not appear to have an objective in mind, nonetheless, one can always be specified—namely, the objective of accepting true answers to questions as true. Accepting a hypothesis *H* would then be equivalent to acting on the basis of *H* relative to that objective.

Resorting to this strategy would be to miss the point of the discussion. To say that accepting a hypothesis is the same as acting on the basis of *H* in order to obtain true answers is tantamount to asserting that accepting *H* is equivalent to accepting *H*. One could not conclude from this alone that the problem of deciding what to believe is an all fours with decision problems in quality control —at least with respect to the value-neutrality thesis. In the latter kind of problem, the objectives are "practical"; in the former, they are "theoretical."

In order to avoid misunderstanding, therefore, an open-ended decision problem will be understood to be a decision problem for which no practical objective has been specified.[14] Consequently, the issue at stake in the debate between Jeffrey and Churchman is whether there is any sense in which a person can meaningfully and consistently be said to accept a hypothesis as true without having a practical objective. The following considerations are offered in favor of an affirmative answer to this question.

(i) Many apparently intelligible questions are raised and answered in the sciences for which practical objectives are difficult to specify. What practical objectives are at stake when an investigator is deciding whether to accept or reject the principle of parity, the hypothesis of an expanding universe, or the claim that Galileo never conducted the Leaning Tower experiment? One could try to show that appearances are deceiving and that practical objectives are always the goals of such decision problems. However, this would be difficult to prove. Furthermore, it would not follow that appearances *must* be deceiving and that practical objectives *must* be operative. Indeed, the cases just cited would normally be considered to be problems of deciding what to accept as or believe to be true regardless of whether practical objectives are involved. This seems to indicate that there is a sense of "accepting a hypothesis" which is meaningfully applied to choices in open-ended situations.

(ii) Even in the case of decision problems where practical objectives are involved, it often seems appropriate to distinguish between acting on the basis of a hypothesis relative to that objective and accepting the hypothesis as true. Suppose that an investor in oil stocks knows that if a certain oil company whose stocks are selling at a low price strikes oil at a certain location the price of the stock will increase one hundredfold. The investor might buy stock in the company while suspending judgment as to the eventual discovery of oil. Here is a case were one would normally say that a person has acted on the basis of a hypothesis and perhaps was justified in doing so without accepting the hypothesis as true or being warranted in so accepting it.

One could reply by saying that the investor refused to accept the hypothesis because such acceptance would have been tantamount to acting on the basis of the hypothesis of an oil strike relative to some practical objective other than making a profitable investment. However, such an objective would not always be easy to find. Furthermore, the situation would normally be considered a case of action without belief regardless of whether the existence of a practical objective could be shown or not.

(iii) There seems to be a sense in which it is possible for a person to believe in the truth of a hypothesis and nonetheless refuse to act on it. He may even be justified in proceeding in this fashion. The Sabin live virus polio vaccine serves as an illustration. The available evidence might warrant belief in the safety and effectiveness of the vaccine without justifying a program of mass inoculation.[15]

(iv) A plausible case can be made for saying that even when a person is deciding how to act in order to realize a practical objective he will have to accept some statements as true in a sense that does not meet the conditions of (5). The evidence upon which he bases his decisions consists of statements which he accepts as true. He might have to accept the truth of statements asserting the degrees to which various hypotheses are confirmed relative to the available evidence. Finally, he will also have to accept the truth of statements that indicate the best actions relative to his objectives given the truth of various hypotheses.[16]

The considerations just advanced suggest that there is a familiar sense in which a person can meaningfully and consistently accept or reject a hypothesis in an open-ended situation. In that sense, (5) is false and Rudner's argument in favor of (4) and against the value-neutrality thesis fails.

This result need not in itself be fatal to the Churchman-Rudner position. Apologists for this view could admit the meaningfulness of this sense of "accepting a hypothesis" and deny that the aim of the sciences is (or ought to be) to accept or reject hypotheses in that sense. They might contend that scientific inferences indicate how one ought to act on the basis of hypotheses but not what one ought to believe. The rejection of the value-neutrality thesis would flow quite naturally from this transmutation of scientific inquiry into a quest for normative principles. Oddly enough, however, it is Jeffrey, an apparent defender of the value-neutrality thesis, who denies that scientists accept and reject hypotheses.

III

Jeffrey proposes a conception of the aim and function of science also suggested by Carnap[17] and Hempel.[18] According to this view, a scientist does not, or at least should not, accept and reject hypotheses. Instead, he should content himself with assigning degrees of confirmation to hypotheses relative to the available evidence. Anyone who is confronted with a practical decision problem can go to the scientist to ascertain the degrees of confirmation of the relevant hypotheses. He can then utilize this information together with his own estimates of the seriousness of mistakes in order to decide upon a course of action.

One consequence of this view is that all non-deductive inference in science consists in assigning degrees of confirmation to hypotheses relative to given evidence. Indeed, Carnap defines inductive inference in this way.[19] Hence, if Carnap is correct in maintaining that degrees of confirmation can be ascertained without consideration of values, the Carnap-Hempel-Jeffrey view supports the value-neutrality thesis.[20] However, the value-neutrality thesis is upheld at the expense of the claim that scientists accept or reject hypotheses. In this respect, the Carnap-Hempel-Jeffrey view breaks as radically with tradition as does the Braithwaite-Churchman-Rudner position.

In his paper, Jeffrey offers an extremely clever argument to show that scientists can neither accept nor reject hypotheses.

> On the Churchman-Braithwaite-Rudner view it is the task of the scientist as such to accept and reject hypotheses in such a way as to maximize the expectation of good for, say a community for which he is acting. On the other hand, our conclusion is that if the scientist is to maximize good he should refrain from accepting or rejecting hypotheses, since he cannot possibly do so in such a way as to optimize every decision which may be made on the basis of those hypotheses. We note that this difficulty cannot be avoided by making acceptance relative to the most stringent possible set of utilities (even if there were some way of determining what that is) because then the choice would be wrong for all less stringent sets. One cannot, by accepting or rejecting the hypothesis about the polio vaccine, do justice both to the problem of the physician and the veterinarian. The conflict can be resolved if the scientist either contents himself with providing them both with a single probability for the hypothesis (whereupon each makes his own decision based on the utilities peculiar to his problem) or if the scientist takes on the job of making a separate decision as to the acceptability of the hypothesis in each case. In any event, we conclude that it is not the business of the scientist as such, least of all of the scientist who works with lawlike hypotheses, to accept or reject hypotheses.[21]

Jeffrey's argument rests upon two lemmas: (a) if scientists accept and reject hypotheses (1), then they must make value judgments (4); and (b) if (1) is true, then (4) is false. The inevitable conclusion is that (1) is false—i.e., that the scientist neither accepts nor rejects hypotheses.

Jeffrey accepts (a) without any question as having been established by Rudner. His argument for (b) may be paraphrased as follows: Deciding whether to accept or reject a hypothesis is an open-ended decision problem—i.e., there is no practical objective in terms of which seriousness of error can be assessed. Hence, if a scientist decides to accept or reject a hypothesis, he cannot be taking the seriousness of error into account. Consequently, if (1) is true, (4) is false.

In spite of its persuasive character, Jeffrey's argument breaks down at several points.

(i) Rudner's argument for lemma (a) has already been shown to hold only if accepting a hypothesis is understood to be meaning*less* in open-ended situations. On the other hand, Jeffrey's argument for (b) depends upon the understanding that accepting a hypothesis is meaning*ful* in such cases. Hence, Jeffrey is guilty of equivocation.

(ii) Jeffrey's argument from the truth of (1) to the falsity of (4) depends upon the assumption that the decision problem is an open-ended one. An open-ended decision problem has been understood to be one that lacks a *practical* objective. However, such problems may still have a theoretical objective. It is at least an open question whether such an objective can serve as a basis for ascertaining the seriousness of mistakes.

(iii) Even if theoretical objectives cannot function in this way, Jeffrey's inference from (1) to the negation of (4) can still be avoided. It has been argued that Jeffrey is correct in asserting and Churchman is wrong in denying that there is a sense of "accepting a hypothesis" that is meaningful in open-ended situations. This does not mean, however, that this sense of "accepting a hypothesis" is meaningful *only* in open-ended situations. A person may decide what to believe only in order to believe true statements. But he may wish to believe statements which are true and which have some other desirable characteristic such as simplicity, explanatory power, effectiveness as propaganda, or a consoling emotive connotation. And the sense in which he accepts a statement as true in attempting to realize one of these objectives will be the same sense in which he might accept statements as true in open-ended situations. Again, it is at least an open question whether a scientist *qua* scientist has such a practical objective in accepting and rejecting hypotheses and, hence, has a basis for determining the seriousness of mistakes.

The failure of Jeffrey's argument does not, of course, imply the falsity of his conclusion. Indeed, another argument can be offered for rejecting (1). Whatever may be the merits of the inference from (1) to (4), empiricists are committed to accepting the inference from (1) to (3)—i.e., the inference from the claim that scientists accept and reject hypotheses to the need for assigning minimum probabilities for such acceptance and rejection. How are such minimum probabilities to be assigned? If no plausible alternative to a procedure that takes the values of the investigator into account is available, then (1) entails the rejection of the value-neutrality thesis.

Defenders of the Carnap-Hempel-Jeffrey view might feel that we are in such a predicament. Not wishing to abandon the value-neutrality thesis, they reject the conception of the scientist as one who accepts and rejects hypotheses.[22] However, following this strategy is like crashing into Scylla in order to avoid sinking in Charybdis. As Jeffrey himself admits,[23] the scientific literature suggests that scientists do often accept and reject hypotheses in a sense incompatible with (5). Furthermore, they often appear to feel that it is at least part of their business to do so. Consequently, an attempt to construct a theory of scientific inference based on the assumption that scientists do accept and reject hypotheses seems warranted.

IV

The question that remains is whether on this assumption the value-neutrality thesis can be maintained. An answer to this question seems to depend upon determining the manner in which minimum probabilities for accepting and rejecting hypotheses are assigned according to the canons of scientific inference. A study of the procedures for assigning minimum probabilities cannot be undertaken in this paper. Nonetheless, two possible outcomes of such an investigation that would support the value-neutrality thesis are worth mentioning. A consideration of these possibilities will serve to clarify the content of the value-neutrality thesis and to focus attention on the issues that must be settled before an adequate assessment of its merits can be made.

(a) The necessity of assigning minimum probabilities for accepting or rejecting hypotheses does not imply that the values, preferences, temperament, etc., of the investigator, or of the group whose interests he serves determine the assignment of these minima. The minimum probabilities might be functions of syntactical or semantical features of the hypotheses themselves. Indeed, they might not be determined by any identifiable factors at all other than certain rules contained in the canons of inference. These rules might fix the minima in such a way that given the available evidence two different investigatorrs would not be warranted in making different choices among a set of competing hypotheses. If the canons of inference did work in this way, they would embody the value-neutrality thesis.

(b) Even if the minimum probabilities were functions of identifiable values, the value-neutrality thesis would not necessarily have to be abandoned. When a scientist commits himself to certain "scientific" standards of inference, he does, in a sense, commit himself to certain normative principles. He is obligated to accept the validity of certain types of inference and to deny the validity of others. *The values that determine minimum probabilities may be part of this commitment.* In other words, the canons of inference might require of each scientist *qua* scientist that he have the same attitudes, assign the same utilities, or take each mistake with the same degree of seriousness as every other scientist. The canons of inference would, under these circumstances, be subject to the value-neutrality thesis; for the value-neutrality thesis does not maintain that the scientist *qua* scientist makes no value judgments but that given his commitment to the canons of inference he need make no further value judgments in order to decide which hypotheses to accept and which to reject.[24]

Thus, the tenability of the value-neutrality thesis does not depend upon whether minimum probabilities for accepting or rejecting hypotheses are a function of values but upon whether the canons of inference require of each scientist that he assign the same minima as every other scientist.

V

The arguments offered in this paper do not conclusively refute the major theses advanced by Rudner or Jeffrey. However, these arguments have justified further examination of the view that scientists accept or reject hypotheses in accordance with the value-neutrality thesis. In particular, it has been shown that Rudner and Churchman have failed to prove that a scientist must take the seriousness of mistakes into account in order to accept or reject hypotheses where the seriousness of mistakes is relative to practical objectives; it has also been shown that even if Rudner and Churchman were correct, the value-neutrality thesis would not entail Jeffrey's abandonment of the view that scientists accept or reject hypotheses; and, finally, it has been argued that even if scientists must take the seriousness of mistakes or other values into account in determining minimum probabilities, they may still accept or reject hypotheses in accordance with the value-neutrality thesis.

The outcome of this discussion is that the tenability of the value-neutrality thesis depends upon whether the canons of scientific inference dictate assignments of minimum probabilities in such a way as to permit no differences in the assignments made by different investigators to the same set of alternative hypotheses. An answer to this question can only be obtained by a closer examination of the manner in which minimum probabilities are assigned in the sciences. This problem will be the subject of another paper.

NOTES

1. C. W. Churchman, *Theory of Experimental Inference* (New York: Macmillan, 1948), Chap. xv.

2. R. B. Braithwaite, *Scientific Explanation* (Cambridge: Cambridge University Press, 1955), pp. 250–54.

3. R. Rudner, "The Scientist *qua* Scientist Makes Value Judgments," *Philosophy of Science*, XX (1953), 3.

4. *Ibid.*, p. 2.

5. The canons of scientific inference can be construed to be normative principles. The value-neutrality thesis does not deny this but does insist that given an initial commitment to these principles, the scientist need not and should not let his values, attitudes, and temperament influence his inferences any further. It is this claim that Rudner appears to deny.

6. Actually Rudner's version of (4) is stronger than mine. According to Rudner, the importance of making a mistake can be construed in "a typically ethical sense." In order to simplify the discussion, this rider will be dropped. The importance of making a mistake will be understood to be a function of the values, attitudes, preferences, and temperament of the investigator or group whose interests he serves regardless of the ethical character of these values, etc. Understood in this sense, (4) is still incompatible with the value-neutrality thesis.

7. Rudner, *op. cit.* pp. 2–3.

8. C. W. Churchman, "Science and Decision Making," *Philosophy of Science*, XXXIII (1956), 248.

9. Perhaps "*A* is believed by *X* to be the best procedure" should replace "*A* is the best procedure." The following discussion does not, however, demand a choice between these two definitions.

10. "The terms 'accepting' and 'rejecting' a statistical hypothesis are very convenient and are well established. It is important, however, to keep their exact meaning in mind and to discard various additional implications which may be suggested by intuition. Thus, to accept a hypothesis *H* means only to decide to take action *A* rather than action *B*. This does not mean that we necessarily believe that the hypothesis *H* is true. Also if the application of a rule of inductive behavior 'rejects' *H*, this means only that the rule prescribes action *B* and does not imply that we believe that *H* is false." [J. Neyman, *A First Course in Probability* (New York: Henry Holt & Co., 1950), pp. 259–60.] In this passage, Neyman does identify accepting a hypothesis *H* with acting on *H*. However, he refuses to identify accepting *H* with believing that *H*. In effect, therefore, he suspends judgment regarding the truth of (5).

11. R. C. Jeffrey, "Valuation and Acceptance of Scientific Hypotheses," *Philosophy of Science*, XXXIII (1956), 242.

12. This expression is due to Churchman (*loc. cit.*).

13. *Ibid.*, pp. 248–49.

14. By a "theoretical" objective, I shall understand any objective that is concerned with selecting true hypotheses from a given list. A practical objective is one that is not theoretical. This dichotomy overlooks distinctions between ethical, practical, and aesthetic objectives by grouping them together. It also treats many objectives as practical that might legitimately be held to be theoretical. The purpose of the twofold partition of objectives, however, is to avoid a trivial interpretation of (5) while permitting Churchman and Rudner as much leeway as possible in their interpretation of this assumption.

15. This claim might seem counterintuitive. There is a widely held view that if a person really believes in a hypothesis he should be ready to act on it. This "put up or shut up" analysis may be understood in two ways: (a) Belief in *H* implies acting on the basis of *H*. R. M. Martin seems to adopt the former view (*Toward a Systematic Pragmatics*, Amsterdam, North Holland, 1959, p. 11). This version of the "put up or shut up" analysis does not seem adequate to at least one familiar sense of "accepting a hypothesis." The very fact that people often think that one ought to act on a hypothesis if one believes it implies that one might not so act. Sense (b) of the "put up or shut up" analysis seems more plausible. Nonetheless it yields results that themselves appear to be counterintuitive. If this thesis demands readiness to act relative to *any* objective, then one would not be warranted in accepting a hypothesis as true unless the degree of confirmation approached certainty. For there is always the possibility that some objectives exist relative to which mistakes are so serious as to demand enormously high degrees of confirmation. Such a requirement seems unreasonable. On the other hand, if the objectives relative to which one should be ready to act are restricted in some way, it is difficult to see how the restrictions could be specified without destroying the initial plausibility of the "put up or shut up" analysis.

16. I owe this observation to Mortimer Kadish.

17. R. Carnap, *Logical Foundations of Probability* (Chicago: University of Chicago Press, 1950), pp. 205–7.

18. C. G. Hempel, review of Churchman's *Theory of Experimental Inference*, in *Journal of Philosophy*, XLVI (1949), 560.

19. Carnap, *op. cit.*, p. 206.

20. The difference between this view and the revised version of the Churchman-Rudner position suggested above is that the latter considers the scientist as a formulator of practical policy whereas the former considers him to be an adviser to the policy maker. This difference reflects itself in differing conceptions of non-deductive inference. According to the revised Churchman-Rudner view, the "conclusion" of a non-deductive inference is a choice of a course of action. According to the Carnap-Hempel-Jeffrey view, the conclusion is an assignment of a degree of confirmation to a hypothesis.

21. Jeffrey, *op. cit.*, p. 245.

22. Hempel (*loc. cit*) comes closer to arguing in this way than either Carnap or Jeffrey.

23. Jeffrey, *op. cit.*, p. 246.

24. It should also be clear that the value-neutrality thesis says nothing concerning the rationale for adopting scientific canons of inference but only about the content of these canons.

The Justification of Induction

Bertrand Russell

ON INDUCTION

In almost all our previous discussions we have been concerned in the attempt to get clear as to our data in the way of knowledge of existence. What things are there in the universe whose existence is know to us owing to our being acquainted with them? So far, our answer has been that we are acquainted with our sense-data, and, probably, with ourselves. These we know to exist. And past sense-data which are remembered are known to have existed in the past. This knowledge supplies our data.

But if we are to be able to draw inferences from these data—if we are to know of the existence of matter, of other people, of the past before our individual memory begins, or of the future, we must know general principles of some kind by means of which such inferences can be drawn. It must be known to us that the existence of some one sort of thing, *A*, is a sign of the existence of some other sort of thing, *B*, either at the same time as *A* or at some earlier or later time, as, for example, thunder is a sign of the earlier existence of lightning. If this were not known to us, we could never extend our knowledge beyond the sphere of our private experience; and this sphere, as we have seen, is exceedingly limited. The question we have now to consider is whether such an extension is possible, and if so, how it is effected.

Let us take as an illustration a matter about which none of us, in fact, feel the slightest doubt. We are all convinced that the sun will rise tomorrow. Why? It this belief a mere blind outcome of past experience, or can it be justified as a reasonable belief? It is not easy to find a test by which to judge whether a belief of this kind is reasonable or not, but we can at least ascertain what sort of general beliefs would suffice, if true, to justify the judgment that the sun will rise tomorrow, and the many other similar judgments upon which our actions are based.

It is obvious that if we are asked why we believe that the sun will rise to-

---

From Bertrand Russell, *The Problems of Philosophy*, by permission of the Clarendon Press, Oxford.

morrow, we shall naturally answer, "Because it always has risen every day." We have a firm belief that it will rise in the future, because it has risen in the past. If we are challenged as to why we believe that it will continue to rise as heretofore, we may appeal to the laws of motion: The earth, we shall say, is a freely rotating body, and such bodies do not cease to rotate unless something interferes from outside, and there is nothing outside to interfere with the earth between now and tomorrow. Of course it might be doubted whether we are quite certain that there is nothing outside to interfere, but this is not the interesting doubt. The interesting doubt is as to whether the laws of motion will remain in operation until tomorrow. If this doubt is raised, we find ourselves in the same position as when the doubt about the sunrise was first raised.

The *only* reason for believing that the laws of motion will remain in operation is that they have operated hitherto, so far as our knowledge of the past enables us to judge. It is true that we have a greater body of evidence from the past in favour of the laws of motion than we have in favour of the sunrise, because the sunrise is merely a particular case of fulfilment of the laws of motion, and there are countless other particular cases. But the real question is: Do *any* number of cases of a law being fulfilled in the past afford evidence that it will be fulfilled in the future? If not, it becomes plain that we have no ground whatever for expecting the sun to rise tomorrow, or for expecting the bread we shall eat at our next meal not to poison us, or for any of the other scarcely conscious expectations that control our daily lives. It is to be observed that all such expectations are only *probable*; thus we have not to seek for a proof that they *must* be fulfilled, but only for some reason in favour of the view that they are *likely* to be fulfilled.

Now in dealing with this question we must, to begin with, make an important distinction, without which we should soon become involved in hopeless confusions. Experience has shown us that, hitherto, the frequent repetition of some uniform succession or coexistence has been a *cause* of our expecting the same succession or coexistence on the next occasion. Food that has a certain appearance generally has a certain taste, and it is a severe shock to our expectations when the familiar appearance is found to be associated with an unusual taste. Things which we see become associated, by habit, with certain tactile sensations which we expect if we touch them; one of the horrors of a ghost (in many ghost stories) is that it fails to give us any sensations of touch. Uneducated people who go abroad for the first time are so surpised as to be incredulous when they find their native language not understood.

And this kind of association is not confined to men; in animals also it is very strong. A horse which has been often driven along a certain road resists the attempt to drive him in a different direction. Domestic animals expect food when they see the person who usually feeds them. We know that all these rather crude expectations of uniformity are liable to be misleading. The man who has fed the chicken every day throughout its life at last wrings its neck instead, showing that more refined views as to the uniformity of nature would have been useful to the chicken.

But in spite of the misleadingness of such expectations, they nevertheless exist. The mere fact that something has happened a certain number of times causes animals and men to expect that it will happen again. Thus our instincts certainly cause us to believe that the sun will rise tomorrow, but we may be in no better a position than the chicken which unexpectedly has its neck wrung. We have therefore to distinguish the fact that past uniformities *cause* expectations as to the future, from the question whether there is any reasonable ground for giving weight to such expectations after the question of their validity has been raised.

The problem we have to discuss is whether there is any reason for believing in what is called "the uniformity of nature." The belief in the uniformity of nature is the belief that everything that has happened or will happen is an instance of some general law to which there are *no* exceptions. The crude expectations which we have been considering are all subject to exceptions, and therefore liable to disappoint those who entertain them. But science habitually assumes, at least as a working hypothesis, that general rules which have exceptions can be replaced by general rules which have no exceptions. "Unsupported bodies in air fall" is a general rule to which balloons and aeroplanes are exceptions. But the laws of motion and the law of gravitation, which account for the fact that most bodies fall, also account for the fact that balloons and aeroplanes can rise; thus the laws of motion and the law of gravitation are not subject to these exceptions.

The belief that the sun will rise tomorrow might be falsified if the earth came suddenly into contact with a large body which destroyed its rotation; but the laws of motion and the law of gravitation would not be infringed by such an event. The business of science is to find uniformities, such as the laws of motion and the law of gravitation, to which, so far as our experience extends, there are no exceptions. In this search science has been remarkably successful, and it may be conceded that such uniformities have held hitherto. This brings us back to the question: Have we any reason, assuming that they have always held in the past, to suppose that they will hold in the future?

It has been argued that we have reason to know that the future will resemble the past, because what was the future has constantly become the past, and has always been found to resemble the past, so that we really have experience of the future, namely of times which were formerly future, which we may call past futures. But such an argument really begs the very question at issue. We have experience of past futures, but not of future futures, and the question is: Will future futures resemble past futures? This question is not to be answered by an argument which starts from past futures alone. We have therefore still to seek for some principle which shall enable us to know that the future will follow the same laws as the past.

The reference to the future in this question is not essential. The same question arises when we apply the laws that work in our experience to past things of which we have no experience—as, for example, in geology, or in theories as to the origin of the Solar System. The question we really have to ask is: "When two things have been found to be often associated, and no instance is known of the one occurring without the other, does the occurrence of one of the two, in a fresh instance, give any good ground for expecting the other?" On our answer to this question must depend the validity of the whole of our expectations as to the future, the whole of the results obtained by induction, and in fact practically all the beliefs upon which our daily life is based.

It must be conceded, to begin with, that the fact that two things have been found often together and never apart does not, by itself, suffice to *prove* demonstratively that they will be found together in the next case we examine. The most we can hope is that the oftener things are found together, the more probable it becomes that they will be found together another time, and that, if they have been found together often enough, the probability will amount *almost* to certainty. It can never quite reach certainty, because we know that in spite of frequent repetitions there sometimes is a failure at the last, as in the case of the chicken whose neck is wrung. Thus probability is all we ought to seek.

It might be urged, as against the view we are advocating, that we know all natural phenomena to be subject to the reign of law, and that sometimes, on the basis of observation, we can see that only one law can possibly fit the facts of the case. Now to this view there are two answers. The first is that, even if *some* law which has no exceptions applies to our case, we can never, in practice, be sure that we have discovered that law and not one to which there are exceptions. The second is that the reign of law would seem to be itself only probable, and that our belief that it will hold in the future, or in unexamined cases in the past, is itself based upon the very principle we are examining.

The principle we are examining may be called the *principle of induction*, and its two parts may be stated as follows:

(a) When a thing of a certain sort A has been found to be associated with a thing of a certain other sort B, and has never been found dissociated from a thing of the sort B, the greater the number of cases in which A and B have been associated, the greater is the probability that they will be associated in a fresh case in which one of them is known to be present;

(b) Under the same circumstances, a sufficient number of cases of association will make the probability of a fresh association nearly a certainty, and will make it approach certainty without limit.

As just stated, the principle applies only to the verification of our expectation in a single fresh instance. But we want also to know that there is a probability in favour of the general law that things of the sort A are *always* associated with things of the sort B, provided a sufficient number of cases of association are known, and no cases of failure of association are known. The probability of the general law is obviously less than the probability of the particular case, since if the general law is true, the particular case must also be true, whereas the particular case may be true without the general law being true. Nevertheless the probability of the general law is increased by repetitions, just as the probability of the particular case is. We may therefore repeat the two parts of our principle as regards the general law, thus:

(a) The greater the number of cases in which a thing of the sort A has been found associated with a thing of the sort B, the more probable it is (if no cases of failure of association are known) that A is always associated with B;

(b) Under the same circumstances, a sufficient number of cases of the association of A with B will make it nearly certain that A is always associated with B, and will make this general law approach certainty without limit.

It should be noted that probability is always relative to certain data. In our case, the data are merely the known cases of coexistence of A and B. There may be other data, which *might* be taken into account, which would gravely alter the probability. For example, a man who had seen a great many white swans might argue, by our principle, that on the data it was *probable* that all swans were white, and this might be a perfectly sound argument. The argument is not disproved by the fact that some swans are black, because a thing may very well happen in spite of the fact that some data render it improbable. In the case of the swans, a man might know that colour is a very variable characteristic in many species of animals, and that, therefore, an induction as to colour is peculiarly liable to error. But this knowledge would be a fresh datum, by no means proving that the probability relatively to our previous data had been wrongly estimated. The fact, therefore, that things often fail to fulfill our expectations is no evidence that our expectations will not *probably* be fulfilled in a given case or a given class of cases. Thus our inductive principle is at any rate not capable of being *disproved* by an appeal to experience.

The inductive principle, however, is equally incapable of being *proved* by an appeal to experience. Experience might conceivably confirm the inductive principle as regards the cases that have been already examined; but as regards unexamined cases, it is the inductive principle alone that can justify any inference from what has been examined to what has not been examined. All arguments which, on the basis of experience, argue as to the future or the unexperienced parts of the past or present, assume that inductive principle; hence we can never use experience to prove the inductive principle without begging the question. Thus we must either accept the inductive principle on the ground of its intrinsic evidence, or forgo all justification of our expectations about the future. If the principle is unsound, we have no reason to expect the sun to rise tomorrow, to expect bread to be more nourishing than a stone, or to expect that if we throw ourselves off the roof we shall fall. When we see what looks like our best friend approaching us, we shall have no reason to suppose that his body is not inhabited by the mind of our worst enemy or of some total stranger. All our conduct is based upon associations which have worked in the past, and which we therefore regard as likely to work in the future; and this likelihood is dependent for its validity upon the inductive principle.

The general principles of science, such as the belief in the reign of law, and the belief that every event must have a cause, are as completely dependent upon the inductive principle as are the beliefs of daily life. All such general principles are believed because mankind have found innumerable instances of their truth and no instances of their falsehood. But this affords no evidence for their truth in the future, unless the inductive principle is assumed.

Thus all knowledge which, on a basis of experience tells us something about what is not experienced, is based upon a belief which experience can neither confirm nor confute, yet which, at least in its more concrete applications, appears to be as firmly rooted in us as many of the facts of experience. The existence and justification of such beliefs—for the inductive principle, as well shall see, is not the only example—raises some of the most difficult and most debated problems of philosophy. We will, in the next chapter, consider briefly what may be said to account for such knowledge, and what is its scope and its degree of certainty.

Max Black

THE INDUCTIVE SUPPORT OF INDUCTIVE RULES

1. THE VIEW TO BE EXAMINED

It is very commonly said that any attempt to "justify" induction by means of an inductive argument must beg the question. The following are characteristic expressions of this view:

> It is impossible that any arguments from experience can prove this resemblance of the future: since all these arguments are founded on the supposition of that resemblance [Hume, *Enquiry*, Section IV].

> All arguments which, on the basis of experience, argue as to the future or the unexperienced parts of the past or present, assume the inductive principle; hence we can never use experience to prove the inductive principle without begging the question [Russell, *The Problems of Philosophy* (London, 1912), p. 106].

> Though it is true that the inductive inference has been successful in past experience, we cannot infer that it will be successful in future experience. The very inference would be an inductive inference, and the argument would be circular. Its validity presupposes the principle that it claims to prove [Reichenbach, *Theory of Probability* (Berkely, 1949), p. 470].

It would be hard to find any writer who disagreed with these statements.

I want to distinguish two views that these writers may be advocating: (a) The first is that *no* inductive argument ought to be regarded as correct until a philosophical justification of induction has been provided; and, hence, induction must not be used in the attempt to provide such a justification. This is not the view that I wish to discuss in this essay.[1] (b) It may be held that if all inductive arguments are conducted according to one and the same inductive rule, *inductive* inference in support of *that* rule is bound to be circular. (If *several* logically

---

independent inductive rules are admitted, there need be no circularity in using one of them to support another.) This second view may seem as obvious as (a), but it is certainly a different contention. (a) arises from sweeping doubts about any and every induction; but a philosopher might take *some* inductive arguments to be correct and still agree with (b). Indeed, (b) looks almost self-evident.

Nevertheless, I shall try to show that the view I have called "(b)" is mistaken. The conclusion I wish to establish is that if all inductive arguments were to be conducted according to one and the same inductive rule, an inductive argument in support of that rule might still be "correct," i.e., might satisfy all the tests that render an inductive argument acceptable.

2. EXAMPLES OF INDUCTIVE ARGUMENTS

The following are simple examples of good inductive arguments:

(i) That egg has been boiling for twenty minutes, so it is bound to be hard by now.

(ii) My car has never failed to start when the temperature was above 30—it is pretty certain to start today.

(iii) There are a thousand tickets in this lottery and only one prize: Almost certainly, this ticket will not win the prize.

(iv) All kinds of acids under all sorts of conditions have invariably turned blue litmus red: Hence, acids turn blue litmus red.

(v) In a random selection of housewives interviewed, four out of five were found to play bridge; probably the proportion of bridge players among all housewives is close to four fifths.

These examples may serve to remind us of the wide variety of arguments that are commonly called "inductive."[2]

3. SOME FEATURES OF INDUCTIVE ARGUMENTS

I hold each of the following remarks to be true, though I shall not defend them here. (For the sake of brevity, I shall assume throughout that the premises, if there are more than one, are amalgamated into a single premise.)

(i) In no inductive argument is the conclusion entailed by the premise.

In other words, the conjunction of the premise and the negation of the conclusion is not a self-contradiction. This follows from the meaning of "inductive argument": that an inductive argument is not a special kind of deductive argument is a necessary truth.[3]

(ii) An inductive argument may indicate the *degree of support* that the premise gives to the conclusion.

Sometimes, the conclusion may simply be presented as following from the premise [see example (iv) of the last section] with no indication of the strength of the premise; but very commonly the "degree of support" is explicitly shown by the use of expressions such as "pretty certainly," "probably," "extremely likely," and so on. When this happens, the conclusion is still asserted *with* a certain degree of support, and in the cases here considered is not *about* that degree of support. (There may, however, be second-order arguments about degrees of support.) In this respect, inductive arguments differ sharply from deductive ones. The latter either establish their conclusion or they do not—*tertium non datur*. An inductive argument, however, may offer stronger or weaker reasons in support of its conclusion.[4]

(iii) Some inductive arguments are "correct," i.e., the conclusion in question is established (perhaps with an indicated degree of support) by the premise.

I use the word "correct" in preference to "valid," reserving the latter for deductive arguments. It is to be presumed that the reader can recognize some instances of inductive correctness; if not, this essay will be of no value to him.

(iv) A correct inductive argument may have a true premise and a false conclusion.

Suppose we argue: "This pack of cards has been shuffled, so probably nobody will get thirteen hearts in the deal." If the pack of cards has been shuffled but somebody does get thirteen hearts in the deal, this does not show that the argument is incorrect. But of course, if this happened very often, the case would be altered.

(v) Every correct inductive argument belongs to a class of arguments (call it the "associated class") all of which are correct, and every incorrect argument to a class all of which are incorrect.[5]

Roughly speaking, correctness and incorrectness are properties of the members of *classes* of inductive arguments. In a limiting case, the *associated class* might be a unit class. I shall not discuss the defining properties of such classes.

(vi) All the members of a class of associated arguments conform to the same *rule of argument*.

The rule has the form: "To argue from such-and-such a premise that (probably, certainly, very likely, etc.) such-and-such a conclusion holds." It will be convenient to say that the argument is *governed by* the corresponding rule.

(vii) In order for an inductive argument to be correct, the rule to which it conforms must be *reliable*.

A rule is reliable if it yields true conclusions in *most* cases in which it governs arguments having true premises. If a rule governs an argument with a true premise and a true conclusion, it may be said to be *successful* in that instance; if it governs an argument with a true premise and a false conclusion, it may be said to be unsuccessful. Thus a rule is reliable when its successes outnumber its failures. Clearly, we can also speak of *degrees* of reliability.

(viii) From any inductive rule there may be derived a corresponding principle.

If the rule is, "To argue from X (with certainty, high or low probability) that Y," the corresponding principle is, "Whenever X, then Y."[6] A principle, unlike a rule, has a truth value, but the principle corresponding to a correct argument need not be true. For we demand only that the rule be reliable (see iv and vii above).

(ix) If a principle corresponding to rule is true, it is a contingent truth, not a necessary one.

For otherwise the argument would be deductive, not inductive.

(x) A sound inductive inference need not involve the formulation of the rule governing the argument used.

I distinguish between an inference and an argument. In the former, the speaker *asserts* the premise and treats it as a reason for asserting the conclusion. He is then using an argument. If his inference is to be sound, the argument must be used *legitimately*, in a sense to be explained later (see section 9). But the conditions of legitimate use do not include explicit formulation of the rule governing the argument. If the rule is formulated, the argument and the inference may be said to be *formal*. Not all sound inductive inferences are formal.

(xi) There is no single rule governing all correct inductive arguments.

The writers we are to discuss deny this, so I shall conduct the discussion on the supposition that it is false. Yet I am inclined to believe that many of the difficulties in this subject can be traced back to the incorrect supposition that there is a supreme inductive principle.

4. THE INDUCTIVE RULES TO BE CONSIDERED

As I have already said, the writers quoted assume that there is a single supreme inductive rule or principle which governs all correct inductive arguments and has to be shown to be true (or at least probable) if any inductions are to be sound. Now it is not easy to find plausible candidates for the role of such a supreme rule or principle, and writers on the foundation of induction have proposed many alternatives. But I hope to conduct the discussion in such a way that it may apply with little or no modification, no matter what formulation is chosen for the alleged first principle or rule. For the sake of definiteness in what follows, I propose to consider separately the following rules alone:

R_1: To argue from *All examined instances of* A*'s have been* B to All A*'s are* B.

R_2: To argue from *Most instances of* A*'s examined in a wide variety of conditions have been* B to (probably) *The next* A *to be encountered will be* B.

The first of these may be taken to express a rule governing arguments by "simple enumeration." It will be remembered that many philosophers of induction hold that the justification of induction depends in the end upon the justification of this unsophisticated type of inductive argument.[7] The second rule has been included so that we may have one example before us of a type of rule governing arguments in which the conclusion receives a greater or lesser degree of indicated support from the premise (cf. the use of the word "probably").

Both R_1 and R_2 are somewhat crudely expressed. Some readers may be troubled by the temporal references in their formulation; and others might reasonably demand something less vague than the phrase "a wide variety of conditions" that occurs in the phrasing of R_2. Such niceties will not affect our argument, however.

5. TWO SELF-SUPPORTING ARGUMENTS

Let us now formulate two inductive arguments designed to support R_1 and R_2, respectively.

(a_1): All examined instances of the use of R_1 in arguments with true premises have been instances in which R_1 has been successful.

Hence:

All instances of the use of R_1 in arguments with true premises are instances in which R_1 is successful.

We may remind ourselves that we have agreed to call a rule successful if it is used in an argument having a true premise and a true conclusion. So the conclusion of (a_1) says that no argument governed by R_1 and having a true premise ever has a false conclusion. This amounts to saying that R_1 is perfectly reliable, and hence we can parallel (a_1) with the following less exact argument:

(a_{11}): R_1 has always been reliable in the past.

Hence:

R_1 is reliable.

The argument in support of R_2 that we propose to consider runs as follows:

(a_2): In most instances of the use of R_2 in arguments with true premises examined in a wide variety of conditions, R_2 has been successful.

Hence (probably):

In the next instance to be encountered of the use of R_2 in an argument with a true premise, R_2 will be successful.

There is no simplified form of this argument that will exactly parallel (a_{11}) above. But the following will serve our purpose:

(a_{22}): R_2 has usually been successful in the past.

Hence (probably):

R_2 will be successful in the next instance.

The arguments have been so formulated that (a_1) is governed by R_1, and (a_2) by R_2. Thus the first maintains in its conclusion the (perfect) reliability of the rule to which that very same argument conforms; and the second maintains in its conclusion the continued success of the rule to which the argument itself conforms. I shall call these arguments *self-supporting.*

Our task accordingly narrows itself down to determining whether and in what sense either (a_1) or (a_2) is guilty of circularity.

6. THE ALLEGED CIRCULARITY IS NOT THAT OF "PETITIO PRINCIPII"

When the charge of circularity is made against an argument, one of two things is commonly alleged to be wrong. It may be that one of the premises is identical with the conclusion, or is, perhaps, that same conclusion in different words. In a more subtle version of circularity, at least one of the premises is such that it is impossible to get to know its truth without simultaneously or antecedently getting to know the truth of the conclusion. "The premise unduly assumed is generally not the conclusion itself differently expressed but something which can only be proved by means of the conclusion" [H. W. B. Joseph, *Introduction to Logic* (Oxford, 1916), p. 592].

Can the charge of circularity against our arguments (a_1) and (a_2) be construed in accordance with this normal meaning of "circularity"? Let us see, first, whether the conclusion of the first argument appears—perhaps in disguise—as a premise. The conclusion of (a_1) is that the rule R_1 is always successful (when applied to arguments having true premises). There is here only one premise, to the effect that R_1 has always been successful in all cases of its application to arguments having true premises *so far examined.* Clearly, this premise is entailed by the conclusion, but not vice versa. Hence (a_1) is free from the first kind of circularity, that arises when a premise duplicates the conclusion.

But could the premise of (a_1) be known to be true, without the conclusion of (a_1) being simultaneously or antecedently known to be true? Well, I suppose that the premise (R_1 has always been successful when applied to arguments having true premises) might easily be found to be true by the use of observation and memory, i.e., without the use of *any* inductions conforming to R_1. But in order to have the most unfavorable case before us, let us suppose that the evidence in favor of the premise of (a_1) was itself obtained by means of inductions conducted according to the same rule R_1. Even so, there would be no circularity. For in none of these arguments could the statement, R_1 *is always successful*—or, what is the same thing, R_1 *is reliable*—have been a *premise.* All of these arguments were *governed by* R_1, but this did not require them to assert that the rule to which they conformed was reliable. (I do not checkmate a king in a game of chess unless my move is legal, but in making the move I do not *announce* that the move is legal, nor is any such announcement required in order to make that move legal.)

Now precisely similar considerations apply to the argument we called (a_2).

Inspection shows that the conclusion is not entailed by the premise; so (a_2) is free from the more obvious type of circularity. And even if we suppose that the truth of the premise was established by means of inductions using R_2 itself, it is by no means required that a statement about R_2's reliability should have been a premise in any of those arguments. Thus (a_2), like (a_1), is free from each kind of circularity.

But we hardly need these details to establish our point. The following considerations are simpler and equally decisive. Any argument that is circular, in the traditional sense here in point, is a *valid deductive argument*, however worthless it may be for the purpose of getting new information from the conclusion. For if one of the premises duplicates the conclusion, it is logically impossible for all the premises to be true while the conclusion is false. And again, if it is impossible to know the premises to be true, without simultaneously or antecedently coming to know that the conclusion is true, this can only be because the conjunction of the premises entails the conclusion. So, if either or our "self-supporting" arguments really did "beg the question," it would have to be a valid deductive argument, and hence not an inductive argument at all. But whatever faults these arguments may be supposed to have, it is plain that neither is deductively valid—in neither case does the conjunction of the sole premise and the denial of the conclusion result in a self-contradiction. Hence, neither of them is deductively valid, and so neither of them can beg the question.

Of course, this is by no means the end of the matter. Our self-supporting arguments admittedly continue to look suspicious. Take (a_1) for instance. Its conclusion is that the very same rule used in arriving at that conclusion is reliable: More argument will be needed before we can be sure that there need be nothing logically reprehensible in this kind of proceeding.

First, let us convince ourselves that our self-supporting arguments can at least be correct, whether they have any cognitive value or not. (The reader will please remember that I am using "correct" as an analogue for inductive arguments of the term "valid," reserved for deductive arguments. Just as a deductive argument can be valid but cognitively worthless, so an inductive argument might conceivably be correct, but useless.)

7. SELF-SUPPORTING ARGUMENTS CAN BE CORRECT

Suppose, for the sake of the argument, that R_1 is in fact perfectly reliable, i.e., that every argument conforming to R_1, and having a true premise, will also have a true conclusion. Then, there is no reason why the premise of (a_1) should not be true; and if this is so, our assumption is that the conclusion of(a_1)will also be true. Thus, in this case, as in all others of the application of R_1, the argument that it governs will never lead from a true premise to a false conclusion, and such an argument will be perfectly correct.

We can perhaps make the situation more vivid by imagining that a machine has been constructed for delivering conclusions according to the rule R_1 from data supplied to it. Then, if the machine is informed that "All examined instances of the working of this machine upon true data have been instances of successful operation of the machine," it will deliver the conclusion "All instances of the operation of this machine upon true data are successful." And, upon our assumption, that R_1 is reliable, this conclusion will be true. There is nothing in the nature of the data fed to the machine upon *this* occasion to prevent it from drawing the right conclusion, here as always. No paradox results from the supposition that the machine makes true predictions about *its own* reliability.

Or again, suppose another machine drawing conclusions, this time according to the rule R_2. When informed "In most of the examined workings of this machine (in a wide variety of conditions) a true conclusion was derived whenever a true premise was furnished," the machine will report the conclusion: "(Probably) the next instance of the application of this machine to a true premise will result in a true conclusion." And there is no reason why this answer should not be true (at any rate, in a majority of such cases, which is all that we need).[8]

We are not machines, however. And we shall not regard the use of arguments (a_1) and (a_2) as justified unless certain other conditions are satified. For an argument may be "correct" in the sense explained above and yet its use by somebody may be held illegitimate. For example, suppose Smith believes some rule R to be usually unsuccessful, but nevertheless deliberately draws a conclusion from a premise in accordance with R. Now Smith might be mistaken in his belief and R might really be reliable, but this is no defense of this procedure. Though the argument he used was a "correct" one, his use of it was unjustified, and his inference was illegitimate. (Cf. the case of a man who inadvertently tells the truth while lying.)

We may say, following W. E. Johnson,[9] that before an inference may be regarded as legitimate or "reasonable," certain "epistemic conditions" must be satisfied by the reasoner. Now it is conceivable that circularity of a kind more subtle than any we have yet considered might arise in the satisfaction of the epistemic conditions for the legitimate use of our self-supporting arguments.

8. WHEN IS THE USE OF A DEDUCTIVE RULE OF INFERENCE LEGITIMATE?

For the sake of comparison, let us begin by considering the epistemic conditions that must be satifised before anybody's use of a *deductive* rule can count as legitimate. One might be tempted to say that no deductive inference is legitimate unless all the arguments governed by that rule are known to be valid.[10] But this is to demand too much: All we are entitled to ask is that the man who applies the rule shall have no good reasons to suppose that the rule ever governs an invalid argument. Many a deductive inference is made by a reasoner who has never consciously formulated the rule to which he was conforming, let alone considered whether there were any reasons for suspecting the rule to be untrustworthy. But consider the most self-conscious and deliberate deductive reasoning that you please: Suppose the reasoner has formulated the rule he is using, finds its validity self-evident, and after weighing any objections against it that he knows, still continues to find it self-evident.[11] Suppose, moreover, that he has gone to the trouble of considering whether it is deducible from, or is at any rate consistent with, other rules whose validity he finds self-evident. If the rule survives this process of critical scrutiny, the reasoner has done all that could be reasonably demanded according to the most rigorous standard of self-consciously rational employment of a deductive rule. But even this extraordinary and unusual procedure of preliminary cross-examination of the rule to be employed does not guarantee that all the arguments that conform to the rule shall be valid; for what seems self-evident after the most searching criticism may nevertheless still be false. Yet this process of prior test is the very utmost that can be demanded of a reasoner (and far less than this would normally be regarded as enough). If a man has done as we have described, then, so far as concerns the employment of the deductive rule in question, he has satisfied all the epistemic conditions.

9. WHEN IS THE USE OF AN INDUCTIVE RULE LEGITIMATE?

Inductive rules are not self-evidently reliable, nor can their reliability be deduced from self-evident truths. Yet there is something in inductive reasoning that is analogous to self-evidence in deductive reasoning. Some inductive rules impress us as intrinsically trustworthy—so much so, that it calls for a very considerable effort of abstraction from the situations in which we use them to be able to entertain seriously the possibility that they may not be reliable after all. Unless the inductive rules we use "inspire confidence," as we might say, we are hardly reasonable in using them. But a severe critic of inductive reasoning may regard such untutored trust in a rule as very likely to lead to error. (For a deductive rule that looks self-evident may, after all, be not a tautology; and an industrious rule that we cannot help trusting may not be reliable—may produce false conclusions from true premises most of the time.) Very well, then—let us try to satisfy the most stringent demands that such a critic might feel called upon to make. Let us suppose we take the rule that initially "inspires confidence" and, not content with that, make the most painstaking examination of a wide variety of cases in which that rule has been used in the past, considering also all the objections that might plausibly be brought against the use of that rule. Suppose, now, that our rule has survived this searching cross-examination and still seems to us, as it did at the outset, worthy of our confidence in its reliability. Then, surely, all the guarantees that could reasonably be asked of us will have been supplied. And our use of that rule will be legitimate, even if that rule should, after all, and in spite of our best efforts, prove unrealible. To demand more than this, e.g., that the rule should be *known* to be reliable before it is used, would be to render all genuine induction impossible. For if the rule were known to be reliable prior to its employment, we could adjoin the assertion that the rule was reliable to our other premises and *deduce* that desired conclusion. Thus, we seem to have stated sufficiently stringent conditions for the reasonable use of an inductive rule. The question that I now want to settle is whether the satisfaction of conditions as stringent as these will lead, in the case of our self-supporting arguments, to any kind of circularity or triviality.

10. CAN A SELF-SUPPORTING ARGUMENT BE CORRECT WITHOUT TRIVIALITY?

We have already seen that (a_1) and (a_2) may be correct without begging the question. Suppose now that somebody using (a_1) tries to satisfy the stringent conditions outlined above. Suppose, in fact, that before he argues, he must first regard R_1 as reliable, must then have taken deliberate steps to check this spontaneous confidence in R_1 by examining previous instances of the use of R_1, possible objections to its use, and its logical connections with other rules that he trusts. If R_1 survives such a process of criticism, the man who uses it will, of course, have excellent reasons for holding it to be reliable. Now will this render the inference (a_1) superfluous?

Well, no doubt, inductive inferences will have been used in the course of finding good reasons for one's confidence in the reliability of R_1 (and many of these inferences may even have been conducted in accordance with that very same rule R_1). If that very same inductive evidence for R_1's reliability is now produced again as the premise of (a_1), (a_1) will indeed yield no new knowledge, and will then, indeed, lead to useless repetition of what was known at the outset. *But there is nothing in the specification of* (a_1) *that requires us to confine ourselves to the evidence that we previously had for* R_1*'s reliability*. Having evidence, *K*, say, that gives us excellent reason to hold R_1 reliable, and thus to make any employment of *R* in an inductive argument legitimate by the most stringent tests that might be demanded, there is nothing to prevent us going in search of further relevant evidence. If we find even one new argument in which R_1 has been successfully employed, and refer in formulating the premise of (a_1) to that new evidence as well as to the evidence we already had, that self-supporting argument may really be of value. For consider what would have been the result of our going in search of a new argument conforming to R_1 and finding that R_1 was in fact unsuccessful in that new instance. Then we should have discovered that although we did previously have good reason (indeed the best) to trust R_1, further experience had shown after all that R_1 was not perfectly reliable as we had supposed. And just because further experience might discredit R_1, the argument to the effect that further experience yielded a result favorable to R_1 has cognitive value.

The case we have made for the use of (a_1) is an artificial one. For experience does *not* show that R_1 is invariably successful and the rules that we use in legitimate inductive inferences have only a greater or lesser *degree* of reliability. And for this reason, (a_2) comes closer than (a_1) to representing the type of argument that we do sometimes have occasion to use in practice.

Suppose, as before, that some exacting critic of inductive reasoning refuses to allow us to use (a_2) until we have previously convinced ourselves by careful investigation that R_2 is trustworthy. We shall be satisfied if such investigation gives us good reason to suppose that R_2 has a sufficiently high degree of reliability. Suppose such evidence assembled. Then, if we simply repeat the evidence in the form of a premise for (a_2), that argument will certainly accomplish nothing except to remind us of what we already knew. But, as before, there is nothing to stop us going in search of further evidence in support of R_2. In favorable cases, such new evidence will not merely support the general conclusion as to R_2's reliability: It may *raise* the degree of the reliability of that rule. And it is precisely in this way that arguments like (a_2) have cognitive value.

For how do we in fact proceed in inductive inquiries? No doubt we are taught to argue inductively before we are ever in a position to reflect upon the reliability of the rules that we find ourselves already using. (And if not, how could there be any matter for reflection?) But as we become self-conscious and critical about the methods that we are using, we come to realize that it is of the essence of inductive method that the rules according to which it operates must be treated as being probably subject to qualifications and restrictions that are not clearly apprehended when those rules are used unreflectively. And it is precisely in rendering clearer such restrictions that inductive arguments in support of *particular* inductive methods find their greatest use in practice. It is by the application of inductive methods to those same methods that induction, in favorable cases, may become "self-correcting." And in all this, there is no more vicious circularity involved than there is in the truism that tools can be used to sharpen themselves. Inductive arguments in support of inductive rules can meet the tests that determine the acceptability of an inductive argument. Anybody who thinks he has good grounds for condemning all inductive arguments will also condemn inductive arguments in support of inductive rules. But he will have no good reasons for singling out what we have called self-supporting arguments for special condemnation.

. . .

NOTES

1. I have discussed the general problem of the justification of induction in Chap. iii of *Language and Philosophy* (Ithaca, 1949).

2. The lack of numerical precision in the examples is deliberate. A good deal of harm has been done in the philosophy of induction by the choice of sophisticated examples from highly developed sciences. We need to bear in mind throughout the discussion the extent to which inductive argument can be imprecise and inexplicit.

3. This point is argued in detail in the earlier discussion referred to in n. 1. See especially pp. 66–68.

4. What is here being called "degree of support" may be the same as what other writers have meant by "degree of confirmation."

5. This assertion involves some degree of idealization. Try formulating the membership of example (i) of section 2, for instance. (Yet that argument, as it stands, is a perfectly good example of inductive argument and does not need the addition of further "assumptions.")

6. Suppose the argument has the form "*X*, so probably *Y*." An alternative way of proceeding would be to consider the statement "Whenever *X* then probably *Y*" as the "associated principle." I do not believe this would make any difference to the conclusions I have tried to draw in this essay.

7. Mill and Nicod are two of the best-known examples of this.

8. I owe the example of the machine to R. B. Braithwaite—see his valuable discussion of "The Justification of Induction," Chapter 8 of his *Scientific Explanation* (Cambridge, 1953). Braithwaite is one of the very few writers who do not think that inductive justifications of induction are viciously circular. But I think he is unnecessarily restrained in his conclusions on the matter and my own treatment follows somewhat different lines.

9. *Logic*, Part II (Cambridge, 1922), p. 8.

10. A rule may be called valid if all the deductive arguments conforming to that rule are valid.

11. As I have said before, this Cartesian attitude toward an argument is not needed in order to make that argument correct. To say otherwise would be to make all arguments, deductive and inductive alike, impossible. If I may take no step until I have proved I am entitled to, it becomes logically impossible for me to walk at all. First comes the correct reasoning—then, if need be, its justification.

P. F. Strawson

DISSOLVING THE PROBLEM OF INDUCTION

. . .

7. We have seen something, then, of the nature of inductive reasoning; of how one statement or set of statements may support another statements, *S*, which they do not entail, with varying degrees of strength, ranging from being conclusive evidence for *S* to being only slender evidence for it; from making *S* as certain as the supporting statements, to giving it some slight probability. We have seen, too, how the question of degree of support is complicated by consideration of relative frequencies and numerical chances.

There is, however, a residual philosophical question which enters so largely into discussion of the subject that it must be discussed. It can be raised, roughly, in the following forms. What reason have we to place reliance on inductive procedures? Why should we suppose that the accumulation of instances of *A*'s which are *B*'s, however various the conditions in which they are observed, gives any good reason for expecting the next *A* we enocunter to be a *B*? It is our habit to form expectations in this way; but can the habit be rationally justified? When this doubt has entered our minds it may be difficult to free ourselves from it. For the doubt has its source in a confusion; and some attempts to resolve the doubt preserve the confusion; and other attempts to show that the doubt is senseless seem altogether too facile. The root-confusion is easily described; but simply to describe it seems an inadequate remedy against it. So the doubt must be examined again and again, in the light of different attempts to remove it.

If someone asked what grounds there were for supposing that deductive reasoning was valid, we might answer that there were in fact no grounds for supposing that deductive reasoning was always valid; sometimes people made valid inferences, and sometimes they were guilty of logical fallacies. If he said that

From Chap. 9 of P. F. Strawson, *Introduction to Losical Theory*. Reprinted by permission of the publishers, Methuen & Company, Ltd., London.

we had misunderstood his question, and that what he wanted to know was what grounds there were for regarding deduction *in general* as a valid method of argument, we should have to answer that his question was without sense, for to say that an argument, or a form or method of argument, was valid or invalid would *imply* that it was deductive; the concepts of validity and invalidity had application only to individual deductive arguments or forms of deductive argument. Similarly, if a man asked what grounds there were for thinking it reasonable to hold beliefs arrived at inductively, one might at first answer that there were good and bad inductive arguments, that sometimes it was reasonable to hold a belief arrived at inductively and sometimes it was not. If he, too, said that his question had been misunderstood, that he wanted to know whether induction in general was a reasonable method of inference, then we might well think his question senseless in the same way as the question whether deduction is in general valid; for to call a particular belief reasonable or unreasonable is to apply inductive standards, just as to call a particular argument valid or invalid is to apply deductive standards. The parallel is not wholly convincing; for words like "reasonable" and "rational" have not so precise and technical a sense as the word "valid." Yet it is sufficiently powerful to make us wonder how the second question could be raised at all, to wonder why, in contrast with the corresponding question about deduction, it should have seemed to constitute a genuine problem.

Suppose that a man is brought up to regard formal logic as the study of the science and art of reasoning. He observes that all inductive processes are, by deductive standards, invalid; the premises never entail the conclusions. Now inductive processes are notoriously important in the formation of beliefs and expectations about everything which lies beyond the observation of available witnesses. But an *invalid* argument is an *unsound* argument; an *unsound* argument is one in which *no good reason* is produced for accepting the conclusion. So if inductive processes are invalid, if all the arguments we should produce, if challenged, in support of our beliefs about what lies beyond the observation of available witnesses are unsound, then we have no good reason for any of these beliefs. This conclusion is repugnant. So there arises the demand for a justification, not of this or that particular belief which goes beyond what is entailed by our evidence, but a justification of induction in general. And when the demand arises in this way it is, in effect, the demand that induction shall be shown to be really a kind of deduction; for nothing less will satisfy the doubter when this is the route to his doubts.

Tracing this, the most common route to the general doubt about the reasonableness of induction, shows how the doubt seems to escape the absurdity of a demand that induction in general shall be justified by inductive standards. The demand is that induction should be shown to be a rational process; and this turns out to be the demand that one kind of reasoning should be shown to be another and different kind. Put thus crudely, the demand seems to escape one absurdity only to fall into another. Of course, inductive arguments are not deductively valid; if they were, they would be deductive arguments. Inductive reasoning must be assessed, for soundness, by inductive standards. Nevertheless, fantastic as the wish for induction to be deduction may seem, it is only in terms of it that we can understand some of the attempts that have been made to justify induction.

. . .

10. Let us turn from attempts to justify induction to attempts to show that the demand for a justification is mistaken. We have seen already that what lies behind such a demand is often the absurd wish that induction should be shown to be some kind of deduction—and this wish is clearly traceable in the two attempts at justification which we have examined. What other sense could we give to the demand? Sometimes it is expressed in the form of a request for proof that induction is a *reasonable* or *rational* procedure, that we have *good grounds* for placing reliance upon it. Consider the uses of the phrases "good grounds," "justification," "reasonable," etc. Often we say such things as "He has *every justification* for believing that *p*"; "I have *very good reasons* for believing it"; "There are *good grounds* for the view that *q*"; "There is *good evidence* that *r*." We often talk, in such ways as these, of justification, good grounds or reasons or evidence for certain beliefs. Suppose such a belief were one expressible in the form "Every case of *f* is a case of *g*." And suppose someone were asked what he meant by saying that he had good grounds or reasons for holding it. I think it would be felt to be a satisfactory answer if he replied: "Well, in all my wide and varied experience I've come across innumerable cases of *f* and never a case of *f* which wasn't a case of *g*." In saying this, he is clearly claiming to have *inductive* support, *inductive* evidence, of a certain kind, for his belief; and he is also giving a perfectly proper answer to the question, what he meant by saying that he had ample justification, good grounds, good reasons for his belief. It is an analytic proposition that it is reasonable to have a degree of belief in a statement which is proportional to the strength of the evidence in its favour; and it is an analytic proposition, though not a proposition of mathematics, that, other things being equal,* the evidence for a generalization is strong in proportion as the number of favourable instances, and the variety of circumstances in which they have been found, is great. So to ask whether it is reasonable to place reliance on inductive procedures is like asking whether it is reasonable to proportion the degree of one's convictions to the strength of the evidence. Doing this is what "being reasonable" *means* in such a context.

As for the other form in which the doubt may be expressed, viz., "Is induction a justified, or justifiable, procedure?", it emerges in a still less favourable light. No sense has been given to it, though it is easy to see why it seems to have a sense. For it is generally proper to inquire *of a particular belief*, whether its adoption is justified; and, in asking this, we are asking whether there is good, bad, or any, evidence for it. In applying or withholding the epithets "justified," "well founded," etc., in the case of specific beliefs, we are appealing to, and applying, inductive standards. But to what standards are we appealing when we ask

*This phrase embodies the large abstractions referred to in secs. 5 and 6

whether the application of inductive standards is justified or well grounded? If we cannot answer, then no sense has been given to the question. Compare it with the question: Is the law legal? It makes perfectly good sense to inquire of a particular action, of an administrative regulation, or even, in the case of some states, of a particular enactment of the legislature, whether or not it is legal. The question is answered by an appeal to a legal system, by the application of a set of legal (or constitutional) rules or standards. But it makes no sense to inquire in general whether the law of the land, the legal system as a whole, is or is not legal. For to what legal standards are we appealing?

The only way in which a sense might be given to the question, whether induction is in general a justified or justifiable procedure, is a trival one which we have already noticed. We might interpret it to mean "Are all conclusions, arrived at inductively, justified?" i.e., "Do people always have adequate evidence for the conclusions they draw?" The answer to this question is easy, but uninteresting: It is that sometimes people have adequate evidence, and sometimes they do not.

11. It seems, however, that this way of showing the request for a general justification of induction to be absurd is sometimes insufficient to allay the worry that produces it. And to point out that "forming rational opinions about the unobserved on the evidence available" and "assessing the evidence by inductive standards" are phrases which describe the same thing, is more apt to produce irritation than relief. The point is felt to be "merely a verbal" one; and though the point of this protest is itself hard to see, it is clear that something more is required. So the question must be pursued further. First, I want to point out that there is something a little odd about talking of "the inductive method," or even "the inductive policy," as if it were just one possible method among others of arguing from the observed to the unobserved, from the available evidence to the facts in question. If one asked a meteorologist what method or methods he used to forecast the weather, one would be surprised if he answered: "Oh, just the inductive method." If one asked a doctor by what means he diagnosed a certain disease, the answer "By induction" would be felt as an impatient evasion, a joke, or a rebuke. The answer one hopes for is an account of the tests made, the signs taken account of, the rules and recipes and general laws applied. When such a specific method of prediction or diagnosis is in question, one can ask whether the method is justified in practice; and here again one is asking whether its employment is inductively justified, whether it commonly gives correct results. This question would normally seem an admissible one. One might be tempted to conclude that, while there are many different specific methods of prediction, diagnosis, etc., appropriate to different subjects of inquiry, all such methods could properly be called "inductive" in the sense that their employment rested on inductive support and that, hence, the phrase "noninductive method of finding out about what lies deductively beyond the evidence" was a description without meaning, a phrase to which no sense has been given; so that there could be no question of justifying our selection of one method, called "the inductive," of doing this.

However, someone might object: "Surely it is possible, though it might be foolish, to use methods utterly different from accredited scientific ones. Suppose a man, whenever he wanted to form an opinion about what lay beyond his observation or the observation of available witnesses, simply shut his eyes, asked himself the appropriate question, and accepted the first answer that came into his head. Wouldn't this be a noninductive method?" Well, let us suppose this. The man is asked: "Do you usually get the right answer by your method?" He might answer: "You've mentioned one of its drawbacks: I never do get the right answer; but it's an extremely easy method." One might then be inclined to think that it was not a method of finding things out at all. But suppose he answered: "Yes, it's usually (always) the right answer." Then we might be willing to call it a method of finding out, though a strange one. But, then, by the very fact of its success, it would be an inductively supported method. For each application of the method would be an application of the general rule, "The first answer that comes into my head is generally (always) the right one"; and for the truth of this generalization there would be the inductive evidence of a long run of favourable instances with no unfavourables ones (if it were "always"), or of a sustained high proportion of successes to trials (if it were "generally").

So every successful method or recipe for finding out about the unobserved must be one which has inductive support; for to say that a recipe is successful is to say that it has been repeatedly applied with success; and repeated successful application of a recipe constitutes just what we mean by inductive evidence in its favour. Pointing out this fact must not be confused with saying that "the inductive method" is justified by its success, justified because it works. This is a mistake, and an important one. I am not seeking to "justify the inductive method," for no meaning has been given to this phrase. *A fortiori*, I am not saying that induction is justified by its success in finding out about the unobserved. I am saying, rather, that any successful method of finding out about the unobserved is necessarily justified by induction. This is an analytic proposition. The phrase "successful method of finding things out which has no inductive support" is self-contradictory. Having, or acquiring inductive support is a necessary condition of the success of a method.

Why point this out all at? First, it may have a certain therapeutic force, a power to reassure. Second, it may counteract the tendency to think of "the inductive method" as something on a par with specific methods of diagnosis or prediction and therefore, like them, standing in need of (inductive) justification.

12. There is one further confusion, perhaps the most powerful of all in producing the doubts, questions, and spurious solutions discussed in this part. We may approach it by considering the claim that induction is justified by its success in practice. The phrase "success of induction" is by no means clear and perhaps embodies the confusion of induction with some specific method of prediction, etc., appropriate to some particular line of inquiry. But, whatever the phrase may mean, the claim has an obviously circular look. Presumably the suggestion is that we should argue from the past "successes of induction" to the continuance of those successes in the future; from the fact that it has worked hitherto to the conclusion that it will continue to work. Since an argument of

this kind is plainly inductive, it will not serve as a justification of induction. One cannot establish a principle of argument by an argument which uses that principle. But let us go a little deeper. The argument rests the justification of induction on a matter of fact (its "past successes"). This is characteristic of nearly all attempts to find a justification. The desired premise of section 8 was to be some fact about the constitution of the universe which, even if it could not be used as a suppressed premise to give inductive arguments a deductive turn, was at any rate a "presupposition of the validity of induction." Even the mathematical argument of section 9 required buttressing with some large assumption about the makeup of the world. I think the source of this general desire to find out some fact about the constitution of the universe which will "justify induction" or "show it to be a rational policy" is the confusion, the running together, of two fundamentally different questions; to one of which the answer is a matter of nonlinguistic fact, while to the other it is a matter of meanings.

There is nothing self-contradictory in supposing that all the uniformities in the course of things that we have hitherto observed and come to count on should cease to operate tomorrow; that all our familiar recipes should let us down, and that we should be unable to frame new ones because such regularities as there were too complex for us to make out. (We may assume that even the expectation that all of us, in such circumstances, would perish, were falsified by someone surviving to observe the new chaos in which, roughly speaking, nothing foreseeable happens.) Of course, we do not believe that this will happen. We believe, on the contrary, that our inductively supported expectation-rules, though some of them will have, no doubt, to be dropped or modified, will continue, on the whole, to serve us fairly well; and that we shall generally be able to replace the rules we abandon with others similarly arrived at. We might give a sense to the phrase "success of induction" by calling this vague belief the belief that induction will continue to be successful. It is certainly a factual belief, not a necessary truth; a belief, one may say, about the constitution of the universe. We might express it as follows, choosing a phraseology which will serve the better to expose the confusion I wish to expose:

I. (The universe is such that) induction will continue to be successful.

I is very vague: It amounts to saying that there are, and will continue to be natural uniformities and regularities which exhibit a humanly manageable degree of simplicity. But, though it is vague, certain definite things can be said about it. (1) It is not a necessary, but a contingent, statement; for chaos is not a self-contradictory concept. (2) We have good inductive reasons for believing it, good inductive evidence for it. We believe that some of our recipes will continue to hold good because they have held good for so long. We believe that we shall be able to frame new and useful ones, because we have been able to do so repeatedly in the past. Of course, it would be absurd to try to use I to "justify induction," to show that it is a reasonable policy; because I is a conclusion inductively supported.

Consider now the fundamentally different statement:

II. Induction is rational (reasonable).

We have already seen that the rationality of induction, unlike its "successfulness," is not a fact about the constitution of the world. It is a matter of what we mean by the word "rational" in its application to any procedure for forming opinions about what lies outside our observations or that of available witnesses. For to have good reasons for any such opinion is to have good inductive support for it. The chaotic universe just envisaged, therefore, is not one in which induction would cease to be rational; it is simply one in which it would be impossible to form rational expectations to the effect that specific things would happen. It might be said that in such a universe it would at least be rational to refrain from forming specific expectations, to expect nothing but irregularities. Just so. But this is itself a higher-order induction: Where irregularity is the rule, expect further irregularities. Learning not to count on things is as much learning an inductive lesson as learning what things to count on.

So it is a contingent, factual matter that it is sometimes possible to form rational opinions concerning what specifically happened or will happen in given circumstances (I); it is a noncontingent, a prior matter that the only ways of doing this must be inductive ways (II). What people have done is to run together, to conflate, the question to which I is answer and the quite different question to which II is an answer; producing the muddled and senseless questions: "Is the universe such that inductive procedures are rational?" or "What must the universe be like in order for inductive procedures to be rational?" It is the attempt to answer these confused questions which leads to statements like "The uniformity of nature is a presupposition of the validity of induction." The statement that nature is uniform might be taken to be a vague way of expressing what we expressed by I; and certainly this fact is a condition of, for it is identical with, the likewise contingent fact that we are, and shall continue to be, able to form rational opinions, of the kind we are most anxious to form, about the unobserved. But neither this fact about the world, nor any other, is a condition of the necessary truth that, if it is possible to form rational opinions of this kind, there will be inductively supported opinions. The discordance of the conflated questions manifests itself in an uncertainty about the status to be accorded to the alleged presupposition of the "validity" of induction. For it was dimly, and correctly, felt that the reasonableness of inductive procedures was not merely a contingent, but a necessary, matter; so any necessary condition of their reasonableness had likewise to be a necessary matter. On the other hand, it was uncomfortably clear that chaos is not a self-contradictory concept; that the fact that some phenomena do exhibit a tolerable degree of simplicity and repetitiveness is not guaranteed by logic, but is a contingent affair. So the presupposition of induction had to be both contingent and necessary: which is absurd. And the absurdity is only lightly veiled by the use of the phrase "synthetic a priori" instead of "contingent necessary."

Wesley C. Salmon

INDUCTIVE INFERENCE

In our capacity as rational beings, we find it imperative to reflect upon our activities. The need for reflection extends to our intellectual as well as our practical behavior. Science, as one of the most important of human activities, thus becomes the object of scrutiny. We can make empirical studies of science much as we study other human activities. We can trace its origins and development, we can examine its relationship to its social context, and we can investigate the motivations which lead to its pursuit. From the need to reflect upon science there arises a group of metascientific disciplines such as history, sociology, and psychology of science.

Unlike certain other human activities, however, science involves a cognitive claim. Whatever historical, social, and psychological factors nave influenced and molded science, the basic fact remains that science is presumed to yield knowledge. Such knowledge may enable us to predict future events, control our environment, explain diverse phenomena, or understand the world around us. It may have great practical import or it may merely satisfy our curiosity. Regardless of the use to which scientific knowledge is put, the cognitive claim is there. It is fundamental to the rationale for the pursuit of science. If science could not produce knowledge, it is hard to imagine what possible reason would remain for doing science.

When we reflect upon the cognitive claim of science, another radically different kind of metascientific problem arises. Can we justifiably maintain that scientific results do, in fact, constitute knowledge? Can we legitimize the cognitive claim of science? Such questions are no longer empirical; they are philosophical or logical. In order to answer such questions, it is necessary to examine the logical structure of science. We must look at the kinds of inference which are employed in science.

It is important to be clear from the outset that the questions I am raising do

From *Philosophy of Science: The Delaware Seminar II.* Reprinted by permission of the University of Delaware Press.

not concern scientific method as a method of dicovery. Questions about the methods by which scientists achieve scientific results, gain insights, or think up hypotheses are not being considered. The problem is, instead, one of justification. Given that a conclusion is well established or highly confirmed according to the accepted canons of scientific justification, on what grounds may we accept this conclusion as embodying knowledge? Another way of putting the same problem is this. Given that a conclusion is well supported by scientific evidence, is the scientific concept of evidence a legitimate one? The concept of scientific evidence is defined by the canons of scientific inference.

These questions about the foundations of scientific inference have, I believe, great import. It is possible in principle to conceive of different nonscientific canons of inference; indeed, in practice, canons which are distinct from and in conflict with those of science have been adopted. Science is not the only human activity for which cognitive claims are made. Astrologists, numerologists, clairvoyants, mystics, and other types of seers affirm that their methods yield knowledge—that their canons of inference are superior to those of science. Furthermore, science evokes a wide variety of attitudes, from blind worship to total revulsion. These conflicting attitudes are often present in the same individual. Some people accept science on six days and reject it on the seventh. Some people accept science as applied to physical phenomena, but reject its application to human affairs. In the face of such conflicts, it is essential to assess the cognitive claims of science. It is a problem of seeing whether there are sound logical reasons for supposing that the conclusions of science constitute more reliable knowledge than conclusions established on the basis of alternative canons of inference.

The view is rather commonly held that science is, at bottom, a matter of faith; the canons of science are accepted on faith. Science is a faith which exists alongside of and on a par with other faiths. Like other faiths, it may be forced upon one by his culture, temperament, glands, or genes. People with differing cultures, temperaments, glands, or genes may be expected to accept different and conflicting faiths. It is not my purpose to disparage faith, for various kinds of faith may have valuable functions in our lives. The point is simply that the grounds for accepting a faith have nothing to do with cognitive claims. If science is fundamentally a matter of faith, then the only valid reasons for pursuing science are noncognitive. Under these circumstances, science cannot legitimately lay claim to a knowledge-producing function.

Science has been called a sacred cow.[1] If the foregoing characterization is correct, it was rightly so-called. Furthermore, it is the most extravagant faith ever to infect mankind. It has paraded under false pretenses. We have poured vast resources into scientific endeavors in the expectation of achieving knowledge. If it should turn out that the cognitive claims of science cannot be legitimized—that science is basically a faith—then, in all reasonableness, this should have a profound effect upon our expectations concerning the results of science and upon the resources we are willing to pour into it. There are, after all, many sorts of divination which are simpler, cheaper, more amusing, and more exciting than scientific experimentation. The only basis for preferring science lies in its cognitive claim. If our devotion to science is to be more than grotesque scientism, we must find a way of showing that science is better suited to establishing knowledge than are its potential rivals.

It might seem, offhand, an easy task to show that the canons of science are suitable for the establishment of knowledge of the world we live in. After all, the accomplishments of science are there for all to see, and we can judge it by its fruits. Unfortunately, David Hume's searching analysis has shown that this way of justifying the method of science is logically fallacious. He has shown, in fact, that the problem of legitimizing the cognitive claims of science is an extremely difficult one—so difficult that even at present there is no generally accepted solution.

THE PROBLEM

In order to understand clearly the nature of the problem we face, let us begin by making an exhaustive distinction between two kinds of inference: deductive (or demonstrative) and nondemonstrative. When we speak of deductive inference, we mean to confine ourselves to valid deductive inferences. The conclusion of a valid deductive inference follows necessarily from its premises; it is impossible for the conclusion to be false if the premises are true. This type of inference is justified by its truth-preserving character. Its use can never lead from true premises to false conclusions, so falsity cannot be introduced by the use of such inferences. The truth-preserving character of deductive inference is a result of the fact that any information given in the conclusion is already present, explicitly or implicitly, in the premises. The following simple syllogism is an example of a valid deductive inference:

All horses are mammals.
All mammals have lungs.
―――――――――――――――
All horses have lungs.

It is not difficult to see that this example has the characteristics claimed for deductive inferences.

Nondemonstrative inferences are simply those which lack the characteristics of deductive inferences. In particular, a nondemonstrative inference may have a false conclusion even though it has true premises, and the conclusion has content not possessed by the premises. The following inference is nondemonstrative:

All observed horses have lungs.
―――――――――――――――
All horses have lungs.

Again, the example obviously exhibits the mentioned characteristics.

We may summarize the distinction as follows. Deductive inference has the truth-preserving characteristic; nondemonstrative inference has a content-extending function. According to the view of logic which I take to be correct—which was, I believe, first clearly stated by Hume—these characteristics are incompatible with each other. Deduction has the truth-preserving characteristic precisely because it sacrifices the content-extending function. Nondemonstrative inference has the content-extending function precisely because it sacrifices the truth-preserving characteristic.

The concept of deductive inference was specified so as to include only valid inferences. Nondemonstrative inferences were characterized without reference to validity or logical correctness. Thus, nondemonstrative inference includes any kind of fallacious inference whatever. Any nondemonstrative inference has the content-extending function. This means only that the conclusion says something not said either implicitly or explicitly by the premises. The fact that the conclusion is the conclusion of a content-extending inference with true premises does not, by itself, provide any ground for accepting it. We must, therefore, attempt to identify those types of nondemonstrative inference, if any, which do provide a reason for the acceptance of their conclusions if they have true premises. Such inferences, if there are any, extend our knowledge. Given that the premises of such an inference are known, the inference extends our knowledge to include the additional content of the conclusion. Such inferences, then, have not only a content-extending function, but also a knowledge-extending function. I would suggest that the term "induction" be reserved to refer, strictly speaking, to any kind of nondemonstrative inference which has, in the sense just indicated, a knowledge-extending function. However, there is no particular harm in using this term less strictly to apply to inferences which are generally regarded as correct, even if they have not been rigorously established as correct. In what follows, I shall feel free to use the term "induction" in this looser sense.

It is important to realize that the characterization of induction given here is very different from certain traditional characterizations. Induction is not defined as inference from the particular to the general; it is not defined as the inverse of deduction; it is not defined as induction by enumeration; it is not defined as a method of discovery. Induction is any kind of nondemonstrative inference which has the property that the truth of its premises, though not necessitating the truth of its conclusion, provides good grounds for the acceptance of its conclusion.

Inductive inference is a species of nondemonstrative inference. In order to distinguish inductive inferences from other nondemonstrative inferences we must establish criteria for determining whether a given type of inference does, in fact, provide good grounds for the acceptance of its conclusion if it has true premises. This, in turn, requires that we consider the problem of what constitutes good grounds for acceptance. In so doing, we find ourselves face to face with Hume's problem of the justification of induction.

We may now say without hesitation that science purports to be basically inductive. Science is not satisfied merely to describe what we observe directly; it is concerned to establish knowledge of (as yet) unobserved facts. This knowledge may be embodied in predictions or postdictions of particular events, or in general statements which apply alike to the observed and the unobserved. No matter how extensively an empirical science may use the deductive techniques of mathematics, its inferences are ultimately inductive, because it claims to fulfill the knowledge-extending function. It purports to establish conclusions on the basis of premises which do not logically exhaust the content of those conclusions.

Hume saw the fundamental import for the logic of science of the problem of justifying inferences from the observed to the unobserved.[2] Although he has (with doubtful justice) been criticized for supposing that induction by enumeration is the only kind of inductive inference in science, we can pose his problem without any such assumption. It can be stated as follows:

(1) It is impossible to prove *deductively* that the conclusion of any nondemonstrative inference with true premises must be true. Since the content of the conclusion exceeds the content of the premises, there is no a priori reason why the premises cannot be true and the conclusion false. If it were possible to prove that the conclusion must be true if the premises are true, the inference would be deductive rather than nondemonstrative, and as such, it could not fulfill the knowledge-extending function of scientific inference.

(2) Any attempt to prove *inductively* that nondemonstrative inferences with true premises will have true conclusions will involve either vicious circularity or vicious regress. Unless we have justified some type of nondemonstrative inference, we have no established inductive principles to use in an inductive justification. In particular, the attempt to legitimize the cognitive claims of science for future applications on the basis of past successes is an attempt to justify induction inductively. Since the justifiability of induction is precisely the question at issue, we cannot legitimately use induction for its own justification.

The conclusion seems to follow inexorably that induction cannot be justified at all, and consequently, that the cognitive claims of science cannot be legitimized.

Hume's argument, in the hands of a clever irrationalist, is devastating. Suppose a fundamentalist wishes to defend his view that the age of the universe is precisely 5966 years. He may admit that the scientific evidence is adverse, but it is only scientific evidence. Leaving out tedious dialectical details, the argument comes down to this. There is no rational basis for the belief that the world was created in 4004 B.C., but this belief is no more irrational than the belief in scientifically established conclusions. The possibility of such defenses of every conceivable irrationality is intolerable.

Needless to say, numerous attempts have been and continue to be made to circumvent the difficulties raised by Hume. They range all the way from frontal attacks upon Hume's arguments to the view that the problem of induction is a pseudo problem. It is impossible to discuss all the solutions, resolutions, and dissolutions, but I should like to treat two approaches which seem to me to have primary contemporary interest. The first is, I believe, unsuccessful, but the second holds some promise of success.

A TEMPTING MOVE

One common way of distinguishing between (correct) induction and valid deduction is to say that a valid deduction with true premises establishes its conclusion as true, while a correct induction with true premises establishes its conclusion as probable. Hume, it has been argued, failed to find a way of justifying induction because his requirements for justification were too stringent. Admittedly, Hume showed that it is impossible to prove that inductive inferences with true premises will have true conclusions. But, it is argued, this point amounts only to the platitude that induction is not deduction. In order to justify induction, the argument continues, it is not necessary to show that inductive inferences with true premises will have true conclusions; it is sufficient to show that they will have probable conclusions.

This is an appealing approach to the problem of induction, but to assess its adequacy we must consider the meaning of the key term "probable." Broadly speaking, there are two senses available for consideration. The first is the frequency sense of the term—roughly, that which is probable is that which happens often. Using this sense of the term, the statement that inductive inferences with true premises establish their conclusions as probable means that a large percentage of such inferences have true conclusions. The frequency sense of "probable" thus provides no help in dealing with the problem of induction, for Hume's argument blocks this move directly. Hume's argument showed not only that it is impossible to prove that *every* inductive argument with true premises will have a true conclusion. This point had been recognized long before Hume. His argument shows, in addition, that it is impossible to prove that *any* inductive inference with true premises will have a true conclusion. For all we can prove, every future inductive inference with true premises may have a false conclusion. In the frequency sense of "probable," Hume has shown that we cannot prove that inductive conclusions from true premises are probable.

The second general sense of "probable" identifies probability with rational belief. This sense includes the logical interpretation of probability and the concept of probability as degree of confirmation. Such conceptions of probability have given rise to what I shall call, with some inaccuracy, the "ordinary language solution" of the problem of induction.[3] The argument runs as follows.

To say that an inductive conclusion is probable means that we would be rationally justified in accepting it on the basis of the available evidence—that is, on the basis of its premises. What determines whether we are rational in accepting a given conclusion on the basis of given premises? The accepted canons of induction, the accepted canons of science. If we want to justify a conclusion, it is sufficient to show that it is related to an adequate body of scientific evidence in terms of the accepted rules of inductive inference. If we ask, further, whether we are justified in accepting these rules of inductive inference, the answer is that we are because this is just what it means to be rational. To ask for any further justification is to ask for a justification for being rational. The justification of induction lies in its rationality. When we are clear about the meanings of words, according to this ordinary language argument, we see that any further request for a justification of induction is tantamount to asking whether it is rational to be rational. Such tautological questions, if they have any sense at all, demand an affirmative answer.

According to the ordinary language solution, then, the answer to Hume's problem is immediately forthcoming when we become clear about the meanings of such key terms as "probable" and "rational." Because of a direct analytic relation between rationality and the canons of induction, the problem of the justification of induction dissolves. The dissolution of the problem shows that we are dealing with ultimate justifactory principles. The analytic connection between the canons of induction and rationality shows that the canons of induction are ultimate principles of justification. The canons of induction can be used in justifying, but they cannot be justified because there are no more ultimate principles in terms of which to carry through the justification.

Appealing as this kind of answer is—and it is widely accepted in one form or another—it is totally unsatisfactory.[4] In order to see this, we must examine the concept of justification itself. Such examination reveals an ambiguity which is fatal to the ordinary language attempt to dissolve the problem of induction. In an extremely illuminating study, Feigl has distinguished two kinds of justification: validation and vindication.[5] Rules, principles, or propositions are validated by deriving them from more basic rules, principles, or propositions. For example, a rule of conditional proof in deductive logic can be validated by proving (the deduction theorem) that any conclusion which can be derived using conditional proof can be proved using only the basic logical rules, without the help of conditional proof. Vindication, by contrast, consists of showing that the adoption of a given rule, principle, or propostion fulfills an accepted purpose. The fundamental rules of deduction would be vindicated, not by deriving them from more fundamental rules or principles, but by showing that they fulfill the truth-preserving function. We want methods of inference which can be used with the confidence that their use will never introduce a falsehood. If we can show that false conclusions can never be deduced from true premises, the rules of deduction are vindicated.

The ordinary language argument shows that the fundamental rules or principles of induction cannot be validated by reference to more fundamental rules or principles. It does not follow that these principles or rules are incapable of vindication of that they are in no need of vindication. This point becomes even clearer when we note that terms like "rational" have ambiguities corresponding to that of the term "justification." To be rational is to behave justifiably, so if there are two distinct senses of "justify" there are also two distinct senses of "rational."

In the sense of "rational" corresponding with validation, the acceptance of a particular inductive inference may be shown to be rational by showing that this inference conforms to the accepted canons of induction. In this sense of "rational," there is no way of showing that the basic canons of induction are rational. In the sense of "rational" corresponding with vindication, it does make sense to ask whether it is rational to accept the usual canons of induction. This amounts to asking whether use of the accepted canons of induction fulfills a certain function, namely, the knowledge-extending function.

Recognizing these two important and distinct senses of "rational," it is easy to see that the apparently pointless question, "Is it rational to be rational?" is not necessarily pointless. If we take the first occurrence of the word "rational" in the sense corresponding with vindication and the second occurrence in the sense corresponding with validation, the question becomes, "Do the accepted canons of induction actually fulfill the knowledge-extending function?" This is precisely the question Hume raised. The ordinary language approach did not dissolve the problem; it only hid it under an ambiguity.

The basic difficulty with the ordinary language dissolution of the problem of induction lies in the fact that pervasive cultural attitudes often become firmly entrenched in ordinary language. It is not surprising to find these attitudes expressed in the ordinary language approach to the problem. "Rational" is a highly honorific word and, in addition, it has an admitted connection with the accepted canons of induction and science. These facts are an expression of the social approval of science in our culture circle. From the ordinary language attempt to dissolve the problem of induction there emerges a kind of vindication of induction. It is this. Science employs certain widely accepted methods of inference. If you want to have the approval that comes from being scientific, then use the accepted canons of induction. Science and its inductive canons are vindicated on the basis of the widespread social acceptability of science. This vindication fails, however, to legitimize the cognitive claims of science. On this vindication, science is, indeed, a sacred cow. There is, perhaps, nothing wrong with seeking social approval, but this goal should not be confused with the goal of trying to find and adopt methods of inference which fulfill the knowledge-extending function.

We began this section by considering an initially attactive approach which involves the concept of probability. The idea was that induction is justified, not by showing that it yields true conclusions from true premises, but rather, that it yields probable conclusions from true premises. We end with a dilemma. If we understand "probable" in a frequency sense, it is impossible to show that induction does produce probable conclusions. If we understand "probable" in a nonfrequency sense, it may be possible to show that inductive conclusions are probable, but we cannot provide any reason for preferring probable conclusions over improbable ones. Given that an event is probable in a nonfrequency sense, what reason is there for believing it will happen? Given that a proposition is probable in a nonfrequency sense, what reason is there for believing it to be true? These questions reformulate Hume's question once more. Introduction of the concept of probability does not dissolve the problem of induction, though it may lead to some interesting reformulations.

PRAGMATIC VINDICATION

A unique approach to Hume's problem, which differs radically from the traditional attempts to refute Hume and from the modern attempts to show that Hume's problem is not a genuine problem, was proposed by Reichenbach.[6] Reichenbach's "pragmatic" justification of induction suffers from well-known difficulties, yet it is, in my opinion, the only promising approach to the problem. In this section I shall present his solution and indicate certain ways of dealing with some of the difficulties.

Reichenbach regarded the problem of the justification of induction as a genuine philosophical problem—one which cannot be dissolved by clearing up a few elementary linguistic confusions. Furthermore, he regarded as sound the conclusion that it is impossible to prove, either deductively or inductively, that inductions with true premises will always, or even sometimes, have true conclusions. Thus he agreed that, in the frequency sense of "probable," it is impossible to show that inductive conclusions are probable. Nevertheless, Reichenbach maintained, we can prove deductively that a certain inductive method is the method best suited to fulfill the knowledge-extending function. Formulating his viewpoint very roughly, he held that we can demonstrate that if there is any method of inference whatever which fulfills the knowledge-extending function, then his rule of induction will also do so. This is not to say that the method of induction will succeed in establishing true conclusions on the basis of true premises, nor is it to say that the method of induction is the only method which will. His thesis is that induction will succeed if any method will. This is more than we can say for any other method.

I shall begin by formulating Reichenbach's argument very loosely, in order to give an intuitive idea of his strategy. A tighter formulation will then be presented.

Hume had pointed out that inductive inference may fail completely to establish knowledge of the unobserved if nature is not uniform. We cannot validly show, either a priori or a posteriori, that nature is uniform, prior to a justification of induction. Therefore, we cannot justify induction. Reichenbach agrees that we cannot, prior to a justification of induction, establish the uniformity of nature. We can, however, examine the two possibilities: that nature is uniform and that nature is not uniform. In either case, we can see what happens when we employ the standard inductive method and when we use other methods. The following table emerges:

	Nature uniform	Nature not uniform
Induction employed	Success	Failure
Other method employed	Success or failure possible	Failure

It is fairly clear that inductive inferences will successfully establish knowledge of the unobserved if nature happens to be uniform, and that they will fail if nature should turn out to be chaotic and lawless. This is not to say that every inductive inference with true premises will have a true conclusion if nature is uniform, but only that persistent use of the inductive method of science will eventually establish such knowledge. If nature is uniform there is the possibility that other methods of gaining knowledge may work, but, even on the assumption of uniformity, we have no proof that any alternative method will succeed. The crucial entry is the last in the table. Reichenbach asserts that even the alternative methods will fail if nature is not uniform. His reason for this assertion is that the continued success of any alternative method would constitute a uniformity, contrary to the assumption of nonuniformity. Another way of putting the same point is this. If an alternative method were to succeed consistently, this would constitute a uniformity to which induction could be applied. Thus, if an alternative method were to succeed, induction would also succeed in yielding knowledge of the unobserved. We have, therefore, everything to gain and nothing to lose by using induction. If induction fails, no other method could possibly succeed.

The foregoing presentation is excessively loose, because it suffers from a failure to specify what is meant by "inductive method," "success," and "uniformity." It fails, in addition, to take account of the fact that there may be varying degrees and kinds of uniformity, and that nature could be uniform in one domain but not in another. Reichenbach has rendered the same argument much more precisely, but to present the more precise version a few preliminaries are required.

Since we are concerned with foundational problems, it will be necessary to consider very rudimentary forms of inference. These elementary forms can be concatenated to produce more complex forms. This procedure is familiar and well established in studies of the foundations of mathematics. Reichenbach selects the rule of induction by enumeration as his fundamental rule of induction. This rule governs inference from an observed sample of a class to a conclusion about the whole class; roughly, the rule permits the inference that the whole class matches the sample. Reichenbach couples this rule with the frequency interpretation of probability, according to which probability is defined as the limit of the relative frequency of an attribute in an infinite sequence. The rule of induction by enumeration permits the inference that the limit of the relative frequency in the infinite sequence equals (or approximates within a small interval) the relative frequency of that attribute in the observed initial section of the sequence. This interpretation of probability demonstrably satisfies the axioms of the mathematical calculus of probability, so the whole machinery of this calculus is automatically available for use. The mathematical calculus, by itself, cannot produce any empirical probability statements, but it can transform given empirical probability statements into other related probability statements. The rule of induction by enumeration governs the establishment of fundamental empirical probability statements. Reichenbach has developed a complex theory of scientific inference, but it all hinges basically upon his justification of the rule of induction by enumeration.

We can now present his justification more precisely, using the foregoing loose presentation as a model. The basic kind of uniformity with which Reichenbach is concerned is the statistical regularity that consists in the existence of a limit of the relative frequency of an attribute in a sequence. Universal laws are special cases of this sort of uniformity. Reichenbach's argument is now easily translated into precise terms; the preceding table can be rewritten as follows:

	Sequence has a limit	Sequence has no limit
Rule of induction by enumeration adopted	Value of limit established	Value of limit not established
Another rule adopted	Value of limit may or may not be established	Value of limit not established

The argument is similar to the preceding one. The entries in the table are verified by examining the definition of "limit" and the rule of induction by enumeration.

The statement that the relative frequency of an attribute in an infinite sequence has a limit means that there exists some number, the limit, such that the observed relative frequency in any sufficiently long initial section of the sequence matches that number as closely as you like. Let us use the symbol "$F^n(A, B)$" to designate the relative frequency with which the attribute B occurs among the first n members of the ordered class A. To say that there exists a limit P of the relative frequency means, more precisely:

> There exists a number P such that, for any $\varepsilon > 0$, there exists an integer N such that, for any $n > N$, $|F^n(A, B) - P| > \varepsilon$.

This is tantamount to the assertion that, if there is a limit of the relative frequency, the persistent use of the rule of induction by enumeration, applied to larger and larger initial sections of the sequence, will establish the limit within any desired interval of accuracy $\pm\varepsilon$.

It is an immediate consequence of the definition of a limit that, if a sequence has a limit, the rule of induction by enumeration will successfully ascertain the value of that limit, in the sense just specified. Other methods may successfully ascertain the value of the limit if one exists. For example, we might write fractions on slips of paper and put them in a hat. By drawing one of these slips from the hat, we might get the correct value of a given limit, but obviously there is no necessity of such success, nor is there any necessity that repeated drawings will yield numbers arbitrarily close to the correct value. On the other hand, if there is no limit, no method can ascertain its value. Reichenbach's justification is this. There is no way to prove, either a priori or a posteriori, prior to a justification of induction, that a given empirical sequence of events will have a limit for the relative frequency of a particular attribute. We cannot know beforehand whether nature is uniform in this respect. Nevertheless, to attempt to establish the value of a limit of this relative frequency, it is advantageous to use the rule of induction by enumeration, for if there is a limit the rule of induction by enumeration will establish its value and if there is no limit no method can establish its value. If any rule will work, induction by enumeration will work.

Many objections have been brought against Reichenbach's attempt to justify induction by enumeration, but the most serious objection is one he was clearly aware of. There is an infinite class of rules—called "asymptotic rules"—which are equally justified by the same argument. Any rule for inferring the limit of the relative frequency can be represented as follows:

$$\text{From } F^n(A, B) = m/n, \text{ to infer } \lim_{n\to\infty} F^n(A, B) = m/n + c$$

where c is a "corrective" term which may be specified as we choose. The rule of induction by enumeration is the one that results by making c identically zero. An asymptotic rule results from any specification of c according to which c converges to zero as n goes to infinity. Obviously, there are infinitely many asymptotic rules, and they all share with induction by enumeration the property of successfully ascertaining the limit of the relative frequency if there is a limit. The problem is to find grounds for preferring one of these rules to the others. Reichenbach's selection on grounds of descriptive simplicity will not do.

I shall illustrate the problems which arise in attempting a selection from among all (not only asymptotic) rules for inferring the limit of the relative frequency by reference to the standard model. Suppose we have an urn containing marbles, and we are interested in the frequencies with which various colors are drawn. As usual, we assume the marbles are indestructible and that each one is replaced after each draw to insure an infinite sequence. Now let us examine several rules for inferring the limit of the relative frequency of a certain color. They are given in the following table:[7]

$$\text{From } F^n(A, B) = m/n, \text{ to infer } \lim_{n\to\infty} F^n(A, B) =$$

(1) m/n	Induction by enumeration
(2) $1/k$	A priori rule
(3) $(n - m)/n$	Counter-inductive rule
(4) $(m/n + 1/k)/2$	Compromise rule
(5) $(n - m)/n(k - 1)$	Normalized counter-inductive rule
(6) $[1/(n + 1)](m + 1/k)$	Vanishing compromise rule

In this table, n is the size of the observed sample, m is the number of times the color in question has occurred in the sample, and k is the number of mutually exclusive and exhaustive color predicates we distinguish.

Aside from the first rule, and possibly the last, these rules are hardly the kind anyone would seriously propose; however, if we can see what is wrong with them it will help in finding principles for justification of an acceptable inductive rule. It seems to me, in fact, that this is an excellent way of posing Hume's problem. Given the infinity of possible rules, how are we to justify the selection of a unique rule as superior to all others? This formulation has a great advantage over the traditional question, "Are we justified in using induction?" When we are faced with the infinite number of candidates, the problem loses much of its apparent triviality.

Rules 1–3 exemplify three basic possibilities. In rule 1, past experience is a positive guide to the future; if red marbles have been drawn frequently, we infer that they will continue to be drawn frequently in the future. In rule 2, past experience is no guide to the future; it is irrelevant. Suppose there are three possible colors: red, yellow, and blue. The a priori rule permits us to conclude that the limit of the relative frequency of red is 1/3, regardless of the frequency with which red has been drawn. According to rule 3, past experience is a negative guide to the future. If red has occurred often in the observed sample, rule 3 sanctions the inference that it will occur seldom in the long run; indeed, it sanctions the inference that the limit of the frequency of red equals the observed frequency of nonred. Rules 4–6 make some improvements upon rules 2 and 3.

Several of the rules in our list are nonasymptotic; namely, rules 2–5, Consider rule 3. Suppose that the limit of the relative frequency of red is, in fact, 1/3. As we apply the counterinductive rule to larger and larger observed samples, our inferred values of the limit of the relative frequency will approach 2/3. This rule has the general property that (leaving aside the exceptional case in which the value of the limit is 1/2) if the relative frequency has a limit, persistent use of the counterinductive rule will lead to inferences which converge to a necessarily incorrect value. Reichenbach's argument shows, I believe, the superiority of his rule over any nonasymptotic rule.

Rule 3 has another disadvantage which is even more serious. Suppose again that there are three possible colors: red, yellow, and blue. Suppose, further, that these three colors occur with equal frequencies in an observed sample. Applying the counter-inductive rule for each color, we get the result that the limit of the relative frequency is 2/3 for each. This means that the limit of the relative frequency of the disjunctive attribute of being either red or yellow or blue is $2/3 + 2/3 + 2/3 = 2/1$. This is a logical absurdity, for relative frequencies, and limits thereof, must lie between zero and one. Furthermore, the sum of the limits of the relative frequencies of a mutually exclusive and exhaustive set of attributes must equal one. Let us use the symbol IV $\lim F^n(A, B)$ to denote the inferred value of the limit of the relative frequency of B on the basis of the sample consisting of the first n members of A. We can lay down the following *normalizing condition*:

Let $B_1, \ldots, B_k$ be any set of attributes mutually exclusive and exhaustive within A. The following relations must hold:

$$\text{IV} \lim F^n(A, B_i) \geq 0$$

$$\sum_{i=1}^{k} \text{IV} \lim F^n(A, B_i) = 1$$

Rules which satisfy the normalizing condition are called "regular."

Rule 3 is the only nonregular rule in our list. In particular, rule 5, the normalized counterinductive rule, is regular. Nevertheless, it shares with rule 3 the feature of making past experience negative evidence for the future. Furthermore, rule 5 is, as previously noted, nonasymptotic. This becomes obvious when we note that rule 5 coincides with rule 3 for $k = 2$.

Rule 2 is a purely a priori rule; results of observation do not enter into the inferences governed by this rule. Rule 1 is a purely empirical rule; no a priori "corrective" term is added to the observed frequency in inferring the value of the limit. Rule 4 is a simple compromise which averages the results of rules 1 and 2. Rule 4, although regular, is clearly nonasymptotic.

Rule 6, the vanishing compromise rule, is similar to rule 4 in combining an empirical factor with an a priori factor, but in rule 6 the a priori factor carries less weight as the amount of empirical evidence increases. In the limit, rule 6 coincides with rule 1; that is, for arbitrarily large samples the difference between the results of these two rules becomes arbitrarily small. For this reason, rule 6 is asymptotic; in addition, it is regular.

It is worthwhile to pause a moment to assay our progress. Starting from an infinite class of possible rules for inferring values of limits of relative frequencies, we found good reasons for rejecting rules which are either nonasymptotic or nonregular. The question is, how many regular asymptotic rules are there? The answer is that there are infinitely many. Furthermore, although for any given sequence the results of all these asymptotic rules converge to the same value, i.e., the limit of the relative frequency, the convergence is nonuniform. In fact, for any sample size and any possible observed frequency in a sample of that size, we may choose arbitrarily any number in the closed interval [0, 1] for our inferred value of the limit and find a regular asymptotic rule to justify that inference.[8] In other words, we have made no effective progress in reducing the possible arbitrariness of inductive inference.

The vanishing compromise rule has, however, a different kind of defect—one which it shares with rules 2, 4, and 5. Consider again the a priori rule. The fact that the number k is mentioned in this rule signals a fundamental difficulty. This number can be considered in either of two ways; it may be the number of color predicates we are using or it may be the number of color properties that actually exist. Construing it in either way, we get into serious trouble.

Suppose, first of all, that k is the number of color predicates we are using. For example, we assumed above that "red," "yellow," and "blue" constituted a mutually exclusive and exhaustive set of predicates for our population of marbles drawn from the urn. Under these conditions, the a priori rule led to the inference that the limit of the relative frequency of red is 1/3. Suppose now that we strengthen our language in a way which does not affect the meaning of "red" at all. We introduce the terms "light blue" and "dark blue" as terms which are mutually exclusive and together equivalent to "blue." As a result of this purely terminological change, which has no bearing on the meaning of "red," the a priori rule now sanctions the inference that the limit of the relative frequency of red is 1/4. The a priori rule thus leads to a contradiction: The limit of the relative frequency of red equals both 1/3 and 1/4. We must introduce a condition to exclude rules of this sort. I have proposed the *criterion of linguistic invariance*:[9]

> Given two logically equivalent descriptions (in the same or different languages) of a body of evidence, no rule may permit mutually contradictory conclusions to be drawn on the basis of these statements of evidence.

Rules 2, 4, 5, and 6 violate this criterion because each of them mentions k.

Since rule 6 is regular and asymptotic, the only objection to it is the violation of the criterion of linguistic invariance. Let us examine this violation. First, let us use a language with only two color predicates, "red" and "nonred"; in this case $k = 2$. Second, let us use a language with the predicates "red," "yellow," "blue," "orange," "green," and "purple"; in this case $k = 6$. We stipulate that "red" means the same in both languages. Let us make five draws from the urn and apply the vanishing compromise rule at each opportunity. Because of the a priori factor, we can make an inference with no empirical evidence. The results are as follows:

n	Color	$F^n(A, B)$	IV lim ($k = 2$)	IV lim ($k = 6$)	Difference
0	—	—	1/2	1/6	1/3
1	red	1/1	3/4	7/12	1/6
2	nonred	1/2	1/2	7/18	1/9
3	red	2/3	5/8	13/24	1/12
4	red	3/4	7/10	19/30	1/15
5	nonred	3/5	7/12	19/36	1/18

The difference converges to zero as n goes to infinity, but it does not vanish for any value of n. For $n = 5$, we have the following two inferences with logically equivalent premises and logically contradictory conclusions:

(1) Of five observed marbles, three are red and two are nonred.

The limit of the relative frequency of red marbles is 7/12.

(2) Of five observed marbles, three are red and two are either yellow, blue, orange, green, or purple.

The limit of the relative frequency of red marbles is 19/36.

We might propose to get rid of such contradictions by insisting upon the adoption of a complete set of color predicates. Such an approach is hardly feasible. It requires that we be able to determine a priori how many distinct colors there are, and that this be some finite number. If there is, in fact, a continuum of colors, this program will be impossible, for there is no nonarbitrary way of splitting up the continuum a priori. This is precisely the kind of difficulty which arises if we construe k to be the number of colors that exist instead of merely the number of color predicates we utilize. We can hardly suppose it to be a truth of pure reason that there are precisely 347 colors. If one wishes to tread the futile path of attempting to justify induction through synthetic a priori truths, he will do better to stick to the principle of uniformity of nature or the principle of limited independent variety.

The kind of contradiction which arose from the vanishing compromise rule is just the sort of contradiction which dealt the death blow to the classical interpretation of probability. The classical interpretation made use of the principle of indifference, a principle which states that two possible occurrences are equally probable if there is no reason to suppose one will happen rather than the other. This principle gives rise to the Bertrand paradox, which is exemplified by the following case. Suppose a car has traversed a distance of one mile, and we know that the time taken was between one and two minutes, but we know nothing further about it. Applying the principle of indifference, we conclude that there is a probability of 1/2 that the time take was in the range of 1 to 1-1/2 minutes, and a probability of 1/2 that the time taken was in the range 1-1/2 to 2 minutes. A logically equivalent way of expressing our knowledge is to say that the car covered the distance at an average speed between 30 and 60 miles per hour. Applying the principle of indifference again, we conclude that there is a probability of 1/2 that the average speed was between 30 and 45 miles per hour, and a probability of 1/2 that the average speed was between 45 and 60 miles per hour. Unfortunately, we have just been guilty of self-contradiction. A time of 1-1/2 minutes for a distance of one mile is an average speed of 40, not 45, miles per hour. On the basis of the same information, formulated in different but equivalent terms we get the result that there is a probability of 1/2 that the average speed is between 30 and 40 miles per hour, and also that there is a probability of 1/2 that the average speed is between 30 and 45 miles per hour. Since it is not impossible that the average speed is between 40 and 45 miles per hour, the foregoing results are mutually incompatible.

Let us again assay our progress. We have examined six rules of inference and have three grounds for rejecting possible candidates. The results are as follows:

Rule	Asymptotic	Regular	Linguistically invariant
(1) Induction by enumeration	Yes	Yes	Yes
(2) A priori	No	Yes	No
(3) Counter-inductive	No	No	Yes
(4) Compromise	No	Yes	No
(5) Normalized counter-ind.	No	Yes	No
(6) Vanishing compromise	Yes	Yes	No

Each of these rules, except the first, can be rejected for failure to be regular or for failure to be linguistically invariant. This result can be generalized. It is demonstrable that every rule for inferring limits of relative frequencies, with the exception of the rule of induction by enumeration, fails to meet one or the other of these conditions.[9*, 9†]

I shall not reproduce the proof here, but the idea behind the proof is, perhaps, philosophically illuminating. Suppose, still, that we are dealing with some set of mutually exclusive and exhaustive attributes. It is a trivial truth of arithmetic that the sum of the relative frequencies of these attributes in any sample is one. It is an equally trivial truth of mathematical analysis that, if these relative frequencies have limits, the sum of these limits must equal one. If we add something to the observed frequency of one attribute in order to get the inferred value of the limit, we must take that amount away from the observed frequencies of other attributes to get the inferred values of the limits of their relative frequencies. On what basis might we decide which observed frequencies ought to be increased and which ones ought to be decreased? If we do it in terms of the words which are used to refer to these attributes, then we pave the way to violation of the criterion of linguistic invariance and to logical contradiction. If we do it in terms of the attributes themselves, we require a kind of synthetic a priori knowledge we could not possibly have. It might be supposed that we could do it strictly in terms of the observed frequencies themselves; for instance, we might try adding to smaller frequencies and taking away from larger ones—a kind of Robin Hood principle of robbing the rich to give to the poor—but this is mathematically impossible. The only admissible "corrective" function, c, which is a function of observed frequencies alone, is that which is identically zero, in short, the one which gives rise to the rule of induction by enumeration. Any deviation from the rule of induction by enumeration leads to some unconscionably arbitrary biasing of the evidence. We can, therefore, select one unique rule of inference from the infinity of possible candidates. In fact, this selection can be made without reference to whether the rule is asymptotic. It turns out that the remaining candidate is asymptotic. Since all others have been rejected, we may use the asymptotic character of induction by enumeration to justify its acceptance. Reichenbach's justification of induction is now cogent.

The justification is as follows. There is only one rule for inferring limits of relative frequencies which is free from contradiction. Every rule except this one permits the establishment of a logical contradiction of the basis of consistent evidence. Such rules are unsatisfactory. We are left with a simple choice. Either we accept the rule of induction by enumeration for purposes of inferring limits of relative frequencies, or we forego entirely all attempts to infer limits of relative frequencies. We cannot prove beforehand that we will be successful in inferring limits of relative frequencies by using induction by enumeration, for the relative frequencies of the attributes we deal with may not have limits. But we can be assured that, if such limits do exist, persistent use of induction by enumeration will establish them to any desired degree of accuracy.

REMAINING PROBLEMS

I am not suggesting that the results presented in this paper succeed in fully legitimizing the cognitive claims of science. At best, I hope we have a beginning in that direction, and I think we do. There are two grounds for optimism. First, the properties of rules herein considered are of fundamental importance. If an antiscientific method can be presented with some degree of clarity, we can examine it from the standpoint of regularity, linguistic invariance, and convergence (asymptotic properties), and perhaps show its inferiority to scientific methods. Even if we cannot completely justify the methods of science, it is useful to be able to discredit various forms of irrationalism. Second, the results so far achieved will, I hope, provide a basis for a full-blown inductive logic whose rules are justified. In conclusion, I should like to indicate what seem to me to be the most important outstanding problems that need to be solved before this program can be completed.

1. The Goodman Paradox

Goodman[10] has shown how the rule of induction by enumeration can lead to paradoxical results if applied in connection with certain peculiar sorts of predicates. This paradox can be turned into an argument to the effect that even the rule of induction by enumeration fails to meet the criterion of linguistic invariance. This paradox necessitates the imposition of certain restrictions upon the kinds of predicates admitted into our scientific language. I have tried, elsewhere, to show how this paradox is to be blocked.[9†] My justification for saying, as I did above, that the rule of induction by enumeration is linguistically invariant hinges upon the successful elimination of the Goodman paradox.

2. The Short Run

It has often been correctly noted that we deal in practice with finite sequences only. Various attempts have been made to assure the applicability of inductive knowledge to finite classes of unobserved events.[11] One approach is to finitize the frequency interpretation of probability, so that probability is identified with the actual relative frequency in a finite sequence. Another approach is to attempt to justify a short-run rule for inferring the relative frequency in a finite sample from the value of the limit of the relative frequency in an infinite sequence. Another approach is to attempt to justify a rule for inferring directly from one finite sample to another nonoverlapping finite sample. None of these approaches to the problem of application of inductive knowledge to finite numbers of unobserved cases has been worked out with complete success, but I know of no reason to regard any of them as hopeless.

3. *Complex Inductive Inferences*

It is obvious that science utilizes much more complex forms of inductive inference than any we have discussed. In particular, an essential feature of scientific inference is the confirmation of scientific hypotheses by means of the so-called "hypothetico-deductive method." More complex inductive rules must be validated or vindicated. They may be capable of validation on the basis of induction by numeration. Following Reichenbach's approach, it may be possible to show that the more complex forms are concatenations of inductions by enumeration. On the other hand, it may be necessary to provide a separate vindication of additional and more complicated rules. Either way, the task has yet to be completed.

NOTES

1. See subnote 1*. I am only borrowing a phrase, not characterizing the content of the book.

 * A. Standen, *Science is a Sacred Cow*, E. P. Dutton & Co., Inc., New York, 1950.

2. D. Hume, *An Enquiry Concerning Human Understanding*, Open Court Publishing Co., LaSalle, Illinois, 1946.

3. P. F. Strawson, *Introduction to Logical Theory*, Wiley, New York-London, 1952, Chap. IX.

4. This point is argued at greater length in subnote 4*.

 * W. C. Salmon, "Should We Attempt to Justify Induction?" *Philosophical Studies*, VIII, No. 3 (April, 1957).

5. H. Feigl, "De Principiis Non Disputandum . . . ?" in M. Black, ed., *Philosophical Analysis*, Cornell University Press, Ithaca, New York, 1950.

6. H. Reichenbach, *Experience and Prediction*, The University of Chicago Press, 1938, and *The Theory of Probability*, University of California Press, Berkeley and Los Angeles, 1949.

7. Any rule in this table can be written so that the inferred value of the limit is represented in the form set out above, $m/n + c$, by the simple expedient of adding and subtracting m/n. For example, in the a priori rule it becomes $m/n + (1/k - m/n)$. This shows, incidentally, that the foregoing characterization of rules is broad enough to include even rules which fail to utilize the empirical evidence resulting from the observation of samples.

8. Proof of this assertion is given in subnote 8*.

 * W. C. Salmon, "The Predictive Inference," *Philosophy of Science*, XXIV, 180 (April, 1957).

9. See subnotes 9* and 9† for discussion of this criterion.

 * W. C. Salmon, "Vindication of Induction," in H. Feigl and G. Maxwell, eds., *Current Issues in the Philosophy of Science*, Holt, Rinehart, and Winston, New York, 1961.

 † W. C. Salmon, "On Vindicating Induction," in Henry E. Kyburg, Jr., ed., *Induction: Some Current Issues*, Wesleyan University Press, Middletown, Conn., forthcoming.

10. N. Goodman, *Fact, Fiction, and Forecast*, Harvard University Press, Cambridge, Mass., 1955.

11. W. C. Salmon, "The Short Run," *Philosophy of Science*, XXII, 214 (1955).

BIBLIOGRAPHY

Black, Max. *Language and Philosophy*, Chap. III. Ithaca, New York: Cornell University Press, 1949.

———. *Models and Metaphors*, Chaps. XI–XII. Ithaca, New York: Cornell University Press, 1962.

———. *Problems of Analysis*, Part III. Ithaca, New York: Cornell University Press, 1954.

Carnap, Rudolf. *The Continuum of Inductive Methods*. Chicago: The University of Chicago Press, 1952.

———. *Logical Foundations of Probability*. Chicago: The University of Chicago Press, 1950.

Feigl, Herbert. "De Principiis Non Disputandum . . . ?" in *Philosophical Analysis*, Max Black, ed. Ithaca, New York: Cornell University Press, 1950.

Goodman, Nelson. *Fact, Fiction, and Forecast*. Cambridge, Massachusetts: Harvard University Press, 1955.

Hempel, Carl G. "Inductive Inconsistencies," *Synthese*, XII, 439–69 (1960).

Hume, David. *An Enquiry Concerning Human Understanding*.

Katz, Jerrold J. *The Problem of Induction and its Solution*. Chicago: The University of Chicago Press, 1962.

Keynes, John Maynard. *A Treatise on Probability*. London: Macmillan and Co., Limited, 1952.

Kyburg, Henry E., Jr. *Probability and the Logic of Rational Belief*. Middletown, Connecticut: Wesleyan University Press, 1961.

Mises, Richard von. *Probability, Statistics and Truth*, 2nd English ed. London: George Allen and Unwin Ltd, 1957.

Nagel, Ernest. *Principles of the Theory of Probability*, International Encyclopedia of Unified Science, I, No. 6. Chicago: The University of Chicago Press, 1955.

Popper, Karl R. *The Logic of Scientific Discovery*. New York: Basic Books, Inc., 1959.

Reichenbach, Hans. *Experience and Prediction*. Chicago: The University of Chicago Press, 1938.

———. *The Theory of Probability*. Berkeley and Los Angeles: University of California Press, 1949.

Salmon, Wesley C. "On Vindicating Induction," in *Induction: Some Current Issues*, Henry E. Kyburg, Jr., ed. Middletown, Connecticut: Wesleyan University Press, forthcoming. Also to be printed in *Philosophy of Science*, 1963.

———. "The Predictive Inference," *Philosophy of Science*, XXIV, No. 2 (April, 1957).

———. "Regular Rules of Induction," *Philosophical Review*, XLV, No. 3 (July, 1956).

———. "The Short Run," *Philosophy of Science*, XXII, No. 3 (July, 1955).

———. "Should We Attempt to Justify Induction?" *Philosophical Studies*, VIII, No. 3 (April, 1957).

———. "Vindication of Induction," in *Current Issues in the Philosophy of Science*, Herbert Feigl and Grover Maxwell, eds. New York: Holt, Rinehart, and Winston, 1961.

Strawson, P. F. *Introduction to Logical Theory*, Chap. IX. New York: John Wiley & Sons, Inc., 1952.

Venn, John. *The Logic of Chance*, 4th ed. New York: Chelsea Publishing Company, 1962.

Wright, Georg Henrik von. *The Logical Problem of Induction*, 2nd ed. New York: The Macmillan Company, 1957.

Postscript

Norwood Russell Hanson

IS THERE A LOGIC OF SCIENTIFIC DISCOVERY?

Is there a logic of scientific discovery? The approved answer to this is "No." Thus Popper argues:[1] "The initial stage, the act of conceiving or inventing a theory, seems to me neither to call for logical analysis nor to be susceptible of it." Again, "There is no such thing as a logical method of having new ideas, or a logical reconstruction of this process." Reichenbach writes that philosophy of science "cannot be concerned with [reasons for suggesting hypotheses], but only with [reasons for accepting hypotheses]."[2] Braithwaite elaborates: "The solution of these historical problems involves the individual psychology of thinking and the sociology of thought. None of these questions are our business here."[3]

Against this negative chorus, the "Ayes" have *not* had it. Aristotle (*Prior Analytics* II, 25) and Peirce[4] hinted that in science there may be more problems for the logician than just analyzing the arguments supporting already invented hypotheses. But contemporary philosophers are unreceptive to this. Let us try once again to discuss the distinction F. C. S. Schiller made between the Logic of Proof and the Logic of Discovery.[5] We may be forced, with the majority, to conclude "Nay." But only after giving Aristotle and Peirce a sympathetic hearing. Is there *anything* in the idea of a "logic of discovery" which merits the attention of a tough-minded, analytic logician?

It is unclear what a logic of discovery is a logic of. Schiller intended nothing more than "a logic of inductive inference." Doubtless his colleagues were so busy sectioning syllogisms that they ignored inference which mattered in science. All the attention philosophers now give to inductive reasoning, probability, and the principles of theory construction would have pleased Schiller. But, for Peirce, the work of Popper, Reichenbach, and Braithwaite would read less like a *Logic of Discovery* than like a *Logic of the Finished Research Report*. Contemporary

---

logicians of science have described how one sets out reasons in support of a hypothesis once proposed. They have said nothing about the conceptual context within which such a hypothesis is initially proposed. Both Aristotle and Peirce insisted that the proposal of a hypothesis can be a reasonable affair. One can have good reasons, or bad, for suggesting one kind of hypothesis initially, rather than some other kind. These reasons may differ in type from those which lead one to accept a hypothesis once suggested. This is not to deny that one's reasons for proposing a hypothesis initially may be identical with his reasons for later accepting it.

One thing must be stressed. When Popper, Reichenbach, and Braithwaite urge that there is no logical analysis appropriate to the psychological complex which attends the conceiving of a new idea, they are saying nothing which Aristotle or Peirce would reject. The latter did not think themselves to be writing manuals to help scientists make discoveries. There could be no such manual.[6] Apparently they felt that there is a *conceptual* inquiry, one properly called "a logic of discovery," which is *not* to be confounded with the psychology and sociology appropriate to understanding how some investigator stumbled on to an improbable idea in unusual circumstances. There are factual discussions such as these latter. Historians like Sarton and Clagett have undertaken such circumstantial inquiries. Others—for example, Hadamard and Poincaré—have dealt with the psychology of discovery. But these are not logical discussions. They do not even turn on conceptual distinctions. Aristotle and Peirce thought they were doing something other than psychology, sociology, or history of discovery; they purported to be concerned with a *logic* of discovery.

This suggests caution for those who reject wholesale any notion of a logic of discovery on the grounds that such an inquiry can *only* be psychology, sociology, or history. That Aristotle and Peirce deny just this has made no impression. Perhaps Aristotle and Peirce were wrong. Perhaps there is no room for logic between the psychological dawning of a discovery and the justification of that discovery via successful predictions. But this should come as the conclusion of a discussion, not as its preamble. If Peirce is correct, nothing written by Popper, Reichenbach, or Braithwaite cuts against him. Indeed, these authors do not discuss what Peirce wishes to discuss.

Let us begin this uphill argument by distinguishing

(1) reasons for accepting a hypothesis H, from
(2) reasons for suggesting H in the first place.

This distinction is in the spirit of Peirce's thesis. Despite his arguments, most philosophers deny any *logical* difference between these two. This must be faced. But let us shape the distinction before denting it with criticism.

What would be our reasons for accepting H? These will be those we might have for thinking H true. But the reasons for suggesting H originally, or for formulating H in one way rather than another, may not be those one requires before thinking H true. They are, rather, those reasons which make H a *plausible type of conjecture*. Now, no one will deny *some* differences between what is required to show *H* true and what is required for deciding *H* constitutes a plausible kind of conjecture. The question is: Are these logical in nature, or should they more properly be called "psychological" or "sociological"?

Or one might urge, as does Professor Feigl, that the difference is just one of refinement, degree, and intensity. Feigl argues that considerations which settle whether *H* constitutes a plausible conjecture are of the *same type* as those which settle whether *H* is true. But since the initial proposal of a hypothesis is a groping affair, involving guesswork amongst sparse data, there *is* a distinction to be drawn; but this, Feigl urges, concerns two ends of a spectrum, ranging all the way from inadequate and badly selected data to that which is abundant, well diversified, and buttressed by a battery of established theories. The issue therefore remains: Is the difference between reasons for accepting *H* and reasons for suggesting it originally one of logical type, or one of degree, or of psychology, or of sociology?

Already a refinement is necessary if our original distinction is to survive. The distinction just drawn must be reset in the following, more guarded, language. Distinguish now

(1′) reasons for accepting a particular, minutely specified hypothesis *H*, from
(2′) reasons for suggesting that, whatever specific claim the successful *H* will make, it will, nonetheless, be a hypothesis of one *kind* rather than another.

Neither Aristotle, nor Peirce, nor (if you will excuse the conjunction) myself in earlier writings,[7] sought this distinction on these grounds. The earlier notion was that it was some particular, minutely specified *H* which was being looked at in two ways: (1) What would count for the acceptance of that *H*, and (2) what would count in favor of suggesting that same *H* initially.

This latter way of putting it is objectionable. The issue is whether, *before* having hit a hypothesis which succeeds in its predictions, one can have good reasons for anticipating that the hypothesis will be one of some particular *kind*. Could Kepler, for example, have had good reasons, *before* his elliptical-orbit hypothesis was established, for supposing that the successful hypothesis concerning Mars's orbit would be of the noncircular kind?[8] He *could* have argued that, whatever path the planet *did* describe, it would be a closed, smoothly curving, plane geometrical figure. Only this *kind* of hypothesis could entail such observation-statements as that Mars's apparent velocities at 90 degrees and at 270 degrees of eccentric anomaly were greater than any circular-type *H* could explain. Other *kinds* of hypotheses were available to Kepler: for example, that Mars's *color* is responsible for its high velocities, or that the dispositions of Jupiter's moons are responsible. But these would not have struck Kepler as capable of explaining such suprising phenomena. Indeed, he would have thought it *un*reasonable to develop such hypotheses at all, and would have argued thus [Braithwaite counters: "But exactly which hypothesis was to be rejected was a matter for the 'hunch' of the physicists."[9] However, which *type* of hypothesis Kepler chose to reject was not just a matter of "hunch."]

I may still be challenged. Some will continue to berate my distinction between reasons for suggesting which type of hypothesis *H* will be, and reasons for accepting *H* ultimately.[10] There may indeed be "psychological" factors, the opposition concedes, which make certain types of hypotheses "look" as if they might explain phenomena. Ptolemy knew, as well as did Aristarchus before him and Copernicus after him, that a kind of astronomy which displaced the earth would be theoretically simpler, and easier to manage, than the hypothesis of a geocentric, geostatic universe. *But*, philosophers challenge, for psychological, sociological, or historical reasons, alternatives to geocentricism did not "look" as if they could explain the absence of stellar parallax. This cannot be a matter of logic, since for Copernicus one such alternative *did* "look" as if it could explain this. Insofar as scientists have *reasons* for formulating types of hypotheses (as opposed to hunches and intuitions), there are just the kinds of reasons which later show a particular *H* to be true. Thus, if the absence of stellar parallax constitutes more than a psychological reason for Ptolemy's resistance to alternatives to geocentricism, then it *is* his reason for rejecting such alternatives as *false*. Conversely, his reason for developing a geostatic type of hypothesis (again, absence of parallax) was his reason for taking some such hypothesis as *true*. And Kepler's reasons for rejecting Mars's color or Jupiter's moons as indicating the kinds of hypotheses responsible for Mars's accelerations were reasons which also served later in establishing some hypothesis of the noncircularity type.

So the objection to my distinction is: The only *logical* reason for proposing *H* will be of a certain type is that *data* incline us to think some *particular H* true. What Hanson advocates is psychological, sociological, or historical in nature; it has no logical import for the differences between proposing and establishing hypotheses.

Kepler again illustrates the objection. Every historian of science knows how the idea of uniform circular motion affected asstronomers before 1600. Indeed, in 1591 Kepler abandoned a hypothesis because it entailed other-than-uniform circular orbits—something simply inconceivable for him. So psychological pressure against forming alternative types of hypotheses was great. But *logically* Kepler's reasons for entertatining a type of Martian motion other than uniformly circular were his reasons for accepting that as astronomical truth. He first encountered this type of hypothesis on perceiving that no simple adjustment of epicycle, deferent, and eccentric could square Mars's observed distances, velocities, and apsidal positions. These were also reasons which led him to assert that the planet's orbit is not the effect of circular motions, but of an elliptical path. Even after other inductive reasons confirmed the truth of the latter hypothesis, these early reasons were *still* reasons for accepting *H* as true. So they cannot have been reasons merely for proposing which type of hypothesis *H* would be, and nothing more.

This objection has been made strong. If the following cannot weaken it, then we shall have to accept it; we shall have to grant that there is *no* aspect of discovery which has to do with logical, or conceptual considerations.

When Kepler published *De Motibus Stellae Martis*, he had established that Mars's orbit was an ellipse, inclined to the ecliptic, and had the sun in one of the foci. Later (in the *Harmonices Mundi*) he generalized this for other planets. Consider the hypothesis *H'*: *Jupiter's* orbit is of the noncircular type.

The reasons which led Kepler to formulate *H'* were many. But they included this: that *H* (the hypothesis that *Mars's* orbit is elliptical) is true. Since Eudoxos, Mars had been the typical planet. (*We* know why. Mars's retrogradations and its movement around the empty focus—all this we observe with clarity from earth because of earth's spatial relations with Mars.) Now, Mars's dynamical properties are usually found in the other planets. If its orbit is ellipsodial, then it is reasonable to expect that, whatever the exact shape of the other orbits (for example, Jupiter's), they will all be of the noncircular type.

But such reasons would not *establish H'*. Because what makes it reasonable to anticipate that *H'* will be of a certain type is *analogical* in character. (Mars does *x*; Mars is a typical planet; so perhaps all planets do the same kind of thing as *x*.) Analogies cannot establish hypotheses, not even *kinds* of hypotheses. Only observations can do that. In this the hypothetico-deductive account (or Popper, Reichenbach, and Braithwaite) is correct. To establish *H'* requires plotting its successive positions on a smooth curve whose equations can be determined. It may then be possible to assert that Jupiter's orbit is an ellipse, an oviform, an epicycloid, or whatever. But it would not be reasonable to expect this when discussing only what type of hypothesis is likely to describe Jupiter's orbit. Nor is it right to characterize this difference between "*H*-as-illustrative-of-a-type-of hypothesis" and "*H*-as-empirically established" as a difference of psychology only. *Logically*, Kepler's analogical reasons for proposing that *H'* would be of a certain type were good reasons. But, logically, they would not then have been good reasons for asserting the truth of a specific value for *H'*—something which sould be done only years later.

What are and are not good reasons for reaching a certain conclusion is a logical matter. No further observations are required to settle such issues, any more than we require experiments to decide, on the basis of one's bank statements, whether one is bankrupt. Similarly, whether or not Kepler's reasons for anticipating that *H'* will be of a certain kind are *good* reasons is a matter for logical inquiry.

Thus, the differences between reasons for expecting that some as yet undiscovered *H* will be of a certain type and those that establish this *H* are greater than is conveyed by calling them "psychological," "sociological," or "historical."

Kepler reasoned initially by analogy. Other kinds of reasons which make it plausible to propose that an *H*, once discovered, will be of a certain type, might include, for example, the detection of a formal symmetry in sets of equations or arguments. At important junctures Clerk Maxwell and Einstein detected such structural symmetries. This allowed them to argue, before getting their final answers, that those answers would be of a clearly describable type.

In the late 1920's, before anyone had explained the "negative-energy" solutions in Dirac's electron theory, good analogical reasons could have been advanced for the claim that, whatever specific assertion the ultimately successful

H assumed, it would be of the Lorentz-invariant type. It could have been conjectured that the as yet undiscovered *H* would be compatible with the Dirac explanation of Compton scattering and doublet atoms, and would fail to confirm Schrödinger's hunch that the phase waves within configuration space actually described observable physical phenomena. All this could have been said before Weyl, Oppenheimer, and Dirac formulated the "hole theory of the positive electron." Good analogical reasons for supposing that this *type* of *H* would succeed could have been and, as a matter of fact, were advanced. Indeed, Schrödinger's attempt to rewrite the Dirac theory so that the negative-energy solutions disappeared was *rejected* for failing to preserve the Lorentz invariance.

Thus reasoning from observations of *A*'s as *B*'s to the proposal "All *A*'s are *B*'s" is different in type from reasoning analogically from the fact that *C*'s are *D*'s to the proposal "The hypothesis relating *A*'s and *B*'s will be of the same type as that relating *C*'s and *D*'s." (Here it is the *way* *C*'s are *D*'s which seems analogous to the way *A*'s are *B*'s.) And both of these are typically different from reasoning involving the detection of symmetries in equations describing *A*'s and *B*'s.

Indeed, put this way, what *could* an objection to the foregoing consist of? Establishing a hypothesis and proposing by analogy that a hypothesis is likely to be of a particular type surely follow reasoning which is different in type. Moreover, both procedures have a fundamentally logical or conceptual interest.

An objection: "Analogical arguments, and those based on the recognition of formal symmetries, are used because of inductively established beliefs in the reliability of arguments of that type. So the cash value of such appeals ultimately collapses into just those accounts given by *H-D* theorists."

Agreed. But we are not discussing the *genesis* of our faith in these types of arguments, only the *logic* of the arguments themselves. *Given* an analogical premise, or one based on symmetry considerations—or even on enumeration of particulars—one argues *from* these in logically different ways. Consider what further moves are necessary to convince one who doubted such arguments. A challenge to "All *A*'s are *B*'s" when this is based on induction by enumeration could only be a challenge to justify induction, or at least to show that the particulars are being correctly described. This is inappropriate when the arguments rest on analogies or on the recognition of formal symmetries.

Another objection: "Analogical reasons, and those based on symmetry are *still* reasons for *H* even after it is (inductively) established. They are reasons *both* for proposing that *H* will be of a certain type and for accepting *H*."

Agreed, again. But, analogical and symmetry arguments could never *by themselves* establish particulars *H*'s. They can only make it plausible to suggest that *H* (when discovered) will be of a certain type. However, inductive arguments can, by themselves, establish particular hypotheses. So they must differ from arguments of the analogical or symmetrical sort.

H-D philosophers have been most articulate on these matters. So, let us draw out a related issue on which Popper, Reichenbach, and Braithwaite seem to me not to have said the last word.

J.S. Mill was wrong about Kepler (*A System of Logic*, III, 2–3). It is impossible to reconcile the delicate adjustment between theory, hypothesis, and observation recorded in *De Motibus Stellae Martis* with Mill's statement that Kepler's first law is but "a compendius expression for the one set of directly observed facts." Mill did not understand Kepler (as Peirce notes [*Collected Papers*, I, p. 31]). (It is equally questionable whether Reichenbach understood him: "Kepler's laws of the elliptic motion of celestial bodies were inductive generalizations of observed fact . . . [he] observed a series of . . . positions of the planet Mars and found that they may be connected by a mathematical relation . . .")[11] Mill's *Logic* is as misledading about scientific discovery as any account proceeding via what Bacon calls "*inductio per enumerationem simplicem ubi non reperitur instantia contridictoria.*" (Indeed Reichenbach observes: "It is the great merit of John Stuart Mill to have pointed out that all empirical inferences are reducible to the *inductio per enumerationem simplicem. . . .*")[12] The accounts of *H-D* theorists are equally misleading.

An *H-D* account of Kepler's first law would treat it as a high-level hypothesis in an *H-D* system. (This is Braithwaite's language.) It is regarded as a quasi-axiom, from whose assumption observation-statements follow. If these are true—if, for example, they imply that Uranus's orbit is an ellipse and that its apparent velocity at 90 degrees is greater than at aphelion—then the first law is confirmed. (Thus Braithwaite writes: "A scientific system consists of a set of hypotheses which form a deductive system . . . arranged in such a way that from some of the hypotheses as premises all the other hypotheses logically follow . . . the establishment of a system as a set of true propositions depends upon the establishment of its lowest level hypotheses . . .")[13]

This describes physical theory more adequately than did pre-Baconian accounts in terms of simple enumeration, or even post-Millian accounts in terms of ostensibly not-so-simple enumerations. It tells us about the logic of laws, and what they do in finished arguments and explanations. *H-D* accounts do not, however, tell us anything about the context in which laws are proposed in the first place; nor, perhaps, were they even intended to do so.

The induction-by-enumeration story *did* intend to do this. *It* sought to describe good reasons for initially proposing *H*. The *H-D* account must be silent on this point. Indeed, the two accounts are not strict alternatives. (As Braithwaite suggests they are when he remarks of a certain higher-level hypothesis that it "will not have been established by induction by simple enumeration; it will have been obtained by the hypothetico-deductive method. . . .")[14] They are thoroughly compatible. Acceptance of the second is no reason for rejecting the first. A law *might* have been inferred from just an enumeration of particulars (for example, Boyle's law in the seventeenth century, Bode's in the eighteenth, the laws of Ampere and Faraday in the nineteenth, and much of meson theory now). It could *then* be built into an *H-D* system as a higher order proposition. If there is anything wrong with the older view, *H-D* accounts do not reveal this.

There *is* something wrong. It is false. Scientists do not always discover every feature of a law by enumerating and summarizing observables. (Thus even

Braithwaite[15] says: "Sophisticated generalizations (such as that about the proton-electron constitution of the hydrogen atom) . . . [were] certainly not derived by simple enumeration of instances . . .") But *this* does not strengthen the *H-D* account as against the inductive view. There is *no H-D* account of how "sophisticated generalizations" are *derived.* On his own principles, the *H-D* theorist's lips are sealed on this matter. But there are conceptual considerations which help us understand the *reasoning* that is sometimes successful in determining the type of an as-yet-undiscovered hypothesis.

Were the *H-D* account construed as a description of scientific practice, it would be misleading. (Braithwaite's use of "derived" is thus misleading. So is his announcement [p. 11] that he is going to explain "how we *come to make* use of sophisticated generalizations.") Natural scientists do not "start from" hypotheses. They start from data. And even then not from commonplace data, but from surprising anomalies. (Thus Aristotle remarks[16] that knowledge begins in astonishment. Peirce makes perplexity the trigger of scientific inquiry.[17] And James and Dewey treat intelligence as the result of mastering problem situations.)[18]

By the time a law gets fixed into a *H-D* system, the *original* scientific thinking is over. The pedestrian process of deducing observation-statements begins only after the physicist is convinced that the proposed hypothesis is at least of the right type to explain the initially perplexing data. Kepler's assistant could work out the consequences of *H'* and check its validity by seeing whether Jupiter behaved as *H'* predicts. This was possible because of Kepler's argument that what *H* had done for Mars, *H'* might do for Jupiter. The *H-D* account is helpful here; it analyzes *the argument of a completed research report.* It helps us see how experimentalists elaborate a theoretician's hypotheses. And the *H-D* account illuminates yet another aspect of science, but its proponents have not stressed it. Scientists often dismiss explanations alternative to that which has won their provisional assent along lines that typify the *H-D* method. Examples are in Ptolemy's *Almagest*, when (on observational grounds) he rules out a moving earth, in Copernicus's *De Revolutionibus* . . . , when he rejects Ptolemy's lunar theory, in Kepler's *De Motibus Stellae Martis*, when he denies that the planes of the planetary orbits intersect in the center of the ecliptic, and in Newton's *Principia*, when he discounts the idea that the gravitational force law might be of an inverse cube nature. These mirror formal parts of Mill's *System of Logic* or Braithwaite's *Scientific Explanation.*

Still, the *H-D* analysis remains silent on reasoning which often conditions the discovery of laws—reasoning that determines which type of hypothesis is likely to be most fruitful to propose.

The induction-by-enumeration story views scientific inference as being from observations to the law, from particulars to the general. There is something true about this which the *H-D* account must ignore. Thus Newton wrote: "The main business of natural philosophy is to argue from phenomena. . . ."[19]

This inductive view, however, ingores what Newton never ignored: The inference is also from *explicanda* to an *explicans.* Why a beveled mirror shows spectra in sunlight is not explained by saying that all beveled mirrors do this. Why Mars moves more rapidly at 270 degrees and 90 degrees than could be expected of circular-uniform motions is not explained by saying that Mars (or even all planets) always move thus. On the induction view, these latter might count as laws. But only when it is explained why beveled mirrors show spectra and why planets apparently accelerate at 90 degrees will we have laws of the type suggested: Newton's laws of refraction and Kepler's first law. And even before such discoveries were made, arguments in favor of those *types* of laws were possible.

So the inductive view rightly suggests that laws are somehow related to infeıence *from* data. It wrongly suggests that the resultant law is but a summary of these data, instead of being an explanation of these data. A logic of discovery, then, might consider the structure of arguments in favor of one *type* of possible explanation in a given context as opposed to other *types*.

H-D accounts all agree that laws explain data. (Thus Braithwaite says: "A hypothesis to be regarded as a natural law must be a general proposition which can be thought to explain its instances; if the reason for believing the general proposition is solely direct knowledge of the truth of its instances, it will be felt to be a poor sort of explanation of these instances . . ." [*op. cit.*, p. 302].) *H-D* theorists, however, obscure the initial connection between thinking about data and thinking about what kind of hypothesis will most likely lead to a law. They suggest that the fundamental inference in science is from higher-order hypotheses to observation-statements. This may characterize the setting out of one's reasons for making a prediction after *H* is formulated and provisionally established. It need not be a way of setting out reasons in favor of proposing originally of what type *H* is likely to be.

Yet the original suggestion of a hypothesis type is often a reasonable affair. It is not as dependent on intuition, hunches, and other imponderables as historians and philosophers suppose when they make it the province of genius but not of logic. If the establishment of *H* through its predictions has a logic, so has the initial suggestion that *H* is likely to be of one kind rather than another. To form the first specific idea of an elliptical planetary orbit, or of constant acceleration, or of universal gravitational attraction does indeed require genius—nothing less than a Kepler, a Galileo, or a Newton. But this does not entail that reflections leading to these ideas are nonrational. Prehaps *only* Kepler, Galileo, and Newton had intellects mighty enough to fashion these notions initially; but to concede this is not to concede that their reasons for first entertaining concepts of such a type surpass rational inquiry.

H-D accounts begin with the hypothesis as given, as cooking recipes begin with the trout. Recipes, however, sometimes suggest, "First catch your trout." The *H-D* account is a recipe physicists often use after catching hypotheses. However, the conceptual boldness which marks the history of physics shows more in the ways in which scientists *caught* their hypotheses than in the ways in which they elaborated these once caught.

To study only the verification of hypotheses leaves a vital part of the story untold—namely, the reasons Kepler, Galileo, and Newton had for thinking their hypotheses would be of one kind rather than another. In a letter to Fabricus, Kepler underlines this:

Prague, July 4, 1603

> Dear Fabricius,
> . . . You believe that I start with imagining some pleasant hypothesis and please myself in embellishing it, examining it only later by observations. In this you are very much mistaken. The truth is that after having built up an hypothesis on the ground of observations and given it proper foundations, I feel a peculiar desire to investigate whether I might discover some natural, satisfying combination between the two . . .

Had any *H-D* theorist ever sought to give an account of the way in which hypotheses in science *are discovered*, Kepler's words are for him. Doubtless *H-D* philosophers have tried to give just such an account. Thus, Braithwaite[20] writes: "Every science *proceeds* . . . by thinking of general hypotheses . . . from which particular consequences are deduced which can be tested by observation . . . ," and again, "Galileo's deductive system was . . . presented as deducible from . . . Newton's laws of motion and . . . his law of universal gravitation . . ."

How would an *H-D* theorist analyze the law of gravitation?

(1) First, the hypothesis *H*: that between any two particles in the universe exists an attracting force varying inversely as the square of the distance between them ($F = \lambda Mm/r^2$).
(2) Deduce from this (in accordance with the *Principia*)
 (a) *Kepler's* Laws, and
 (b) *Galileo's* Laws.
(3) But particular instances of (a) and (b) square with what is observed.
(4) Therefore *H* is, to this extent, confirmed.

The *H-D* account says nothing about how *H* was first puzzled out. But now consider why, here, the *H-D* account is prima facie plausible.

Historians remark that Newton's reflections on this problem began in 1680 when Halley asked: "If between a planet and the sun there exists an attraction varying inversely as the square of their distance, what then would be the path of the planet?" Halley was astonished by the immediate answer: "An ellipse." The astonishment arose not because Newton *knew* the path of a planet, but because he had apparently deduced this from the hypothesis of universal gravitation. Halley begged for the proof; but it was lost in the chaos of Newton's room. Sir Isaac's promise to work it out anew terminated in the writing of the *Principia* itself. Thus the story unfolds as an *H-D* plot: (1) from the suggestion of a hypothesis (whose genesis is a matter of logical indifference—that is, psychology, sociology, or history) to (2) the deduction of observation statements (the laws of Kepler and Gallieo), which turn out true, thus (3) establishing the hypothesis.

Indeed, the entire *Principia* unfolds as the plot requires—from propositions of high generality through those of restricted generality, terminating in observation-statements. Thus Braithwaite[21] observes: "Newton's *Principia* [was] modelled on the Euclidean analogy and professed to prove [its] later propositions—those which were confirmed by confrontation with experience—by deducing them from original first principles . . ."

Despite this, the orthodox account is suspicious. The answer Newton gave Halley is not unique. He could have said "a circle" or "a parabola," and have been equally correct. The general answer is: "A conic section." The greatest mathematician of his time is not likely to have dealt with so mathematical a question as that concerning the possibility of a formal demonstration with an answer which is but a single value of the correct answer.

Yet the reverse inference, the retroduction, *is* unique. Given that the planetary orbits are ellipses, and allowing Huygen's law of centripetal force and Kepler's rule (that the square of a planet's period of revoution is proportional to the cube of its distance from the sun), the *type* of the law of gravitation can be inferred. Thus the question, "If the planetary orbits are elipses, what form will the gravitational force law take?" invities the unique answer, "An inverse square type of law."

Given the datum that Mars moves in an ellipse, one can (by way of Huygen's law and Kepler's third law) explain this uniquely by suggesting how it might follow from a law of the inverse square type, such as the law of universal gravitation was later discovered to be.

The rough idea behind all this: Given an ellipsodial eggshell, imagine a tiny pearl moving inside it along the maximum elliptical orbit. What *kind* of force must the eggshell exert on the pearl to keep the latter in this path? Huygen's weights, when whirled on strings, required a force in the string, and in Huygen's arm, of $F_{(k)} \propto r/T^2$ (where r signifies distance, T time, and k is a constant of proportionality). This restraining force kept the weights from flying away like stones from David's sling. And something like this force would be expected in the eggshell. Keplers' third law gives $T^2 \propto r^3$. Hence, $F_{(k)} \propto r/r^3 \propto 1/r^2$. The force the shell exerts on the pearl will be of a kind which varies inversely as the square of the distance of the pearl from that focus of the ellipsoidal eggshell where the force may be supported to be centered. This is not yet the law of gravitation. But it certainly is an argument which suggests that the law is likely to be of an inverse square type. This follows by what Peirce called "retroductive reasoning." But what *is* this retroductive reasoning whose superiority over the *H-D* account has been so darkly hinted at?

Schematically, it can be set out thus:

(1) Some surprising, astonishing phenomena $p_1, p_2, p_3 \ldots$ are encountered.[22]

(2) But $p_1, p_2, p_3 \ldots$ would not be surprising were a hypothesis of H's type to obtain. They would follow as a matter of course from something like H and would be explained by it.

(3) Therefore there is good reason for elaborating a hypothesis of the type of H; for proposing it as a possible hypothesis from whose assumption $p_1, p_2, p_3 \ldots$ might be explained.[23]

How, then, could the discovery of universal gravitation fit this account?

(1) The astonishing discovery that all planetary orbits are elliptical was made by Kepler.

(2) But such an orbit would not be surprising if, in addition to other familiar laws, a law of "gravitation," of the inverse square type obtained. Kepler's first law would follow as a matter of course; indeed that kind of hypothesis might even explain why (since the sun is in but one of the foci) the orbits are ellipses on which the planets travel with nonuniform velocity.
(3) Therefore there is good reason for further elaborating hypotheses of this kind.

This says something about the rational context within which a hypothesis of *H*'s type might come to be "caught" in the first place. It begins where all physics begins—with problematic phenomena requiring explanation. It suggests what might be done to particular hypotheses once proposed—namely, the *H-D* elaboration. And it points up how much philosophers have yet to learn about the kinds of reasons scientists might have for thinking that one kind of hypothesis may explain initial perplexities; why, for example, an inverse square type of hypothesis may be preferred over others, *if* it throws the initially perplexing data into patterns within which determinate modes of conection can be perceived. At least it appears that the ways in which scientists sometimes reason their way *towards* hypotheses, by eliminating those which are certifiably of the wrong type, may be as legitimate an area for conceptual inquiry as are the ways in which they reason their way *from* hypotheses.

Recently, in the Lord Portsmouth collection in the Cambridge University Library, a document was discovered which bears on our discussion. There, in "Additional manuscripts 3968. No. 41, bundle 2," is the following draft in Newton's own hand:

> And in the same year [1665, twenty years before the *Principia*] I began to think of gravity extending to ye orb of the Moon, and (having found out how to estimate the force with which a globe revolving within a sphere presses the surface of the sphere), from Kepler's rule . . . I deduced that the forces which keep the planets in their Orbs must be reciprocally as the squares of their distances from the centres about which they revolve . . .

This manuscript corroborates our argument. ("Deduce," in this passage, is used as when Newton speaks of deducing laws from phenomena—which is just what Aristotle and Peirce would call "retroduce.") Newton *knew* kow to estimate the force of a small globe on the inner surface of a sphere. (To compare this with Halley's question and our pearl-within-eggshell reconstruction, note that a sphere can be regarded as a degenerate ellipsoid—that is, where the foci superimpose.) From this and from Kepler's rule, $T^2 \propto r^3$, Newton determined that, whatever the final form of the law of gravitation, it would very probably be of the inverse-square type. These were the resaons which led Newton to think further about the details of universal gravitation. The reasons for accepting one such hypothesis of this type *as a law* are powerfully set out later in the *Principia* itself; and they are much more comprehensive than anything which occurred to him at this early age. But without such preliminary reasoning Newton might have had no more grounds than Hooke or Wren for thinking the gravitation law to be of an inverse-square type.

The morals of all this for our understanding of contemporary science are clear. With such a rich profusion of data and techniques as we have, the arguments necessary for *eliminating* hypotheses of the wrong type become a central research inquiry. Such arguments are not always of the *H-D* type; but if for that reason alone we refuse to scrutinize the conceptual content of the reasoning which precedes the actual proposal of definite hypotheses, we will have a poorer understanding of scientific thought in our time. For our own sakes, we must attend as much to how scientific hypotheses are caught, as to how they are cooked.

NOTES

1. Karl Popper, *The Logic of Scientific Discovery* (New York: Basic Books, 1959), pp. 31–32.

2. Hans Reichenbach, *Experience and Prediction* (Chicago: University of Chicago Press, 1938), p. 382.

3. R. B. Braithwaite, *Scientific Explanation* (Cambridge: Cambridge University Press, 1955), pp. 21–22.

4. C. S. Peirce, *Collected Papers* (Cambridge, Mass.: Harvard University Press, 1931), Vol. I, Sec. 188.

5. F. C. S. Schiller, "Scientific Discovery and Logical Proof," Charles Singer, ed., *Studies in the History and the Methods of the Sciences*(Oxford: Clarendon Press, 1917), Vol. I.

6. "There is no science which will enable a man to bethink himself of that which will suit his purpose," J. S. Mill, *A System of Logic*, III, Chap. I.

7. Cf. *Patterns of Discovery* (Cambridge, Mass.: Harvard University Press, 1958), pp. 85–92; "The Logic of Discovery," in *Journal of Philosophy*, LV, 25, 1073–89, 1958; More on "The Logic of Discovery," *op. cit.*, LVII, 6, 182–88, 1960.

8. Cf. *De Motibus Stellae Martis* (Munich), pp. 250ff.

9. *Op. cit.*, p. 20.

10. Reichenbach writes that philosophy "cannot be concerned with the first, but only with the latter" (*op. cit.*, p. 382).

11. Reichenbach, *op. cit.*, p. 371.

12. *Inid.*, p. 389.

13. Braithwaite, *op. cit.*, pp. 12–13.

14. *Ibid.*, p. 303.

15. *Ibid.*, p. 11.

16. Aristotle, *Metaphysics* 982b, 11ff.

17. Peirce, *op. cit.*, Vol. II, Book III, Chap. ii, Part III.

18. Cf. John Dewey, *How We Think*, (London: Heath & Co., 1909), pp. 12f.

19. Newton, *Principia*, Preface.

20. *Op. cit.*, pp. xv, xi, 18.

21. *Op. cit.*, p. 352.

22. The astonishment may consist in the fact that p is at variance with accepted *theories*—for example, the discovery of discontinuous emission of radiation by hot black bodies, or the photoelectric effect, the Compton effect, and the continuous β-ray spectrum, or the orbital aberrations of Mercury, the refrangibility of white light, and the high velocities of Mars at 90 degrees. What is important here is *that* the phenomena are encountered as anomalous, not *why* they are so regarded.

23. This is a free development of remarks in Aristotle (*Prior Analytics*, II, 25) and Peirce (*op. cit.*). Peirce amplifies: "It must be remembered that retroduction, although it is very little hampered by logical rules, nevertheless, is logical inference, asserting its conclusion only problematically, or conjecturally, it is true, but nevertheless having a perfectly definite logical form" (*op. cit.*, I, 188).

BIBLIOGRAPHICAL ESSAY

This bibliography is intended as an introduction to, rather than a comprehensive survey of, the literature on the philosophy of science. The reader who wants further bibliographical information should consult the bibliographies of the books mentioned below.

The best introduction to the history of the philosophy of science is R. Blake, C. J. Ducasse, and E. Madden's *Theories of Scientific Method* (University of Washington Press, 1960). In connection with this book, the reader should consult the selections in the following historically oriented anthologies: B.A. Brody and N. Capaldi's *Science: Men, Methods and Goals* (W. A. Benjamin, 1968) and J. J. Kockelman's *Philosophy of Science: The Historical Background* (Free Press, 1968).

Several good texts introduce the reader to contemporary work in the philosophy of science. These include: C. G. Hempel, *Philosophy of Natural Science* (Prentice-Hall, 1966), E. Nagel, *The Structure of Science* (Harcourt, Brace, and World, 1961), A. Pap, *An Introduction to the Philosophy of Science* (Free Press, 1962), I. Scheffler, *The Anatomy of Inquiry* (Knopf, 1963), and J. J. C. Smart, *Between Science and Philosophy* (Random House, 1968). In connection with these texts, the reader may find it useful to look at some of the articles collected in A. Danto and S. Morgenbesser's *Philosophy of Science* (Meridian Books; 1960), H. Feigl and M. Brodbeck's *Readings in the Philosophy of Science* (Appleton-Century-Crofts, 1953), H. Feigl and G. Maxwell's *Current Issues in the Philosophy of Science* (Holt, Rinehart and Winston, 1961), and E. Madden's *The Structure of Scientific Thought* (Houghton Mifflin Co., 1960).

Most of the important current work in the philosophy of science appears in journals and collections of articles. Two journals that specialize in the philosophy of science are *Philosophy of Science* and *British Journal for the Philosophy of Science* (*B. J. P. S.*). The following collections of articles appear periodically: *Boston Studies in the Philosophy of Science* (Humanities, 1963, 1965, 1968), Vols. I, II, III; *Philosophy of Science: The Delaware Seminar* (Interscience, 1963) Vols. I, II; *Minnesota Studies in the Philosophy of Science* (University of Minnesota Press,, 1956, 1958, 1962), Vols. I, II, III; *University of Pittsburgh Series in the Philosophy of Science* (University of Pittsburgh Press, 1962, 1965, 1966) Vols. I, II, III; and *Logic, Methodology, and Philosophy of Science* (Stanford University Press, 1962), Vol. I; (North Holland, 1956), Vol. II.

The covering-law models of explanation and prediction are defended by Hempel in many of his articles collected in C. G. Hempel's *Aspects of Scientific Explanation* (Free Press, 1965). Hempel's claim that explanations and predictions have the same structure is critically discussed in N. R. Hanson's "On the Symmetry between Explanation and Prediction," *Philosophical Review* (1959); N. Rescher's "Discrete State Systems, Markov Chains, and Problems in the Theory of Scientific Explanation and Prediction," *Philosophy of Science* (1963); I. Scheffler's "Explanation, Prediction, and Abstraction," *B. J. P. S.* (1957); and M. Scriven's "Explanation and Prediction in Evolutionary Theory," *Science* (1957) and "Explanations, Predictons, and Laws," *Minnesota Studies*, Vol. III. It is defended by A. Grunbaum in "Temporally Asymmetric Principles, Parity Between Explanation and Prediction, and Mechanism vs. Teleology," *Philosophy of Science* (1962). The problem of laws of nature is discussed in H. G. Alexander's "General Statements as Rules of Inference," *Minnesota Studies*, Vol. II, W. Kneale's "Universality and Necessity," *B. J. P. S.* (1961); Chapter 4 of E. Nagel's *The Structure of Science;* Chapter 16 of A. Pap's An *Introduction to the Philosophy of Science;* H. Reichenbach's *Nomological Statements and Admissible Operations* (North Holland, 1954); and Chapter 3 of S. Toulmin's *Philosophy of Science* (Hutchinson, 1953). Scriven's alternative analysis of explanations is developed in his articles "Definitions, Explanations and Theories," *Minnesota Studies*, Vol. II, and "Truisms as the Ground for Historical Explanations," in P. Gardiner's *Theories of History* (Free Press, 1959), while Bromberger's analysis is developed further in "An Approach to Explanation" in R. J. Butler's *Analytical Philosophy Second Series* (Blackwell, 1965). Many of the main articles on functional and teleological explanations have been collected in J. Canfield's *Purpose in Nature* (Prentice-Hall, Inc., 1966). Finally, the problem of historical explanation is discussed in many of the articles collected in P. Gardiner's *Theories of History* (Free Press, 1959); S. Hook's *Philosophy and History* (New York University Press, 1963); and W. Dray's *Philosophical Analysis and History* (Harper and Row, 1966).

The logical positivist conception of theories is discussed in R. Braithwaite's *Scientific Explanation* (Cambridge University Press, 1953), Chapters 1–4; R. Carnap's "Testability and Meaning," *Philosophy of Science* (1936, 1937); "The Methodological Character of Theoretical Concepts," *Minnesota Studies*, Vol. I; C. G. Hempel's "The Theoretician's Dilemna," *Minnesota Studies*, Vol. II; and E. Nagel's *Structure of Science*, Chapters 5 and 6. The observational-theoretical distinction is attacked by M. Spector in "Theory and Observation," *B. J. P. S.* (1966). The problem of the role of models is extensively discussed by M. Hesse in her book *Forces and Fields* (Nelson, 1961) and in her articles "Operational Definition and Analogy in Physical Theories," *B. J. P. S.* (1952), "The Role of Models in Physics," *B. J. P. S.* (1953), and "Theories, Dictionaries and Observation," *B. J. P. S.* (1958). These last two problems are also discussed by P. Achinstein in his new book *Concepts of Science* (Johns Hopkins Press, 1968). Feyerabend presents his views in the following articles: "Explanation, Reduction, and Empiricism," *Minnesota Studies*, Vol. III; "Probems of Microphysics," *University of Pittsburgh Series*, Vol. I; and "Problems of Empiricism," *University of Pittsburgh Series*, Vol. II. His views are criticized by P. Achinstein in "On the Meaning of Scientific Terms," *Journal of Philosophy* (1964), and by D. Shapere in "Meaning and Scientific Change," *University of Pittsburgh Series*, Vol. III. Finally, there is an important symposium containing papers by Feyerabend, Putnam, Sellars, and Smart in the second volume of the *Boston Studies*, and Smart summarizes these discussions in Chapter 3 of *Between Science and Philosophy*.

A very basic but good introduction to the problem of confirmation is B. Skyrms's *Choice and Chance* (Dickenson, 1966), whereas H. Kyburg's "Recent Work in Inductive Logic," *American Philosophical Quarterly* (1964), is a good summary of recent work in this area. Most of the important classical articles in this area are collected in M. Foster and M. Martin's *Probability, Confirmation, and Simplicity* (Odyssey, 1966). Hempel's discussion of confirming instances is critically evaluated in J. Canfield's "On the Paradox of Confirmation," *Metrika* (1962) and in R. Carnap's *Logical Foundations of Probability* (University of Chicago Press, 1950). The paradoxes of confirmation are discussed in J. Agassi's "Corroboration versus Induction," *B. J. P. S.* (1959); I. J. Good, "The Paradoxes of Confirmation," *B. J. P. S.* (1960, 1961); J. L. Mackie's "The Paradoxes of Confirmation," *B. J. P. S.* (1963); and J. W. N. Watkin's "Between Analytical and Empirical," *Philosophy* (1957). Much of the discussion of these issues is summarized in Part III of I. Scheffler's *The Anatomy of Inquiry*. Carnap's theory of degrees of confirmation is presented in his books *The Logical Foundations of Probability* and *The Continuum of Inductive Methods* (University of Chicago Press, 1952), and in "The Aim of Inductive Logic" in the first volume of *Logic, Methodology, and Philosophy of Science*. The theory is discussed in several articles in P. Schilpp's *The Philosophy of Rudolf Carnap* (Open Court, 1963), and Carnap also presents a reply to these criticisms. Goodman's problem has recently been discussed in a symposium in the *Journal of Philosophy* (1966), with contributions by R. Jeffreys, J. J. Thomson, J. Wallace, and a reply by Goodman. It is also discussed by R. J. Butler in "Messrs. Goodman, Green, and Grue" in his book *Analytical Philosophy, Second Series* (Blackwell, 1965) and by I. Scheffler in Part III of his *Anatomy of Inquiry*. The fullest discussion of inductive acceptance rules, with complete references to the whole literature on this problem, is I. Levi's *Gambling with Truth* (Knopf, 1967). The problem of justifying induction has been much discussed in recent years. Salmon's most elaborate exposition of his approach is his article "The Foundations of Scientific Inference" in Vol. III of the *University of Pittsburgh Series*. This approach is criticized by I. Hacking in "Salmon's Vindication of Induction," *Journal of Philosophy* (1965); their disagreement is discussed by I. Levi in "Hacking and Salmon on Induction," *Journal of Philosophy* (1965). Black's justfication is criticized by Salmon in "Should We Attempt to Justify Induction," *Philosophical Studies* (1957) and by Achinstein in "The Circularity of a Self-Supporting Inductve Argument," *Analysis* (1961–62). Black replies to these criticisms in Chapter 12 of his book *Models and Metaphors* (Cornell University Press, 1962). A position similar to Strawson's is defended by P. Edwards in "Bertrand Russell's "Doubts about Induction," *Mind* (1949), and is criticized by J. O. Urmson in "Some Questions Concerning Validity" in *Essays in Conceptual Analysis* (Macmillan, 1959) and by Salmon in "Should We Attempt to Justify Induction."

Finally, the following books offer an introduction to areas of the philosophy of science not covered in this volume: P. Benacerraf and H. Putnam's *Philosophy of Mathematics: Selected Readings* (Prentice-Hall, Inc., 1964); M. Capek's *The Philosophical Impact of Contemporary Physics* (Van Nostrand, 1961); A. Grunbaum's *Philosophical Problems of Space and Time* (Knopf, 1963), and M. Brodbeck's *Readings in the Philosophy of the Social Sciences* (Macmillan, 1968).

by Salmon in [illegible] and by Achinstein in "The Concept of [illegible] Evidence" [illegible] (1961-62) [illegible] (Cornell University Press, 1963) [illegible] "Derived [illegible] Questions [illegible] Validity" [illegible] also by Salmon [illegible]

The following books [illegible] philosophy of science [illegible] (Prentice-Hall, Inc., 1966) [illegible] (Macmillan, 1953).